VOICE INTERACTION DESIGN

The Morgan Kaufmann Series in Interactive Technologies

Series Editors:
- Stuart Card, PARC
- Jonathan Grudin, Microsoft
- Jakob Nielsen, Nielsen Norman Group

VOICE INTERACTION DESIGN

Crafting the New Conversational Speech Systems

Randy Allen Harris
University of Waterloo

ELSEVIER

AMSTERDAM • BOSTON • HEIDELBERG • LONDON
NEW YORK • OXFORD • PARIS • SAN DIEGO
SAN FRANCISCO • SINGAPORE • SYDNEY • TOKYO

Morgan Kaufmann Publishers is an imprint of Elsevier

MORGAN KAUFMANN PUBLISHERS

AN IMPRINT OF ELSEVIER SCIENCE

Publishing Director	Diane Cerra
Publishing Services Manager	Simon Crump
Project Manager	Daniel Stone
Editorial Assistant	Asma Stephan
Cover Design	Ross Carron Design
Cover Image	(left image) © Thinkstock, Inmagine; (middle image) © Brandix Pictures, Inmagine; (right image) © Lonley Planet Images, Getty Images.
Design Manager	Cate Rickard Barr
Composition	SNP Best-set Typesetter Ltd., Hong Kong
Technical Illustration	Dartmouth Publishing
Copyeditor	Gary Michael Spahl
Proofreader	Phyllis Coyne Proofreader Services
Indexer	Linda Caravelli
Interior printer	The Maple-Vail Book Manufacturing Group
Cover printer	Phoenix

Morgan Kaufmann Publishers is an imprint of Elsevier.
500 Sansome Street, Suite 400, San Francisco, CA 94111

This book is printed on acid-free paper.

Permissions may be sought directly from Elsevier's Science & Technology Rights Department in Oxford, UK: phone: (+44) 1865 843830, fax: (+44) 1865 853333, e-mail: *permissions@elsevier.com.uk*. You may also complete your request on-line via the Elsevier homepage (*http://elsevier.com*) by selecting "Customer Support" and then "Obtaining Permissions."

Library of Congress Cataloging-in-Publication Data
Application Submitted

ISBN: 1-55860-768-4

For information on all Morgan Kaufmann publications,
visit our Web site at www.mkp.com or www.books.elsevier.com

Printed in the United States of America
04 05 06 07 08   5 4 3 2 1

This book is dedicated to my brothers and sisters

Jeff
Debbie
Nirm
Logendra
Naveen
and
Avashinee

Table of Contents

CHAPTER 14

Scripting   423

CHAPTER 15

Iterative Evaluation   473

CHAPTER 16

Conclusion — Pursuing Habitability

Preface

We need to be able to work at a level of abstraction concrete enough to provide leverage within the task-artifact cycle, yet abstract enough to cumulate and develop as a theory base.
— John M. Carroll

There are new machines abroad, talking machines. They are very impressive devices, especially to people who understand the magnitude of the problems involved. But the making of machines that talk is dominated by concerns for recognition rates, vocabulary size, and processing capacity, with comparative inattention to the interaction. There has not been, up to this point, sufficient focus on what these machines should actually say and listen. The simple premise of this book is that if the effort for the human-factors aspects matched the recognition and processing aspects of these devices, then we would all be happier with our talking machines.

You

I wrote this book to serve a wide range of speech-related interests, and you are the best judge, by far, of whether it serves yours. But I can help, I hope, by sketching out the sort of people I developed the book to engage, to assist, and even occasionally to entertain.

If you are a **speech-system interaction designer**, or an interaction designer aspiring to work with speech systems, you are the primary audience, especially if you work with (or aspire to work with) systems that have a conversational style. I have laid out for you a soup-to-nuts overview of the linguistic, pragmatic, and conversational principles that undergird speech interaction, and coupled that with a detailed map of the process for developing a voice interaction, cramming examples in at every turn and steadily outlining the necessary instruments.

If you are a **multimodal designer**, for whom speech is not the sole interaction channel, but one of several possibilities, I wrote this book for you as well. The speech aspects of multimodal design work are different from those of speech-only interaction design — more complicated in some ways, simpler in others — but they depend just as heavily on the fundamentals of speech, conversation, and interface development explained throughout this book.

If you are part of the **human-computer interaction** community generally, a practitioner or researcher who is broadly interested in the various terrains interfacing between humans and machines — their respective constraints, and affordances — you are also in my audience. Speech-system designers read good interaction-design books regularly, most of which are predominantly about graphic interfaces, not because we design such systems (though some do), but just for ideas, and information, and new perspectives. Now that there is a body or voice interaction books appearing, graphic-design people and human-computer folks generally will surely profit from reading them — for ideas, information, and perspectives — and I have kept such readers in mind while developing this book. Most pertinently, the conversational focus I adopt relentlessly in this book is especially rich for human-computer professionals, all of whom understand that the field is defined at almost every level as a type of conversation between human and computer agents.

If you are a **manager of speech projects**, you will also benefit from this book, and reading it can help you identify the necessary personnel for such projects, provide them with the appropriate resources, and help you understand the specific voice interaction elements your project demands.

Similarly, if you are a **client of a speech company**, *Voice Interaction Design* will help you to know the questions to ask, the features to request, the data you can gather to guide and facilitate such projects, and the sorts of interactions you can expect the system to support.

Me

I am an academic, with input/output in linguistics, literature, rhetoric, and technical communication. I am a teacher, with input/output in linguistics, rhetoric, human factors, usability, documentation, and information design. I am a corporate hack, with input/output in quality assurance, usability, and graphic and voice interaction development. If this sounds like unstable and opportunistic careering around a group of loosely related fields and professions, you're right. It is. But opportunistic careering has its virtues. On the academic front, it has given me routine exposure, from a broad range of perspectives, to the best that has been thought and said about the issues and problems of language — the most intimate and gregarious and human-defining singularity we share. On the teaching front, it has given me routine exposure to keen and challenging minds who don't settle for easy answers, unless, as sometimes proves to be the case, the answers really are easy. On the corporate front, it has brought me into contact with innovative, dedicated, practical minded folk, and their remarkable, obtuse machines. My work on all three fronts is integrated into this book.

This Book

Voice Interaction Design is not a Book of Pristine Theory, in the coherent-system-of-explanation-and-prediction sense. There is no Unified Theory of Voice Interaction here. It is also not a Book of Messy Practice, in the first-we-tried-this-and-then-we-tried-that sense. There is no Story of Speech Application X here.

Voice Interaction Design is a book of *theorized practice*, a book that brings a diverse body of theory to a growing body of practice and elaborates the ways the first can inform the development of the second.

The body of theory, while not elegant, is robust. It comes from very smart people, who have thought very hard about the features, patterns, and problems of communication for thousands of years. I have brought the results of all this thought into new relationships and done some relabeling, grafting and pruning. The merit of the theory in this book is not mine though it belongs to my sources. The body of practice, while not unitary, is also robust. The data is good. It comes from decades of work, with an especially furious burst of activity in the nineties through the turn of the century — work by smart and hardworking people, who strive to design and build machines that speak and, when spoken to, respond helpfully and congenially. Its substantial merit, too, belongs to my sources.

The body of theory in this book is a potpourri, assembled for utility from the results, speculations, and research programs of a variety of overlapping and interpenetrating fields: human–computer interaction, conversational analysis, philosophy of language, cognitive psychology, social psychology, computational linguistics, rhetoric, technical communication, chatterbot theory, and interface design. There is, in short, nothing sacred about what has been assembled here. Frankly, it is profane, in the old-fashioned sense of the term. It is outside the temple, all the temples. I am proceeding without allegiance to, or reverence for, any particular body of doctrine. My methodological model, the model of theorized practice, is the crafts.

The applied research that has fed this book is considerably more unified than the theory. The sources are diverse: conference proceedings, books, technical articles, academic papers, observation, expert advice, hallway chit-chat, and long hours stuck in automated telephony hell. But the motivations behind all of these sources orbit around one overwhelming theme: how to get these stubborn, brilliant, ubiquitous instruments — computers — to behave cooperatively through the medium of speech.

The Research

A huge amount of the theory and practice I draw on, especially involving the human–computer interaction, telephony, and computer speech communities, I collected via the Web, which occasions both gratitude (to my sources) and an apology (to my readers). I am tremendously grateful to the researchers who have made their work so available on the

Internet, and more generally to the many sources of opinion and information that populate this vital, virtual world. This is an explosion of open-source information that in many ways recalls the development of public science in the 17th century out of secretive, information-hoarding practices like alchemy. While the quality can be highly variable, much of this available research is superb, and the movement to make it so widely available is almost entirely salutary.

But — here's the apology — I'm sorry that I have not always been able to document those sources as well as I would like. In particular, many of the documents that I have worked with either have unreliable pagination (those published in conference proceedings but seen by me only as downloaded PDF or PS files), or no pagination at all (those published only in HTML or XML); some sites disappeared after I used them, so their URLs became useless for citation purposes; and some have multiple addresses, so that picking a single authoritative URL to cite became a bit of a guessing game. In general, I have prioritized the citations according to authors, titles, and original venues, trusting that anyone with a search engine will be able to track them down without too much trouble.

There likely remains both linkrot and vagueries in the citations, however, and I know how aggravating it is to see a quotation tagged by something like "(Derf, 1996)," and discover that the source is 36 pages long in an unsearchable file format, if locatable at all. But I've done the best that I can, and the alternative in many cases was either eliminating many sources altogether, or dedicating my remaining years to pursuing final, ultimate authoritative citations for them all, while the manuscript for this book lay moldering in my computer.

I have also minimized the number of footnotes in the text. There still a handful in almost every chapter, offloading concerns from the (already brimming) main text, or addressing counter-arguments and counter-themes to my assorted proclamations. Part of being an academic, of course, is being obsessive. Earlier versions of this book were encrusted with notes full of qualifications, ramifications, and preoccupations. All but a narrow group of the readers found the footnotes mostly a bother. I have relented, hacking them back savagely. To readers as anal-retentive as myself, I apologize. For everyone else, I hope the book has increased fluidity.

Dialogue Inclusions

In general, I have played somewhat fast and loose with the snippets of dialogue I incorporate in this book from other sources — capitalizing, punctuating, eliminating distractions — always in the hopes of clarifying what is going on, or of minimizing transcription weirdnesses. For instance, a great many speech researchers use English orthography for reporting their dialogue data (and even their scenario data) without using its standard conventions, or using them very irregularly — not capitalizing names of days or months, for instance, but capitalizing proper names and first-person singular pronouns. I'm not sure at all what the motivations for this practice are, though it may be related to the widespread belief among some researchers that spoken language is "ungrammatical" (they're wrong; it's not). In any case, my apologies to any authors who take offense. I am confident that I haven't introduced any distortions, and rarely does any point hinge on one of my adjustments. I am not doing this editing out of a sense of grammatical correctness, to be a smart-ass, or for aesthetics, but simply to clarify parts of the dialogue (use/mention distinctions, for instance, and self-quotations), and to eliminate textual suggestions that spoken language is deficient.

In general, it is the human/human dialogues that are most heavily edited in this book (removing much of the transcription machinery of conversation analysts in particular); with human/machine dialogues, my alterations are mostly a matter of capitalization, punctuation, agent-labeling, and line breaks.

Some Acknowledgments

One pleasant evening at the beginning of this century, I was having dinner at La Toscana, with Rick and Linda Serafini, and Rick was holding forth on investment opportunities with a company that designed software for talking to your appliances. A week later, Douglas Wright, former president of the University of Waterloo and all-round technology *bon vivant* called to tell me about a company that he was working with that was developing speech-recognition products. He invited me to drop by, on the off-chance that someone who worked in linguistics might have some advice for them about the finer points of language. I accepted. *Ipso novo* — thank you Rick and Doug — I had a whole new research interest.

Many, many people since then have helped me to bring this book together.

I am especially grateful to Scott Brave, Jennifer Lai, Clifford Nass, and Nicole Yankelovich for sharing their unpublished work with me.

Lai and Yankelovich also read and reviewed versions of this book, as did Daryle Gardner-Bonneau, Martha Lindeman, Chris Schmandt, and, most generously, Ellen Isaacs, whose early-draft review deserves special praise — a meticulously close reading that was a model of collegial helpfulness. Other members of the dialogue and speech-system communities have also been kind — providing tips, suggestions, advice, responses to inquiries, and general goodwill, starting chronologically and decisively with Victor Lee. Ian Mccallum, Sunny Mendes, Richard Rosinski, Greg Sanders, Laura Sasaki, Phil Shin, Steve Shepherd, Flora Shiu, and Rakesh Tailor were also generous with their time and expertise. Not all of these folks agree with what I have to say here, and some of them disagree quite violently with my overall perspective or with some specific claims. But the importance of their feedback to the quality and integrity of this book is immeasurable.

My colleagues at the University of Waterloo have been supportive and creative allies — especially, at a critical junction, Harry Logan, Andrew McMurry, and Glenn Stillar — for which I thank them heartily. I have also had the benefit of several discussions with

Chrysanne DiMarco. Fakhri Karray, along with Otman Basir of the University of Guelph, have been particularly generous in allowing me to work with and explore the technology that they are developing.

My contact with students at the University of Waterloo has been the most sustained source of joy I have had in my professional life. I thank all of them collectively, but a small subset needs to be singled out for helping keep me engaged and challenged about voice-interaction design: Heather Calder, Gabriel Chan, Dewlyn D'Mellow, David Flett, Opal Gamble, David Gillis, Kim Honeyford, John Jong-Suk Lee, Teresa Winky Mak, Sheila McConnell, Kim McMullen, Sarah Mohr, Maria Andrusiak Morland, Amy Oulette, Robert Shanks, Mike Truscello, Aliya Walji, Phil Wang, Michelle Willer, and Karl Wierzbicki. Zarsheesh Divecha and Kateryna Zolotkova helped with the bibliography. Thank you all — with special mention to Dewlyn, for her bank project and for many lively discussions; to Gabriel for sending me useful materials; to the stellar '04 class (Heather, David, Kim, and Sarah); and to Mike.

Diane Cerra, Belinda Breyer, Mona Buehler, and Daniel Stone have been wonderfully supportive in the iterative creation of this book, and Matt Wagner gave me some very useful advice early on. The experience working with Morgan Kaufmann has been all that I could have hoped for.

Antepenultimately, I am deeply obliged to Galen, for inspiration, insatiable curiosity, and intense dedication; for writing, and illustrating nine books to my one, five of them in one week; and for protecting me from "the forces of evil" that sought to interfere with my writing time. The badges, too, were a big help. Penultimately, I am grateful also to Oriana, for dancing, singing, loving, sowing joy, slipping notes and pictures into the study, getting down off my stacks of books when I asked, and for, sometimes, not pounding on the study door. Ultimately, I thank Indira, always Indira, for everything but the mornings.

While [interactive speech] technology improves constantly, it is unlikely that, in the foreseeable future, it will approach the robustness of computers in science fiction movies.
    — Jennifer Lai and Nicole Yankelovich

You've seen it in Star Trek, Mission Impossible *and* James Bond *— futuristic technologies based on voice recognition and verification. Long the stuff of science fiction fantasy, voice technologies are now a business reality thanks to Nuance.*
    — Nuance Communications (www.nuance.com, 2002)

Toto . . . I have a feeling we're not in Kansas anymore.
    — Dorothy

Introduction

The user interface to an interactive product such as software can be defined as the languages through which the user and the product communicate with each other.
— Deborah Mayhew

Interfaces

User interfaces (hereafter, mostly just "interfaces" with various specifying adjectives, like "voice" and "graphic") are the media by which people interact with digital systems. Although it is an abstract and quite shallow image (in particular, an interface is part of the system, not outside it, and strictly speaking, so is the user), it is nevertheless useful to picture the interface in the terms of Figure 1.1: A user performs some physical actions according to an established protocol, effecting input, which trigger electronic operations. Almost always, these operations include indications for the user about the effect of her actions; that is, feedback.

Voice interfaces are interactive media in which the input is primarily or exclusively speech, and so is the feedback. They are new phenomena, especially in the conversational style that dominates the approach of this book, and their design draws on wide bodies of language research, from linguistics, philosophy, sociology, and psychology. The results of this research, in turn, need to be interpreted, then applied, through the disciplines of computer science, telephony, and interaction design.

Voice interfaces are so new, indeed, that their properties and possibilities are still being worked out, daily. So new, that there is still only a partial awareness among designers that those properties and possibilities need to be investigated in a way that is independent of specific systems and technologies. So new, that — even though there is widespread

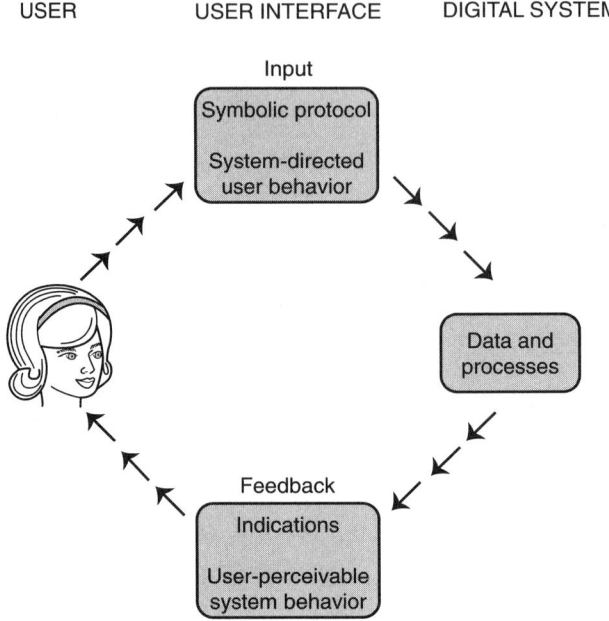

USER   USER INTERFACE   DIGITAL SYSTEM

Input

Symbolic protocol

System-directed user behavior

Data and processes

Feedback

Indications

User-perceivable system behavior

FIGURE 1.1 A rough schematic of user interfaces

recognition that building usable speech systems overwhelmingly implicates the field of human–computer interaction — the word *interface* is not often used by the very people who are developing them. "Spoken Language Systems" is more common, along with a variety of terms featuring "dialogue" — like "Automatic Telephone Dialogues" and "Spoken Dialogue Systems." These labels reflect much the same attitude that characterized the early years of graphic interface development, which was subsumed under general application development and often designed by the same software engineers who designed the system mechanics.

Voice interfaces are among the growing range of interfaces populating the modern landscape, from the rapid, virtually invisible interfaces of digital calculators to the awkward on-screen interfaces of digital televisions, most of them exploiting multiple modes (chiefly, sound, vision, and touch). But three specific interface categories are significant for the evolution and design of voice interfaces. Two are familiar from computer interaction: the nearly archaic command-line interface and the ubiquitous graphic interface. The third comes from telephony: the keypad interface, which provides interaction with messaging applications, automated reception systems, telephone banking, and the like. It is in these domains, where keypad interfaces have dominated for a decade or more, that voice interfaces are beginning steadily to appear. An inevitable and near-total eclipse of keypad interfaces is around the next bend (how far ahead that bend is depends on economic and technological developments; but it *is* the next bend).

Since all three of these interface types are implicated in voice-interaction design, and since I will be drawing analogies from them throughout the book, we'll look at them briefly in turn, and also glance at the related area of multimodal interfaces, where voice may come to play an increasingly important role.

Command-line Interfaces

149. Implementation restriction: The "stringrange" on-condition cannot be enabled when the "substr" pseudo-variable is used. Note that this restriction does not apply to the "substr" built-in function.
　　— A PL/I Multics error diagnostic

Command-line interfaces were usually just called "Man-Machine Interfaces" (MMIs) in their period of dominance, the 1970s. The name reflected both the gender imbalance that characterized computer development and the fact that they were the only game in town. As in Figure 1.2, they work on a linguistic paradigm (strings of alphanumeric "words" arranged in a determinate syntax).

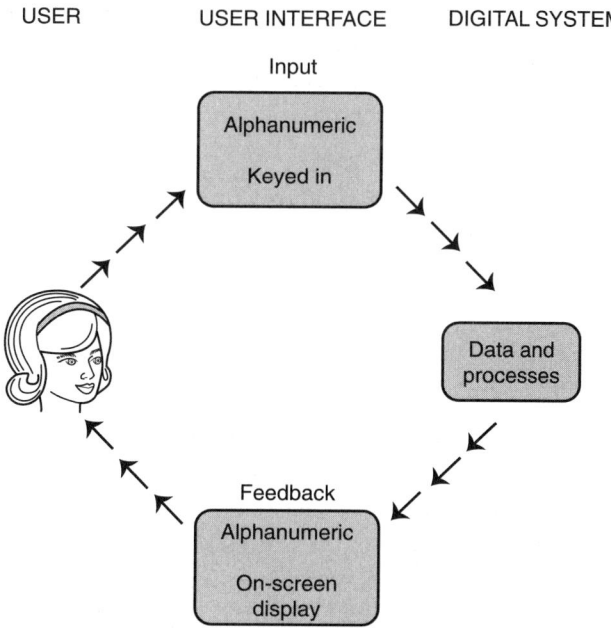

FIGURE 1.2 A rough schematic of command-line user interfaces

Users issue typed commands and receive textual feedback, resulting in interactions of the following sort:

User $MESSAGESYSTEM RETRIEVE

MTS   Mailbox EYVQ: 1 new, 4 old messages

That particular specimen is the command users once issued to MTS (the Michigan Terminal System) to get a list of email, followed by MTS's response.

Command-line interfaces like MTS or the better-known Multics (Multiplexed Information and Computing Service) could be tremendously efficient, compressing a whole range of operations into one sleek line of text, so long as the user knew their narrow vocabularies and rigid syntaxes. But these interfaces incorporated very little sense of the user, and were brutally unforgiving. In particular, there was comparatively negligible attention to the clarity of the feedback. The system either did what you told it to, with minimal confirmation of its actions, or it generated an error message, often cryptic in the extreme, to account for why it wasn't doing what you thought you told it to do. Here is a from-the-trenches characterization of the user's relation to the system, during the heyday of command-line systems:

> *The user is often placed in the position of an absolute master over an awesomely powerful slave, who speaks a strange and painfully awkward tongue, whose obedience is immediate and complete but woefully thoughtless, without regard to the potential destruction of its master's things, rigid to the point of being psychotic, lacking sense, memory, compassion, and — worst of all — obvious consistency.*
> (Miller and Thomas, 1977: 172)

They weren't completely lacking in goodwill, however, and some subsystems even showed a measure of personality. I was once fortunate enough to get, upon issuing the following command, the next-following response:

Me $MESSAGESYSTEM RETREIVE NEW

MTS   Didn't your momma ever tell you: I before E, except after C?

The last bastions of this era are MS DOS (Microsoft Disk Operating System), which has been overlain by various incarnations and generations of Windows, and Unix (named by way of an arcane, nose-thumbing pun on *Multics*), which undergirds (among other graphic interfaces) Linux and Mac OS X. While command-line interfaces still have active user populations in some specialized communities, and while the graphic overlays sometimes have to be peeled back to the command-line level when things go wrong, or when higher efficiencies are needed, they are powerful yet lumbering dinosaurs; graphic interfaces have clearly inherited the earth.

Graphic Interfaces

Those buttons, graphics, and words on the screen through which we control information.
— The explication of "interface" in the jacket blurb to Steven Johnson's
Interface Culture

Graphic interfaces are often just called "interfaces" these days, so much have they become the default, but also GUIs (Graphical User Interfaces). They work on an object-manipulation paradigm (representations are moved, altered, or engaged — sometimes all three — through the user's wielding of a pointing device, usually augmented by physical buttons or keys) as outlined in Figure 1.3.

Users interact with systems, usually in clusters, represented by dedicated areas of the screen looking something like the object in Figure 1.4. That particular specimen is from the earliest web browser, Mosaic. It has the now-familiar doodads: buttons along the top for back, forward, home-page, refresh, and save; along with a scroll bar; links; a text field for keyboard input; and a primitive menu (the box at the top with the down-facing arrow head provides for navigation among previously visited sites). The display is graphic, the

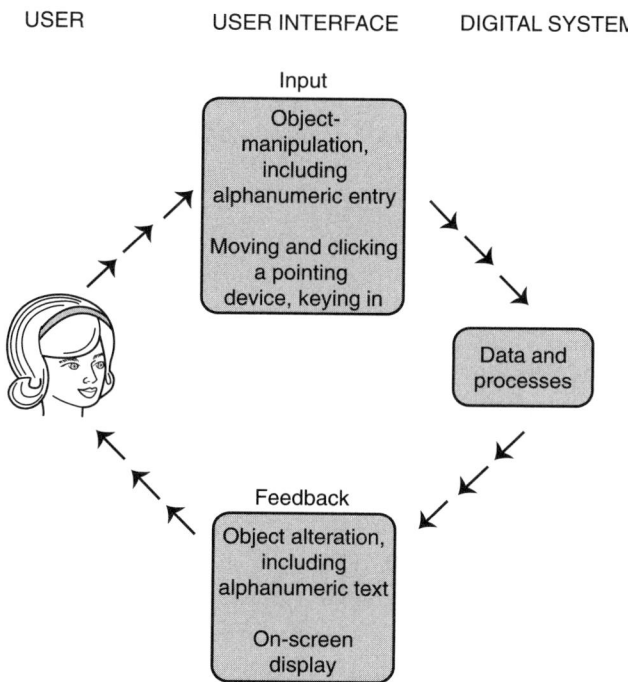

FIGURE 1.3   A rough schematic of graphic user interfaces

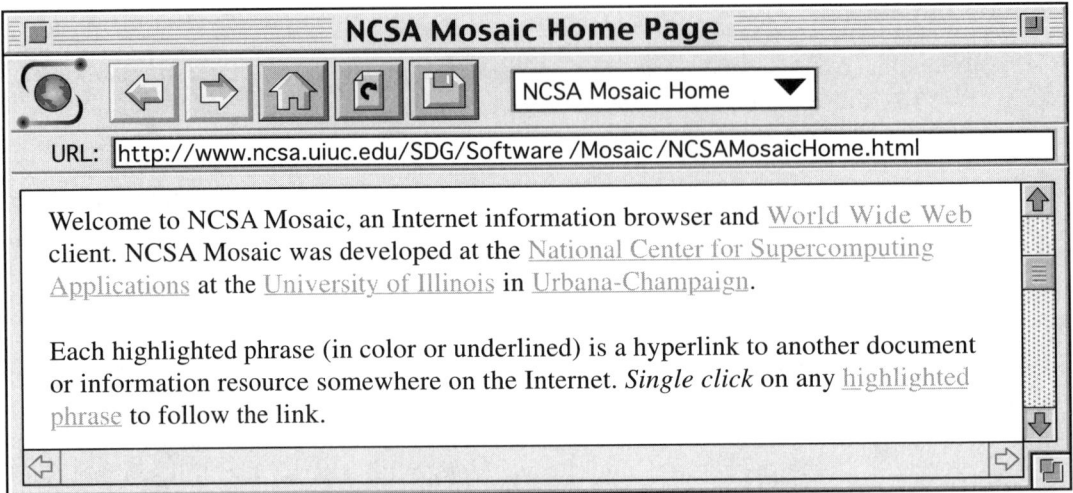

FIGURE 1.4   A typical early graphic-interface window

input is direct manipulation. When graphic interfaces were still widely viewed as second rate in the (much smaller and snobbier) computer community, they were awarded the acronym WIMP, which, ignoring its derisive connotations, still captures the interaction style fairly well: Window-Icon-Mouse-Pointer. (Now that other input/output modalities are under widespread exploration, the acronym has come back to life; the most familiar adjective for alternative interfaces is "non-WIMP.")

Command-line interfaces have not, of course, disappeared under the dominance of WIMPs. As I mentioned, a few have survived in peripheral communities (Unix is the most notable of these). But, more significantly, command-line interfaces can be seen in broken and spectral form haunting some functionalities of graphic interfaces. One of the most familiar elements of graphic interfaces is the menu, which includes the (naturalized, rationalized, and contextualized) names of commands; buttons, too, are commands; field labels cue parameter entry; and so on. Graphic interfaces have added imagery and a more expansive sense of space, separating and distributing the functions of the command-line interface, but bits and pieces of the latter live on in every cranny and nook of the former.

Graphic interfaces are about to undergo a similar fate.

Multimodal Interfaces

During multimodal communication, we speak, shift eye gaze, gesture, and move in a
powerful flow of communication that bears little resemblance to the discrete keyboard and
mouse clicks entered sequentially with a graphical user interface.
    — Sharon Oviatt and Philip Cohen (2000)

The era of the graphic interfaces is nearing an end, but they too will live on, as part of multimodal systems with increased physical inputs and a wider range of outputs — in both cases, chiefly motion and sound. On the input side, gesture and speech will become increasingly important; on the feedback side, video and sound (prominently including speech).

While multimodal interfaces will soon be ubiquitous, and while the most important incorporated modality will certainly be voice, I don't explore the combination of voice with other input/feedback modes in this book, except in passing. Many of the principles of pure voice interfaces we take up are of course applicable to speech in multimodal design, especially the discussion of language that occupies Part I. But I mention multimodal interfaces here largely to explain their absence from my explicit concerns in this book.

I concentrate on voice-only interfaces, with a few casual remarks about voice as a modality in multimodal systems, for two reasons. Designing speech interaction as one modality among several is too easy. And it is too hard. Speech as a modality is too easy because the redundancy of context puts far less pressure on speech design. As one simple example, response delays in speech when that's the only channel are confusing and highly annoying. Think of listening on a telephone to seconds and seconds of silence after you've asked a question. You ask again, but still there's no response. Being put on hold to the caramel tones of a further-homogenized Bee Gees tune only makes it worse. But if you ask a question of someone in person, and they hold up their hand to indicate you should please hang on for a moment — while they catch your eye to suggest they just have to finish up with this other task first — you're far more liable to be patient. I am.

Multimodal systems have this characteristic. They don't require you to focus all your attention on one channel for feedback, a channel that is thereby blocked to further activity while you can only sit and grind your teeth. A graphic display can show that processing is going on, or that an external data source is causing a logjam — some explanation and some temporal indication as to when you can return to the activity — while the multitasking capabilities inherent in such systems let you do something else in the meantime. The other modalities augment speech.

If you can design a voice-only interface, designing the voice elements of a multimodal system is, if not exactly a piece of cake, certainly much easier than the reverse: moving from voice-modality design to voice-only design. Similarly, concerns such as conversational style, continuous recognition, and speaker-independence are all much less urgent for multimodal systems, since there are enough other channels available that individual words or short phrases can do efficient work for the user without her concerning herself much with nuance.

But designing speech as one modality among several is also too hard, in another way — because of the incredible context-sensitivity of language. In a multimodal, multitasking environment, in an office where there are phones ringing, other people talking, and your attention may be widely distributed, something as (seemingly) simple as the referent for a

pronoun, or as (seemingly) determinate as the topic of a discourse, can be extremely difficult for humans to peg down, never mind for machines.

On that note, let's return to our rapid survey of interface styles.

Keypad Interfaces

81#	*Menu of the Touchtone Teller*
51#	*Current Certificate and Savings Rates*
52#	*Current Loan Rates*
11#	*Savings Account Balance (01)*
12#	*Checking Account Balance (02)*
15#	*Specific Account Balance (Plus 2 digit Account Number)*
16#	*Loan Balance (Plus 3 digit Loan Number)*
9#	*Last ATM or Electronic Withdrawal or Transfer*
19#	*Interest Paid on Loan*
20#	*To Access Another Member Account*
66#	*Change Access Code*
69#	*Card Activation*
80#	*Repeat Last Response*
*#	*Cancel Transaction*
99#	*End Call*

Some codes from Touchtone Teller Star One Credit Union

Keypad interfaces, shown in Figure 1.5 and sometimes called TUIs (Touchtone User
Interfaces), work on a linguistic paradigm, employing "words," usually in isolation, though
sometimes in a determinate syntax. They consist of recorded instructions or queries that
users respond to with (literal) button pushes.

A typical exchange looks like this:

Phone: To make your withdrawal from your savings account, press "1." From
        drafts, press "2." From shares, press "3." From a loan advance, press "4."

User: beep

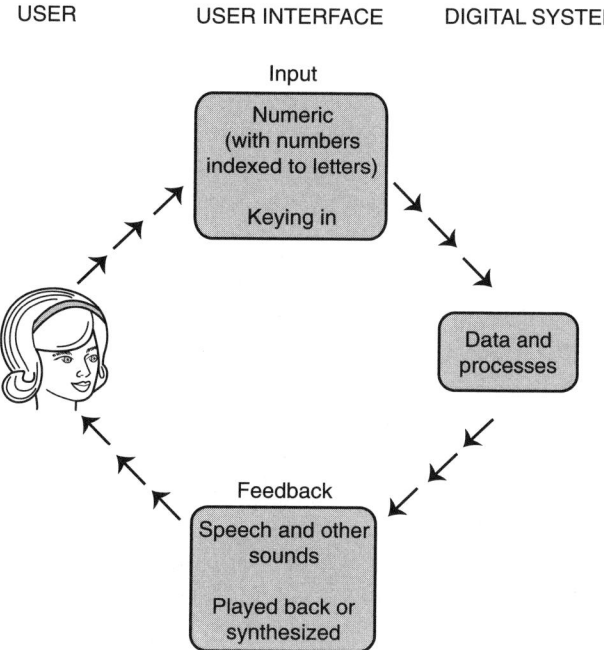

USER   USER INTERFACE   DIGITAL SYSTEM

Input

Numeric
(with numbers
indexed to letters)

Keying in

Data and
processes

Feedback

Speech and other
sounds

Played back or
synthesized

FIGURE 1.5   A rough schematic of keypad user interfaces

As hard as the designers work, as ingenious as some of their solutions are, and as diligent as their user research is, these systems are the bane of most telephone users' existence, and their poor satisfaction rating is one of the principal motivations behind the development of voice interfaces. The catalogue of their annoyances include:

- Arbitrariness of key-function relationships: why does 1 mean "savings," 2 "drafts," 3 "shares," 4 "loan advance"?

- Transience of the key-function relationships: with the exception of a few stable key-function match-ups (* might always mean "main menu" in a given system; 0 might mean "get me a real human"), for each successive exchange in the encounter the keys have different functions (1 means "banking," then it means "withdrawals," then it means "savings," . . .).

- The size, number, and arrangement of the keys: there are too few (only twelve active commands can be available at any given time), too densely packed, in a small matrix rationalized only for short, infrequent, and unidirectional numeric input, not for even brief interactions.

- The location of the keys: most keys are on the handset nowadays, which has to be held up to the head for feedback, then away from the head for input, then up to the head again for feedback.

- The seemingly interminable, hierarchical navigation trees ("menus"): after the user presses 1 (or 2 or 3), there is another branch, and another; then the user might be forced to climb back up the tree and take another branch — a gagged and bound (but for one finger) monkey, trying to follow the mindless, one-note commands of a ruthless task-master.

Speech combats all of these annoyances by its very nature. Words have meaning. That's their job. Meaning doesn't have to be assigned. (Alas, however, early voice systems did indeed have prompts like "for transportation, say 'one,'" sadistically avoiding the natural semantics of words.) And there are a lot of words to choose from, not just twelve. And the user's head is deployed for both the production of input and the reception of feedback. And language is largely independent of conceptual structure; it can instantiate hierarchies, or level them.

The specific solutions to most of these problems, however, are not automatic with the use of language. Aside from the fortuitous production/reception properties of the human head, overcoming these problems falls to implementation and design. While languages have huge numbers of words, systems don't. Their vocabulary space is limited, and one of the most critical jobs of voice interaction design is allocating that space efficiently. Also, while hierarchies can be apparently flattened, they remain very important architectural strategies in system design. Voice interaction design often involves not an organizational structure which avoids hierarchies, but an interaction style which doesn't overly determine the users' paths through them.

Voice Interfaces

No longer the sovereign property of humans, speech has become an ability we share with machines.
— Sarah Borruso

Voice interfaces have a range of labels — as above, many of them are configurations of "spoken," "dialogue," and "system," with each other or related words, usually reduced to acronyms (see the glossary for the most common terms). *VUI* (Voice User Interface) has recently emerged as the leading short-hand term. They work on a linguistic paradigm (word strings), and consist of utterances, plain and simple: speech in and speech out, as outlined in Figure 1.6. Niels Ole Bernsen and Laila Dybkjær define a pure voice interface as "a *uni-modal speech input/speech output* system which conducts a dialogue about a *single task* in a *single language* with a *single user* at a time" (2000). (Their preferred term, actually, is SLDS — for Spoken Language Dialogue System — but it overlaps so closely with what I am calling a "voice interface" that I am comfortable poaching their definition.)

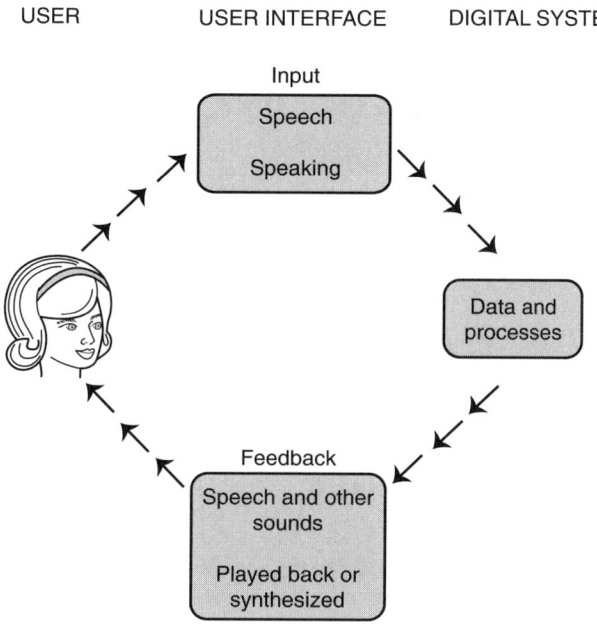

USER   USER INTERFACE   DIGITAL SYSTEM

FIGURE 1.6   A rough schematic of voice user interfaces

While we humans are pretty good at working in multiple modes, and many of us are good at multitasking as well, the voice interface's unimodality is one of its chief sources of power. The other is the fact that the relevant modality, speech, is one to which humans are very highly adapted.

Speech in, speech out: the caller says something, the system says something back. Here's an example:

SpeechActs   Welcome to SpeechActs. Please say your full name.

Nicole   Nicole Yankelovich.

SpeechActs   You currently have "Lunch with Kate Ehrlich" until 2 pm. At 3 pm
             you have "Brainstorming meeting in the lab." At 6:30 pm you have
             "Dinner with Ellen."

Nicole   What do I have tomorrow?

SpeechActs   Tomorrow, you have "Dentist appointment."

Nicole   What about Bob?

SpeechActs	On Tuesday, November 16th, Bob has "Visit to Georgia Tech."
Nicole	And on Wednesday?
SpeechActs	On Wednesday, November 17th at 2:15 pm, Bob has "Return to Boston." At 5:15 pm, Bob has . . ."
Nicole	Stop. Tell me what I have.
SpeechActs	On Wednesday, November 17th you have no appointments.

(Yankelovich, 1994)

This specimen comes from a first-generation conversational voice interface developed at Sun Microsystems in the early 1990s to provide telephone access to basic computer tasks such as e-mail and calendars. (In addition to Yankelovich, 1994, see also Yankelovich, 1995, 1996, and Martin et al., 1996.) While this was an in-house system and more of a test-bed than a product, and consequently primitive in many respects, you can probably see some of its virtues at a glance:

- The caller has much more initiative in the flow of the interaction. Nicole is in control.

- The application allows for implicit reference. Nicole does not have to say "What about Bob's schedule tomorrow?" A simple "What about Bob?" does it.

- The logic of the application is consistent but not bullheaded. Close to the end of this sequence, for instance, SpeechActs starts telling Nicole about Bob's Wednesday appointments, which is reasonable, given that the topic of the exchange at this point (raised by Nicole in her previous turn) is Bob's appointments. Once Nicole realizes she and SpeechActs aren't on the same page, though, she can repair the problem quickly.

- The hierarchical structure is opaque (though architecturally still there): Nicole isn't forced to navigate the structure stepwise. She really wants to know about her own appointments, for instance, when SpeechActs starts telling her about Bob's, and the application lets her get elegantly to her own calendar. She gets there not by explicitly going back up to a "calendar" menu fork, and then down the "Nicole Yankelovich" branch. She just says "Stop" and SpeechActs drops this line. In the same turn she adds "Tell me what I have" and the system — retaining both the topic "calendar" and the focus on Wednesday, November 17th — responds with information about her schedule.

And, most importantly, the feature that the others all hang on:

- The language the caller can deploy is far more natural. It's command-like in places, more abrupt than most of us would probably talk to another person ("Stop. Tell me what I have."). But SpeechActs is not a person, and does not take offense. It's not the tone ("brusque" here, rather than "polite"), however, that makes the interaction linguistically natural. It's the ability to collapse a sequence of system actions (or states) into one ordinary-language utterance (not a specialized syntax of isolated operators), which gets its meaning from context as much as from dictionary entries for the words. And, of course, there are also utterances that would be perfectly at home in a human-human dialogue ("What about Bob?").

Conversational voice interfaces are not the only type of voice interface; they are the preferred voice interface for most phone-based interactions, and they are the focus of this book, but they are computationally expensive and creatively demanding, and are barely known among the general public. If you ask the general public about a "speech application" they might have encountered, they are likely to think of menu-driven voice systems that share the structure of keypad interfaces, and they are likely to curse.

These far-more-common speech applications behave like this:

Telefónica	For information regarding transportation, say "transportation." For information regarding entertainment, say "entertainment." For information regarding weather, say "weather."
Caller	Transportation.
Telefónica	For information regarding air travel, say "airport." For information regarding bus travel, say "buses." For information regarding rail travel, say "trains."
Caller	Airport.
Telefónica	For arrivals, say "arrivals." For departures, say "departures."

And so on. And so on. And so on.[1]

These applications — voice-response systems — have their virtues, and involve many design subtleties. Done right, they can eliminate the arbitrariness and transience of function-semantics that plague keypad systems. They have available vocabularies substantially bigger than 12 keys. And they efficiently capitalize on the human head's dual input-output capacities. For these reasons, voice-response interfaces are an important advance over

---

1: This dialogue is extrapolated from an early application of AT&T's Intelligent Network service deployed in Spain (called, as in the text, Telefónica), reported in Jacob et al. (1992) and Wilpon (1994). The extrapolation takes some poetic license, but to the advantage of the original system, which was even more annoyingly tied to the keypad design legacy. The actual prompts in fact directed the caller to say "uno" for transportation, "dos" for entertainment, and so on.

keypad interfaces, but their call-and-response, one-function-at-a-time interaction, and the hierarchical structure that determines the interactive sequence, makes them tedious to use.

Voice-response systems are not evil, despite the rants I have heard from users, and occasionally given vent to myself. Even the dreaded keypad systems are not wholly without merit (if you know them well, or have a good cheat sheet, they can be very efficient). Indeed, designing conversational voice interfaces needs to build upon voice-response interactions, not start from the ground up, and sometimes conversational interactions will proceed indistinguishably from voice-response interactions. Sometimes conversational breakdowns require voice-response remedies. Sometimes, even the keypad should be integrated into a conversational voice interface (for security, in particular, or robustness when signal quality is low).

But the advantages of conversational interaction are so substantial, because the available resources of human–human vocal communication are so rich, that it is far-and-away the best design style. And it is the topic of this book.

Why Speech?

Admittedly, the keyboard option might be less practical at this moment. . . .
— Caption on a 2002 ad for Vbox (a small, hand-held, voice-and-keypad input device), on a small poster mounted above a urinal.

Implementing voice interfaces to eliminate the scourge of the keypad and counteract the curse of the hierarchy has far-reaching implications, and not just for our collective sanity as a user population. The increasingly competitive call-center sector, for instance, is a commercial area with an economic meter ticking in the billions every year. Voice systems will inevitably become more and more important in this sector, as they already have (along with keypad systems) in phone platforms of most other commercial operations, and ease of use will quickly separate the successful systems from next week's boat anchors. Conversational design is the (potential) road to usability.

Why speech? Largely because the alternative modalities for portable information devices lead to interactive abominations. Telephones are ubiquitous, information is abundant, and there are many times when the former is the best (most convenient, most portable, safest) way to access the latter. But even souped up with liquid-crystal displays, wheels, soft keys, memory cards, features, features, and more features — whatever their truly ingenious designers can glue on — there is an input/output bottleneck that hardware can never solve. The screens, if they are small enough to be genuinely portable, are too small for much information output, and the keys are too small for our gross digits to punch in much information. As *New York Times* technology reporter, Katie Hafner, laconically put it, "browsing the Web with a five-line screen and a tiny number pad is not a very gratifying experience" (Hafner 2002).

The Japanese word for the hoards of mobile informavores roaming our landscapes is *oyayubizoku*, which translates as "the tribe of the thumb," for the digit they are always poking at their portable devices. Thumbs are good. Their opposability has served us hominids very well. But as organs of communication, they fall vastly short of mouths. Given the choice, most people would go with voice over appendages for input into itsy-bitsy devices, any day.

As for output, the screens, even at their small dimensions, certainly have potential. We humans are very efficient visual processors. Information fed to us along that modality can be very rich. But there are still considerable difficulties. Not all information is best in visual format. I would rather hear "warm and hazy with a thirty percent chance of afternoon thundershowers" than try to puzzle out a series of pictograms, especially if I had to squint at a miniature screen to make them out. And even information that might be best presented visually, I would still prefer to hear if it meant I could keep my eyes on something more urgent, like an eighteen-wheeler coming up fast in my rear view mirror.

There are also significant barriers to graphic interfaces that voice interfaces can overcome, both individual and situational. Not everybody is equally facile with graphic interfaces. Some people can't see, or can't see very well. Some people can't point and click and type (a category which includes not just the severely disabled but the increasing legions of computer users with repetitive strain injuries). Some people, often called "children," don't have the dexterity, or even the hand size for keyboards and pointers. And not every environment is equally amenable to direct-manipulation input. Computer systems would be better served by voice input and output when they are used in situations where hands and eyes might otherwise be better occupied — repairing equipment, for instance, or sorting through inventory, or just (in the case of hands) jammed into pockets or mittens for warehousing operations in a North Dakota January.

Speech is, further, very resilient as a side channel, making it the ideal mode for what Salvucci (2001) calls *secondary-task interfaces*. These are interfaces for systems or functions when the computational activity is not the primary task (for instance, supporting an installation where the user is busy handling equipment but still needs to check part numbers, follow procedures, and the like). The issue of safety can become quite important for secondary-task interfaces when the primary activity is potentially dangerous, such as driving. As long as the guy screaming past me on the highway and switching lanes with abandon insists on using his telephone, I would be happier if he had voice dialing. As long he wants traffic or weather or stock information on demand, while he is switching lanes, I would rather have him talking and listening than trying to push buttons or squinting at a tiny display.

Speech also has the potential to reduce the mode confusion common in everyday digital living. One of the most compelling attractions of *Enterprise*-style life-in-the-future is the ability to tell the elevator what deck you want to go to, to tell the holodeck to play sultry music or end program, to tell your replicator "Tea. Earl Gray. Hot." Sure, the holodeck and the replicator would be handy, but it is homogenization of the interfaces — just talk — which makes that world so seamless. As most speech-system researchers will tell you, shortly after they say hello, the world of *Star Trek* is a long way off (it does take place in the 24th century after all). But the possibility of interacting with, say, entertainment appliances "without first scanning a cryptic array of choices on any of several remote control devices" (Hafner, 2002b) is imminently possible.

Speech also opens up more than just input and output channels. It can open up an additional cognitive dimension. Affective computing is becoming a significant research pursuit, and voice is a more reliable indicator of emotion than anything available to a graphic interface. Word choice can signal mood quite directly ("great" suggesting more enthusiasm than "good," for instance). But the pace and style are even more reliable. Rapid volume changes suggest anger or urgency; slower, more deliberate speech correlates with seriousness, chirpy speech with pleasure. To the extent that computers will make headway either interpreting or inducing emotion, they will need a fuller understanding of speech.

Why Conversation?

> *The brain is actually a very shitty computer.*
> — Richard Wallace

We're good at conversation, we humans. True, some people are more entertaining at dinner parties than others, and some people know how to make you feel more interesting when you talk to them than other people do. Some people you enjoy conversing with, some you don't. But all of them are "good at conversation" in the same fundamental way of knowing that a greeting pairs with a greeting, but a question pairs with an answer; that drawn-out intonation means the speaker isn't ready to relinquish his turn to talk just yet, even though he may have run out of words; that specific routine strategies of coherence and cohesion weave notions into topics, words into discourse. There is variability in our capacities to deploy this conversational know-how, just as there is variability in the velocity and grace with which we move, but we are all masters of conversation in very much the sense that we are all masters of bipedal locomotion. It's something we do, after early childhood acquisition, instinctively.

Computers, thankfully, are fairly good at actuating our conversational dispositions, in limited, task-specific ways. It is not part of their essential structure, which really only carries out computational instructions — changing states, shuttling information around, setting conditions, and the like. The facility of computers to work with our conversational intuitions is the result of decades of difficult, user-centered interface crafting; of developing systems that accommodate our native human strategies and talents rather than (just) requiring humans to accommodate the single-minded, instruction-processing dispositions native to computers. At the very outset of real-time, individual-user, responsive computing, programmers quickly adopted a conversational metaphor for this new mode of computer behavior (Orr 1968, vi). Interactive design begins with this metaphor. One of the foundational articles in the evolution of graphic interfaces (Foley and Wallace, 1974) charted the significant linguistic analogs in graphic interfaces, stressed the critical importance of feedback, and argued for a design philosophy anchored in the concept of "Natural Graphic Man–Machine Conversation." The importance of this philosophy for the success of graphic interfaces is hard to overestimate; it governs the systematic tendency to treat most user actions not just as instructions to be processed but as contributions toward a goal, contributions requiring a semantic response.[2]

But it is a metaphor. We pass the pointer over a tool and get the semantic pop-up response "Text Box" or "Insert Picture." We key command+S and get the semantic response of a highlighted File menu and little jingly sound. On the other side of the interaction, our machines await instructions disguised as semantic input like pressed "Open" or "Print" action buttons. The computer doesn't care about those labels, only about the microcode they get translated into. For the machine's purposes, they could be reversed, or "Geöffnet" and "Druck," or "Archie" and "Jughead." The semantics are for the humans.

Our dealings with computers are rife with pieces of natural language, not just in button labels and menu-item names, but in "dialogs," often defined by the conversational turn structure of standard utterance-pairs like question and answer. "Do you want to save the changes you made?" my word processing application asks me when I quit, soliciting any of three responses, "Save," "Don't save," and "Cancel" (and biasing for "Save"). This sort of interaction would never be mistaken for a chat; the metaphor only goes so far, and most of the time you wouldn't want it to go any further. Graphic interaction is *similar* to conversation in important ways, but it is not a *type* of conversation, and it shouldn't be.

Now, however, we have computers that can process speech and respond in kind, whose input-output semantics therefore come almost exclusively from natural language (rather than partially and metaphorically), and whose ear-and-mouth (rather than hand-and-eye) interactivity triggers our conversational dispositions more immediately and insistently than any other technology we currently have. The question is, then, should we thwart these dispositions or court them?

The options in the design of these new interactive creatures are subtle and varied, and we will explore them in some depth, but the question of an overall design philosophy can be oriented with respect to two poles: the keypad model, where the machine presents options, constrained by the extremely limited input options to resemble a multiple-choice exam; and the conversational model, where the interaction occurs between agents working towards a mutual goal.

2: By *semantic response* here, I mean only that the system responds not just with an internal change of states but with a meaningful signal to the user about the change of states. In particular, I do not mean "semantic" in the Foley and Wallace sense of specifications of system functions.

Should these voice systems, that is, lead us about by the tongue, utterance by utterance, up and down the ladders of an obscure hierarchical network? Should they demand that we repeat isolated little noises they dictate to us, a staggeringly atypical pattern of talk? Should they respond to us with repetitions of their own, parroting our words back to us incessantly for confirmation, like a four-year-old kid trying to drive her brother crazy? Or should they function negotiatively — making suggestions, but also following our lead, initiating but also responding to communication repairs, interacting in humane and sensible ways? Interaction design is a tradeoff, always, between machine and human. But the keypad model favors the machine in that tradeoff; the conversational model favors the human.

The interactions that result from good implementations of the conversational model lose their metaphorical quality. They are not *like* conversations, they *are* conversations. They are not freewheeling dinner party conversations. They are more on the order of service-encounter conversations, where one person helps another with some information or assists their purchase of a hammer.

"Before too long," Jef Raskin quotes a mobile computing authority, "you may not have to worry about an interface at all. You may find yourself simply speaking to your computer" (1993: 122, 2000: 2). Raskin reproves this authority for failing to understand that voice interaction *is* an interface. Perhaps the authority doesn't understand that (so much context gets lost in isolated quotation that it is hard to tell). But he doesn't say "there will be *no* interface." He says "you won't have to *worry* about the interface." It's a fair bet he's not talking about the if-you-hate-voice-systems-say-"yes," menu-driven, hierarchical, voice interaction — which you *do* have to worry about. He is talking about talk, about speaking, about conversation.

What voice interfaces require, as Hayes and Reddy put it very early on in the development of these systems, is the capacity for "graceful interaction" (1983); conversation is the model for graceful interaction using voice.

Talking to Machines

There is evidence that people's communication with computers differs [from] their communication with humans.
  — Niels Ole Bernsen, Hans Dybkjær, and Laila Dybkjær

Talk is not all the same. It occurs within *genres*, just like text (hand-scribbled notes, email, junk mail, novels) and television (sitcoms, documentaries, award shows, dramas) occur within genres. For that matter, locomotion (walking, running, driving, sitting in a rickshaw) and shopping (for hardware, for groceries, for lingerie) have distinct genres. Activity has contexts, governing conditions, strategies; and talking is a hugely distributed activity, broaching many contexts, conditions, and strategies. Hardware-store talk is different from lingerie-store talk, which is different from ordering a hamburger or greeting a neighbor. Moreover (to admit just a little bit of complexity before retreating again to simplicity) these categories can overlap and interpenetrate: it may be that our neighbor sells hamburgers, or that we are shopping in a department store for a hammer and a Valentine's Day gift at the same time.

Looking only at the narrow matter of who-talks-when, genres of talk can

be viewed as a continuum ranging from the relatively unconstrained turn-taking of mundane conversation, through various levels of formality, to ceremonial occasions in which not only who speaks and in what order, but also what they will say, are pre-arranged (for instance, wedding ceremonies).

(Hutchby and Woffit, 1998: 147; they are paraphrasing the groundbreaking but less-succinct Sacks, Schegloff, and Jefferson, 1974.)

One might also locate genres of talk on other continua, defined by, say, content (hardware, lingerie, hockey), context (store, phone, street corner), or purpose (commercial transaction, information exchange, social maintenance). The variables are extensive. And that's the point: when we talk about talk we have to (1) remain sensitive to genre, and (2) remember the variables.

The subject of this book (talk with machines) is an emergent genre. More accurately, it is a range of emergent genres. This subject has been explored seriously for over forty years, and imagined for hundreds more, but it is only now coming into existence. Only now are the contexts and conditions, conventions and strategies, being fashioned at both ends of the telephone. We don't yet know how people prefer to talk to machines. We definitely don't know how they will be talking to them in twenty years. But there are plenty of hints, lots of studies, and we all have hunches. My strongest hunch (but certainly not *only* mine), and the belief that drives this book, is that people will want to piggy back their speech-dealings with machines on the cognitive and social evolution of language; that they will, for instance, prefer continuous speech to isolated words; that they will want to manage their inputs, in response to the machine's outputs, much the way they manage conversational turns with other people; that they will be happier with a system which allows them to participate actively rather than one that leads them through a series of navigational obstacles.

This belief is *not* that people will treat talking machines exactly the way they treat talking organisms. On average, there will surely be less concern for social graces, along with far more willingness to issue direct commands, to interrupt, and to be generally more abrupt. For instance, people often use prerequests to get attention and set up an information query, but utterances like "Can I ask you something?" are not likely to be commonly directed to a voice system; or, if they are, they would far more likely be simple, direct questions (rather than prerequests) by a user who wants to check on the interactive possibilities of the system.

On the other hand, there will surely be more accommodation made for conceptual and performance errors; people speak differently to children, foreigners, and persons with cognitive deficits than they do to adults who are cognitively adept and fluent in their own language. They won't have the same kindness and tolerance, surely, towards the voice-interface front ends to powerful multinational corporations that they might show towards a mentally challenged young woman asking for directions. But people do have the capacity to alter their articulations and expectations as a function of their partners-in-talk, and this capacity is called upon by voice interfaces.

But these actual and potential differences do not argue, as some people take them to, that the path to better voice interfaces should not run through the wide territory of natural speech interaction research. On the surface, it might seem to make more sense to study how people interact with voice systems now. That research is clearly important, but it is nowhere near sufficient, and used exclusively will only result in dull, unimaginative systems. Since (1) voice systems have manifold technological limitations and design flaws, and (2) speakers have a remarkable capacity to adapt to language behaviors, even primitive and bizarre language behaviors, studying current voice system interactions is of quite limited use in understanding what the optimal design principles are for voice interfaces. That's like studying how people shuffle around in manacles in order to understand the optimal design principles for running shoes.

We don't know the specifics, because these interfaces are so new. But to generalize from how people interacted via speech with computers as recently as five years ago to how they will interact with a system built in the next five years is a highly questionable enterprise. To generalize from how they speak to each other, however, while not without its risks, provides a more promising foundation.

The idea is not to build machines that can converse fluently on 19th century aesthetics or the implications of iridium layers for paleobiology, though such machines will surely come in this century. The idea is simply to build machines that allow people to exploit their naturally cultivated linguistic and conversational skills, rather than developing an entirely new repertoire of language behaviors because the interface designers think it is perfectly reasonable for a speech system to talk like this:

TOOT   Hi. This is AT&T Amtrak Schedule System. This is TOOT. How may I help you?

Caller   I'd like to get a train from Philadelphia to New York.

TOOT   I heard you say "go to New York from Philadelphia." When do you want to leave?

Caller   Sunday.

TOOT   I heard you say "leave on Sunday." What time do you want to leave?

Caller	10:30 p.m.
TOOT	I heard you say "around 10:30 p.m." Do you want me to find the trains from Philadelphia to New York on Sunday around 10:30 p.m. now?
Caller	Yes.
TOOT	I am going to get the train schedule for you . . .

(Walker, Kamm, and Litman, 2000: 6)

Never mind the bizarre, and bizarrely iterated, phrase "I heard you say 'X.'" In every single case that "X" is wrong, it is not what the caller, in fact, said. The caller in this exchange is remarkably cooperative, but TOOT is not; in fact, it violates a well-known conversational maxim such that providing more information than is required during your turn is surly. A humanely designed speech system, one that operated in a way that learns from how humans engage in dialogue, would talk more like this:

Caller	Sunday.
TOOT-2	Sunday? What time?

There is confirmation here, which speech systems require far more often than humans; the hypothetical TOOT-2 does not act just like a person. But it doesn't fall into rote confirmation, either. And TOOT-2 takes the next conversational move (seeking a departure time) efficiently.[3] The systems don't have to be human, just — to use Jef Raskin's beautiful design term — humane.

What Isn't in This Book

You can't have everything. Where would you keep it?
— Steven Wright

This book is largely unconcerned with the mechanical specifics of speech systems, or their implementational languages. I follow the advice of Michael Norman and Peter Thomas about the application of conversational human factors principles to speech-system

---

3: I am acutely aware of the arrogance involved in playing armchair designer, as I do here and elsewhere in the book — sitting back and proposing changes on paper to a real, working system that inventive and hardworking people have sweated to develop — and, in fact, the current Amtrak speech system (1-800-USA-RAIL) is somewhat less dogmatic than the original TOOT. But the progressive road for interaction design, as everyone who works in the field understands fully, is iterative, and I am just trying to use this iterative spirit to investigate design possibilities and to push voice interface development forward.

development, which they say "should be *concerned with* technology, but must be *independent of* technology" (1990: 55).

There is a little bit here and there about recognition and synthesis and natural language understanding, but only at the most basic level, and only from the design perspective (to understand, for instance, why recognizers are so error prone, or what characteristics of synthetic speech can be altered to influence the perception of emotion). There are already good books on speech technologies, by more knowledgeable people. (As we go to press, the best ones in my estimate are Robert Rodman's [1999] *Computer speech technology,* for elegance, clarity, and patient explanation; Douglas O'Shaughnessy's [2000] second edition of *Speech communications: Human and machine,* for comprehensiveness and detail; and Christopher Schmandt's [1994] *Voice communication with computers: Conversational systems,* for careful attention to the constraints these technologies place on voice interaction design. The field is developing quite rapidly, however, so there may be more up-to-date books within a year or two of the publication of this book, *Voice Interaction Design* — perhaps even new editions of Rodman, O'Shaughnessy, or Schmandt.)

The speed of technological advances, in fact, is the principal reason for this book's indifference to speech technologies. The advances in signal processing algorithms, memory, and brute power have far outstripped design in speech-system research. This book is an attempt to push design.

While there is an undeniable developmental reciprocity between technology and design principles, the latter are a good deal more stable and always need to take the driver's seat. Those of us who were around for the frustrating early years of graphic interfaces remember a lesson that we can take for speech-system development — from the many awkwardly designed systems that sprang up and died while technology (the applications) ruled supreme. When graphics experts and human factors specialists began to develop controlling interests in graphic interface design, things changed very much for the better. Speech system development currently needs more language experts and human factors specialists than coders. This book aims to help push the field in that direction.

My Approach

Contrary to popular opinion in at least some circles, applications of speech technology don't generally fail because a speech recognizer's accuracy is 93% instead of 97%. . . . They fail because human factors concerns are not addressed.
— Daryle Gardner-Bonneau

I have a broader conception of voice interaction design than many current practitioners, who see voice interaction design and development as challenging and important work, but still as something of an add-on to the actual, true, and real engineering object, the speech system — that complex of circuits and code that defines recognition, parsing, language-modeling, and speech synthesis. In this view, a common one, "the interface" is isolated to one module, often called the dialogue manager. In my view of human–computer interaction (by speech in specific, but by any modality in general), there is no such segregation. A voice interface is the sum of all human-factors concerns in the design and development of a speech system (including, but not limited to, "dialogue management"). My arguments to this end are bound to be controversial in some quarters; they have already rubbed some practitioners the wrong way.

I not only insist at every opportunity on the primacy of human factors concerns in speech system development — exactly what you would expect, I would hope, from someone writing on user interface design — I insist on the necessity of a thorough grounding in the specific, rich human factors research base on how people use language interactively. This base has been developing for millennia, and our travels sometimes take us back to Ancient Greece, but it gathered incredible momentum in the last half of the 20th century, in a range of overlapping fields, chiefly linguistics, philosophy, sociology, and psychology. I also insist that heavyweight resources — both technological and human — be put behind the application of this research to voice interface development. This insistence especially has been called extravagant, excessive, idealistic — not by everybody, but by a noteworthy contingent of old-guard speech-system folk.

Yet, these arguments really only come down to the low-level application of common sense. Language is a fundamentally human instrument, with a distinct and complex character. Its composition and use is certainly related to general cognitive principles (which I also advocate voice interface design respect, of course), but it also has a wide range of unique elements, relations, and protocols; moreover, even general-purpose cognitive principles have somewhat distinct effects in the realm of linguistic interaction. Attention, for instance, and salience, have quite specific realizations in language.

Designing speech-based interactive systems with only a loose understanding of this character dooms those systems to mediocrity, if not outright failure. And thorough understanding does not come cheaply. Understanding the nature of linguistic interaction in a theoretical way takes energy and time. But understanding it in an applied way takes more. It takes energy and time too, but also money, machines, and methodology.

To understand the way a community uses language to accomplish a set of goals, like booking tickets or ordering pizzas or doing some banking, requires closely studying that language domain. It will have its own terminology, its own utterance structures, and its own interactive patterns. These need to be collected into two interdependent databases — a general one, called a *corpus*, which houses and organizes a wide sample of language use, especially dialogues; and a specific one, called a *lexicon*, which houses and organizes the characteristic words of that domain. Building and then using such databases requires people and computers. My advocacy of them (and more particularly, of corpus linguistics) for voice interface design is a hallmark of the extravagance for which my approach has been criticized.

But these tools, or tools very much like them, are mainstays of the other areas of speech-system development, and all I am ultimately advocating is a resource allocation model that puts the same effort — technological and human — behind interface development as behind recognition, parsing, and language understanding. That seems perfectly reasonable to me. Moreover, since, as Roni Rosenfeld notes, conversational systems "require a lengthy development phase," anyway, "which is data and labor intensive" (quoted in Green, 2001), it is deeply misguided not to foreground human-factors concerns throughout that process. Leaving user issues to the end doesn't work with graphic interfaces; leaving them until the end with a system as dependent on user dispositions as an interactive speech application is an even worse idea. The speech-system development cycle, as I propose in the book, should be a voice interface development cycle.

I also argue for resource-heavy testing, for the use of other digital tools, like digital design specifications, the use of knowledge-representation scripts as the basis of the interaction model, and the model of the voice interface as an expert system. From the human-factors language research base, I draw on the findings of lexicography, conversation analysis, pragmatics, computational linguistics, and the collaborative-action framework of social psychology, as well as a range of more local disciplines, like phonetics, syntax, and semantics. Some of the components of this mélange are more accepted than others — but collectively they nowhere receive the kind of attention I give them here.

My terminology, too, is occasionally nonstandard — perhaps most egregiously for many traditional speech-system people in my use of the word *vocabulary* for what some of them call a *dictionary*, and most call a *grammar*; that is, for the storehouse of interrelated acoustic models that defines the "sublanguage" of the interaction: what the users can say, what the system can hear. But I have also adopted or developed words from a range of disciplines to cover the overlaps and amalgams that are necessary when dealing with such a large and diverse body of research. In all cases, I've done this solely in the interest of conceptual clarity. Individuals who have investments in one bag of terminology or another may not take kindly to the way I have altered or ignored their terms, but I hope that I have served the design community's best interests.

And, while there is no shortage of opinions and recommendations in this book, I have avoided a formulaic connect-the-dots approach that would falsely suggest that you just have to connect up X, Y, and Z if you want a great voice interface, only X and Y for a good one. Such design books, and there are many, trivialize the richly creative, collaborative process of crafting intricate, interactive artifacts for human use. And they are misleading. "Design is complex," as Deborah Mayhew, among many others, has noted, "and there is no cook-book approach that can rely on general principles and guidelines alone" (Mayhew, 1999: 4).

The point of this book, simply put, is to provide voice interaction designers with the knowledge and the strategies to craft language-using applications that behave the way a language user expects other language users to behave. But that knowledge and those strategies do not provide a prefabricated speech system.

The Rest of This Book

[In] the fusion of art and technology we call interface design . . . the inventors and practitioners have blurred into one holistic unit, like a science lab hosting a creative writing seminar.
— Steven Johnson

A major reason that good graphic interfaces (integrated with useful tools) are so effective is that they capitalize on the contextually embedded, your-turn-then-my-turn, mutually reinforcing, collaborative strategies of human-human interaction. Using computers in the 21st century is far closer to the way we interact conversationally with other people than it is to the way we use a shovel or a bicycle or a chain saw. All user interfaces are dialogic — turn-by-turn communicative exchanges of information and intention, more-or-less similar to the dialogues between humans. But with non-voice interfaces this dialogic quality remains metaphorical.

Voice interfaces are literally dialogues: two agents communicating in one coherent, spoken discourse. So, while the principles of discourse have lessons to offer the design of all human-computer interactions, those lessons are much more direct and solid and authentic for voice interface design.

There are still figurative elements to all this. The "agent" at one end of the line does not have the cognitive and social capacities of the agent at the other end. Notions like intention and personality and even communication can be used at the machine end of the line only under an explicit fiction, where the voice represents a system, a company, its designers and developers, the way a particular main window (with its menus, bars, and palettes) represents Illustrator, Adobe, Rick Boyce, Jeff Bradley, Paul George, and so on. But because the system, institution, and people are condensed into a voice, these figurative aspects are fading, and it may not be long before terms like "ticket agent" and "travel agent" primarily denote synthetic creatures.

The extent to which the figurative aspects recede, and the speed with which they do so, will depend on one thing, and one thing only: the facility these agents can acquire with spoken language. That facility is the focus of the rest of the book.

First, in **Part I: Talk**, I outline some of the features and principles of speech, from acoustic wave form to discourse. The material is unabashedly opportunistic, and for the most part oblivious to theoretical nuance. There have been, for instance, a good many exchanges among representatives of speech-act theory (philosophers and linguists, mostly), conversational analysis (sociologists), and discourse analysis (sociologists again, with literary theorists and refugee linguists thrown in). Some exchanges are little more than embarrassing turf wars, some are serious attempts at syntheses, or outright absorptions, and some are just dismissive hand-waving. Some are useful, some trivial. I have simply plundered this literature for what speech applications need, and ignored the counter-arguments about, say, whether speech-act theory makes sense in the context of a conversation, because (for instance) strict versions of the theory require seven components of illocutionary force, none of which determines the appropriate conditions for a reply to that act, and conversations depend on understanding reply conditions. If I can jerry-rig two approaches together sufficiently to generate something useful for voice interaction design, that's good enough for me. And there is more jerry-rigging going on in Part I than trying to reconcile speech-act theory and conversational analysis and text linguistics. The whole chapter is a stew of results from fields and approaches that sometimes view each other with antagonism. Tough. Let them fight it out. We'll take what we can use.

The chapters in this section are:

Speech, an introduction to the section.

Sound and Meaning, an overview of the mechanics of language, follows the conventional (and fruitful) approach of four specific focal ranges: the phonological, which attends to sound; the lexical, which attends to the words built out of those sounds, when they are linked to meanings; the syntactic, which attends to how words travel in certain acceptable packs (those packs usually labeled "grammatical"); and the semantic, which attends to the way those packs make composite meanings ("Bart poked Lisa" and "Lisa poked Bart" have all the same elements, but different composite meanings). The chapter also briefly takes up the more distributed linguistic notion of prosody, the rhythms that give speech its life, and sketches out the processes of speech synthesis.

Doing Things with Words, a survey of the impact of context on language use, draws mostly on the traditions of ordinary language philosophy, which opens up the focal range even further to include circumstances and purposes ("Fire!" is a very different utterance in a crowded theatre than on the battlefield). Chief among the circumstances and purposes considered in this chapter are the ones that govern conversational exchange.

Conversation, a review of the principles guiding what some theorists call talk-in-interaction, continues to draw on ordinary language philosophy, particularly for the way in which it reveals that speech is a form of action ("yes," for instance, doesn't just signal agreement; it might initiate, or finalize, a purchase). The organizing insights of the chapter are from a branch of sociology called *conversation analysis*, and a branch of psychology which doesn't have a snappy label but which takes a perspective on talk that articulates the single most important attitude for voice interaction design, almost a mantra: that talking is a species of *collaborative action*.

Glue, an overview of the networks of reference and relations that bind dialogues (and all discourses) together. The words, utterances, turns, and exchanges consolidate into dialogues because they invoke the same entities and they support each other, on the two elemental levels of language, content and form. Those networks effect the coherence and the cohesiveness of the dialogue. Coherence is a function of the conceptual networks, cohesion of the formal networks.

Diction, borrowing a somewhat quaint-sounding word from rhetorical theory, centers on the single most critical element of voice interface design, choosing the optimal word for a given audience, purpose, and context. We draw most fully here on the work of corpus lexicographers, people who figure out how words function by looking very broadly at how they are used.

In the second section, **Part II: Design,** I outline more specifically, and in greater detail, the processes, principles, and practices of crafting voice interfaces. The natural model for comparison is the design of graphic interfaces, which transfers very well in one way, and quite poorly in another. The overall graphic interface design *process* carries over very well: up-front task and user analysis, iterative development with lots of testing, and situated design. There are differences in the development cycle — in the depth of domain analysis, especially the use of natural dialogue studies, and in the importance of Wizard of Oz studies — but the most significant differences are in the specific crafting of the interface. The design medium, spoken language, means a much different palette of tools to work with, and a much different set of constraints to work under.

The chapters in this section are:

Crafting Voice Interfaces, an introduction to the section and a treatment of several basic issues, starts with the first question to ask, at the outset of the design process, whether or not the target service is appropriate for voicing. Of particular concern, because there is both so much potential and so much hype, is voicing web sites, a highly visual medium which often has text that is poorly suited for voicing. We also take up two sets of design contaminants — the attempt to apply graphic design sensibilities to a verbal medium, and the use of menus in a speech system. The chapter rounds to a discussion of the guiding sensibility of this book generally, and the design section specifically, the pursuit of habitability, the property of a discourse model to provide users the room they need to achieve their goals in linguistic comfort.

The Team and the Process presents a roll call of the job functions necessary to develop a habitable voice interface, and the iterative process in which they can do it. The roll call is controversial, chiefly in its overall size and its inclusion of a Lexicographer. But all the roles are unquestionably necessary for the development, even if several of them are combined in one individual: the leader is an Interaction Architect, the Lexicographer is responsible for the research necessary to build the discourse model, Dialogue Writers put the speech system's utterances together, the Soundscape Designer creates the nonspeech audio, the Quality Assurance Prime oversees the development process and enforces the milestones, and the Usability Prime guides the heavy user testing necessary for voice interface development; Subject-matter and Speech-technology Experts are also important team members. The development process this team follows is the familiar spiral model of iterative expansion, from prototyping through release.

Users, Tasks, an outline of data-gathering considerations and strategies for understanding the users and their tasks, starts with suggestions on how to capitalize on whatever existing data there might be on the people and processes in the target domain for the voice service, but spends most of its time outlining primary research techniques for observing and interviewing the users, as well as charting and analyzing their tasks.

Building the Discourse Model revisits habitability in terms of the goals necessary to achieve a coded understanding of the register that defines the voice interface, in terms of lexical, syntactic, and functional criteria. This chapter advocates the use of natural dialogue studies and strongly recommends the adoption of two powerful design instruments in carrying out and profiting from those studies — a dialogue-rich, register-specific corpus, and a digital lexicon.

Agents is a discussion of the principal considerations in designing the character(s) of a voice interface. There is, unaccountably in my view, a controversy about whether voice interface agents should be human-like or not (*anthropomorphic* is the preferred term of the controversy, though I prefer *personified*). I review the controversy, mostly to demonstrate that voice interface agents, by their very linguistic nature, are already personifications, and the design questions concern how one comes to grips with that fact. The chapter outlines the best ways to come to grips: issues of branding, aesthetics, personality, gender, emotion, and ethical character are all investigated, as well as the reasons for recorded-voiced and/or synthetic-voiced agents, and the reasons for using a single or using multiple agents in the interface.

Dialogue Matters is largely an outline of strategies and concerns with respect to communicative slippages. Things always go wrong with speech systems — in part because of the inherent fragility of the technology, in part because of the immense difficulty of working with a medium as slippery as spoken discourse — and this chapter focuses on how to minimize those slippages. It offers a taxonomy of errors and slippages. It recommends preventative measures such as effective prompting, careful vocabulary building, and certain resolution strategies. It recommends how to fix the slippages that do occur, through strategies that guide the user in the repair process. And it treats other significant dialogue matters related to prevention and repair — managing initiative, tapering and expanding system output, as well as issues of working with preexisting text, especially from graphic sources, and concerns that arise because of the legacy often inherited by speech systems that replace human or keypad interactions.

Scripting is the development of a detailed design specification, with two principal components, writing the dialogue and planning the call flow, both of which this chapter charts out thoroughly. It develops the idea that a conversational voice interface is a form of expert system and begins with a conventional knowledge script for a take-out/delivery order service, elaborating that script in the direction of dialogue acts and conversational turns, on the one hand, and of an interactive architecture on the other. This chapter also advocates the adoption of a digital design spec, rather than (just) a paper-based document.

Iterative Evaluation, an account of the testing procedures needed throughout the development cycle, pays particularly detailed attention to a method highly suited to voice-interface development, Wizard of Oz testing. Oz testing has a human (the Wizard) simulating a speech system, which allows for the early and rapid testing of design ideas. The human, and the necessary support resources (computers, other humans, possibly a vocoder), are unique to this methodology, but otherwise the process is of a piece with usability testing, so I spend comparatively less attention on such testing, but I also review heuristic evaluation methods, and beta-testing, field studies, and what I call the pluralistic talkthrough.

    Conclusion — Pursuing Habitability, briefly wraps up the themes of the book.

    Glossary; there is also a comprehensive glossary of relevant terms from linguistics, philosophy, sociology, telephony, human-computer interaction, and speech system research.

Summary

The need for conversational agents has become acute with the widespread use of personal machines with which to communicate.
— Yorick Wilks

"Today, you have to understand the systems," W.S. "Ozzie" Osborne (general manager of IBM's Speech Systems division) said at the turn of the century, "but soon the systems will understand you. That's the transition over the next couple of years, from you learning machines to machines learning you" (quoted in Hellweg, 1999). That's what this book is about: helping the designers of the machines build ones that better understand and accommodate people's speech behaviors.

    In this chapter, I have outlined the various user interfaces relevant to the evolution and design of voice interfaces, including a general description of voice interfaces themselves. I have also argued for the importance of speech as an interaction modality in a range of circumstances — chiefly eyes-and/or-hands-busy circumstances — and for the centrality of human–human conversation in voice interaction design. I have sketched out the main components of this book. And I have articulated a vision of the driving importance of human-factors research to the design of voice interfaces, as well as conceiving that research very broadly, to encompass the work done in a range of (sometimes mutually hostile) disciplines on the fundamentally human phenomenon of interactive language use.

Talk

Speech is evanescent.
— Herbert C. Clark and Deanna Wilkes-Gibbs

Think of the tools in a tool-box: there is a hammer, pliers, a saw, a screwdriver, a rule, a glue-pot, nails, and screws. The functions of words are as diverse as the functions of these objects.
— Ludwig Wittgenstein

Speech is the telephone network, the nervous system of our society, much more than the vehicle for the lyrical outbursts of the individual soul. It is a network of bonds and obligations.
— J. R. Firth

What we're dealing with is the technology of conversation.
— Harvey Sacks

Speech

Every workman in the exercise of his art should be provided with proper implements.
For the fabrication of complicated and curious pieces of mechanism, the artisan requires
a corresponding assortment of various tools and instruments.
　— Peter Mark Roget

One fact that all consumer-end technology companies either realize — or get driven from the market by failing to realize — is that fidelity to human factors research is what keeps their products from ending up in the commercial dustbin.

Human factors in speech-system development originates in linguistics, philosophy, sociology, and psychology — the disciplines that concern themselves with what humans do when they use language.

There will be resistance among some readers, perhaps maybe even a touch of horror, at the level of detail in this book about the workings of language. Technologists are especially susceptible to this horror. Everybody talks, pretty much, and everybody does it well enough to get other people to pass them the salt when they want it, or to give them information about the weather, or to transfer funds between two accounts. That native ability should be enough, shouldn't it? Who cares about coherence relations or conversational implicature?

Ultimately this attitude comes down to a prejudice against the technical understanding of human activity. Some people, for instance, can see the point behind a university degree or two and an entire career based on the technical understanding of mechanical activity, like robotics, but not human activity, like the use of natural languages. At some point, we just have to wish such people good luck and go back to our business. But there are some responses that can serve to highlight the prejudice. Such folks usually don't complain, for instance, about the amount of detail in a book about artificial languages, like C++

or Perl or Java — all of them comparatively trivial linguistic constructs when set cheek by jowl with natural languages, like Japanese or Mandarin or English. Nor do they pause to reflect that everybody, pretty much, walks, runs, and jumps, yet the technical knowledge required to build machines that can perform these activities is vast. (Activities, we might add, which a huge range of organic forms share, while language is unique to humans.) And, in fact, the few chapters in this section on language are little more than a primer to a range of important concepts. There is much more which might be said; there are books longer than this one just on *verbs*.

In This Section

In a language, one can communicate about communication.
— Charles F. Hockett

One of the properties that distinguishes natural languages from every nonhuman communication system we know, is the ability to talk about talk. Vervet monkeys, for instance, have an extensive repertoire of communicative hoots — among them, different warning sounds for avian predators (like eagles), lateral-attacking terrestrial predators (like leopards), and low-attacking terrestrial predators (like pythons). Apparently, they can even lie with these signals, to gain the upper hand in a skirmish, or to divert attention from food they don't want to share (see, for instance, Seyfarth, R. M. and Cheney, 1992, or Hauser, 1997). What they can't do is say "Oops! I meant *avian* predator," or "That was a fine articulation of the lateral-attacking terrestrial predator cry, Dick," or "Pardon me, Jane, but did you just say *low-attacking terrestrial predator*?"

Vervets can talk, more or less. But they can't talk about talk. We can. We do. Incessantly. We constantly use language directly, of course, but we also paraphrase it, quote it, spell it, teach it, criticize it, philosophize it, talk, talk, talk about talk. We even have pieces of language whose primary function is to refer to other pieces of language (pro-forms, like "he," "she," "it," and "does so"). Talk about talk, as I'm sure you suspect, has a catchy linguistic label: *reflexiveness*. Just telling you that word, of course, is an example of that word.

There are two crucial points to make about voice interfaces and reflexiveness, the first perfectly obvious, the second less so:

1. Without reflexiveness, we wouldn't be able to *build* voice interfaces, which requires talk about talk in great detail while planning out lengthy, interactive, highly variable stretches of multiparty talk.

2. Our *systems* need reflexiveness, too. Without it, they wouldn't be able to function beyond a level so primitive that they would even annoy and frustrate vervet

monkeys. The systems need to clarify, query, contextualize, and negotiate — using language about language to ensure successful, quality interactions.

In this section we get reflexive — in large part to satisfy the first point, but also to work towards systems that increasingly satisfy the second point. Our talk about talk over the next five chapters, while giving due respect to the complexity of language, is focused entirely on taming that complexity in ways which draw on speakers' intuitions and inclinations about talk, and which help us avoid frustrating those intuitions and inclinations. The idea is to predict and shape users' behaviors, without dictating to them, or insulting them, or punishing them for reasonable moves. It's not easy, but the more you know about how people talk, the more manageable it becomes.

For readers who are new to speech-system work, or who have been doing it for a while and want now to ground their work more fully in scholarship and research, and thereby expand their range of strategies, these chapters will in large part offer a new vocabulary. For readers who are already familiar with the concepts and terminology of this scholarship, these chapters offer something more in the nature of a stalking ground, in which to hunt for new connections and perspectives among the various fields they bring together.

Both groups of readers (and the several shades between) I expect and encourage to read with determined self-interest — the beginners with broad opportunism, the experienced practitioners with narrower opportunism — to see the ways in which this material can be turned into strategies and understandings they can deploy in voice interaction design. In both cases, the most important effect of this section should be an attitude, fostered or reinforced, of not taking language for granted, of attending consistently to its deeply ramified nature, of always thinking out the implications of language choices.

The knowledge we explore in this section comes in five packages, briefly sketched below.

Sound and Meaning

This chapter follows the very well-worn traditional path between the two defining cruces of language: sound and meaning. We make noises; other humans extract meaning from them. How? Minimally, by the command of four domains: the phonological, which concerns the features and combinatorics of elemental language sounds (wee noises that correspond very roughly to letters); the lexical, which concerns the structure of the words, those primal sound-meaning units built out of wee speech noises (like "the" and "peevish" and "greenhouse"); the syntactic, which concerns the combinatorics of word clusters ("twenty small cigars" is an okay English word cluster; "small twenty cigars" isn't); and the semantic domain, concerned with the isolated meaning of those word clusters (that "Rocky smelled bad," for instance, makes an assertion about Rocky's aroma, while "Rocky smelled badly" makes an assertion about his olfactory abilities). The simplest descriptor of the material in this chapter is "linguistics."

Doing Things with Words

The simplest descriptor for this chapter is "pragmatics," a word which derives from a Greek verb (*prassein*) meaning "to do, act, perform." As important as the four base-level linguistic domains are to the functioning of language, we rarely if at all just talk, generating isolated and motiveless word clusters of the sort that linguistics investigates. I'm not sure such talk is possible, even by Dadaists, but if so, it is extremely uncommon. We talk, even in widely roaming chats that seem to defy the notion of a topic, with purposes and within contexts. We talk to do. Pragmatics investigates the way people act through speech. "We're out of oatmeal" *is* an assertion about the contents of the cupboard, but as an action, over the phone to a spouse at the supermarket, what it *does* is request her to pick some up so the kids will have breakfast tomorrow. In particular, this chapter focuses on the way people act through reciprocal speech, or conversation.

Conversation

Pragmatics is a field of study that develops mostly out the work of ordinary-language philosophers, with some linguists thrown into the mix. This chapter takes up two other fields that have made very significant contributions to the scholarly talk about talk. Language is fundamentally social (it is the primary vehicle of cooperation) but it is just as fundamentally psychological (we have it in our heads; it works by affecting the state of other people's heads). So, both sociologists and psychologists have crucial insights into talk for voice interaction design. In particular, the work of a subfield called *conversation analysis* is extremely helpful for understanding (and designing) the turn-by-turn mechanics of everyday talk (that greetings come in pairs, that answers follow questions, that both parties usually ensure that people get the chance to correct their own slippages, rather than getting corrected by others). But perhaps most important of all for voice interaction design is the subfield of psychology that explores how talkers work collaboratively to build networks of reference and association, how they don't just operate within a pre-existing context, they construct new contexts through their talk.

Glue

Here we draw rather heavily on the research of text linguists into how utterances combine with each other (for instance, how and when a paraphrase functions, or a justification, or a contrast, within an exchange) into discourses that are internally coherent and cohesive. This research has been largely neglected in voice interface design, leading to awkward and seriously disjointed interactions, and to complete breakdowns in coherence.

Diction

The last chapter in this section draws on one final research field, corpus analysis. Every voice interface project should be grounded by a lexicon, a collection of the words and phrases that the speech system can deploy (use *and* recognize; a cardinal rule of voice interface design is that the system should not "speak" anything it cannot also "hear"). Many projects currently assemble this collection with a grab bag of ad hoc methods — introspection, collective brainstorming, possibly some user interviews or surveys. But the project lexicon is the absolute bedrock of a speech system, and it should be built in a systematic and coherent way, in order to assure the interface has proper diction. I know that phrase — "proper diction" — sounds prissy and archaic, like the sort of thing one would find in an ancient handbook about what to do if you're invited to tea with foreign dignitaries. But it just means proper word choice, and nothing is more crucial for voice interaction design than using the right word at the right time. This chapter investigates dictionaries, thesauri, and lexicon-building for a fuller understanding of diction.

All five of these chapters are heavily peppered with examples, drawn from both human–human talk and human–machine talk.

Sound and Meaning

Grammar is a tricky, inconsistent thing. Being the backbone of speech and writing, it should, we think, be eminently logical, make perfect sense, like the human skeleton. But, of course, the skeleton is arbitrary, too. Why twelve pairs of ribs rather than eleven or thirteen? Why thirty-two teeth?
— John Simon

"Since in human speech, different sounds have different meaning," Leonard Bloomfield wrote, "to study the coordination of certain sounds with certain meanings is to study language." Bloomfield was one of the most influential linguists of the twentieth century, and he is being somewhat imperious here, as linguists tend to be. There are many ways to study language, and we will sample from a variety of them in this section. But studying the coordination of sounds and meanings has always been the mission of linguistics, and that's where we'll start.

You don't have to know linguistics in great depth to design good interfaces. You can forget all the horror stories you may have heard about Government/Binding Theory or Head Driven Phrase Structure Grammar or split infinitive constructions. You don't need to lose sleep over all the diacritics in the International Phonetic Alphabet.

But you need to know enough about the linguistic basics — a general vocabulary of terms and concepts — to follow the relevant literature and the drift of the speech/language technologists on the project. You need some level of comfort talking about talking; language is your medium.

We need to start by expanding our image of the voice interface we developed in Chapter 1, separating out the main conceptual divisions of a speech system. Figure 3.1 is still roughly schematic (it could be rendered in much finer detail), and it is conceptually, not physically, representational (different speech systems will have different arrays of modules

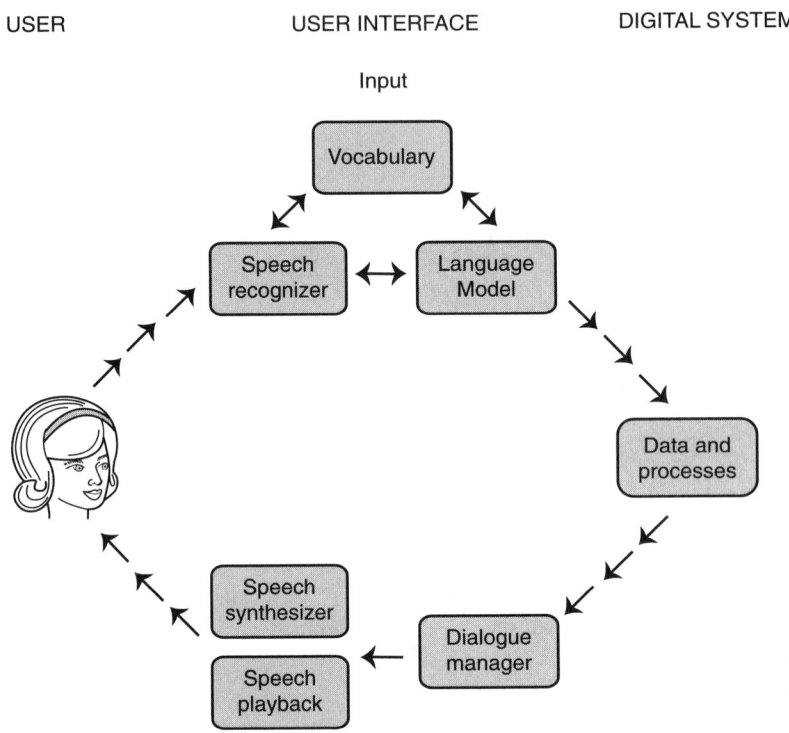

FIGURE 3.1   A conceptual diagram of speech systems

and there is no claim here that the conceptual divisions here correspond to those modules). But the concepts, and the general flow of processing among them, outlined in Figure 3.1 are all necessary elements of a speech system: a recognizer, to deal with the raw sounds; a vocabulary, to correlate those sounds with words; a language model, to find the structures, syntactic and semantic, that words are assembled into; and a dialogue manager, to decide on system utterances; as well as a synthesizer and/or playback component to turn the dialogue manager's decisions into speech.

The notion "voice interface" begins to dissolve somewhat into the notion of "speech system" at this level of discussion. From the user's perspective, especially if it works well, there is no "interface" at all, nor any "speech system." There is only a mode of getting information. Just as graphic interfaces work best when they are "transparent," when they disappear psychologically into the task, the best speech systems are characterized by the user's perception that she is merely talking directly to an application or a database. From the development perspective, however, the interface is distributed throughout the functionality of the system; my own definition of *voice interface*, the one that I develop throughout this book, is "all the human-factor concerns associated with a given speech-system."

Figure 3.1 is abstract, since it represents the "speech recognizer" as distinct from the "vocabulary," the storehouse of acoustic models that the recognizers seeks to match up with the speech input, and from the "language model." But the recognizer is really a signal-processing pattern matcher, which tries to assign an identity to a pattern clump, using interpretive rules based on the context, on the syntax, and on the presence of other pattern clumps in the input. Where the "recognizer" leaves off and the "vocabulary" or "language model" begins is not as clean and simple as boxes and arrows suggest. And, of course, different recognition systems behave differently. But the abstraction allows us to isolate, in particular, the talk about words and word choice that is so fundamental for voice interface design.

I should also point out that my use of the word *vocabulary* here is mildly controversial. I'm certainly not the only one who uses it in this context, but the traditional word for the acoustic-model store house in a speech system is *grammar*. For a variety of reasons, mostly to do with how substantially it distorts other uses of the word, I prefer to avoid that use of *grammar*.

The bulk of the interaction design effort will focus on the vocabulary (what the system should listen for) and the dialogue manager (what the system can and should say), but the success of the interface depends on collaboration with language model development (which phrases and structures are most common to the user population and the discourse genre), recognizer development (which pronunciations are relevant), and speech synthesis/playback development (which types of voices and personalities are the most credible and the most satisfactory to the user).

In this chapter, focusing on linguistics, we explore findings that bear mostly on recognition, synthesis, parsing, and vocabulary. Later chapters will devote more attention to findings that bear more fully on the vocabulary, and that bear on dialogue management.

Most of the progress that linguistics has made has come from the methodological principle known as "reductionism." That word is often used as an accusation in ordinary speech, an accusation of taking something complex and filtering off or ignoring the complexities — which is exactly what reductionism does. Linguists have faced such accusations, too, for good reason. Their work is founded on reductionism and language is incredibly complex. But reductionism has its virtues. A cake is a mildly complex substance, but one can't understand it, let alone build it, unless one reduces it to ingredients (flour, sugar, eggs, . . .) and procedures (sift, stir, pour, bake).

Reductionism has allowed linguists to attend to four core perspectives, all investigated somewhat independently of each other: sounds (phones, phonemes, and syllables, the domain of phonetics and phonology), words (morphemes, the domain of lexicography); expressions (phrases and sentences, the domain of syntax), and meaning (propositions, the domain of semantics). A further linguistic phenomenon, the speech rhythms effected by variations in length, loudness, and frequency — that is, prosody — commingles with the other dimensions so completely that it cannot be studied profitably by reductionism on the same scale. Table 3.1 charts this mapping of language into the fields and foci of linguistics.

		Language			
		Sounds	**Words**	**Expressions**	**Meanings**
Linguistics	Elements studied	Phones, phonemes, syllables	Morphemes	Phrases, sentences	Propositions
		P r o s o d y			
	Fields of study	Phonetics, phonology	Lexicography	Syntax	Semantics

TABLE 3.1 Language and Linguistics

A Note about "Grammaticality"

Most of the occasions for the troubles of the world are grammatical.
—Michel de Montaigre

While we are on the general topic of linguistics, there is one issue we should probably deal with before going any further, a canard. There is a piece of folk wisdom that circulates widely among dialogue designers that speech is consistently "ungrammatical." It can be, but so can written text, and, in any case "grammatical" is one of those loaded terms that can often mean little more than "good" (in this case, "good according to my sixth-grade teacher"), so that its use systematically disparages spoken discourse (and privileges written discourse). That's a very dangerous attitude for a dialogue designer to have. There are lots of false starts, hesitations, and other phenomena in speech that get filtered out in writing, and edited out in publishing. But they have little bearing on grammaticality, and are just part of the natural order of things that speech systems have to deal with (filtering and editing as they process, in the way humans filter and edit as they listen).

Grammatical, in linguistic terms, does not concern "good" or even "according to the rules in textbooks that specify how people should talk and write." It has nothing (at least nothing very directly) to do with issues like split infinitives or dangling modifiers or the like. *Grammatical*, in linguistics, means "consistent with a specific grammar"—either an empirical grammar derived from investigating some language, or a formal grammar that some linguists have built for computational or experimental purposes. If that definition is too awkward, then just remember (1) that *grammatical* in linguistics does not mean "good" at all, (2) that spoken language is not particularly ungrammatical, and (3) that you need to maintain the utmost respect for spoken language.

Sound

We move our Tongue with ease, and can readily diversifie the sound of our Voice in different manners. For this reason Nature has disposed Man to make use of the Organs of the Voice to give sensible signs of what he wills and conceives.
   The disposition of these Organs is wonderful.
    — Thomas Hobbes

The sound of language is the province of two closely related linguistic sub-disciplines, phonetics and phonology. Phonetics studies the raw wave form, the noises; phonology studies the specific ways languages exploit these noises. For instance, if you put your lips together, build up a little air behind them, and then quickly open them, you've made a sound that lots of languages put to use; in English, we make versions of that sound (a voice-less bilabial stop, if you must know; [p] in the International Phonetic Alphabet, or IPA) at the beginnings of words like *pin*, the ends of words like *lip*, and in assorted other locations (*spin, camper, liposuction*). Notice that we're not talking about the *letter* here (which can stand for other sounds altogether, as in *phone*, and even for no sound at all, as in *psycho*) but about the specific *sound* that you make when you put your lips together and then open them up to let the acoustic signal out. Try another one: put your tongue up close to the little bump just behind your top, front teeth, and force air through it quickly (that's an alveolar fricative; [s] in the IPA). When your attention is drawn to the physical maneuvers behind these sounds, they can seem peculiar, but of course you do them automatically, often thousands of times a day, very much the way you walk, turn your head, or punch buttons on the remote control.

Speech sounds, from the raw-wave-form perspective, are called *phones* (literally, "sounds" in Greek). From the exploited-by-specific-languages perspective, they are *phonemes*. It is at this point — very, very early in the attempt to link sound (a material thing) with meaning (a mental thing) — that the job of linguistics gets messy. "It would be convenient," Wallace Chafe has said, "if linguistic units could be identified unambiguously from phonetic properties . . . however, the physical manifestations of psychologically relevant units are always . . . messy and inconsistent" (1994: 58).

Firstly, teasing the phones apart is problematic. Chafe's analogy is eggs in a frying pan; you can find the centers easily enough, but finding where one ends and the other begins is impossible. They merge at the boundaries.

Secondly, the material blurs into the mental here, at a very immediate level of linguistic analysis, before sentences or even words enter the picture. Phonemes, that is, are not really *sounds* at all, but abstract mental representations of *sets* of sounds.

I know this seems weird, and it routinely keeps linguistics students up at night, especially before exams, but it is absolutely crucial to understanding language generally, and speech recognition issues specifically: phonemes are abstract sets of speech sounds (therefore, sets of phones). What this comes down to is that when you hear or say things you perceive as the same sound in your language (like two voiceless bilabial stops, two [p]s), they can be acoustically quite different; in another language they might even be treated as different sounds. The fact is, phones aggregate into *different* sets (phonemes) in different languages. In Japanese, the phones that begin the English words, *right* and *light*, are part of the same aggregate, the same phoneme, which is why Japanese speakers mix up their pronunciation of them and of words using them (*rock* and *lock*, *rip* and *lip*, *rice* and *lice*) when they pronounce English. Conversely, we don't even *have* the original sound in the middle of the words we borrowed from Japanese as *tofu* and *typhoon*. The f-sound we pronounce them with is a gross approximation of the native Japanese sound (which is actually somewhat closer to the sound which begins *when* and *which*).

This all strikes many people as unutterably pedantic and even painful, but it is precisely because of this ability — the ability to make a constrained range of sounds, in constrained sets of relationships with other sounds, and the dependent ability to hang meanings on specific clumps of those sounds (like *pin* and *lip*) — that we can get people to pass the salt or inquire about their opinion of *Citizen Kane* or ask them why on earth they would put a pin through their lip.

Now this is where it goes from pedantic to scary in speech-recognition terms. These speech sounds, like snowflakes and fingerprints, are never the same, *never*, not even when they come out of your mouth twice in a row, let alone when they come out of different people's mouths — men's, women's, children's, baritones', tenors', sopranos', Bluto's, Popeye's, Julia Child's. These differences can be a headache for speech-recognition machines. James Glass and his MIT colleagues, for instance, report that their JUPITER system has particular trouble with children's voices (Glass et al., 1999), which operate at higher frequencies than adult voices; the telephone cuts off some of the higher ranges in the first place, eliminating parts of the signal, but recognizers are just not trained for kids' voices. Accents, dialects, kids — all these input-problems are tractable in principle, but in practice they throw so many marbles into the recognition problem space that they can drive engineers out twelve-story windows.

There are similarities among these different sounds of course, powerful ones. There are the similarities shared by all of one speaker's utterances. People speak in unique ways, but not randomly unique ways. One Julia Child utterance can be recognized as coming from the same person as another Julia Child utterance. This uniquely similar characteristic has two benefits for voice systems. Firstly, it allows for speaker verification — for machines to singularly identify a speaker — a technology that paves the way for very individualized user profiles, and that can provide high degrees of security with speech applications. Secondly, it allows speech applications to learn individual speech patterns, so that accuracy improves with usage.

It is the between-speaker similarities, of course, that allow us to communicate, the resemblances that my sounds have to your sounds when we both say "Hope the rain don't hurt the rhubarb," but in the speech-recognition business we can never lose sight that they are similarities, not identities. We do all right by speech noises most of the time, because our brains are specially tuned for speech noises, and because we get lots of practice. But this incredible diversity of the "same" sounds, compounded by the number of "different" sounds overall (American English has around forty distinct phonemes), gives speech recognizers a very hard time.

It gets worse for speech recognizers. The page you're looking at has physical manifestations of psychologically relevant units, too. It has a bunch of little marks on it — informally, we call them *letters*, but really they are specific letter images. The letter "t", for instance, is not, strictly speaking, a mark on a page. It is a set of possible marks on a page; the psychologically relevant unit is that set. Figure 3.2 is a brief illustration of this point. It serves up a few examples of the letter "t", drawn from different typefaces. None of them is a pure letter "t", but all of them manifest that letter. Linguists call the set of possible marks a *grapheme*, and graphemes have some very strong parallels with phonemes. But a letter-image on a page is a much simpler perceptual item than a phone in a sound stream. The letter image is discrete. It has space around it, allowing us to see it fully. It also allows machines to see it fully. Optical character recognition, and text-to-speech engines, have it very easy on this score, vastly easier than speech-to-text systems. Phones in a sound stream are continuous. They overlap each other. They influence each other. It is not only impossible to tell where one leaves off and another begins, it often doesn't even make sense to say there is a sequence of phones at all. Rather, there is an acoustic clump that contains phonetic information.

Speech recognizers don't yet resolve phones (or extrapolate phonemes) out of an acoustic speech clump. Rather, they store patterns that share phonetic values with acoustic

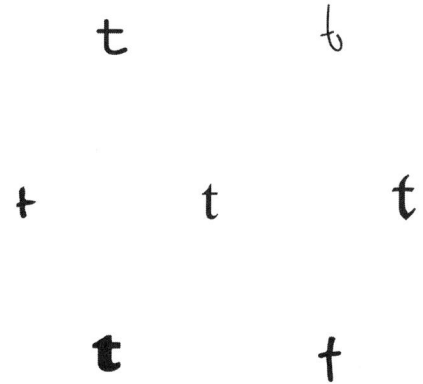

FIGURE 3.2   Some letterforms of the grapheme "t"

speech clumps, and match those clumps to interpretations. A more operative notion in the speech clump (as opposed to the phone) is the syllable. Think of Chate's pan of fried eggs. The analogy works even better for syllables than for phones. You can't tell where one egg leaves off and others begin, but you can spot the centers easily enough. The yolk is the center of the syllable. There are two basic classes of phones, consonants and vowels, which work together to make syllables, and syllables constitute stable enough patterns, especially when they are stressed (that is, given relatively more duration, volume, and intensity), to provide fairly reliable patterns to match; syllables are the yolks in the pan. Vowels are sounds which sustain the same frequency and pitch long enough to structure the syllable yolks; acoustically, they are periodic sounds, with sustained recurrent patterns: tones. Vowels are often called "musical." They are the speech elements that go on for very long times in arias. Consonants tend to be abrupt (like the bilabial stop, [p]) or harsh (like the alveoloar fricative, [s]), and their job is usually to separate the vowels from each other, occurring at syllable boundaries; acoustically, they tend to be aperiodic, with random buzzes or bursts.[1]

But there's a really lovely part in all this that helps us and our machines resolve speech noises more efficiently: vowels are quite sensitive, so that the consonants around them leave distinct acoustic impressions on them.[2] Vowels, in short, are good. And stressed vowels are the best of all, a maxim that is useful when designing system utterances that will be kind to your recognizers. If vowels are the yolks in the speech-clump frying pan, stressed vowels are yolks from ostrich eggs. The American speech-technology company, HeyAnita,[3] may have a very opaque name in terms of coding its business domain. The company could be a cleaning service, or a flapjack emporium, so meaningless is the name. But the phonology of the name is pure genius: if users of the company's voice-portal service wants to alert the agent, "Anita," they say the acoustically very robust, two-stressed-vowel phrase — you guessed it — "Hey! Anita!" To stretch Professor Chafe's analogy a bit, it is easier to distinguish a group of two fried-eggs-in-a-pan from some other group of two fried-eggs-in-a-pan,

---

1: Speech sounds are really much better differentiated by way of a continuous scale than by the binary consonant/vowel division. Some consonants, for instance, can be periodic when they are sustained, and can even hold center position in a syllable. And there is a class of sounds (sometimes called semi-vowels) that are in all respects vowel-like except for their duration (which is consonantally brief).

2: This information is represented in vowel subpatterns called formants (the most intense frequency regions in the sound). If you hear engineers talking about "formant structure" they're talking about two things: first, that vowels have different frequency patterns from each other (the formant structure of [i] is different from the formant structure of [o]), and second, that those frequency patterns change as a function of the sounds around them (the vowel [i] has a somewhat different formant structure in the sequence [tin] than in the sequence [nit]). What that means is that if the recognizer gets a distinct resolution of a vowel it not only has information about the vowel itself, but about its neighbors, the consonants (most of which are otherwise much tougher to detect).

3: *http://www.heyanita.com/*, last accessed April 2004.

than it is to tell single fried eggs apart. The centers could be closer together in one group than the other, or one could have the eggs aligned horizontally, the other vertically, or diagonally, and so on. There are more possible patterns with two centers than with one.

Words

Each word may not unfitly be compared to an invention.
  — William Dwight Whitney

Sounds don't have meaning; they get meaning. There are some interesting sound/meaning correlations in language, like the presence of [s], [r], and [l] in moving-liquid words — *river, stream, rivulet, rill, spring, splish, splash, sploosh.* But you needn't look any further than another language to see that the sound clumps we link to meanings overwhelmingly get that linkage not because of anything about the sounds, but just because of conventions and habits of the sound-clump users: speakers. A yappy, hairy quadruped is denoted by the sound-clump *chien* in the conventions of French speakers, *Hund* for German speakers, *dog* for English speakers. There's nothing inherent in any of those sound sequences that means yappy, hairy quadruped. They just do, for the speakers of those languages.

Sound-clumps get paired with meanings in languages in a variety of interesting ways, but the pairings are ultimately *arbitrary* (to use the standard linguistic and semiotic term for this trait); that's how symbols work. From the perspective of the language users, of course, the sound/meaning linkages are not arbitrary. A dog *is* a yappy, hairy quadruped to English speakers, and if that's not enough, its participation in a whole network of other terms (*hound-dog, dog-catcher, doggy,* . . .) proves it is not arbitrary. But from the abstract perspective of sound and meaning, words are inescapably arbitrary: why *these* sounds with *those* meanings? Just because.

Sound-clumps paired to meanings are called *words*, if they occur freely, *affixes* if they are bound to other sound-clump/meaning pairs. The name of the superordinate category for words and affixes is *morpheme*. After that, the terminology (appropriately) goes all to hell. I'm just going to pick the word *lexicography* for the study of words and their parts, and plow blithely on. I will also pretty much ignore affixes from here on out, except to note that while they play less of a role in English than in many languages, they can still be big trouble. For recognizers, their low perceptual salience combined with potentially high semantic consequences can be a big headache: acoustically the differences between *expensive* and *inexpensive* are very slight; semantically, they are major. For vocabularies, they can entail big space allocations; unlike our heads, where we seem to have lexical rules which provide some storage economy, most speech systems need separate entries for all relevant inflections of a word: *ship, ships, shipped, shipping, shipper,* . . . .

Words (and, what are almost the same, catch phrases like "Hey Anita!" or "sports utility vehicle" or "SUV") are the most important linguistic elements, by far, in the voice-interface designer's kit bag. The principal elicitation task of an interface utterance is to get the caller to speak in the way that best serves the recognizer. Even the very best speech recognition systems don't get anywhere close to clear resolution for *everything*, and while they are sure to improve, 100% reliability is a long way off (if ever; *humans* aren't 100% reliable for signal resolution). The signal might be weak mechanically, but more significantly, much of the signal is by nature low-res (the unstressed syllables). And, even if the recognizer could resolve everything, in the sense of confidently returning a phonetic transcription of the wave form, the voice system can't *understand* everything. I'm betting you don't know what to do with this word: *fenks*. English is just too darn big — too big for your mental dictionary. And for mine: I didn't know the word either, until I rummaged around for an obscure and preposterous word in *Mrs. Byrne's dictionary of unusual, obscure, and preposterous words* (Byrne, 1994). But it doesn't take a word like *fenks* (leftover whale blubber, by the way, especially when it is used as fertilizer or as an ingredient in the pigment Prussian blue) to throw voice systems for a loop. Their vocabularies are far smaller than yours and mine. And, what makes it worse, their size is not just a memory issue. The vocabularies are smaller in principle, an increase in vocabulary size means an increased demand for processing. George Furnas and his colleagues (1983, 1987; see also Brennan, 1998) called this tradeoff, suitably enough, "the vocabulary problem," and it implicates the most significant technical issue confronting voice-interface design: which words get included?

But in both cases — signal resolution and vocabulary capacity — a prime component of the solution involves diction: *knowing the right words*. You can't do anything directly about noises or gaps, but you can work to get the caller to use the most resilient terms, the ones that give the recognizer its best odds. On the output side the dialogue system has to use words that strike the caller forcefully, that she will be more likely to speak back; that is, system output must effectively constrain the caller to remain within the system's vocabulary. On the input side, the better the system vocabulary reflects the users' discourse patterns (what they are likely to say in a given context), the better the system's chances of dealing with the utterances.

From the output perspective, there is a variety of ways to induce users to say specific words (therefore, of constraining their speech), which we will explore throughout the book, but one of the most important is priming. It is a very well established result of psycholinguistic research that hearing (or seeing) words dramatically increases the likelihood of producing those words and their confreres. Hearing (or seeing) the word *rock*, for instance, disposes us to say the word *rock*. Moreover, it disposes us to say words like *stone* (a semantic relative), *crock* (a phonological relative), *rocking* (a morphological relative), and *music* (a sometime neighbor, or *collocate*). Like many aspects of language that you need to keep in mind for voice interaction design, priming happens whether you want it to or not.[4] Your job is to constrain it in the directions that aid recognition and the overall coherence of the system.

Lexical priming may be a manifestation of the more general sociolinguistic phenomenon known as *convergence*: people have a tendency to speak in the way their interlocutors speak (phrasing, clause structure, word use, and even pronunciation). It is a form of tacit cooperation, and speech is a highly cooperative enterprise. We all do it, all the time, but convergence is most clearly seen in those people we all know who have gone to England for a two-week vacation and come back talking like Michael Caine. It isn't *all* affectation; some of it is the natural impulse for linguistic conformity. Moreover, convergence may operate even more powerfully in human–computer interaction (see Bretan et al., 1995).

From the input perspective, knowing what users are likely to say in a given context is not as daunting as it might first appear. It *is* daunting, but not hopelessly so. And the reason is "a given context." Some words will turn up frequently in a conversation about air travel (*booking, aisle, Tuesday*), some won't (*antifreeze, scuttle, woodpecker*). In fact, "given contexts" entail quite limited word lists. In studies of human–human dialogues about planning air travel, the vocabulary clocked in between 1000 and 1200 words (Kowtko and Price, 1989; Peckam, 1993). A similar study into telephony-services dialogues returned a similar number (Karis and Dobroth, 1995), and found the total number of word types in one twelve-minute call about installing phone service was only 188. If you add a computer (or the assumption of a computer) to the dialogue, the numbers are even more encouraging. Robin Woofit and his colleagues, for instance, found that both tokens and types drop dramatically when people assume they are talking to a computer: 5966 types in (travel-register) human–human dialogues dropped to 4005 types when people thought they were talking to a computer; 1004 tokens fell to 399 (Woofit et al., 1997: 50).

Words have been subject to intense scrutiny for thousands of years in the Western intellectual tradition, and they are such complicated little beasts that lexicographers have been known to say things like "all words have their own grammar" (Stubbs, 1996: 37). There is no shortage of things to say about words, and much of this book concerns them at some level or another; several chapters look at them almost exclusively.

But for the purposes of this brief survey of linguistics, there are only three more words I want to use about words: *synonymy, homonymy,* and *ambiguity*. They all concern the sound/meaning linkage.

4: Byrne (2001), for instance, reports on an example of inadvertent priming: ". . . in one recently released application, subscribers consistently began to say MOVE ON to get to the next domain even though they were never explicitly told to do so. As it turned out, the application repeatedly used the same phrase in a slightly different context. In other words, the application itself primed them for MOVE ON as opposed to, for example, NEXT DOMAIN or some other equally likely choice."

Synonymy, Homonymy, and Ambiguity

Too caustic? To hell with the cost. If it's a good picture, we'll make it.
    — Samuel Goldwyn

Synonymy is the situation where words have different sound clumps, but the same (or, not quite, but highly similar) meanings; such words are synonyms (for instance, *attorney* and *lawyer*). Homonymy (also *homophony*) is the situation where words have the same sound clumps, but different meanings; such words are homonyms (for instance, *bare* and *bear*). Ambiguity is the general condition of uncertainty about meaning, of not being able to nail down the specific significations of specific sounds. These concepts are probably not news to you — they are among the few linguistic ideas that make their way into most people's general education — but they all require some attention here because of their implications for design.

Since ambiguity is the biggest of the three — a nest of phenomena, rather than a single phenomenon — let's take it first. Here is an example in a dialogue:

Ford: You'd better be prepared for the jump into hyperspace. It's unpleasantly like being drunk.

Arthur: What's so unpleasant about being drunk?

Ford: You ask a glass of water.

(Adams, 1979: 59)

Like most ramified concepts, there is good news about ambiguity, and there is bad news about ambiguity. The good news is that there is a lot less of it to deal with than you would expect with the incredible litter of words strewn throughout English. A table might potentially be something for eating breakfast off, or something for arranging data in, but listeners in a given context (furniture, mathematics) know which of those potentialities is intended. A dog may be a yappy, hairy quadruped or a promiscuous man, but context will tell. The words that other words hang out with (*collocation*, we'll soon be calling it) go a long way to reducing ambiguity:

Orlando was a dog. He chased rabbits.

Orlando was a dog. He chased women.

Words like *table* and *dog* and *bank* are all, more or less, instances of homonymy: cases of (potential) ambiguity, those cases where multiple meanings can't be resolved because they are attached to the same sound sequence: is it *Mary*, *merry*, or *marry*? Again, context usually will clarify. In this example, we have a noun, an adjective, and a verb, none of which show up in the same syntactic frame, and more than one of which is unlikely to show up in the same discourse. For an interactive speech system, it would be unlikely in the extreme that *bank*, the side of a river, and *bank*, the financial institution would be part of the same vocabulary (and in a dictation system, it's largely irrelevant). But even in dialogues where homonymy genuinely leads to ambiguity, it is often easy to handle. This case shows a user utterance asking for a quote in a stock application:

Caller: How much is Cisco trading at?

Ignoring spelling (which is present in this example only because we are between the covers of a book, and not, say, talking on the phone), the relevant sound-clump could refer to either of two publicly traded companies. But that's not a problem. The system would know this, and come back with a clarification request:

System: There are two companies with that name. Do you want Cisco Systems or SysCo foods?

(Balentine and Morgan, 1999: 190)

These situations can be fairly common — in voice-dialing or email or directory-information services, for instance, when there might be conflicting names; in travel planning, where there may be more than one chain hotel in a city; in banking applications, where there will be lots of products corresponding to *mutual fund*. But they are all resolved in effectively the same way, with a clarification (or disambiguation) request.

If the good news about ambiguity is that there's not very much in ordinary language use, and that it's easy to resolve by just asking the source, then what's the bad news? The bad news is that we aren't dealing with ordinary language use directly. We're dealing with it as strained through a speech recognition engine, which can find ambiguity in places that wouldn't occur to those of us with a neocortex. This problem is generally known as the wreck-a-nice-beach problem (wreck a nice beach . . . recognize speech . . . get it?). Speech recognizers not only have to deal with the same *train*, *drain*, *crane* sorts of misrecognitions that humans confront, but also with their own fascinating construals. Here is what my dictation software thought I said one day:

The food makes everything I feel putter.

Here is what I really said:

Indian food makes everything feel better.

Here are some examples from an interactive speech system:

Spoken: I want to fix this circuit.

Recognized: power a six a circuit

Spoken: There is no wire on connector one zero four.

Recognized: stays know wire I connector one zero for

(Smith, 1997)

Not weird enough for you? Try this other-worldly stew, a passage from the automated transcription of a marketing phone call:

> *that has sweat what you have a minus for the one year before that you you look have all along are right you feel that has performed for you right-now one term I would say average before if there's I would still say to go over average top ten we what you what his you consider than man yeah time you want my name and then I'm-sorry a blue-chip fund number's one because the middle bond fund over ten year period has returned an IRA over a five year period of five point eight are.*
> (Cooper et al., 2001: 3)

Ambiguity doesn't seem big enough to label for this kind of lexical pandemonium. It may look like the ambiguity of words like the two *banks*, and *bear/bare*, and even *Cisco/SysCo* is "in the language," while the problems in this word salad are "in the technology," but it is important to notice that ambiguity is not a property of language; it's a function of interpretation. Human brains map the sound-clump (represented orthographically here as *bank*) into "financial institution" or into "land immediately contiguous to a river" depending on the context; and they fail to resolve it, or resolve it inappropriately, to the extent that context doesn't constrain the interpretation sufficiently. The sound-clumps that produced the transcription example above clearly had other lexical potentialities; namely, the ones the speaker intended. Another example from the same data illustrates the point well: a sound-clump the speaker meant as *JavaOne* was routinely transcribed as *jowl one* (Cooper, 2001: 7).

You should not, at this point, close the book in despair and become a Luddite. The input stream for that example was part of a human–human dialogue, with the speaker making no special concession for the recognition engine, having no interaction with the system at all, and not even getting feedback to monitor how well the system was catching him. And even with all those handicaps, the machinery did quite well. It delivered this text only as a raw first pass. The text was then submitted to a number of algorithms that ended up producing a quite impressive information summary.[5]

5: This summarization was not part of a voice interface, and not carried out in real time (so would not even be available to a voice interface), but it is fascinating all the same and highly suggestive about what the near future holds for dealing with tricky voice input. The authors submitted the text to a timing algorithm, breaking it into hypothetical utterance units. They used boundary cue words (like *yeah, well,* and *OK*) to further refine the utterance groups. They replaced the low-certainty recognitions with place holders and removed the nonspeech elements

The morals of the that-has-sweat-what transcript are simple ones: raw speech recognition is less than flawless, ambiguity is an ever-present menace, and clarification and repair routines must be a driving concern of voice interface design.

The flip side of ambiguity is synonymy, and its implication for voice interfaces is also inverse: synonyms are the saving grace of speech systems, the best friend of the voice interaction designer. They can cause the engineers headaches. (Remember the Vocabulary Problem? The more synonyms in the system, the more strain is placed on the vocabularies and processors.) But a well-chosen range of synonyms gives the user exactly the kind of navigational freedom that nonconversational voice systems are notorious for denying.

The first problem is English. It has a *lot* of synonyms. We have this plethora, this surfeit, this largesse firstly because our language is the product of violent intercourse between two Medieval languages, a German one and a French one. The bulk of our concepts ended up with words from both sides of the family (*big* and *large*, *lawyer* and *attorney*, *dresser* and *bureau*, . . .). It is no coincidence, by the way, that the more elegant-sounding words are from the French side, since the Norman French ran the show during the years they commingled with the Anglo Saxons — all the naughty peasant words for body parts and functions are Germanic in origin, for instance, while all the polite and clinical terminology is French. But that was just the start. After the Anglo-Norman period, English became an aggressive, ambitious, and well-traveled language, borrowing hither and yon as it went. Indeed, given the political, economic, and cultural hegemony that its speakers have historically pursued, James Nicoll is a bit closer to the mark when he says "We don't just borrow words; on occasion, English has pursued other languages down alleyways to beat them unconscious and rifled their pockets for new vocabulary" (1990).

Even *without* that inheriting, borrowing, and rifling, it's hard to believe English vocabulary would not still be an incredible tangle. English is the linguistic vehicle of a culture (or set of cultures) that is so geographically distributed, so economically stratified, so professionally diverse, and so socially heterogeneous that it is routine for linguists to talk not of English in this context, but of Englishes (e.g., Weiner, 1990: 501). All languages have these tendencies, but English is practically pathological in its range of dialects, jargons, and slangs.

Dialects, jargons, and slangs are usually held to be bad things by people with limited understandings of how languages function (mostly because they've been taught by other people who possessed limited understandings of how languages work, but who had evangelic senses of "correctness"). All three *can* be bad, of course; all three are motivated by

---

(the *ums* and *uhs*) altogether. Then they performed a content extraction, pulling out meaningful lexical and phrasal items based in part on the discussion context. In the end, they produced a searchable transcript with high reliability for the key terms: "while the word recognition accuracy of these transcripts was in many cases fairly low, the salient term accuracy was quite high and made these searchable summaries extremely useful" (Cooper et al., 2001: 9).

the dual impulses to foster inclusion among their users, and to enforce exclusion of their nonusers. Whether *telly*, *CTI*, or *fubar* are useful words is highly contingent. It depends on who you are, who you are talking to, and what your mutual circumstances are; that is, they depend on absolutely the most critical factor in communication: context.

The second biggest irritation of most phone-line speech systems, after their hierarchical tree structure, is the requirement to use the system's terminology. But using synonymy is an art, a skill; in application, it is a study in investigative lexicography. You have to find out how (and often when) your users choose one term over another. As Brennan (1998: 3) points out, "to remove a file, possible terms include *remove*, *delete*, *erase*, *expunge*, *kill*, *omit*, *destroy*, *lose*, *change*, *rid*, and even *trash*." But these terms are not equally weighted for any given community. Unix users would be biased toward *remove*; DOS folks would lean toward *delete*; MTS dinosaurs, if their memory held up, might go for *destroy*.

Sometimes the system's vocabulary is familiar enough or inevitable enough that it coincides with user expectations (*checking*, *savings*, *deposit*, *withdrawal*), and often there are reasons to encourage one word over another (*zero* has more acoustic bite than *oh*), but neither of those arguments is sufficient to ignore the alternatives. People might want to "pay a bill," or "pay some bills," rather than the phrase preferred by the system, "make a payment." They might want to "see how much they owe" on their credit card, not "check their credit card balance." They might just want to transfer "funds," rather than "money." Or the reverse. And if all users have only one set of terminological options, even if they are statistically the most frequent terms for the relevant actions, some users are still going to stub their toes on the system's vocabulary. Moreover, when there are a number of activities during one interaction, the chances are that some of the caller's words and some of the system's words are not going to line up. A system with *funds* <u>and</u> *money* in its vocabulary, coded to the same value, *pay . . . bill* <u>and</u> *make . . . payment* both coded for the same action will satisfy more users more often.[6]

There are, however, no true synonyms — two phonologically different, semantically and functionally identical words. Why would there be? They would be a waste of cognitive resources, an unproductive redundancy. The differences of meaning (or, more properly, function) can be subtle, but they are there. Take a clear case: *myopic* and *near-sighted* both describe the same physical condition, and both might occur in ophthalmologist's office or in the kitchen over a cup of coffee. They vary with individuals as well as situations and registers. But English speakers all know that one is a technical term, the other is an

---

6: Bruce Balentine argues the other side of the coin: "it is probable that small, task-specific vocabularies will always be more robust than those that include 'synonyms'" (1999: 232). Yes, it is very probable. But robustness is not the only issue in interaction design. Isolated speech is more robust than continuous speech. Trained recognizers are more robust than speaker-independent recognizers. Dedicated microphones are more robust signal detectors than telephones.

ordinary-language word, and overall they are more disposed to use, and to expect, *myopic* at the ophthalmologist's, *near-sighted* in their kitchen. Context rules.

What this means for voice interface design is two things. Firstly, even in a sea of synonyms, some words are better maps to the meanings and functions you need than others (or, usually, some *set* of words are better maps to the sets of meanings and functions you need). Voice-interface design has suffered from the arbitrariness legacy of keypad–function matches; for instance, in verbatim prompts ("For news say *news*. For sports say *sports*. For weather say *weather*," . . .). What this style may gain in redundancy, it loses in naturalness, grace, and overall user quality. What it misses, as Bruce Balentine points out, is the absolutely essential fact that "words already have their own intrinsic meaning" (1999: 215). "Would you like news, sports, or weather," especially with a suitably weighted intonation, is just as likely to return the appropriate function-tied word, and is much less likely to seem imperious, or just plain bone-headed, to the user. If you don't allow words to carry their semantic weight, you might as well go right back to the arbitrary keypad-emulating prompt style that originated this pattern ("For news press one," . . .).

Secondly, it can often take a good deal of thinking and investigating to find the most appropriate words to populate the recognition vocabularies with. Every voice interaction team needs a lexicographer.

Words have deep affections for one another, along with certain antipathies. As lexicographers will tell you, words travel in recognizable packs. They *collocate*. One of John Rupert Firth's prize examples of collocation was the company that *ass* habitually kept with *silly*. J. R. Firth was a linguist who argued for the importance of the "habitual company" that words keep (Firth, 1957: 14), and his work became very influential for the development of late 20th century computer-mediated lexicography. But his fondness for "silly ass" identifies him as an early 20th century Brit (his use of the example dates from the 1930s). You and I can nod pleasantly about the quaintness of the collocation and move on, but first let's extract the lesson. If we fed some of Firth's recorded speech into a speech recognition engine, and it was having trouble from the acoustic properties alone figuring out if a given pattern matched up with *bass* or *glass* or *pass* or *ass*, but managed to get the preceding *silly* successfully from the input, knowledge of that collocation could prove definitive in resolving the signal.

These company-keeping habits also tell us about a good deal about function and meaning, which brings us, almost, to syntax. Syntax would be impossible without a fundamental distinction between word classes, between words that do mostly semantic duty, the content words, and those that do mostly syntactic duty, the function words. The former are nouns, verbs, adjectives, and adverbs; the latter are, well, everything else (prepositions, participles, intensifiers, pronouns, proverbs, articles, demonstratives, . . . there's a million of 'em). These two main word categories differ on a number of important criteria, but the critical difference is just that the content words carry the overwhelming bulk of the meaning. The function words help the individual word-meanings string into composite, propositional meanings. In this sentence, the function words are single underscored, and the content words are double underscored.

On, then, to syntax.

Syntax

peel, teaspoon, tablespoon, fry, finely, salt, pepper, cumin, freshly, and ginger
    — The ten most frequent words in *Madhur Jaffrey's Cookbook*

Words, even when traveling in packs, aren't nearly enough most of the time. From this list, we can tell the general source pretty easily, but there's no message here, just a general semantic invocation of delectables. As Mark Liberman points out, the issue is not a matter of quantity.[7] Adding more words doesn't help. Here are the ten next most frequent from *Madhur Jaffrey's cookbook*:

stir, lemon, chicken, juice, sesame, garlic, broth, slice, sauce, and *chili*

We still don't know what to do (and I, for one, am getting hungrier). What's missing is structure, the systematic arrangement of words to communicate a sum greater than its parts; what's missing is syntax. Take these two very different sequences of words, different from each other, and not even similar to the last two sequences:

1.   The cook tasted the chicken.

2.   The chicken tasted the cook.

They have all the same words, and even the same general structure (Subject-Verb-Object), but word/syntax mappings make for very different sentences. Most natural language systems wouldn't have too much trouble parsing these two sentences, figuring out how the words go together. Being able to parse sentences 1 and 2 is what lets us tell the taster from the tastee. But, because the inputs are so uncertain much of the time, mapping words into syntax is not a particularly easy job for speech systems.

I once did an experiment with language-damaged patients, getting their responses to sentences like these:

3.   The man showed her baby the pictures.

4.   The man showed her the baby pictures.

7: The example is from his exemplary course page for Linguistics 101 at the University of Pennsylvania, at least the version of it he offered in the Fall of 2000; in particular, it's from his textbook-quality overview of syntax: *http://www.ling.upenn.edu/courses/Fall_2000/ling001/syntax.html*, last accessed April 2004.

It's a snap to tell these sentences apart, right? Not for the people I was working with. They responded pretty much at random between two options when I asked them to match those sentences. Their job was to point to the best-matching picture from an array of drawings that included a depiction of a man showing baby pictures to a woman, another of a man showing pictures to a baby, and two unrelated depictions (as red herrings). They matched either sentence with either (non-herring) depiction, at very close to chance level. What this means is that these people — with a type of damage sometimes called *Broca's aphasia*, sometimes *agrammatism* (for the two most common labels; there are others) — were effectively working with anagrams of the biggest sound-meaning chunks from the signal (*man*, *showed*, *baby* and *pictures*).

They assembled the chunks into the syntactic configurations that made sense to them, and then mapped those assemblages back against the pictures; half the time the assemblages were of the type in sentence 3 above, half the time they were of the sentence 4 type. Never, by the way, did they choose the red herrings — depictions of, say, a man showing hats to some girls. They were getting the words, just not the syntax (see Harris, 1985, 1988).

And here's the point: this is roughly how most conversational voice systems will work. There are basically two ways that speech recognizers work in voice systems. Speech recognizers are pattern matchers, and they either match the entire input patterns (utterances) or the component patterns (words and brief phrases) of that input. Utterance-matchers (the term is mine; they are often just called *grammars*, a label that is highly misleading) have been the most common, but they have a very limited potential in the move toward increasingly conversational voice systems. An utterance-matcher works by comparing the input to a list of allowable utterances. If it matches sufficiently well, the system gets the go-ahead for the appropriate response (and/or other associated behaviors). The main complication with this method, of course, is that it stringently constrains what the user can say. Even if a word is "known" to the system (it's worked for the user before, the system output has included it, and so on), if it shows up in the wrong place in the utterance, the recognition fails.

In contrast, the other recognition scheme — extracting component patterns from the input — is much more flexible in what it can accept, and therefore licenses greater system responsiveness. This architecture is most commonly called *wordspotting*, because the recognition engine gets words (or small word-like phrases) out of the sound stream, and then hands them off to the natural-language-understanding unit for mapping into syntactic structure (and, thereby, into meaning).[8] Not quite. They talk to each other more than that. But that's roughly the allocation of responsibilities.

8: I'm fudging a bit here, since it is the logical syntax that the system is interested in, not, strictly speaking, the English syntax. For instance, the system might use an expression like "tasted (the-cook, the-chicken)" rather than "the cook tasted the chicken." But the distinction is completely moot here.

You might be thinking, as an interface designer, "Yes, well, I sympathize with the natural-language-understanding crews, building machines to work inferentially on a pile of words to assign the most reasonable structure. It sounds darn hard. But what can I do?" Quite a lot, actually, since you're the one most familiar with the discourse domain.

There may be situations where the speech system will have to confront clumps like *man*, *showed*, *baby* and *pictures,* and have to decide between sentences 3 and 4, or even something like sentence 5.

5.   The baby showed the man her pictures.

That is, there may be situations in which any one of several candidate arguments might equally be doing the showing, being the audience, or even getting shown. But such ambiguities are rare.

It is far more likely that a processor will get *cook, chicken,* and *taste,* with the job of deciding between sentences 1 and 2. Since we know the domain of cooks and chickens, we can tell the parser to bias the analysis towards a tasting cook (hence, subject) and a tasted chicken (hence, object). In the domain of travel, for instance, the discourse research should be able to provide weightings for destinations, departure cities, perhaps even months, days, and times. And even with clumps like *man, showed, baby,* and *pictures,* we can rule out some syntactic readings. This one, for instance, is unlikely in the extreme:

6.   The pictures showed the man her baby.

The really sneaky part about syntax, actually, is not figuring out how the words you have should go together, but figuring out what to do with the words you don't have. Take the word *it*. It's trouble. Acoustically, it's big trouble, because it almost never carries stress. But let's assume the recognizer catches it. Now what? It's a place-holder, a sort of abbreviation, an arrow pointing somewhere else. And it's ubiquitous:

A   What is it doing there? Whose is it?

C   It's sitting there.

A   Is it yours?

C   It's Dave's.

A   It's your husband's, huh? He isn't a police officer?

C   No.

A   He just has one?

C   Mm-hmm. It — ah — everyone does, don't they?

(Pomerantz, 1986: 225)

It manifests a syntactic phenomenon known as *anaphora*, a word from ancient Greek that translates as "carry back up the stream", and that's the big problem with pronouns: you have to swim back up the stream looking for the antecedent to figure out what they really mean. In the case of this little piece of dialogue, from a call to a suicide prevention line (A is the counselor, C the caller), all of the *it*s (and a *one*, in the penultimate turn) go back up the discourse stream to the preceding sequence:

A   Do you have a gun at home?

C   A forty-five.

(Pomerantz, 1986: 225)

Anaphora has two important implications for voice interface design. First, you need to use anaphora in the system-side utterances, which requires a good deal of foresight. There are very few language usages that sound more unnatural, more down-right alien, than the dogged use of full noun phrases. Conversely, there is little in language that makes an exchange more fluently cohesive than the consistent use of anaphora (see Halliday and Hassan, 1976). The dialogue snippet above may not be elegant. It may not be artistic. But it is clear that both speakers are in absolute agreement about what the immediate topic is, the forty-five. Other instances of anaphora participate in the exchange, as well — two *he*s reference *Dave*, *your* references the caller, *they* references *everyone*. Try replacing all the pronouns with full words, and you'll get an exchange that would try the patience of Saint John Chrysostom, patron saint of speaking.

Second, as you can probably imagine, anaphora creates nightmares for the language-understanding side of the system. Anaphora resolution has been one of the dominant problems in computational linguistics for decades. Most natural language processing tasks are easier with written texts, because they are much more stable, they are punctuated, they can be annotated, and so on. Anaphora resolution, however, is potentially easier for speech systems. It's not easier because of recognition (proforms are often quite negligible acoustically), nor because of logic (which faces exactly the same problems that text systems do). It's easier because the system has a partner; it can always just ask.

There is one other element of words-not-there that is central to voice interaction design, another critical factor in dialogue naturalness, ellipsis. Here's a case:

Caller: What time is *Twelfth Night* playing tonight?

System: It starts at 8:10 p.m. tonight.

Caller: And *Hamlet*?

(Churcher et al., 1997: 5)

In terms of its semantic specification, there's a great whack of material missing from the caller's second utterance. The utterance *means* "And <u>what time does</u> *Hamlet* <u>start tonight</u>?"

The underlined material is elided, left out, because it can be filled back in, by the hearer, from the context. People like to speak economically as a rule, and ellipsis is a very common feature of ordinary talk, which speech systems therefore have to accommodate (a duty which falls on inference engines of the language model).[9] Equally, the *system* needs to be able to use ellipsis:

MailCall: You have seven messages from Gina-Anne Levow, two from Stuart Adams, and one each from Nicole Yankelovich, Eric Baatz, and Andrew Kehler.

(Marx, 1995: 69)

The system's use of ellipsis is cognitively economical for the user; it is wearying to hear well-understood information specified out. Ellipsis helps avoid tedium. It also just sounds more natural, bringing fluency to the dialogue. And it increases interaction rate (full phrases take longer to utter).

Semantics

Let us calculate.
— Gottfried Wilhelm von Leibniz

The study of meaning has traditionally been conducted under the label *semantics*, though (suitably enough) that word has meant something different in different contexts. For linguists, it usually meant the study of word meaning, and we have already incorporated some of the fruits of this research: homonymy, ambiguity, and synonymy are all aspects of word meaning. For philosophers, it usually signaled sentential (sometimes called *propositional* in this context) meaning. We have already seen some of the results of this line as well, in our brief looks at parsing, anaphora, and ellipsis.

Semantics, however, is not really the locus of meaning. It couldn't be. Meaning is not one thing, and it is not localized, certainly not in the word or the sentence. Meaning is an ever-shifting product, fashioned out of the resources of a language by a speaker, construed by a hearer, and shaped their context. It is produced, moment-to-moment, usually as speaker becomes hearer and then becomes speaker again (the situation that concerns voice interaction design).

There are three important topics to take up under the banner of "semantics," *truth conditions, paraphrase*, and *intention*. But keep in mind that we are only poking at the skin

---

9: The system, however, must prepare robust repair and grounding strategies to help sort out what's missing when the parser is confused about antecedents, or when it makes the wrong inference. Recall the ellipsis in the example from SpeechActs in the first chapter, where the system inferred that the caller wanted to know about Bob's schedule on Wednesday, when she wanted to know about her own.

of meaning here, not probing it in any depth. In subsequent chapters we will get under the surface a bit further, as we explore context and interaction.

Truth conditions are historically the defining theme of semantics. As the imposing *Encyclopedia of Philosophy* puts it, a syntactic system "becomes" a semantic system when rules for establishing truth (or falsity) are added (Kalish, 1967: 350–1). One of the reasons that sentences have been such an intense focus of research into meaning is that they are the bearers of truth (and therefore falsity), and to know what an utterance means is effectively to know what conditions determine its truth. Usually these are specified by translations of the utterance into a peculiar notation. Excuse me while I perform such a translation:

5.   The dog bit the postman.

6.   ∃d & ∃p & B(d,p)

Sample 6, translating yet again, comes out something like this:

There exists a dog. [= ∃d]

and

There exists a postman. [= ∃p]

and

The action "biting" occurred, such that the dog performed it on the postman. [= B(d,p)] Tedious, yes; trivial, to a degree. But the conditions are fully specified: for sentence 5 to be true, there must have been a dog, there must have been a postman, and biting had to transpire, with the dog doing it and the postman having it visited upon him.

Still, the whole translation process would not be worth much more than a stifled yawn if it wasn't for one critical attribute of this system: you can use this algebraic language to link sentences up, an especially useful property in the context of a dialogue. Recall the exchange over a gun on the suicide-prevention line:

A   Do you have a gun at home?

C   A forty-five.

(Pomerantz, 1986: 225)

Strictly speaking, the caller did not answer the counselor's question, which requires a confirmation or denial — "Yes, there is a gun at my home," or "Uh-huh" or "Nope," or something — but the question in this exchange is answered by entailment. It would be tedious (a hard word to avoid when you're talking about taking perfectly obvious pieces of language interaction and turning them into calculus) to chart out the linkage in detail. But, roughly: having a forty-five entails having a gun, and saying you have a gun entails an affirmation of the question "Do you have a gun?" Some sentences are true (or false), in other words, because other sentences, related sentences, are true (or false); so asserting one entails asserting another; and your hearers know this.

Voice interfaces need to be able to work out entailments, too. Indeed, your speakers will often prefer you to entail, rather than to assert directly. For instance, speakers of the following question do not usually want answer A_1. They usually want A_2 (or possibly A_3, but largely because of the information, not because of the affirmation):

Q   Do you have the weather for Springfield?

A_1 Yes.

A_2 Warm and hazy with a chance of afternoon drizzle.

A_3 Yes. Warm and hazy with a chance of afternoon drizzle.

The natural-language engineers, of course, will be happy to go on about logic, entailment, inference, and existential operators until the cows come home, cheerfully translating your request for two lumps of sugar in your coffee into a daunting string of formulas. Your job is just to figure out how entailments and inferences like these can benefit the interactions.

Speaking of truth, we come to *paraphrase*, which we have already seen in a smaller context. Paraphrase is the situation where two (or more) structures mean (more or less) the same thing; minimally, for instance, they have the same truth conditions. "Galen fed Silverback" and "Silverback was fed by Galen" are paraphrases largely because if one is true, so is the other; if one is false, so is the other. Paraphrases entail each other. That is, paraphrase is a species of synonymy, for pieces of language bigger than words. (Paraphrase is sometimes called "structural synonymy," though it can capitalize on lexical relationships as well.) There are two lessons here, both of them familiar by this point. On the input side, just as you need to allow for synonyms, you need to provide for paraphrases. Let's say your own transaction accounting system requires the dates to be mm/dd/yy, but your user says "the 15th of October 1994." You can punish him for saying it that way, and force him to restate it in your format, or you can accept the paraphrase and translate it for your accounting system. Which way will be more satisfying for him?

But here is the really messy part about semantics: you can't just hear (or look at) a sentence and know what it means. Take sentence 5 again ("The dog bit the postman."). It could, of course, mean that the yappy, hairy quadruped applied his teeth to the flesh of the professional deliverer of mail. But it might also mean:

• Hi, I'm Agent Smart [if the sentence was a special passphrase].

• Be careful [if you're about to enter a yard where there is a dog].

• We don't get mail anymore.

• Our dog was put down.

• Etcetera, etcetera, etcetera.

Sentences, that is, can mean almost any damn thing the speaker wants them to mean, in given circumstances, with receptive hearers. Irony, for instance, is when the speaker intends for the hearer to know that he means the literal opposite of what the words + syntax + truth conditions indicate that he means.

This isn't as disastrous as it sounds — people operate with shared "repertoires" of meanings, as the man who first wrote about these issues noted (Grice, 1968; see also Grice, 1957). If you access those repertoires, you can make meaning with them — but if you want to know what a piece of language means, you have to attend not only to "linguistic meaning" (words + syntax + truth conditions) but also to "speaker meaning" (what the speaker intends to communicate). And pinning down speaker meaning can, as you might guess, be a hairy mess.

"Abstruse philosophical nonsense," you're thinking? "Pedantic hogswaddle?" "Needless hairsplitting for someone who just wants to get a voice interface to function properly?" None of the above, unfortunately. It's true that one can get by without having to handle irony or other types of overtly figurative language. But more than anything else in the domain of "content," a speech system needs to come to grips with the speaker meaning.

What if your system, like SpeechActs in the dialogue snippet we looked at in Chapter 1, clung to the last topic when the caller wanted information framed by a more global topic? It would miss speaker meaning, and fail. What if sentence 5 comes in the middle of a transaction with a voice interface for AcmeBook.com? If the logic of the system tries to figure out who is biting whom (words + syntax + truth conditions), it will miss the point entirely. If, however, the system interprets the utterance as naming the title of a children's book (within a given context), and recognizes that the speaker wants to know how much it costs, or whether it is still in print, or if there has been a translation of it into Urdu, or whatever is appropriate to the context, then it has done its job.

As Lochbaum begins her influential computational treatment of speaker meaning, "Agents engage in dialogues and subdialogues *for a reason*" (1994: 1; I added the italics). Knowing the reason(s) gets you most of the way to knowing the speaker's meaning.

All of this brings us to the last critical notion of semantics we will take up, intention. (It is, of course, only shallow intentions, as a communicator, that are relevant. For instance, a speaker may intend to deceive you or to make you feel good, or both with "I love what you've done with your hair." Those are personal and social intentions. Her communicative intention, however, is just that you come to believe that she feels strongly positive about your new hair configuration. That's the level of intention we need to care about.)

Lochbaum continues: "[Agent's] intentions guide their behavior and their conversational partner's recognition of those intentions aids [their understanding of those] utterances" (1994: 1). From the voice interaction perspective, that means the inference engine has to be alive not just to what the users say but what they intend, what the purposes behind their utterances are. The most influential approach to the matter is Grosz and

Sidner's (1986, 1990) notion of "intentional structure."[10] Effectively, this is a formal computational account of the speaker's purposes in saying what he says, and of the relationship between those purposes and the prior dialogue. Any interface designer worth her salt who looked at Grosz and Sidner's work in this area for more than ten seconds would immediately strike her forehead and shout "Aha! A task model!," which should signal to you clearly that this is an area for heavy reciprocity between the natural language folk and the interface folk.

The speaker's intentions for any contribution to a dialogue reflect his purposes in engaging that dialogue, and in the dialogues that we are concerned with, his purposes are to complete specific tasks (get information, make purchases, solve problems). He is not just disinterestedly producing true or false claims about dogs and postmen. He is trying to do things with his words.

Prosody

The vocal apparatus is in fact an extraordinary instrument which easily beats the synthesizer in the amount of distinct sound qualities it can produce. It is an instrument of which we can change the size and shape at will and instantaneously. We can lengthen it and shorten it, widen it or narrow it, increase or decrease the size of its air outlets, change the shape of its resonating chambers.
— Theo van Leeuwen

Overlaid on spoken language, or emanating from it — certainly distinct from its other components, but completely integrated with them — is prosody, the rhythm of speech, the variations in frequency, duration, and volume that give spoken language its life. Prosody gives speech its music, its spunk and zest, its emotional qualities. All of these factors are largely beyond the reach of speech systems. Prosody is too subtle for recognizers to detect anything from it yet — or even to catch it — and is too subtle for speech synthesizers to encode with any sophistication. But prosody is an area of growing research in computational linguistics, especially in connection with agent personality and affective computing.

Prosody is often considered a phonological phenomenon (even by linguists, who should know better; some of them call it *suprasegmental phonology*) because it concerns sound, but it really has no more (nor any less) to do with phonemes than it does with words or syntactic structures or semantics. It plays indiscriminately among all the "levels." It can, for instance, distinguish between word uses. Many words can be either nouns or verbs, depending on use; some of them are distinguished by their stress pattern. Table 3.2 gives a few examples (where the capitalized syllables are the stressed ones — longer and louder than the other).

10: Intentional structure is actually only a third of their approach, which also includes *linguistic structure* (representing the discourse segments), and *attentional state* (an account of the speakers' focus of attention — the objects, properties, and relations that are salient at given points in the discourse).

Word	As noun	As verb	Usage
permit	PERmit		You need a permit to pluck the chicken.
		perMIT	I permit you to pluck the chicken.
contest	CONtest		We held a chicken-plucking contest.
		conTEST	I contest your right to pluck the chicken.
survey	SURvey		He conducted a survey of chicken-pluckers.
		surVEY	I am the master of all the chickens I survey.

TABLE 3.2 Word usage distinguished by stress (capitalization represents longer, louder syllables)

Similarly, sentence type (that is, syntactic information) can be distinguished by prosody. A question, for instance, is a question (in speech) more by virtue of the up-swing in pitch at the end than it is by word order. You can request information with an "assertion" pattern, so long as you swing the pitch up at the end. Orthographically, the upswing is signaled by a question mark.

7.   The cook tasted the chicken?

Other common aspects of sentential prosody are emphasis (indicated here by upper case):

8.   The cook tasted the CHICKEN.

9.   The cook TASTED the chicken.

Sentence 8 is appropriate to a situation in which there is some lack of clarity about what she tasted, and the sentence is here to ensure we know it was in fact the chicken (not, say, the eggplant); sentence 9 suggests a context in which there is a lack of clarity about what she did to the chicken, and insists we recognize it was tasting (not, say, seasoning). And sentence 10, with both upswing and emphasis, expresses incredulity that the cook tasted the chicken (perhaps she is a vegetarian):

10. The cook tasted the CHICKEN?!

But prosody is so unutterably part of speech that it is almost misleading to show a few specific examples like these, as if prosody merely had a few well defined functions in otherwise unprosodic speech. Nothing could be further from reality. Like color in visual depiction, prosody is never absent from speech. Colors may be neutral, prosody may be flat, but if there is no color, there is no visual depiction. If there is no prosody, there is no speech. Even speech synthesizers have prosody, though they are sometimes said not to. Their prosody can be very strange — like a gray peacock or a neon-chartreuse-spotted sheep — but it is not absent.

An Interlude on Speech Synthesis

Machines which, with more or less success, imitate human speech, are the most difficult to construct, so many are the agencies engaged in uttering even a single word — so many are the inflections and variations of tone and articulation, that the mechanician finds his ingenuity taxed to the utmost to imitate them.
    — Scientific American (1871)

There are three basic methods of generating output for speech systems — play-back, wave-form synthesis, and concatenation synthesis — each of which has different implications for acoustic naturalness; that is, for prosody. The first of these methods is just a matter of playing back an utterance recorded by a human. In terms of acoustic naturalness, this one is a function of the talent and the production. Some professional voicers, and/or some producers, can produce fairly weird recordings, but we can just ignore that possibility here. Humans, unsurprisingly, produce the most human-sounding speech. The limitation of play-back, of course, is that the system can only generate output that has been put in the can. Change is problematic, especially frequent or rapid change; imagine a stock-ticker system, in which the prices have to be updated every half hour or so. A great many speech systems have no choice but to use synthetic speech, at least for part of their functionality.

The two forms of speech synthesis have opposite implications for prosody. Wave-form prosody tends to be flat, the gray peacock; concatenative prosody tends to lurch about wildly, the neon-chartreuse-spotted sheep. Everyone is familiar with examples of both, often heard over the telephone.

Wave-form synthesis produces a largely undifferentiated, mechanical sounding stream of language, which is often called "unemotional." But the bigger problem is that it is un-everything; there's little or no variation. It would be more accurate to say that wave-form synthesis encodes one emotion, and only one: boredom. The emotional content of the utterance doesn't matter to the way it sounds, but neither does word placement or utterance type. Wave-form synthesis works by mixing a collection of essential speech parameters (which is why some engineers prefer the term *parametric coding synthesis*), according to a body of rules (which is why other engineers prefer *rule-based synthesis*), to generate a digital representation of speech wave forms. Core parameters include pitch, amplitude, duration, and frequency, which all influence the fidelity of the representation — how much the vowels and consonants sound like the vowels and consonants that humans deploy.[11] They're pretty good at fidelity. Super-ordinate parameters operate more globally, deter-

---

11: Amongst the variations in this research that I'm sliding past is that the parameters and rules might be based on different models — chiefly on either articulatory models or acoustic models (often called formant models, because of the crucial importance of vowel formants for speech perception, not just for the perception of vowels themselves, but also for the perception of their neighboring consonants).

mining how quickly or slowly the syllables and words follow each other (for the speech-rate parameter), or whether all the vowels sound like they were spoken by a male or a female (for gender), or an older person, or a child, or a breathy speaker, or a clipping speaker, and so on. What the rules are not very good at yet (though there have been, and will continue to be, small improvements) is prosody, the utterance rhythms that give speech its liveliness.

The number of parameters these synthesizers can manipulate, and the range over which they can manipulate them, could handle prosody effectively, but formulating reliable rules with respect to syntax (and context) has proven elusive. Linguists have a raft of hoary jokes based on prosodic differences. I'll only subject you to one, the contrast between sentences 11 and 12, which simple punctuation and word spacing conveys to any (literate) native speaker of English, who can read them with subtly different but communicatively distinctive prosodies, but wave-form synthesizers tend to render them exactly the same.

11. What's that in the road ahead?

12. What's that in the road, a head?

One has to say "tend to render them exactly the same" in these contexts, because an engineer could easily build a rule to alter the parameters in order to make these two sentences come out quite naturally and distinctly. But the rule would have to be sensitive to just these two sentences; finding a set of *general* rules that would automatically generate the appropriate intonations for sentences 11 and 12 is quite another matter. As Robert Rodman puts it, "the production of naturally sounding speech is no less complicated than the performance of a symphony" (Rodman, 1999: 196); or, in another image, "the principles that govern natural prosody have proven as elusive as the perfect soufflé" (Rodman, 1999: 184).

Concatenative synthesis has the opposite problem, not a flat soufflé, but a wildly three-dimensional crêpe. Concatenative synthesis takes isolated bits of speech (words, syllables, phones) and stitches them together. If speech bits were all the same, the effect would be like fridge-magnet poetry, a line of neatly uniform words; instead, the effect is more like an old-style ransom note, with different sizes, colors, fonts, and typefaces strewn through the message. The effect is notorious for numbers, and everyone over the age of thirty has heard something over the phone like the following.

Telephone   For what name, please?

Caller   Fred Derf

Telephone   Please hold, . . . that number is

NINE six OH SEven TWO thrEE six OH SEven TWO

That's a mild representation, restricted only to case changes; the grating shift from a recorded voice to badly concatenated word strings is worse than that. The main problem comes from our old friend, the phoneme, which, as you recall, is not instantiated acoustically as a sound but as a set of sounds. Which member of the set shows up depends on what the phonetic neighborhood is (a /p/ next to a /u/ is different from a /p/ next to an /e/). Crude concatenation takes the sound out of its neighborhood, sticks it in another one, and the ear can tell very quickly — jarringly — that it doesn't belong. Add to that the problem of prosody, that strings of words have prosodic relationships as delicate as a soufflé, and you have a recipe for crazily unnatural sounding speech.[12]

Or, you used to have such a recipe. Concatenation synthesis has improved dramatically in recent years, by learning from wave-form synthesis and developing rules for smoothing out concatenated sounds. And right now, concatenation synthesis has the lead in naturalness: the bits and pieces come from real human throats, and they can now be assembled in less grating ways. Not in ways anyone would yet call natural, but less unnatural. In the future, wave-form synthesis is more likely to dominate, as articulatory and formant models are brought together and the prosodic rules become more fully understood. But the perfect synthetic soufflé is some years off, perhaps decades.

Summary

If everyone always agreed on what to call things, the user's word would be the designer's word would be the system's word.
— George W. Furnas, Thomas K. Landauer, Louis M. Gomez, Susan T. Dumais

Linguistics has made tremendous headway in sussing out the way people trade noises to function interpersonally (cooperate, fight, gossip, . . .). But it has done so largely by ignoring the "personal," and the "inter-" as well — the situated motives and the general mechanisms of everyday conversation. That's not a slight on linguistics, or shouldn't be. Language is not a tidy little phenomenon like, say, playing chess, that can be fully specified, but it does have "pieces" and "rules," and you can't get very far toward either understanding or modeling the cognition of language without knowing those pieces, and the rules that move them. The successes of linguistics have come from following the divide-and-conquer methodology so fruitful in the natural sciences: divide sound off from the rest, and examine the hell out of it; ditto for morphemes and words, phrases and sentences, and to a very restricted degree, meaning.

12: Nor do prosody and phonetic variation exhaust the difficulties confronting concatenation. There is a general chunking problem. There are no silences in a speech signal corresponding to the spaces between words. The perceptual cues that one word has ended and another is beginning are very subtle and vary according a range of imperfectly understood variables (such as word and morpheme type). The concatenated elements have to be abutted, which scrambles these perceptual cues.

Dividing sound off has led to the insights of **phonetics** and **phonology** — that there are primary units, **phones**, extractable (more or less) from the raw acoustic wave form, which combine into groups corresponding to language-dependent psychological units, **phonemes**. The raw acoustic wave form is highly variable in the way it maps phonemes into sound clumps, not just because of this set-theoretic relationship, but because the instruments that do the encoding — human vocal chambers — are highly variable themselves, as are the conditions of coding (the effects of a stiff cup of coffee can alter the encoding process substantially, as can a late night, or the presence of an attractive third party). So, while linguistics has discovered a great deal about speech signals and their psychological correlates, speech recognizers have a very difficult job, and they fail regularly.

Dividing **affixes** and **words** off has led to the insights of **morphology** and **lexicography**. We didn't touch on either very much in this chapter, though we'll take lexicography up in a fair amount of detail later on, when we come to Chapter 7 on diction. Managing words is the alpha and omega of interface design — discovering and utilizing the user's terminology, in the given context of use, and inducing the user to adopt the system's chosen slice of that terminology — and a great deal of this book concerns words quite directly; even more of it concerns words indirectly. In the meantime, there are two important principles we noted in this chapter, the **principle of collocation** and the **principle of lexical priming**. The first — the strong propensity that words have to form collectives (so that *educate*, for instance, travels with a different group than *train*) — is important for understanding discourse patterns. And understanding these discourse patterns is crucial both for choosing system terminology, and for the lexical inference aspect of speech recognition (which word or words are most likely to match up with a given acoustic pattern). The second principle — the remarkable fact that people tend to pick up each other's words, even if they have "better" words for a concept in their heads — is important for inducing users to use the system's vocabulary. We also considered **synonymy**, which gives the users flexibility (at a price), **homonymy**, which is something of a pseudo-problem for recognizers (largely because of collocations), and the general notion of **ambiguity**, which is a minor problem for humans and a major headache for recognizers.

In **syntax**, the divide and conquer method, has led (among much else) to the understanding of **anaphora** and **ellipsis**, both of which are critical for the development of voice interfaces. In **semantics**, the method has led to notions like **paraphrase** and **entailment**, again critical for voice interfaces. On the input side of things, the parsing components, and the inference engine, cannot do without an understanding of these elements, especially since they have to be reconstructed from faulty or partial input. We looked at this exchange earlier:

T_1 System: How may I help you?

T_2 Caller: I was trying to place a call and must have dialed the wrong number, can
         I get credit for that?

T$_3$ System: Do you need me to give you credit?

T$_4$ Caller: Yes.

(Boyce, 1999: 49)

(The T$_n$ convention just means Turn-n, which I occasionally use when we need to talk about specific contributions to a dialogue. We will take up the notion *conversational turn* in a bit more detail in Chapter 5. For now, the common sense notion of turn — as it is used in a game, for instance — is sufficient; in fact, it is more or less sufficient throughout the book.)

It is highly unlikely that the exchange went quite like this, from the system perspective. In particular, the system more probably registered something like this:

Caller: blah blah blah blah **place a call** blah blah blah **dialed** blah **wrong number** blah blah blah **credit** blah blah

Or possibly even:

Caller: blah blah blah blah **place** blah blah blah blah blah blah blah blah blah blah blah blah **credit** blah blah

From the term(s) caught, the system has devised the lexico-syntactic hypothesis that the caller means "I placed a call but I dialed the wrong number and I want credit from you for that call," or just "I want credit from you." Whatever the system caught, we know that it filtered enough signal out of the noise to make the offer in T$_3$. In order to make that offer, it has reconstructed the anaphoric argument "I," the elliptical argument "you" (or "AT&T," which is the same in this context).

Another possible scenario has the recognizer catching nothing but "... dialed blah wrong number ... ," in which case it has to draw on entailment possibilities, from a logical space that includes an axiom something like, "if the customer dials a wrong number, the customer may want a credit," or, more directly, "if the customer dials a wrong number, offer the customer a credit." Whatever the specifics of this How-may-I-help-you exchange, we can be sure that, recognizers being what they are, the system did not catch everything in the user's T$_2$, and that an understanding of both syntax and semantics was deployed in generating the correct output.

On the output side, these syntactic and semantic concepts — anaphora, ellipsis, paraphrase, and entailment — are all equally important for a fluid and efficient dialogue. Notice that several of them are also at play in the How-may-I-help-you exchange above. The system's T$_3$ uses anaphora (you, me) and ellipsis (for the wrong number). Anaphora isn't required here (or anywhere) from a strict exchange-of-meanings perspective, but the alternative sounds either like a very stiff butler, or something written by an engineer:

System: Does the customer need AT&T to give the customer credit for the wrong number?

The system's T_3 also uses paraphrase. It is, in fact, a reverse paraphrase of what it has extracted from the caller's input. Many confirmations/queries have this structure, echoing back to the user what she has just said (that is, what the system thinks she has just said). And there is a significant entailment in T_3 as well, that the credit is being offered *because* the user dialed a wrong number. (Notice that this entailment goes through irrespective of what the system detects. It is drawn by the user, and the user has just added "I dialed a wrong number" to the discourse — she has just, as we shall soon be saying, *grounded* that information.)

The remaining linguistic factor sketched out in this chapter, **prosody**, however, is not very available to divide-and-conquer reductionism. It is too fully integrated with phonetics/phonology, lexicography, syntax, and semantics. The only real lesson the study of prosody offers for speech recognition is one that lays bare its limitations, since detecting prosody is a long-range goal, not a current possibility.

We also looked briefly at **speech synthesis** — in particular at the two methods called wave-form synthesis and concatenative synthesis. For speech synthesis, the lessons of prosody are somewhat more immediately encouraging — although **wave-form synthesis** is still some years away from natural speech rhythms, **concatenative synthesis** is becoming more human in its prosody. Even with concatenative synthesis, however, the more subtle effects of prosody (emphatic contrast, emotional expression) are a way off.

The investigation of prosody reveals the shortcomings of linguistics' reductionism for understanding the full sweep of language, especially as a vehicle interpersonal action. It's misleading to say that linguistics ignores context. Linguistics just severely constrains context. The notion of a phoneme as a set of sounds, for instance, would get nowhere without some sense of context — different elements of the set show up in different phonetic contexts. What people usually mean when they say that linguists ignore context is that linguists ignore extra-linguistic context, but even that claim is off center. In particular, semantics has been the subject of pity and scorn for excluding context, but what are truth conditions if not extra-linguistic context? "A cook tastes the chicken" is true in exactly the context in which there is a cook, there is a chicken, and said cook uses his gustatory receptors on said chicken.

But the constraints on context need to be increasingly relaxed to provide an understanding of language sufficient for voice interaction design. The *personal* and the *inter-* have to enter the picture, which is precisely what they will be doing over the next two chapters.

4

Doing Things with Words

Words are deeds.
— Ludwig Wittgenstein

Language is the most powerful instrument of cooperative action that humankind has, that *any* species we know of has. Long before we used it to visit the moon or split the atom or deploy ubiquitous wireless communication devices, Isocrates praised such possibilities this way:

> *In the other powers which we possess we are in no respect superior to other living creatures;*
> *nay, we are inferior to many in swiftness and in strength and in other resources; but,*
> *because there has been implanted in us the power to persuade each other and to make clear*
> *to each other whatever we desire, not only have we escaped the life of wild beasts, but we*
> *have come together and founded cities and made laws and invented arts; and, generally*
> *speaking, there is no institution devised by man which the power of speech has not helped*
> *us to establish.*
>
> (Nicocles, 5–6)

I include this passage, one of the countless paeans in the history of thought to the civilizing, human-defining power of language, not so we can pat ourselves on the back (we do some pretty despicable things with language, too), nor so we can look down our snoots at other species (none of whom, lacking language and technology, have had proportionally even a sliver of the unwholesome impact we have had on the ecology we all share), but merely to emphasize the tremendous disparity between what we do with language and what linguistics can describe.

Linguistics describes the necessary building blocks — we need to know about phonemes and words and truth functions if we are to understand how language operates — but the description is incredibly impoverished as an account of how we put language to work founding cities, making laws, and getting each other to pass the salt, as an account of how we do things with words.

Pragmatics

We looked!
Then we saw him step in on the mat!
We looked!
And we saw him!
The Cat in the Hat!
    — Theodor S. Geisel (Dr. Seuss)

In linguistics and philosophy, the overt inclusion of context in the study of language and its functions is called "pragmatics" (e.g., Stalnaker, 1972). Take a look at sentence 1:

1.   The cat is on the mat.

Pragmatics asks "What *is* that sentence?" and, more particularly, "What is it *doing*?" The answer depends on context. It could be a simple assertion, in the context of the query, "Where's the cat?" It could be a warning — from a cat owner to a friend with an allergy; from one mouse to another in a cartoon where they are scheming to get some cheese; from a parent to a young child when the cat in question is ornery. It could be an assurance — from a house sitter with a reputation for losing cats. It could be a solution — for someone who has been working on a riddle. It could be a linguistic example. In fact, in the environment of this book, sitting up there with the number 1 in front of it, that's just what it is.

What *is* that sentence? It is an *utterance*, a chunk of language functioning in relation to other chunks of language, in a context of use. I've been using the term *utterance* systematically for a while now. It's time to account for that usage. A sentence is an abstract object, a specific syntactic configuration of lexical items conforming to the combinatoric rules of a grammar. An utterance is a physical object, a sequence of sounds or characters conforming to the demands of a communicative situation.

The notion of utterance is an important one because of the emphasis it puts on particularity and context. Utterances are, in principle, unique. They can be loud or soft, quick or slow, complete or incomplete, grammatical or not, a word or a long string of words. In the way *utterance* is used in the pragmatics and discourse analysis communities, however, it usually implicates a sentence, and is rarely larger than a sentence. Take this exchange, for instance:

T_1 Franklin: Who is on the mat?

T_2 Freddy: The cat.

The first utterance in this exchange is a sentence, the second is a pair of words (comprising a noun phrase), but the second, in this context of use, is more than just a randomly chosen pair of words (or random noun phrase). It is an assertion about the spatial relations between the cat and the aforementioned mat, an assertion that effectively is equivalent to sentence 1. The critical notion, clearly, is context of use. Without T_1, we wouldn't have a clue what T_2 is doing.

And when you start thinking about contexts of use, you quickly realize just how powerful and indispensable that notion is, not just for the function of an utterance, but even for the constituent words within it. *Cat*, for instance, might be a feline, but might also be a knickknack shaped like a feline, or a jazz aficionado, or a piece of heavy equipment (the term genericized from the manufacturer, Caterpillar). The *mat* could be a little carpet for wiping your feet, or a clump of hair, or an exercise pad, or the border between a picture and frame (which, in some contexts, might be spelled *matte*). Even the lowly *on* has a range of contextually-determined jobs. Let's say there's a practice of smuggling marijuana by weaving it into mats, and the practice becomes well known enough that *the mat* becomes slang for marijuana; in that context, an utterance like "the mat" could be an answer to a query heard at many a party in the sixties, "Man, what is that cat *on*?"

Context is not just something that one sometimes resorts to for clarification, the way it might seem in isolated discussions like this one. It is absolutely indispensable in ordinary language use. Take ellipsis, for example. We looked at it as a linguistic phenomenon, which involved the omission of later words or phrases based on their earlier presence in an utterance:

2a Franklin wants a burger, Phineas a taco, and Freddy four burritos.

b Franklin wants a burger, Phineas [wants] a taco, and Freddy [wants] four burritos.

The verb *wants* is omitted in the last two clauses in 2a because it "carries over" from the first clause; we know that 2a "really means" 2b because of linguistic ellipsis. But, as William Watt (1968: 347) points out with respect to speech systems, there is *pragmatic* ellipsis, too, a lot of it, in which the context (and not just a previous utterance) licenses the ellipsis. Sentence 3a, for instance, is more properly rendered as 3b (where the bracketed material represents the omissions):

3a Akron, one way, 2 p.m.

b [I would like to buy an airline ticket for a flight to] Akron [on this airline, such that the flight is] one way [and such that the flight leaves at] 2 p.m.

The ellipsis in 3a is not a linguistic matter, depending on earlier phrases in the exchange (though it *might* be related to earlier phrases). It is pragmatic, because the only way to recover the information is by context.

There are two general lessons about context for voice interfaces (and they are lessons that have been well-learned, if not always well incorporated by language engineers). First, if you ignore context, your system will be hopelessly impotent. Second, if you attend to context, you can usually constrain the functions and meanings of an utterance sufficiently to figure out how to deal with it: 3a, for instance, is a snap to resolve as 3b in the context of airline ticket booking. The more specific lessons of context for voice interfaces will occupy us until the end of the chapter.

Conversational Pragmatics

> *Our talk exchanges do not normally consist of a succession of disconnected remarks, and would not be rational if they did. They are characteristically . . . cooperative efforts.*
> — H. Paul Grice

H. Paul Grice, whose work in semantics we saw briefly in the last chapter in our consideration of semantics (he was the first person to raise the issue of speaker meaning), realized that meaning is a two-way street. Communication depends on what the speaker intends the hearer to understand, but it depends equally on the ability of the hearer to infer the speaker's intentions; it is bilateral. Therefore, he turned his attention to conversations.

Grice revealed a set of principles in the way we swap meanings — *conversational maxims*, he called them — that are indispensable to dialogue design. They are simple enough, and so deeply engrained in all of us that they often strike people as staggeringly boring and trivial. Boring they may be to some (though, if you fall in this category, you might be in the wrong business), but they are not trivial. Grice's maxims go a long way to defining how it is the linguistic noises we make at each other lead to cooperative social action. There are four clusters as follows (Grice, 1989: 26–28):

The Maxims of Quantity

- Make your contribution as informative as necessary.

- Do not make your contribution more informative than is necessary.

The Maxims of Quality

- Do not say what you believe to be false.

- Do not say that for which you lack adequate evidence.

The Maxim of Relevance[1]

- Be relevant.

The Maxims of Manner

- Avoid obscurity.

- Avoid ambiguity.

- Be brief.

- Be orderly.

And, I might as well get it over with now, the blindingly obvious principle behind all of them:

The Cooperative Principle

- A conversational utterance should be "such as is required, at the stage at which it occurs, by the accepted purpose or direction of the talk exchange in which [it occurs]" (Grice, 1989: 26)

These maxims, and their governing cooperative principle, have two driving virtues: descriptive and predictive. They describe the default content-guiding procedures people follow when they converse; or, at least, they describe an assumed baseline that guides exchanges. (A baseline for *good* communication, of course; there are liars, paranoids, wind-bags, and other species of deviant conversants abroad.)

Grice's maxims are output-writing guidelines for voice interface writers. They tell your system how to talk. And, even more importantly, they predict what people are up to when they deviate from this baseline. They are inferential guidelines for voice system input. They tell your system how to listen. Grice was interested in trickier exchanges than voice systems need to deal with, but there is a powerful trickle-down effect.

Quantity

The maxims of quantity are tremendously important for speech systems, a fact that has been recognized almost from the beginning. Matt Marx, for instance, uses the don't-say-too-much maxim to warrant MailCall's use of first names only, unless the last name is

---

1: Grice actually calls this the maxim of *relation*, taking his cue from Immanuel Kant's categories of understanding. Many commentators find this label opaque (which is somewhat ironic given Grice's technical understanding of clarity), and I have followed the introduction-to-pragmatics convention of relabeling it with the term *relevance*.

necessary for disambiguation (1995: 77; see also Marx and Schmandt, 1996; Yankelovich, Levow, and Marx, 1995; Chu-Carrol, 1996; and Brennan, 1998: 1, 9).

Quantifying information, though, is not quite like quantifying at-bats or hitting averages or on-base-percentages-against-left-handed-relievers. There are no clear units to count and perform operations on. And humans are not automata: what is enough information for one person is too much for another and not enough for a third. You need to have a Goldilocks "j-u-s-t right" target in mind for information provision, but there is no infallible formula that determines the "j-u-s-t right" for everyone. What is clear is that when you provide too much information, it is fatiguing for listeners; when you don't provide enough, it is debilitating. Here's what happens when you don't provide enough:

System: How may I help you?

Caller: I'd like to make a call and charge it to my calling card.

System: I'm sorry. How may I help you?

Caller: I want an operator.

(Boyce, 1999: 56)

An open prompt (if the context is clear, adequately circumscribing the range of topics) is an excellent initial strategy, but repeating it without some elaboration other than an omnibus failure indicator (as at T_4) is a poor follow-up strategy. You need to say more, providing a fuller indication as to how the caller can proceed. Suzanne Boyce tested this verbatim-repetition strategy (the example is from her research), finding that it did rather poorly, and that even a marginally more instructive "I'm sorry. Please briefly tell me how I may help you," improved performance (1999: 56).

Yankelovich and her SpeechActs collaborators call this strategy of increasing detail *progressive assistance*. Here is one of their progressions, each subsequent recognition failure triggering a more explicit directive:

Sorry?

Sorry. Please rephrase.

I didn't understand. Speak clearly, but don't overemphasize.

Still no luck. Wait for the prompt before speaking.

(Yankelovich et al., 1995)

Now, this series concerns only recognition — not, say, vocabulary or domain — and the project was a testbed. Few commercial users would have the patience to sit through four total recognition failures, no matter how patiently the system was trying to help them enunciate. But the idea of scaling up information under various conditions is critically important to voice interface development, a general practice known as *expansion*.

Progressive assistance is one way of getting at the not-enough-information problem. The inverse, *tapering*, gets at the not-too-much problem. This term denotes a strategy of scaling back the detail in system utterances as a function of discourse history. When users first encounter the recording feature in SpeechActs, for instance, they hear:

Please record your message after the tone. Pause for several seconds when done.

But in subsequent uses of the function, they hear only:

Record then pause.

(Yankelovich et al., 1995; Martin et al., 1996: 39)

Anyone who has listened to thirteen words of synthetic speech when three would do, will recognize the importance of tapering to the overall experience of a user.

The notions of expansion (saying more, as required) and tapering (saying less, as required) both hinge on the maxims of quantity. In particular, they are a matter of knowing how much you should say *now*, given what you said in previous turns (that is on what has been *grounded*, to anticipate a term we will investigate in more detail in the next chapter). And both are extremely useful strategies in voice interface design.

Quality

The maxims of quality are equally significant — more significant, depending on your perspective — not just for voice interface design, but for the design and deployment of any product and service with nontrivial communicative dimensions. But they are the most clearly ethical of Grice's dicta (*all* the maxims are ethical, of course, subsumed under the principle of cooperation; these ones just have the strongest and most explicit ethical implications), and they only implicate usability indirectly. So they are best saved for a more appropriate location; we'll take them up again in connection with agent design (Chapter 11).

I'll pause here only to make two points. Firstly, trust and credibility are considerable attributes of information outlets and retail outlets. Both trust and credibility are tied inextricably to a commitment to warranted truth: to not uttering falsehoods, and to not making unsubstantiated claims. In fact, there are laws about such things. Be good.

Secondly, truth is not enough — nor relevance, quantity, or manner. The maxims interact in necessary ways. Suppose someone asks "Is there a flight from New Orleans to Miami on the morning of the 16th?" The answer "yes" might be qualitatively impeccable, but if the "flight" requires changing planes in Chicago, there may not have been enough information for the caller — quality has been satisfied, but the answer is problematic and possibly misleading because quantity has been violated.

Relevance

The maxim of relevance similarly need not detain us long here, either. In fact, it is a hard maxim for voice interfaces to avoid. They are doggedly relevant. The trouble comes, if it comes, from a mismatch between what the user regards as relevant, and what the system regards as relevant, and again that is best taken up elsewhere (in this case, when we get to the notion of grounding, in Chapter 5).

Manner

The maxims of manner largely concern clarity. Only one of them invokes clarity very directly (and even that is somewhat roundabout, by enjoining us to avoid its opposite, obscurity), but all of them fall into the standard notions of communicative clarity: avoid rare or arcane words and phrases (be direct); avoid loose or indeterminate words and phrases (be specific); don't be longwinded (be brief); don't jumble things together (be orderly).

Not everyone, however, including some of the people distributing them, realize how thoroughly pragmatic these guidelines are. Clarity, directness, specificity, orderliness — none of them is in itself objective or independently moored; all of them are functions of the discourse in which the utterances participate. What is obscure in one discourse is efficient in another.

Take the term, *OOV*. It could mean just about anything, and it is hopelessly unclear if you don't have the key. *Observatoire Océanologique Villefranche-Sur-Mer* is the first corresponding response I got from checking it out on a web search engine, for a marine institute in what looks to be a stunningly pleasant little village on the Riviera. I also turned up *On-Orbit Verification*, which seems to have something to do with satellite tracking, *Out Of Vehicle*, for rally races, and *Out Of Vision*, for television broadcasting. In the discourse of voice interaction design, it means *Out-Of-Vocabulary*, in reference to user utterances that are not in the system vocabulary. Your service is providing weather information and the client asks for what it will be like tomorrow in Eastern Passage, but *Eastern Passage* is not in your vocabulary.

There are three scenarios for the use of OOV, in the speech-system sense, with respect to the Maxims of Manner: (Case 1) OOV is obscure for people who don't participate in that discourse; and (Case 2) its ambiguity could make it misleading to oceanographers, rally drivers, television folk, and people who deal with satellites; but (Case 3) using it is a really efficient way to talk about certain speech-system rejection errors, for people who know the term. The moral here, of course, is that people have vocabularies, too, just like speech machines; some words are OOV for them (Case 1), some words have multiple senses, or other senses than intended, for them (Case 2), but some words are spot-on, high-efficiency terms which are both comprehensive and precise (Case 3).

With a term like *OOV* this communication problem is usually known under the word *jargon*, which is almost always a pejorative term. With *fenks*, the problem might be known under the word *pedantry*, also pejorative. People, like speech systems, don't appreciate it when you use words they don't know. But jargon and pedantry are a function of discourse. If you know speech-systemese, *OOV* is very useful, and there may even be communities, in Greenland perhaps, for whom *fenks* is very useful.

The two remaining maxims of manner — Be brief and Be orderly — implicate syntax more directly than words.

"Be brief" recalls a maxim of quantity ("Do not make your contribution more informative than is necessary"), with which it is sometimes confused. But brevity in this case is established by a sheer word-and-syllable count. The maxims of quantity (somewhat opaquely and loosely) concern the amount of *information*; the manner-maxim of brevity just concerns the amount of *language*.

MailCall's use of ellipsis illustrates this point well. Without ellipsis, listening to the report of a new email stack would be aggravating in the extreme, leaving the caller with an impression of the system as mechanical and thick. Without ellipsis, you would have a highly redundant message.[2] Compare the original version with one in which the elided material is put back in (they both have the same amount of *information*, so the maxim of quantity is not involved, but the second one has more *words*, so the brevity maxim is implicated):

MailCall: You have seven messages from Gina-Anne Levow, two from Stuart Adams, and one each from Nicole Yankelovich, Eric Baatz, and Andrew Kehler.

MailCall: You have seven messages from Gina-Anne Levow. You have two messages from Stuart Adams. You have one message from Nicole Yankelovich. You have one message from Eric Baatz. You have one message from Andrew Kehler.

(Marx, 1995: 69)

The second one might not *look* so bad, but that's because we process text more efficiently than speech. Read it out loud. Of, if you really want to torture yourself, get a text-to-speech

---

2: Redundancy has its uses, of course; in particular, it increases information robustness. And, of course, there are obvious connections between lexical quantity and information quantity. "George Jetson" has both more words and more information than "George," at least the first time it is used. Still, as the MailCall example shows (1) redundancy is unrelated to information quantity, and (2) sheer word quantity quickly becomes burdensome to the user.

device and have it read the second one to you. You will appreciate ellipsis; you will appreciate the injunction to be brief. So will your users.

Being orderly is every bit as important as being brief. Spoken language operates in time, which argues for brevity, and through sequence, which argues for orderliness. Language has certain natural orders, something that engineers don't always believe, because truth conditional semantics tend to ignore linearity. Here's my favorite example (because it yields a long formula in predicate calculus that completely misses the point):

8   It's always the same at parties: either you get drunk and no one will talk to you or no one will talk to you and you get drunk.

(D. Wilson, quoted in Blakemore, 1992: 80)

Aside from indicating D. Wilson's dismal social life in the early 1990s, this utterance shows that natural-language concatenation is not the same as predicate calculus concatenation; the commutative principle does not apply. Natural-language concatenation, that is, like most natural-language phenomena, is orderly. What precedes and what follows is rarely incidental. Take a look at these familiar clauses, all of which would sound extremely awkward in reverse:

cause and effect, hit and run, trial and error, wait and see, park and ride, . . .

Here, the natural ordering is temporal (just as it is in D. Wilson's utterance): languages have a strong preference for using their natural linearity to maintain temporal order. That's one of the reasons that narrative is such a dominant form of discourse in oral cultures, and why even such abstract literary genres as the experimental report work from methods through results to discussion. But there are other orders — the tendency, for instance, of words that have strong conceptual bonds to nestle together, for instance:

9a   racy Italian model

 b   Italian racy model

10a   big pepperoni pizza

  b   pepperoni big pizza

11a   three blind mice

  b   blind three mice

There is nothing syntactically wrong with the b versions here (nor, for that matter, with phrases like "error and trial"), but nationality, kind, and blindness are more essential attributes than appearance, size, and quantity. (These matters implicate an important term in J. R. Firth's linguistics, *colligation* — the joining together of words — which informs an important aspect of voice-interface discourse modeling that we will revisit in Chapter 7.)

Given Information, New Information

> *If I asked you, "Why did you hit that guy?" your answer might be "I hit that guy because he insulted me." However, in this context you certainly wouldn't say "Because that guy insulted me I hit him." This is because "I hit that guy" is now the old or presupposed information and should come in the first part of the sentence, while "because he insulted me" is the new or asserted information and should come at the end.*
> — Bill Byrne

In an utterance, in a conversation, there is information that is known to both agents — that is *grounded* — and there is information just showing up now with the utterance: information that is the *point* of the utterance. Take this exchange:

Caller   Who is its provider?

ARTIMIS   The provider of 36 68 00 00 is Météo-France.

(Sadek and De Mori, 1998: 556–558)

Cropping to just ARTIMIS's response (similar observations hold for the caller's utterance), some of the information it conveys is already grounded (known as "old" or "given" information in the business — see Clark and Haviland, 1977), and some of it is being proposed for grounding ("new" information). ARTIMIS[3] accepts the query by responding to it directly (thereby ratifying *provider* as part of the ground), then presents its assumption that the *it* linked to *provider* is *36 68 00 00* (thereby explicitly advancing that identification to the common ground), then gives the new information: *Météo-France*. Given information (if it comes at all) tends to come first, new information tends to follow it.[4]

3: This tendency is not absolute, and can be overridden by an intonation pattern that foregrounds the new information and backgrounds the given. It is also regularly disregarded within some formal written registers (engineering reports and legal discourse, for instance), which is among the factors that make them so difficult to read until you get used to this specific hobbling pattern. (Learning to read them in this respect is like learning to use crutches; one can become quite adept, but it requires cultivating different muscles and motor programs than regular walking.)

4: ARTIMIS is an acronym for Agent Rationnel à base d'une Théorie de l'Interaction mise en oeuvre par un Moteur d'Inférence Syntaxique (Rational agent based on Interaction Theory implemented on a syntactic inference engine). AGS is an acronym for Audiotel — Guide des Services (Audiotel Guide to services).

This powerful tendency for given information to precede new information is one of the more systematic instances of Grice's orderliness maxim. This arrangement conforms to the notion that utterances are *about* something, that they have a point, a focus. The focus is new information. The rest of the utterance, the given information, is there to explicitly contextualize the new contribution to the dialogue. The linguist Talmy Givón calls the given information in an utterance the "address" for the new information, its "storage locus" in the network of knowledge that constitutes the conversational ground (1990: 899). From a more dialogic perspective, Clark and Schaefer (1987, 1989) identify it with what they call the "acceptance" phase of a conversational contribution, which indicates that the speaker accepts or rejects, or at least acknowledges, the relevant aspects of the previous turn (see also Cahn and Brennan, 1999).

Here is a directory-assistance example that violates the expected given/new order very gratingly, in a way we have all suffered through:

Operator   Hi, this is Joan. What city?

James   Mountain View

Operator   All right?

James   CostPlus

Synth   The number you requested, five, five, five, nine, six, one, six, zero, six, six, can be automatically dialed by pressing one now. An additional charge will apply.

(Cohen, Giangola, and Balogh, 2004: 148)

What James wanted here, what most of us want in these circumstances, is the number. That's the new information he is after, the reason he made the call. But here it is embedded in the old-information slot, the "storage locus," the "acceptance" part of the utterance — stuck in a relative clause attached to the subject. Worse yet, the assistance-system is foisting other information on us (which may or may not be new; most of us know the shill is coming), by putting *it* in the focal, new-information slot. That faux-new information swamps our buffers most of the time, unless we're writing down the number as it comes, or hang up before the swamping starts. We forget. It's easier just to push 1, not coincidentally leading to higher profits for the information-assistance provider.

ARTIMIS's violation is not so egregious, nor so ethically questionable, but it is somewhat unnatural. ARTIMIS accepts a bit too much, doing far more than providing an address. It is, as a recognition system, somewhat insecure. It is making clear what its own map of the neighborhood is, so the caller can reject or adjust the map (or, by saying nothing, accept the whole map). The redundant incorporation of given information is useful for error correction, but it needs to be handled judiciously. A steady diet of too much old information sounds clumsy and mechanical. A human respondent is more likely to just say "Météo-France" or possibly "The provider is Météo-France." Here's a typical human/human exchange in the same register:

Amex   What city are they traveling to?

Caller   To LA.

SRI/Amex (1989; tape 17, call 2)

Speech systems should follow this focal strategy as well, depending on the purpose of the utterance and the level of grounding confidence. That is, unless the confidence level is below some reasonable threshold, given information should be generally pruned back far more than in current voice-interface design practice.

Register

The language we speak or write varies according to the type of situation.
— M. A. K. Halliday

How we speak — which strongly affects how we can follow the Gricean maxims — comes in various genres or styles. The term for situationally effected language styles is *register* — "a set of meanings that is appropriate to a particular function of language, together with the words and structures which express these meanings" (Halliday, 1978: 195).

Computational linguists independently developed a very similar notion, *sublanguage*, defined by Grishman as "the specialized form of a natural language which is used within a particular domain, . . . characterized by a specialized vocabulary, semantic relationships, and in many cases specialized syntax" (2001). The definitions amount pretty much to the same thing, but the term *register*, because of its associations with a pragmatically infused approach to language, is a much more wholesome one for voice interface design.[5]

Registers have three defining dimensions: field, tenor, and mode (Halliday, 1989: 38–9). Field focuses on activities people use the language to accomplish within the register — inquiring, buying, booking a flight. Tenor focuses on the agents who carry out those acts, not as individuals and personalities, but as roles played by participants in communicative events — information seekers and information providers, shoppers and sellers, travelers

---

5: *Sublanguage* also has an unfortunate history of which it is best to divest design work. (In particular, it is inherited from the highly misleading set-of-sentences notion of language so dominant in computational linguistics; Harris 1968 formalizes sublanguage as a subset of the general language.) Too, the prefix *sub-* also gives the word connotations of primitiveness and social subordination, suggesting pidgins and slangs.

and travel agents. Mode is effectively the same as it is in other areas of interactive design, the input/output channels — writing, speaking, text messaging.

In a voice interface, there is one input mode, speech; though when background noise is high, or in other times of desperation, such systems may have to fall back on manipulative input, or button poking. Voice interfaces have two output modes, speech and nonspeech sound (which might be further divided into music and representational sounds). All the registers that a voice interface participates in, then, are constrained to audio modes.

Registers are populated with specialized terms and structures. A weather register, for instance, has terms like *front* and *drizzle* and, for afficionados, *Doppler radar*. It has structures like (among others) template 4, illustrated here by sentences 5–7:

4   TIME-NOUN-PHRASE AUXILIARY-VERB be CONDITION-ADJECTIVE(S), with PRECIPITATION-NOUN-PHRASE PREPOSITION TIME-NOUN-PHRASE.

5   Today's weather will be mild, with occasional showers in the late afternoon.

6   Tomorrow will be cold, with light flurries in the morning.

7   Wednesday should be hot and hazy, with a seventy percent chance of thunder storms towards the evening.

There are sports registers, which differ mildly according to sport; nautical registers; financial registers, which shift with income bracket and time of year; health-care-professionals' registers, and nonprofessionals' health registers; and so on. They interpenetrate, and they all depend on a large set of the same standard linguistic resources, and the same people often engage a collection of registers, but the beauty of it all for voice interfaces is just that they exist: they can be studied, understood, and emulated.

Register and Grice's Maxims

Registers are *manners* of speaking. They implicate the maxims of quantity, quality, and relevance as well. Amount, topicality, and even accuracy of information are related to register. If Fred arrives at a specific point at 7:00 and Barney arrives at 7:01, the utterance "Fred and Barney arrived at the same time," for instance, might be accurate in a gossip register, when reporting about their temporal proximity in showing up at a party. In a sports register, when reporting about their temporal proximity in getting over the finish line in a race, it is highly inaccurate.

But registers are most closely associated with the maxims of manner, and the only way to satisfy the maxims of manner in voice interaction design is to investigate thoroughly the registers of your service. The maxims of manner, again, concern clarity. Language is clear or obscure not on some absolute scale, but as a function of the mode (what is clear in text can be obscure in speech, what is clear in speech can be obscure in text), the agents engaged (what is clear between health care professionals can be obscure between a professional and a layperson), and the activity (what is clear when you are banking — using such terminology as *deposit*, *withdrawal*, and *transfer* — can be obscure in the context of another activity, like baking or gardening). Ditto ambiguity.

Registers also have orders they prefer. Sometimes these orders are reflective of general cognitive and linguistic tendencies of the sort we saw above. Departures before destinations in travel arrangements reflect the temporal bias, for instance, as does from-accounts before to-accounts in bank transfers. There is also a tendency in general observations to precede specific ones — a tendency that the weather structure (4, preceding) follows closely: a general statement about the whole day is followed by a more specific condition at a more specific time ("Warm and muggy with possible thundershowers in the evening"). Sometimes register orders are determined by more local preoccupations (winning team before losing team in sports scores, for instance). Being orderly, being brief, being specific, and being clear are not context free; they are a function of register.

Voice interface design is impossible without a thorough understanding of the relevant registers; in turn, those registers determine the satisfaction criteria for the maxims of manner.

Listening

Hear diligently my speech.
    —Job 21

Grice's maxims are deeply important in designing the system utterances. They are equally important for disposing the system to listen to user utterances. The maxims describe the general strategies for *producing* talk, and therefore they work inversely as strategies for *understanding* talk (which was, in fact, Grice's prime concern, linking this work with his interest in speaker meaning).

We frequently say things to each other that a third party, outside our context, has great difficulty with. Take this example, a perfectly mundane exchange which appears on the surface to violate all of Grice's maxims, and a few that never occurred to him:

Phineas: Take the trash out, will you please, Freddy?

Freddy: The cat is on the mat.

Freddy's response seems utterly unconnected with the utterance that precedes it. But if we want to understand it, as an unremarkable and appropriate response to Phineas's request, we have to begin with the assumption that Grice's maxims are in operation, relative to the context. If we assume that relevance is not violated here, for instance, it only takes a moment to come up with a context in which the cat being on the mat is relevant to Phineas's request. It might be that the mat is in front of the door where the trash will be taken out, that the cat is on it, and that there is a local convention such that moving the cat is not done (at least in Freddy's view of things). Freddy's utterance, then, is a *reason* for not satisfying Phineas's request, possibly an *excuse*.

Similarly, within the context, if Freddy's response is communicatively successful, it would need to be sufficient (quantity), accurate and warranted (quality), and clear (manner).

These matters, as you can probably imagine, are not easy to code. And while there was a good deal of early affection for Grice's work in computational linguistics, the vagueness of his maxims, their interrelations, and the range of behaviors they license have led to some discouragement in various quarters. But, treated as interaction design guidelines, rather than as natural-language processor axioms, they can be extremely useful. Let's say your system needs shipping information at some point, which involves getting information in six categories for the addressee:

- the person's (or the business's) name;

- the street name and number (possibly with an apartment or suite number);

- the city;

- the state (or province);

- the zip (or postal) code; and

- the country

The easiest way to handle this problem from a system perspective is for the interface to lead the user on a question-and-answer interview, and such a subdialogue might not be completely heinous to the caller (who is accustomed to, if not precisely this sort of interrogation, then facsimiles of it on various forms and in rough parallels to it in various human–agent encounters). But let's look at how this might work.

AcmeVoice   What is the name of the recipient?

Caller   Fred Derf

AcmeVoice   Street name and number?

Caller   2 Nass Street

AcmeVoice   City?

Caller   Kitimat

AcmeVoice   Province?

Caller   B.C.

AcmeVoice Postal code?

Caller V8C 2G3

AcmeVoice Canada?

Caller Yes, Canada

All of this assumes flawless recognition, no need for confirmation, and a kindly system-logic that infers to ask for province and postal code (rather than state and zip code) because of originating area code, infers the country for the same reason (strengthened to certainty because of positive answers to "Province?" and "Postal code?"), and so on. But it is *still* tedious.

Now, if the interface builds on Grice's maxims, it can expect (by Quantity) that the caller will supply the address only, and nothing else; (by Manner) that the address will be given in a clear and brief and orderly way (following the specific-to-general conventions of address recital); and by counting on its truth and relevance. This faith in Grice and the caller is *not* enough, of course. The appropriate repair strategies need to be waiting in the wings, and it is still wise to follow the natural chunking tendency of name + address. But, in terms of user satisfaction, this faith is the best place to start.

AcmeVoice What is the name of the recipient?

Caller Fred Derf

AcmeVoice And the address?

Caller 2 Nass Street

Kitimat, B.C.

V8C 2G3

The recognizer logic has to be tuned to:

- expect numerals early in the stream;

- expect letters and numerals late (for postal code, numbers alone for zip);

- expect that the longest inter-utterance gaps will cue the street-chunk / city-province chunk division, and the city-province chunk / postal code chunk division;

- use a geographical database with cross-checking (for instance, does the postal code match the province, the city, the street address? If not, what adjustments are the most likely?);

- anticipate that the user might stop after the street address, or forget to add the postal code, and be prepared with "City?", "Province?", or "Postal code?" as necessary;

- anticipate that the caller might add the country even though it may be redundant by that point in the interaction;

- be prepared to query *for* the country if there's still substantial ambiguity by that point;

- and so on.

Especially since this transaction is over the phone (with low bandwidth, and a narrow frequency range), especially because an address is several chunks long, and especially because the consequences of getting it wrong are so high — this interaction needs to be thought out very, very well. (It also must be verified and, if need be, adjusted, before the transaction is finalized.) But none of that means the user should have to sit through a lengthy inquisition.

Grice's maxims can go a long way to help you know what to listen for. In some cases, they're necessary assumptions anyway. You have to assume, for instance, that users' utterances are honest and warranted (i.e., that they satisfy the maxims of quality), at least to the extent that truth is pertinent. Similarly, you have to assume that their utterances will be relevant (i.e., they will satisfy the maxim of relevance). Users call because they have a task they want to carry out, so their utterances will be relevant to that task, and you just have to build that expectation in as a presupposition of the system. If they *aren't* relevant, that will almost inevitably make the bulk of their utterance out-of-vocabulary, and you need always to have repair strategies on hand for vocabulary failures, steering the callers toward the topic and perhaps enunciation strategies. In any case, it is virtually impossible to build a workable system that can handle irrelevance (that is, as some kind of special case, distinct from handling out-of-vocabulary utterances). Anyone who is so motivated can blow up a voice interaction with very little imagination. Here is radio reporter David Kestenbaum picking on Thrifty Car Rental's interactive voice system:

Thrifty   ... Please say the pick-up date, or say "help" for more information.

David   May 26th, in the year 2010.

Thrifty   And picking up around what time ... ?

David   Around midnight.

Thrifty   You're returning the car to the same airport, right?

David   What are you doing after work?

Thrifty   On what date?

David   Today.

Thrifty   Sorry, I didn't understand. Please say the return date of the car.

David   I want to take you out for dinner.

Thrifty   Sorry, I still didn't understand. Please say the return date of the car —
for example, "June 7th."

David   When am I going to return *you*? It'll just be, just an evening.

Thrifty   I'm sorry. I'm having trouble understanding you. Please hold while I
transfer you to one of Thrifty's sales agents.

(Kestenbaum, 2002).

You can usually count on users to be relevant, unless they want to provoke confusion; in which case, they get their wish. As Robert Rodman points out, "speech recognition will not work if users don't want it to" (1999: 148). It is often overlooked, though, that this is equally true of other input-modes and other interface styles as well. Kestenbaum could have made a command-line interface look pretty goofy too, through syntax or vocabulary violations, or a keypad interface, and even a graphic interface has input conditions that are easy to run afoul of. When I first encountered pointing devices, it was the mid-1980s, in a lab at Rensselaer Polytechnic, an institute with very technologically astute students. Some of them were waving their mice in the air. Some were running them off the edge of the desk. Most were busy looking around for someone they could copy who was doing it right. Kestenbaum would have had a field day; oh, except his medium, radio, has a rather severe limitation with visuals.

As for the maxims of quantity and manner, you can really only do two things: hope users will follow them, and study the discourse to know well what "following them" means. As we have seen, different registers have different notions of directness, orderliness, and so on; they may also have different notions of appropriate information quantity. In order to cue for, and tune for, the best input, you need to know the register well.

Putting your faith in users should not be a big worry, though. You can generally trust them. Anyone can pull a Kestenbaum if they want to; being unclear, irrelevant, prolix, or other forms of pragmatic uncooperativeness are only some of them (talking like Donald Duck or whispering or using a mixture of Latvian and Urdu are also possibilities). But you should feel no allegiance to such people (unless they happen to be testers, who are trying to blow up the system on principled grounds). There's no shame in a car-rental, rate-checking voice system that can't provide David Kestenbaum with a date, or witty banter.

You simply cannot prepare for deliberately hostile users and tricksters. No one sincerely using the system to accomplish a task, however, will be seeking to disable it. They'll be attempting to cooperate, and expecting cooperation back. The maxims outline the forms that cooperation is likely to take.

Speech as Action

The issuing of an utterance is the performing of an action.
— J. L. Austin

Grice focuses on how people cooperatively develop meaning, by observing mutually understood optimality principles. But meaning is not, contrary to widespread belief, even among language professionals, the acme of language. People want to get their meanings across, true enough — they want to be understood — but for a *reason*. If I say "Please pass the salt," I want you to recognize the concepts the words invoke. I want you to operate under the assumptions that I have delivered the appropriate amount of information, that I spoke truthfully, relevantly, and so on; I want you to understand my intentions. But those are all incidental to my primary desire: for you to *do* it. I want the damn salt. Grice's maxims concern information in one way or another: the amount of information, its relevance, its quality, its clarity. And people certainly do a lot of information swapping when they talk, but it is very often subordinate to some action.

Speech is often — maybe always — a form of action. In my salt craving example, the action is to issue a request, in principle no different from pointing to the salt shaker and looking at you hopefully. And speech systems need to have a very good handle on the functions being performed and invoked in every interaction. Functions are the effect of utterances, not the utterances themselves. That's probably easiest to see with an utterance that serves multiple functions, like sentence 12.

12   Your pants are ugly.

If someone says 12, she is making a statement, expressing an opinion, probably insulting you, and possibly persuading you to change.

These acts — stating, opining, insulting, persuading, and lots more — we will call *dialogue acts*, since we are interested in how speech functions in interaction.[6]

The study of speech functions goes back at least as far as Prodicus's classification of sentence types in 5th century BCE Greece. They aren't exactly dialogue acts — their functions are syntactic, rather than pragmatic, a difference we'll get to shortly — but they are the place to start. There are four, and they all come with ten-dollar Latin labels:

6: The more common label in North America is *speech acts* (Searle, 1969), in the U.K., it is *illocutionary acts*. I have adopted *dialogue act* from Walker 1996, though it dates to Bunt's more constrained usage (1979). The term is gaining popularity, especially among speech-system scholars, though it is used somewhat inconsistently, researcher to researcher. I have stretched the term a bit, and my use, unfortunately, is therefore not identical to Walker's or to Bunt's, but the coverage I give it here should help to indicate the range of phenomena to which I apply it.

Constituent Interrogatives

13a What is on the mat?

14a Where did Freddy put the cat?

15a Why is that damn cat always on the mat?

Yes/No Interrogatives

16a Is the cat on the mat?

17a Did Freddy put the cat on the mat?

Imperatives

18a Put the cat on the mat!

19a Vacuum up those hairballs!

Along with, of course, the ever-popular:

Assertions

20a The cat is on the mat.

21a That is one lazy cat.

22a Freddy stepped over the cat.

This set is the absolute core taxonomy of functions that everyone who deals with language needs to be aware of, and the primary functions they perform are critical for most voice interfaces. So we will do a quick overview, but it is utterly crucial to keep in mind that these are sentence *types*, with syntactic functions, not dialogue acts, with pragmatic functions. The latter often supercede the former — sentence 23, for instance, is a yes/no interrogative, which ostensibly seeks a "yes" or "no".

23 Would you please pass the salt?

But the person who utters sentence 23 would rarely be waiting for a "yes" or "no." He is waiting for the salt. Syntactically, it is an *interrogative*; pragmatically, it is a *request*.

Assertions

Assertions are claims about the world (from the Latin *asserere* "to claim rights over" or "state"): about the relation between objects (like cats and mats — 20a), about the relation between objects and states (like cats and laziness — 21a), or about activities (stepping over the cat, performed by Freddy — 22a). These things are the default notion when people think about sentences, and their chief activity is to encode information, assessed in terms of its accuracy (that is, its truth or falsity), though notions like suitability and specificity are often just as relevant. They are the chief structures implicated when the system provides a user with data (weather, sports scores, phone numbers), and vice versa (name, address, credit card information), though the utterances used are almost always truncated versions of full assertions ("Canada 5, USA 4" rather than "The score was Canada 5, USA 4," "Fred Derf" rather than "My name is 'Fred Derf'").

Imperatives

These functions (from *imperare* "to command" or "to requisition") are direct attempts through speech to get activities performed, usually by other humans, but in the last five decades or so by machines as well. MESSAGESYSTEM RETRIEVE NEW is an imperative. It first calls the subsystem it is addressed to, and then issues the order: retrieve (the) new (messages). It is entirely like calling to a boy in a crowd and ordering him to go buy the big prize turkey from the poulterers at the corner. Imperatives are the cornerstone function of keypad interfaces (though, unlike command-line systems, it is the machine that issues the bulk of the orders, in the form of conditionals: "For transportation, press '1'"). In multimodal interfaces, they will likely be the primary form of user-input for a good while. Aside from a few constituent interrogatives ("What day is it?" and "What time is it?"), for instance, virtually all the licensed utterances in Apple's Speakable Items are direct orders "Get my mail," "Open my browser," "Copy this to the clip board."). And imperatives offer effective functionality for speech-only interfaces as well. For instance, voice-enabled email applications usually support commands like "Skip this message," "Delete it," and "Record a response."

Interrogatives

A great deal of the interactive work in voice interfaces also involves interrogatives (*interrogatio*, "a question, questioning"), of both constituent and yes/no varieties.

Constituent interrogatives are labeled thus because their primary job is to seek information that languages package into constituents. They begin (in English) with what are usually called Wh-words (though they include the non-wh *how*): words that signal the category of the queried information. *What* and *who* query noun phrases; *why*, *when*, *how*, and *where* query adverbial phrases. Often the constituents occur on their own (implicating a fuller proposition), but occasionally they come packaged into a full sentence. For instance, functionally suitable answers to the constituent interrogatives, 13a–15a (given above and repeated here for convenience, would look like the ones in 13b–15b (where the constituent is italicized):

13a   What is on the mat?

  b   *The cat.*

      An old chew toy.

      It looks like *oregano.*

14a   Where did Freddy put the cat?

  b   *On the mat.*

      Outside.

      He put the cat *in the tub.*

15a   Why is that damn cat always on the mat?

  b   *He likes it there.*

      It's the only place he can get fresh air.

      Because Franklin spilled a bag of catnip on it.

Yes/no interrogatives, on the other hand, prospect for a simple confirmation (yes) or disconfirmation (no):

16a   Is the cat on the mat?

  b   Yes.

      Yep.

      Yeah, man, he is.

17a   Did Freddy put the cat on the mat?

  b   No.

      Nope.

      Uh-uh.

Both types of interrogative are absolute staples of speech systems. Watch how they drive this interaction:

T$_1$ ARTIMIS   Welcome to the ARTIMIS/AGS server. How can I help you?

T$_2$ Caller   I'm looking for a job in the Calais area. Are there any servers?

T$_3$ ARTIMIS   No there aren't any employment servers for Calais. However, there is an employment server for Pas-de-Calais and an employment server for Lille. Are you interested in one of these?

T_4 Caller   For Nord, not for Calais.

T_5 ARTIMIS   The phone number of the server of employment advertisements for Nord is 36 68 74 44. Do you want any other information?

T_6 Caller   Is it accessible from Calais?

T_7 ARTIMIS   Yes, it is accessible from Calais. Do you want any other information?

T_8 Caller   I also want the weather forecasts.

T_9 ARTIMIS   Are you interested in general weather forecasts or in specialized weather forecasts?

T_{10} Caller   Specialized weather forecasts.

T_{11} ARTIMIS   A server for which specialized weather forecasts?

T_{12} Caller   What can you suggest?

T_{13} ARTIMIS   There are eight kinds of specialized weather forecasts. The two first are atmospheric weather forecasts and coastal weather forecasts. Which specialized weather forecasts are you interested in?

T_{14} Caller   Coastal, for the Nord-Pas-de-Calais area.

T_{15} ARTIMIS   There aren't any servers of coastal weather forecasts for the Nord-Pas-de-Calais region. However, there are two servers of coastal weather forecasts for the departments of Nord-Pas-de-Calais and a server of general weather forecasts for the region of Nord-Pas-de-Calais. Are you interested in one of these?

T_{16} Caller   Yes, the latter.

T_{17} ARTIMIS   The phone number of the second one is 36 68 00 00. Do you want any other information?

T_{18} Caller   Who is its provider?

T_{19} ARTIMIS   The provider of 36 68 00 00 is Météo-France. Do you want any other information?

T_{20} Caller   No.

T$_{21}$ ARTIMIS　Goodbye.

T$_{22}$　Caller　[click]

(Sadek and De Mori, 1998: 556–558)[7]

There are two things in particular to notice with this dialogue. First, interrogatives dominate it. There are thirty sentences or subsentential utterances (e.g., "yes, the latter," "Goodbye"); of this group, almost half are interrogatives. Or, if we count by turn-units (i.e., the elements marked T$_n$) — a far more relevant metric with respect to dialogues — two thirds of them include interrogatives. Second, their syntactic functions are almost always completely subordinate to their pragmatic functions, their dialogue acts, when real business is getting done.

While ARTIMIS is very responsible in answering constituent interrogatives with constituents, yes/no interrogatives with "yes" or "no," the caller is almost completely indifferent to this protocol. Rarely does the caller respond with a "yes" or "no" (or a synonym, like "OK," "sure," or "uh-uh"); on two occasions, ARTIMIS not only gets neither a confirmation nor a disconfirmation back, but rather, another interrogative. And the opening constituent interrogative gets a response that does not supply, nor even implicate, an appropriate constituent. Much of this indifference (and ARTIMIS's success in the face of the indifference) can be accounted for by Gricean inferences and a general focus on intention over literal meanings.

Let's isolate the introductory exchange:

T$_1$ ARTIMIS　Welcome to the ARTIMIS server. How can I help you?

T$_2$　Caller　I'm looking for a job in the Calais area. Are there any servers?

The system asks an open-ended constituent interrogative (T$_1$) which is answered by an independent assertion (T$_2$). Even more strangely, the system seeks direction, even a command, but gets, in return, a statement that focuses on the *caller's* activity ("I'm looking . . ."), not on anything *ARTIMIS* should be doing. A textbook response would be more on the order of (with the answering constituent italicized):

T$_{2*}$　Caller　You can help me *by telling me if there are any job servers in the Calais area.*

　　　　Or just

T$_{2\dagger}$　Caller　*Tell me if there are any job servers in the Calais area.*

7: The dialogue occurred in French, and the version here suffers a bit from the translation. By the way, I have no idea what the difference is between the *region* of Nord-Pas-de-Calais (for which there are no coastal weather servers), and the *department* of Nord-Pas-de-Calais (for which there are apparently two).

The caller, though, does not serve up the constituent on a textbook platter because it is unnecessary. The caller is not responding to the *form* of the interrogative. The caller is responding directly to the *purpose* behind the interrogative, by providing a context and directing an inquiry on that basis. ARTIMIS's T_1 effectively says "I'm ready for business. You can start asking me for information now," and the caller has proceeded accordingly.

Dialogue acts in formal genres like this (a service encounter) frequently come in clusters. The opening turn here involves three principal dialogue acts. It begins with a greeting ("Welcome..."), blended in with an identifier ("the ARTIMIS server"), followed quickly by an offer ("How...?"). Since it is the offer that is primary, not its form (an interrogative), the caller quite reasonably chooses not to address the form, and responds directly to the offer.

Take another interaction from the dialogue:

T_5 ARTIMIS   The phone number of the server of employment advertisements for Nord is 36 68 74 44. Do you want any other information?

T_6 Caller   Is it accessible from Calais?

Again the caller does not reply to the *syntax* — does not reply *yes* or *no* — but responds directly to the intention. ARTIMIS has declared itself ready for another query, and the caller provides one (to which ARTIMIS then responds without batting a processor). The confirmation that yes, indeed, there is further information the caller wants, is entailed by the query itself (and by the lack of a "no"). ARTIMIS also assumes, of course, that the request is relevant; in particular, that the *it* of the caller's question is the phone number just mentioned.

What we are getting at here is the nature of dialogue acts, and we're seeing that they exploit the core taxonomy of interrogative, imperative, and assertive, but do so for a range of communicative activities far broader than that shallow and formal taxonomy suggests. Interrogatives are always interrogatives, imperatives are always imperatives, but their syntactic needs are often satisfied tacitly while the conversants respond more directly to higher-level dialogue acts. The cases where the form maps directly to the act are actually quite rare.

Cutting to the chase, this ancient core taxonomy is syntactic, and dialogue acts are purposive. There is a big difference; purposes employ syntax, but are not in any way determined by it. My solicitation for salt above, for instance, took the form of sentence 23, but it might have taken any number of syntactic forms:

24   Please pass the salt.

25   Is that the salt there?

26   How would you like to pass me the salt?

27   Give me the salt?

28 This soup could use a little salt.

29 Why is it that you assume *I*, of all people, would tolerate a salinity deficiency of this magnitude in my avgolemono when you have a sodium-chloride-dispensing canister immediately due east of your plate?

The last version may strike you as far-fetched, depending on your dining companions, and some of these examples surely strike you as rude, but they are all effectively (in the appropriate context) requests, despite a diversity of syntactic forms. It is the combination of Gricean maxims (especially quantity and relevance), the intentional semantics behind the utterances (that is, the purposes of caller and system), and the notion that speech performs acts (in this case requests) that gets us through what would otherwise be a linguistic log jam.

Not only is there no specific verbal instantiation for a dialogue act — the same act might be phrased any number of ways, like sentences 24–29, which all instantiate a request, and the same one at that — but any individual utterance may serve more than one dialogue act. This utterance, for instance, from the inimitable Foghorn Leghorn, is both a description and an insult:

30 You're about as sharp as a bowling ball, kid.

Dialogue Acts

"Hallo!" said Pooh, in case there was anything outside.
"Hallo!" said Whatever-it-was.
"Oh!" said Pooh. "Hallo!"
"Hallo!"
"Oh, <u>there</u> you are!" said Pooh. "Hallo!"
"Hallo!" said the Strange Animal, wondering how long this was going on.
    — A. A. Milne

Here, with very substantial interdependence, turn by turn, is the way human–human natural dialogues work:

Phone   ring

Nancy   Hello?

Hyla   Hi.

Nancy   Hi!

Hyla   How are you?

Nancy   Fine. How are you?

Hyla   Okay!

Nancy   Good, What's doing?

(Hutchby and Woofit, 1998: 97)[8]

When humans begin telephone calls, there are systematic, multiturn rituals. Nancy and Hyla are involved in an established protocol, a chain of interrelated utterances that illustrate what Emmanuel Schegloff (1986) calls the "core sequences" that routinely begin many classes of telephone calls: the ringing phone, triggered by Hyla, is a *summons*, which Nancy responds to with an *answer*; Hyla's subsequent "Hi." is the *identification* utterance, to which Nancy's "Hi!" serves as the *recognition*; and so on. This opening pattern defers the focus of the conversation until specific role negotiations have been worked out.

"Would the initiation of a typical human–computer conversational telephone call unroll like this?" you ask. That's an empirical question, which can't be fully answered without extensive and focused testing with a conversational system (and/or Wizard of Oz testing). In the absence of data, however, I'm happy to risk a guess: no, not on your life.

This type of utterance swapping use is so utterly social, so divorced from information-exchange that Malinowski (1923: 315) called it *phatic communion*, "a mere exchange of words" (the *phatic* part) whose primary function was to effect "a tie of social sentiment" (the *communion* part).

We know enough about conversational human–computer interaction to know that people don't talk to voice systems the way they talk to other people. People exploit their human–human conversational competence, but they don't treat the machine as another human. What this means is that (1) a typical conversational dialogue with a computer is unlikely, in the extreme, to begin with a series of multiple greetings and phatic inquiries into well-being, but that (2) it may well have other sequences of identification and role-verification, probably as a function of the field (a banking system will prioritize security, for instance, while a traffic-information system will prioritize location). But the actions performed by speech in these systems, as in human–human exchanges, are embedded in a back-and-forth negotiative context.

Dialogue acts are incredibly varied, from the relatively inconsequential and highly mundane acts that we all engage in all the time (routine greetings, requests, assertions); to the more consequential and less common acts that most of us engage in, but more rarely

---

8: From the perspective of conversation analysis, from whence it comes, and upon whose results I now increasingly draw, I have mangled this dialogue data pretty badly, stripping away indications of tone, rate, overlap, pronunciation, and the like. But none of those traits are relevant to the present discussion, and, in general, the conversation analysis transcription conventions are very overdetermined for our purposes. I will continue to treat the data I draw from those sources in this crude, ham-handed way.

(bets, promises, proposals); to the highly consequential acts that only specially-sanctioned people can perform (baptisms, judicial sentences, declarations of marriage, bankruptcy, war). The subset of these acts that are relevant to voice interfaces is considerably narrower (including, at least for some time, none of the specially-sanctioned acts).

Table 4.1 is a chart of dialogue acts, organized around two major categories — the ones that manage the tasks and the ones that manage the dialogue — and their principal sub-categories.[9] It's a dense table, so I'll give you the naked categorical breakdown, too:

- Task management acts

  - Constituitive acts

    - Expressives (e.g., complimenting someone)

    - Declaratives (e.g., sentencing someone to prison)

  - Informative acts

    - Assertives (e.g., stating a fact)

    - Interrogatives (e.g., asking someone for an address)

  - Obligative acts

    - Directives (e.g., requesting someone to pass the salt)

    - System Directives (e.g., calling the help system)

    - Commissives (e.g., offering someone the salt)

- Dialogue management acts

  - Flow-regulating acts (e.g., beginning an exchange)

  - Grounds-keeping acts (e.g., clarifying a point)

There are three things to notice about this scheme, and dialogue acts generally (well, a lot more than three, but three that we will take up immediately). First, the terminology can get a bit screwy, largely because of the idiosyncratic relationships that English nouns

---

9: The main distinction was suggested by Bunt (1989), which is the first systematic separation of informative acts from control acts that I am aware of (his terms for the functions are *information transfer* and *dialogue control*). The task-management side of the table follows Rene Dirven's cognitive typology of speech acts fairly closely (Dirven and Verspoor 1998: 164–167), itself a slight repackaging of Searle's (1979) better known five-category taxonomy. The dialogue-management side builds on the insights of various dialogue theorists, especially Harry Bunt and David Traum, and is influenced strongly by the contribution model of Herbert Clark and his associates.

Dialogue acts							
Task management						Dialogue management	
Constitutive		Informative		Obligative		Flow regulating	Grounds-keeping
Expressives	Declaratives	Assertives	Interrogatives	Directives	Commissives	Flow regulating	Grounds-keeping
praise, agree, disagree, apologize, greet, mitigate, thank, deplore, excuse, congratulate, regret, condole, recognize, swear, lament, protest, boast, compliment, welcome, commiserate, condone, condemn, forgive, complain-that, wish, take-leave-of, approve-of, accept-that, reject, acknowledge-that	marry, wed, sentence, arrest, call-to-order, move, adjourn, dismiss, fire (dismiss-from-work), pronounce, baptize, pardon, christen, name, define, call, permit, dub, abbreviate, appoint, resign, quit, approve, nominate, excommunicate, renounce, endorse, bless, deputize, authorize, summon (legally; i.e., "issue a summons"), comply, terminate task	assert, state, describe, affirm, negate, deny, assume, swear-that, hypothesize, guess, claim, assess, opine, announce, introduce (self), introduce (other), insist, forecast, predict, notify, argue-that, deny, discount, answer, hint-that, explain, correct, cite	constituent-question, yes/no-question, inquiry	summon, advise, propose, ask-if, ask-that, beg, entreat, command, bid, order, forbid, recommend, cajole, suggest, invite, challenge, dare, direct, instruct, request, caution, hint-to, warn, bet, approve, accept-of	promise, pledge, swear-to, volunteer, offer, vow, threaten, book, accept-to, decline-to, guarantee, agree-to, warrant, bid, refuse, bet	initiate-dialogue, initiate-exchange, terminate-exchange, terminate-dialogue, take-turn, keep-turn, release-turn, assign-turn, decline-turn, overlap-turn, recognize **System directives** help-request, orientation-request, navigation-request, command-request, status-request, halt-process, pause, resume, checkpoint, rollback, present, skip, repeat, jump, jump-back, accelerate, decelerate, configure	backchannel, acknowledge, request-acknowledgment, verify, request-verification, self-repair, other-repair, clarify, request-clarification, paraphrase, confirm, request-confirmation, repeat (self), echo (repeat other), request-identification, identify-self, identify-role, identify-institution

TABLE 4.1 A taxonomy of dialogue acts

have to their corresponding verbs. Sometimes the relationship is identity (or effective identity — same spelling, same pronunciation, but different deployment). *Butter* is like this: you <u>butter</u> your toast with <u>butter</u> (the first instance is a verb, the second a noun). And some dialogue acts pattern this way. You <u>request</u> the butter, and what you've done is make a <u>request</u>. You <u>offer</u> the butter, and — *ipso facto* — you've made an <u>offer</u>. But a number of dialogue acts pattern according to more peculiar noun-verb relationships. If you <u>assert</u> something, you've made an <u>assertion</u>. If you <u>acknowledge</u> a remark, you've made an <u>acknowledgment</u>. If you <u>inquire</u> about something, you've made an <u>inquiry</u>. Some of these patterns are pretty systematic, and the nomilization affixes, *-tion* and *-ation* are fairly common among the nouns that correspond to performative verbs (*describe, description*; *negate, negation*; *introduce, introduction*; *notify, notification*). But the peculiarities of English morphology are in high evidence among these terms, and the taxonomic conventions of dialogue acts are to focus on the verbs, while tagging those acts in a specific dialogue requires nouns.[10] I hereby apologize for English.

The second thing to notice about this scheme is that the central division in Table 4.1 functions largely like an old style two-from-column-A, one-from-column-B Chinese restaurant menu. Most utterances have both a task-management dimension and a dialogue-management dimension, and engage multiple actions; some utterances perform three or four dialogue acts. (The principal exceptions are when there are no task management functions, at the beginnings and ends of dialogues, for instance, and when there is an interruption before dialogue-management functions, like turn assignment, can kick in.)

Every utterance, therefore, usually performs at least two dialogue acts simultaneously, the two dimensions of which need to be considered carefully in voice interface design. Every utterance routinely both manages the current task, and manages the dialogue. Task-management acts are more overt and align directly with the pragmatic tradition. Dialogue-management acts are covert and align more with work in sociology, psychology, and computational linguistics.

Task management is the gas pedal, which propels users towards their goals; dialogue management is the steering wheel, which allows them to make the turns and avoid the telephone poles. Task management carries the goods, while dialogue management writes the waybill; checks it at the freight yards, borders, and loading docks; and signs it at the destination. Task management conducts the business; dialogue management keeps the books.

10: The labeling complications don't end here, of course, since English tends to lump together closely acts that are quite distinct. Take the broad notion of acceptance, which, if we specify it functionally, participates in three different types of dialogue acts. One might <u>accept that</u> the Queen is his monarch (an expressive act, since it concerns a cognitive/emotional state), <u>accept</u> her offer <u>of</u> a knighthood (directive, since once it is accepted the obligation is on her to carry out the conference), and <u>accept</u> to follow her commands (commissive, since accepting a command or directive is obliging oneself to carry it out). And now things get really ugly, since if you accept <u>to</u> follow a command, you signal your acceptance <u>of</u> that command.

Task-management acts and dialogue-management acts are, that is, copresent dimensions of every utterance. Let's look again at the successful How-may-I-help-you exchange we saw earlier, in terms of task- (TM) and dialogue-management (DM) acts:

T$_1$ System   How may I help you?

        DM   **exchange initiator**

        TM   **offer, constituent question**

T$_2$ Caller   I was trying to place a call and must have dialed the wrong number.

        DM   **acknowledgment**

        TM   **acceptance-of, assertion**

        can I get credit for that?

        DM   **turn assignment**

        TM   **request**

T$_3$ System   Do you need me to give you credit?

        DM   **acknowledgment, paraphrase, turn assignment**

        TM   **offer, yes/no question**

T$_4$ Caller   Yes.

        DM   **confirmation, turn release**

        TM   **acceptance-of**

(data from Boyce, 1999: 49)

This analysis is fairly shallow, but it is sufficient both for our purposes (illustrating the bidimensionality of dialogue utterances) and for design purposes (charting out the significant functions of the turns).

The system's T$_1$ initiates the exchange (the dialogue has already been initiated by the caller's summons, by dialing in), and initiates it by getting right down to business, offering assistance. The substance of that offer is embodied in an interrogative in the constituent mode, prospecting for an answer something like "by giving me a credit." So we also identify its dialogue action as "constituent question." But it is important to notice that the interrogative is subordinate to the offer.

The caller's T$_2$ acknowledges the offer, accepts it, and makes an assertion (justifying the immediately following request); then, still in T$_2$, but in another clause, comes the request that assigns the turn back to the system. The system acknowledges the request, doing so in a paraphrase of the request that is also (because it is phrased as a question) an offer (T_3). The caller confirms and accepts the offer (T_4). Notice that the subordinate constituent question (of T_1), just as we noticed in some of the ARTIMIS exchanges, is never answered directly, though the dialogue satisfies it (over the course of T_2–T_4).

The third thing to notice about Table 4.1 is that it is very, very full. It uses many categories and terms that are probably opaque at this point. I will introduce and define the major ones over the remainder of this chapter, and into the next, but many of the specific dialogue-act labels won't be seen again in this book. I am not attempting to be exhaustive with this list, which is impossible, but the list is the result of a responsible census. Speech is one of the primary ways, perhaps *the* primary way, in which people act, and people have a huge variety of actions. I hope to give you a sense of that variety, a feeling for the patterns into which it falls, and a terminology to tag the acts in which your interfaces and your users reciprocally engage.

Task-management involves three major categories of dialogue acts, each having two further subcategories: constituitive acts are utterances, like apologies, that are generally recognized socially as more formal; obligative acts are utterances, like promises, that enjoin some future activity; and informative acts are utterances, like assertions, that traffic primarily in data. In terms of the register, the appropriate range of task-management dialogue acts is what comprises the field; they accomplish the activities the register concerns.

Dialogue-management involves two major categories of dialogue acts, both of which help ensure that the progression of the discourse is sufficiently orderly: flow-regulating acts ensure that the speakers exchange opportunities to speak efficiently and cooperatively; groundskeeping acts ensure that the speakers agree about relevant bits of information, commitments, and interpretations that build and shift as the dialogue proceeds.

In the remainder of this chapter I take up the task-management dialogue acts quite systematically (though far from comprehensively), with respect to voice interfaces. The following chapter takes up dialogue-management acts in a more general fashion, along with coherence, cohesion, topic, and focus.

Task Management

Conversation is the fundamental site of language use.
— Herbert H. Clark and Deanna Wilkes-Gibbs

If you didn't know it before you looked at Table 4.1, you know it now: language is a big, sloppy, stew. There is no periodic table of dialogue acts. Table 4.1 is full, but the illustrative verbs are only a sampler. They sketch out a range of acts performed in dialogue, and all of them have been scavenged from various discussions in the literature. Some of the labeling verbs are vague or ambiguous (possibly both), and some may be redundant. Even so, many themes are represented only by a few words (just think, for instance, of all the possible types of motions and rulings that are available to lawyers and judges — people who act, except for the occasional gavel bang, almost exclusively through speech — to get a sense of how much is missing).

This collection of dialogue acts illustrates, in part, the variety of ways we act through our words. The collection is big, certainly bigger than speech systems will be able to accommodate for quite some time. But scaling it radically back, in an attempt to identify only those acts relevant for voice interaction design, is sure to be wrong, if not now, then in a year or two. None of the categories can be excluded in principle, ruled out of the domain of voice interfaces.

We'll discuss all the main categories, and a few representative acts, to get a sense of their descriptive power in accounting for some of the most prevalent functions of language.

Constitutive Acts

> *And they ran to us fast.*
> *They said "How do you?*
> *Would you like to shake hands*
> *With Thing One and Thing Two"?*
>   — (Dr. Seuss)

Defined

Constitutive acts are the speech functions where the connection between speaking and acting is the most obvious, because the talk constitutes — effects, actualizes, *is* — the act, and often names that act in the bargain. If I say "I apologize for using an unwieldy term like *constitutive*" (under the appropriate conditions, like those operating here and now; I've just used the word, you have the potential to be aggravated by this sort of jargon), then, *ipso facto*, I've apologized for using an unwieldy term like *constitutive*. If a chairman says "I adjourn this meeting" (again, under the appropriate conditions — a phrase I will elide from here on in, so long as you promise to remember that acts are incredibly context-dependent), then he has adjourned the meeting.

In a very real sense, of course, *all* dialogue acts are constitutive, since uttering them constitutes the act; if you state that the cat is on the mat, then, by golly, you've constituted the act of stating. So, why is *stating* classed as an informative dialogue act, while *apologizing* and *adjourning* are constitutive? It's somewhat arbitrary, of course, as all categorizations are if you push them a bit. But the acts we call constitutive are prototypical dialogue acts, the clearest cases of acting through speech, where both the act and its conditions are easily identified. Often, in cases like christening and sentencing and marrying ("pronouncing marriage") and wedding (saying "I do" or "I will") and the like, the acts have a great lot of social machinery around them — laws, licenses, sanctioned authorities — and the event does not go through without the utterance.

The other two categories also have very sharply functional definitions, one in terms of information, the other in terms of obligation, so that the constitutive category sometimes has the feel of an elsewhere case, the place to put task-management acts that are neither informative nor obligative.

The best way to understand these categories — indeed, most categorizations in language, even apparently simple ones like "noun," "verb," and "phrase" — is through prototypical examples. Constitutive acts come in two flavors: expressive and declarative. A prototypical expressive act is apologizing, which conveys a sense of regret for some previous act; or forgiving, which conveys a sense of accepting someone else's regrettable act. A prototypical declarative act is a judge or cleric saying "I now pronounce you husband and wife," which effects a new social reality, or the same authorities declaring a pardon — people, who weren't just a moment before, are now married; people who were in legal or religious disgrace a moment before no longer are.

Expressive dialogue acts. Expressive acts, in short, are performed by utterances conveying a mental state for social reasons. I know that's an awkward definition, but the notion of expressives is intuitively clear from most of its specific acts: forgivenesses, congratulations, thanks, greetings, and so on, where the speaker's feelings are expressed for the maintenance, repair, or initiation of social relations. A greeting, for instance, which can be perfunctory and even dismissive, between certain people on certain occasions, is — in its truest, most essential form — an expression of pleasure at encountering someone. Thanking expresses gratitude, congratulations expresses pride-for-another, forgiving expresses forgiveness — all for social purposes (or they wouldn't be expressed). Mitigation (usually just the "please" tacked onto a request, to lower the sense of demand it makes) expresses a touch of humility, a really-I-don't-want-to-bother-you-with-this-but-I-want-X-anyway to accompany an attempt to get X.

Declarative dialogue acts. The other flavor of consitutive dialogue acts — declaratives — are prototypically public pronouncements by specifically sanctioned people: christenings, marryings, arrestings, sentencings. Expressives are mildly ritualistic, usually calling for specific linguistic conventions, with specific words spoken under quite narrow felicity conditions. For instance, an apology calls strictly for words like "sorry" or "apologize," and occurs when the speaker has somehow injured the hearer and wishes to convey regret over the injury. Declaratives are more conventionalized and often ritualized. A marrying calls for a very constrained utterance like "I hereby declare you husband and wife," and occurs at the climax of a ceremony (sometimes highly elaborate), spoken by someone civilly (and perhaps ecclesiastically) ordained to utter them. The wrong person cannot say those words, even at the right moment, and perform the act of marrying. The right person cannot say them at the wrong time and accomplish the act of marrying.

In Voice Interaction Design

Expressive dialogue acts. Speech systems don't, of course, have either mental states or concerns about social relations. But a few expressives are a staple of voice interaction design, which might seem to pose philosophical problems. They are problems we can safely ignore. It doesn't matter to cartoonists or authors, for instance, that characters like Elmer Fudd and Bilbo Baggins don't "really" have mental states; voice interface agents are fictions in very similar ways (as do graphic interface elements — that's not "really" a file folder on your desktop). The same considerations apply to concerns about social relations, though it perhaps aids clarity in commercial situations to rephrase *social relations* as *customer relations*.

For reasons of customer relations — either for the sake of naturalness, or for genuinely affective reasons, or both — voice interface designers have regularly incorporated a small range of expressives from the very beginning (welcoming, greeting, thanking). This small group of expressives is important (more to some customers than to others) for maintaining the sense of respect and courtesy that all customer relationships depend upon at some level. I don't see any reason to expand the use of expressives beyond these few simple courtesies; and, it must be noted, some people may find even this small group annoying and time-consuming from a speech system. I don't say that this list won't grow. It is sure to. But I don't see it growing in any wholesome way. My fear is that the use of expressives will increase in the direction of junk mail and web pop-ups, which offer an empty sort of congratulations for winning, or being selected to win, or being preselected to apply for, something or other. One can imagine automated phone calls a decade or so from now which begin, "Congratulations on the birth of your son, Tristan. Have you considered an education savings plan yet?" or "Lamenting the recent loss of your mother, Gladys, we would like to discuss options for honoring her memory."

A few expressives are likely to come from the user as well — some thanks here, a greeting there, a leave-taking to end the call, perhaps an acknowledgment of something or other. But probably nothing beyond the most routine civilities will make up user expressives, and only out of habit; the ARTIMIS dialogue we looked at earlier, for instance, has zero user expressives (and only two system expressives, a welcome and a leave-taking). User expressives should, by the way, be monitored. In general, people speak as they like to be spoken to; if you catch *pleases* and *thank-yous* from your callers, that is a good indication you should be equally courteous back.

Declarative dialogue acts. The culturally invested rituals that accompany declaratives do not translate to machinery very well. So, declaratives are at the moment almost entirely outside the range of voice systems, though a number of factors might open up this institutional range of dialogue acts in the future. With voice-verification and with increasing administrative automation and self-service, one can easily foresee when at least minor bureaucratic functions might be handled by phone systems (or speech-enabled mall kiosks) which represent the appropriate agencies and have the formal role to utter declaratives: license renewals, home-arrest maintenance, tax-collection notifications. Similarly, a speech system for credit approval might one day authorize someone to use a card (or an utterance; if voice systems take over the world, it's not a far stretch to think of voice-verification components which access credit or payment transfers on the basis of specific utterances by specific people). In the realm of multimodal systems, a Las Vegas-based marrying machine — which recognized "I do," could spray confetti, and had a built-in video camera — would not be far from the current chapel offerings.

There is a noninstitutional set of declaratives, however, that are linked directly and exclusively to discourse. Expressions like "Hereafter, I will call speech acts, *dialogue acts*" are declarative: in the same way that "I hereby declare you husband and wife" enacts social relations, these utterances enact linguistic relations.[11] Definitions (in the sense of stipulating a definition), abbreviations, namings, dubbings, and callings are all in this category (though disentangling which ones are which is not always an easy, or rewarding, task).

User declaratives are rare now, and very likely to remain that way; certainly the prospect of institutional sorts of declaratives from person to machine is a highly bizarre thought. But users might sometimes need the possibility of using some of the narrow class of speech-about-speech declaratives. In particular, the ability to add some words to the system vocabulary could be very useful (perhaps defining it through synonymy). Even naming might be a good feature in terms of customization ("I'll call you Audrey.").

Informative Acts

> *And our fish came down, too.*
> *He fell into a pot!*
> *He said, "Do I like this?*
> *Oh no! I do not."*
> — Theodor S. Geisel (Dr. Seuss)

Defined

Informative utterances deliver or pursue data. They interact most clearly with conversational maxims. In fact, the main role of informative dialogue acts is to calibrate data — information — with respect to the maxims.

Most of them are assertive, committing the speaker to some level of certainty toward some class of facts: statements, descriptions, assumptions, claims, guesses. All of them

---

11: Leech (1983: 180) says that these items should not be considered dialogue (speech) acts because "they are conventional rather than communicative acts," an argument that mystifies me. They *do* communicate, very clearly, an intention to use a given vocable in a given way. They *are* conventional, but not nearly as conventional as a greeting or an objection in court or a declaration of marriage.

establish the speaker's stance with respect to the quality of the information (or, what is roughly the same, the speaker's confidence in the quality of the information). A few are interrogatives, admitting the speaker's lack of information or understanding in order to get the hearer to remedy that lack. Most immediately, they concern the quantity of information. There isn't enough, so they ask for more. But they may also concern quality ("Are you sure?," "How do you know?"), relevance ("What do you mean?"), or manner ("Is that *South* Weber or *North* Weber?").

Although informatives are most explicitly concerned with the movement of data, all of the dialogue acts have at least some informational content. A declarative like "You're fired!", for instance, has the effect of creating information as it is uttered, though its primary job is creating a new social reality; a person who just had a job now doesn't.

In Voice Interaction Design

Informatives are, we all know, absolutely central dialogue acts for speech systems, usually implemented with a very significant asymmetry: the system has the information, and the user wants it. In email or calendar applications, for instance, the system knows when the meetings are, who sent the email, what the subject lines are, what the contents are, and so on; the whole point of the interaction is to transfer that information to the user. Traffic, weather, sports, financial applications, and the like are precisely the same, except that the information is more public domain. And even banking or retail applications, in which specific transactions occur, traffic systematically in informatives (account balances, prices, shipping addresses). Sometimes, the asymmetry is on the other foot, and the user has information the system solicits — names, addresses, credit card information, and the like.

This very heavy reliance on informatives, however, does not mean that the entire gamut of informatives are used in speech system dialogues. Quite the contrary.

Assertives. People select from a wide range of assertives when talking with each other, usually in order to establish a psychological stance with respect to some proposition, from absolute certainty to worrying doubt. The sort of information swapping that goes on with voice systems, and computers generally, rarely involves issues of doubt and commitment; for their part, computers have little call to position their ego with respect to the truth of an utterance (aside from objective probability statements: "ten percent chance of precipitation," and the like).

Certainty is the default stance for people who are asserting, and for speech systems. People just make claims like sentence 1, and the default understanding (by Grice's maxim of quality) is that it is certain and true.

1   The cat is on the mat.

Notice, too, that utterances like 16b and 17b are assertives. Utterance 16b is an affirmative, and is equivalent in assertive force to utterance 16c; 17b is a negation, equivalent to 17c.

16a   Is the cat on the mat?

  b   Yes.

  c   The cat is on the mat.

17a   Did Freddy put the cat on the mat?

  b   No.

  c   Freddy did not put the cat on the mat.

If speakers or their hearers have some level of doubt, it gets marked for degree of commitment, in dialogue acts like the following:

31   I insist the cat is on the mat.

32   I suppose the cat is on the mat.

33   Geisel claims the cat is on the mat.

The negative assertives work almost exactly the same way (you can just slip a *not* before *on* in all of those examples to see that; or, for a different negation, a *do not* before the verb), but there are also a few special cases like the following:

34   I deny the cat is on the mat.

35   I reject (the claim) that the cat is on the mat.

(Remember that these verbs are here just to be explicit. The context — including such factors as intonation, gesture, and dialogue history — is usually sufficient to count an utterance as, say, an insistence, or a suggestion, or a report of someone else's assertion, in normal, person-to-person interaction.)

These, as you might guess, are not the sorts of utterances one wants to hear from a voice interface system:

Speech Acts   I insist you currently have "Lunch with Kate Ehrlich" until 2 pm. At 3 pm, I suppose that you have "Brainstorming meeting in the lab." At 6:30 pm Martin claims you have "Dinner with Ellen."

ARTIMIS   I deny that there are any servers of coastal weather forecasts for the Nord-Pas-de-Calais region. However, I assume that there are two servers of coastal weather forecasts for the departments of Nord-Pas-de-Calais, and Clousseau reports there is a server of general weather forecasts for the region of Nord-Pas-de-Calais. Are you interested in one of these?

System assertives are overwhelmingly of the deadpan-certainty variety. Even when using information with explicitly another source, speech systems do not generally "report" the source, the way humans do, to detach themselves from a commitment about truth values. They "cite" the source as a reflection of responsibility or as to borrow authority (and, hence, *increase* the commitment), or as a form of embedded advertising.

Answers, very common dialogue acts in voice interface interactions, are always assertives, and pair up naturally with acts from the other informative category, interrogatives. That is, they always satisfy an assertive purpose, even when assertion is subordinate to another dialogue act, as in this exchange:

T_1 Amex   So, would you like me to go ahead and set this up to ticket today?

T_2 B.C.   Yes.

SRI/Amex (1989; tape 1, call 1)

The first act here (T_1) is dominantly an offer, subordinately a yes/no question; T_2 is dominantly an acceptance, but it is also an assertive, providing information that amounts to "I would like you to set up the ticket today."

Interrogatives. This is the smallest set in the taxonomy, with only three members, so we can treat it exhaustively. There are yes/no questions and constituent questions, and the slightly mitigated question or query we call an *inquiry*: if yes/no and constituent questions are effectively information-seeking commands (and they are), inquiries are information-seeking requests.

It's the smallest set, but it might also be the ripest for confusion, given the strong syntactic and phonological form interrogatives have, and their ubiquity, and the fact that the terms have overlapping syntactic and pragmatic senses. But, as we noticed with the How-may-I-help-you? and the ARTIMIS exchanges, the pragmatic purposes — that is, the interrogative dialogue acts — are often subordinate to other purposes, like requests ("Could I have some salt?"), offers ("How may I help you?"), or invitations ("Would you like to come to my party?").

While system interrogatives are not as common as system assertives, they are still fairly widespread; in fact, they are way too damned widespread. They can be the bane of users:

System   Authorization number, please.

User   two three one sixty-wa-

System   Please speak the digits of the authorization number, pausing after each digit.

User	two [pause] three [pause] one [pause] six [pause] one [pause] oh [pause] oh
System	I heard two three nine six one oh oh. Is that correct?
User	No, two three *one*.
System	Please answer yes or no. Is the first digit two?
User	Yes.
System	Is the third digit nine?
User	No.
System	Please say the third digit again.
User	One
System	Is the third digit one?
User	Yes.
System	Is the authorization number two three one six one oh oh?
User	Yes.
System	Checking authorization number two three one six one oh oh. Authorization number accepted. You may place your call now.

(Kamm, 1994: 435)

In conversational voice interfaces, system-originating task-management interrogatives should be confined largely to a corner of merchandizing systems, which need credit card details, shipping and contact information, and the like; or distributed sparsely throughout systems. They should not in general be rolled out in long sequences, if that can be avoided, because long strings of interrogatives tend to take (what we will soon be calling) *initiative* away from the user. Sometimes, for instance, the system needs to take direction from the user, to set parameters and orient its conduct, as in the Programming Enhancement Advisor system's early-dialogue query for determining how to shape its advice to the user:

PEA	What characteristics of the program would you like to enhance?
User	Readability and maintainability.

(Moore, 1995: 193)

But once the parameters are set, the system moves directly into advising:

PEA   You should replace (SETQ X 1) with (SETF X 1).

(Moore, 1995: 193)

One of the most important roles for interrogatives is in tuning the task. Speech-system dialogues often need to be calibrated (as do human dialogues) for mutual understanding. In a banking transaction, for instance, the caller might say "I'd like to transfer $50.50 from my checking account." The system would need to come back with "Certainly. Into which account?" (or, if the caller had only one other account, "Certainly. Into savings?").

Wait. There's another (syntactic) question type, isn't there? Yes, of which the previous sentence is an example: the tag question. Syntactically and phonologically (it has the telltale rising intonation), it is a question. But pragmatically it is less an interrogative than a confirmation request (a groundskeeping, dialogue-management act).

There is no guaranteed form-function map in pragmatics, of course, so we are talking only about tendencies, not certainties, but tag questions are usually just assertions masked as questions. That's why lawyers like them so much. Tag questions effectively "allow the questioner to do the testifying" (Walker, 1999: 49). They are also very common with children, who are forming impressions about the world that they want confirmed by adults. Here's a few more tag questions:

36a   The cat was on the mat, was it not?

 b   The cat is on the mat, isn't it?

 c   The cat's on the mat, right?

 d   The cat isn't on the mat, is it?

Sometimes the whole utterance is called a tag question, sometimes it's just the little phrase at the end. It depends on who the syntactician is and which gods she prays to. But in pragmatic terms, it's clear that utterances like 36a–d prototypically perform two dialogue acts, in two stages: asserting (everything before the comma; sometimes these chunks are called the "anchors") and requesting confirmation about that assertion (everything between the comma and the question mark; sometimes called the "tags").

Tag questions are extremely useful devices for voice interfaces, which have generally high groundskeeping demands. Tag questions can briefly and naturally get confirmations that other forms can manage only much more awkwardly. Take this exchange:

T$_1$ TES   Hello, This is Train Enquiry Service. Please speak after the tone. Which information do you need?

T$_2$ Caller   I want to go from Torino to Milano.

T$_3$ TES   Do you want to leave from Trento? Yes or No?

T$_4$ User   No.

T_5 TES Where do you want to leave from?

T_6 User I want to leave from Torino.

T_7 TES Do you want to leave from Torino? Yes or No?

T_8 User Yes.

T_9 TES Do you want to go to Milano? Yes or No?

(Danieli and Gerbino, 1996)

The system builds the ground collaboratively, with confirmation requests at T_3, T_7, and T_9, but in a mechanical, somewhat dictatorial way. Or, take this exchange:

T_1 Danluft: How many people will travel?

T_2 Caller: Two adults and two children.

T_3 Danluft: Two people.

Please state the ID number of the first person.

T_4 Caller: Change.

T_5 Danluft: Two people.

Is this correct?

(Bernsen et al., 1998: 179)

Again, the system is managing the ground. It is alert to the slippage and solicits confirmation, or disconfirmation, from the caller at T_5 about which piece of data needs to be calibrated. Now, take the following snippets of a dialogue, which build the ground more efficiently and easily by deploying tag questions (T_1 and T_7; notice also the casual structure of the constituent question at T_5):

T_1 Thrifty You are flying in, correct?

T_2 David Uh-huh.

. . .

T_5 Thrifty And picking up around what time . . . ?

T_6 David Around midnight.

T_7 Thrifty You're returning the car to the same airport, right?

(Kestenbaum, 2002)

Interrogatives generally are very useful devices for dialogue management issues — repairing ambiguities, soliciting missed input, dealing with out-of-vocabulary terminology, and so on.

Obligative Acts

"You SHOULD NOT be here
When our mother is not.
You get out of his house!"
Said the fish in the pot.
    — Theodor S. Geisel (Dr. Seuss)

Defined

Obligative dialogue acts are utterances that either entreat the hearer (directive) or enjoin the speaker (commissive) to some future action (a future which might start as soon as the utterance is over). Directive acts seek an undertaking from the hearer: prototypical directives are requests, orders, and proposals. Commissive acts are utterances in which the speaker undertakes to do something: promises, offers, threats. Most obligatives are unidirectional, though a few, like bets, oblige both parties. One only bets if, and because, the other party bets. (Notice, however, that obligatives of all sorts can combine in ways which enjoin both speaker and hearer in a linked way: "Gimme that ball or I'll pulverize you" [order + threat]; "Marry me and I will treat you like a goddess" [proposal + promise].)

In Voice Interaction Design

On the surface, obligatives may seem to be the antithesis of human–computer interaction. We humans behave together to accomplish goals by promising to do this, if you do that; offering to do that, if you'll do this; warning each other about the consequences of not getting these done on time; and so on. Obligatives are the essence of *social* coordination, not *mechanical* coordination; surely they're too morally laden to enter into our dealings with *appliances*. OK, we may threaten and command and plead with our machinery regularly, especially when it breaks down or misbehaves, but we don't seriously regard our language output in such cases as dialogue acts, meant to effect a purpose. It's venting.

As they say in the self-help books, we are undergoing a paradigm shift. Obligatives are fundamental to voice interactions, and their use is one of the clearest examples of the way in which computers are becoming social actors in the modern landscape.

Both types of obligatives are crucial to the effective functioning of most speech systems, although there is a very marked asymmetry in their use: directives come overwhelmingly from the user; commissives from the system. We looked at the ARTIMIS dialogue earlier as an example of how interrogatives can drive an exchange, but the interrogatives are syntactic in that dialogue. When you bring dialogue acts into the picture, it becomes clear that obligatives are in the driver's seat. Take this interrogative, asked repeatedly by the system:

ARTIMIS   Do you want any other information?

This is an interrogative, certainly, and requires an answer, but recall that the answers to it are consistently *implied* in the dialogue, only rarely given outright. Most of the caller's responses are like this one, itself an interrogative:

Caller   Who is its provider?

The affirmative to ARTIMIS's interrogative, that is, comes inferentially, because the caller responds directly to the dialogue act, an offer (a commissive) to supply more information. The caller's interrogative, in turn, gets the following response:

ARTIMIS   The provider of 36 68 00 00 is Météo-France.

The syntactic logic of constituent interrogatives gets at the *form* of this exchange (*Météo-France* is the constituent that satisfies *who*), but it is the flow of intentions that tells us what is going on: the system offers some information from a general pool, the caller requests specific information he figures is in that pool, and the system then makes good on its offer, supplying that specific information.

In short, people need to coordinate obligations with speech systems and speech systems need to coordinate obligations with people, and the basis of that coordination is borrowed from the language tools of social interaction. Once again there is a marked asymmetry in most speech-based, human–computer interactions, but lots of human–human obligative coordination is asymmetrical as well: employer/employee, teacher/student, customer/clerk. The fact that speech systems are effectively servants of the caller (however relentless and singleminded in method some of them are) does not reduce the human–human analogy. Indeed, it may strengthen the analogy.

There *are* moral and emotional overtones to many obligatives that make them unlikely to be used by speech systems, or to be used on speech systems; it is difficult to think of someone begging a speech system, or vice versa. But the range of applicable obligatives for human–computer speech interaction is still quite substantial. Directives available to the system include summons (in the ringing-up sense), recommendations, advice, directions, requests, suggestions, warnings, even bets.[12] Available system commissives include offers, refusals, bookings, promises, and guarantees (about information security, for instance). User

---

12: Betting has the relatively unique status of being simultaneously directive and commissive for humans: when people bet they enjoin themselves and each other (to some action, usually a payoff, contingent on some outcome). With machines (slot machines, video terminals, automated race-track tellers), the situation effectively becomes directive: people place the bet by paying up front, so that only the machine is enjoined. If you're uncomfortable

directives mostly take the form of requests for information, orders for merchandise, and a few others — approvals and rejections, for instance (a shipping system that notes the product is going to an address other than the caller's, for instance, might ask "Would you like that purchase gift wrapped?," which the user would then approve or reject). User commissives are largely constrained to promises, refusals, and guarantees.

One set of directives to the system requires special notice: directives related specifically to its functions. These are, simply, *system directives*. While the idea of a conversational voice interface is to emulate a human–human conversation as much as possible, at base we still have a computer and a user, and computers are not humans. If a human is reading out a list or otherwise performing a dialogue-based function for us, and we miss something, we might say something like "What was that last one again?" If we can't pay attention for a moment, we might say "Hold on a moment, please." With a speech system, commands like "Repeat" and "Pause" and "Jump back" are more likely, especially with expert users. With speech systems, we might even command a faster rate of speech, or a slower one, or ask "What can I say here?" These are system directives. While they are technically obligatives, in the sense of obliging the system to behave in a certain way — in fact, they are commands — they concern the flow of the dialogue more than the task, so they are housed in the dialogue management side of Table 4-1, a side we explore more in the following chapter.

Summary

The situation in which words are uttered can never be passed over as irrelevant to the linguistic expression.
— Bronislaw Malinowski

We have taken up three main topics in this chapter, all situation-driven and all strongly interdependent. We have explored conversational maxims, dialogue acts (along with a few syntactic structures), and the over-arching notion of register.

Register is a specific variety of language, which draws on the main language for much of its resources, but which has its own characteristic structures and vocabulary, shaped by

---

with my drifting into the notion that putting a coin into a slot machine is a "dialogue act," don't be. A better term yet for speech act is *communicative act*, since it makes evident that a wave and a "hi," a middle finger held aloft and an obscene utterance, a fist shaken in a face and a "get lost" uttered gutturally, are communicatively synonymous. See Geiss (1995: 12–15) for some discussion. The overall appropriateness of *communicative act* notwithstanding, I prefer *dialogue act* in this book to engage the developing speech system convention, and to keep the focus on language. As for whether the directive of betting is possible with speech systems, given that people are willing to dump money into machines, effectively betting against them (or, more properly, their designers) that randomness will, this time, favor them, and given that such people (unaccountably) often have access to the electronic flow of funds, it is not at all unlikely that some voice interfaces will soon be digital bookies, taking bets and perhaps even initiating them.

three factors — field, tenor, and mode. The **field** of a register is the body of activities it contains (activities which map very tightly onto the dialogue acts). The **tenor** of a register is the collection of available dialogue roles that it includes (information seeker and information provider, buyer and seller, and so on). The **mode** of a register is the medium through which its dialogue acts travel (text, speech, gesture, pictograms).

Dialogue acts are the specific functions that people perform when they produce utterances — greeting each other, promising to call, betting on a game — under the appropriate conditions. There are two main categories, **task management acts** and **dialogue management acts**. We only took up task management acts in this chapter. They have three main subdivisions, each of which has two further divisions.

Constitutive acts are ones in which the acts and the conditions of the acts are most explicit. Constitutive acts can be **expressives**, which correlate with mental states (apologizing, lamenting, congratulating), or **declaratives**, which correlate with institutional rituals or other formal conditions (sentencing, christening, deputizing).

Informative acts are the mainstays of voice-interfaced systems. They consist of either serving up information (**assertives**) or seeking information (**interrogatives**).

Obligative acts, which implicate further acts (linguistic or physical), are also staples of voice interface interactions. Often obligative acts occur in tandem with informative acts. For instance, a request (a **directive**, because it seeks an action from the hearer), might be a request for information: "Do you have snow conditions for Aspen this weekend?"). An offer (a **commissive**, because it proposes an action by the speaker), might be an offer of information: "Would you like to hear about hotel availability?".

Dialogue acts involve a core group of syntactic structures — **assertions**, **imperatives**, and **interrogatives** — but they are tied to them in no absolute way. Offers, for instance, are often syntactically interrogatives ("Would you like to come to dinner Friday?"), but might also be assertions ("You're coming to dinner Friday") or even commands ("Come to dinner Friday"), depending on speaker intention and hearer interpretation. Most confusing terminologically, assertions (as dialogue acts) might be assertions (syntactically), but might also take another syntactic form. "Do I look like Bill Gates?" in the appropriate circumstances might mean "I don't know the answer to your obscure Windows-related inquiry."

Dialogue acts are specific behaviors, specific actions performed by utterances. We also looked at general strategies for language behavior, H. Paul Grice's conversational maxims. They come in four classes, all of which are subsumed under the super-maxim: **Cooperate**.

Cooperation in terms of **quantity**, involves coordinating the amount of information (not too much, not too little, "j-u-s-t right"); in terms of **quality**, cooperation involves trafficking only in reliable information; in terms of **relevance**, cooperation is a matter of sticking to the topic; in terms of **manner**, cooperation concerns the clarity of the message, how easily the hearer can figure out what it means.

While these maxims are somewhat indeterminate (which is often a source of criticism), their determinacy can be ratcheted down quite firmly indeed when they are viewed by the light of a given register. They are general, but yield specific results in specific situations, and an understanding of register helps to bring specificity to situations. Just as register circumscribes the dialogue acts available for an interaction, and even the source of those acts (the tenor identifies the available participant roles, and some roles are more disposed to offers and requests, some to commands), it also circumscribes matters of quantity, relevance, manner, and even quality.

The appropriate quantity of information, for instance, is tied closely to the tenor of an interaction, to its communicative roles. In two helpful-but-limited user stereotypes of interaction design, an "expert" often requires semantically richer, more informative, acts than a "beginner." But these roles are malleable. Someone might be an expert in one context and a beginner in another. Beginners can become experts — even, depending on the task (or field), within the course of one encounter. And, of course, there are lots of intermediate stages.

Since obligatives have informational content as well (not just informatives), two very significant design principles therefore fall into the domain of the quantity maxims, linked closely to tenor and field: **tapering** and **expansion** (with the latter's specific local version, **progressive assistance**). With tapering, the interface starts a given interaction with fully detailed instructions, perhaps combined with informative acts like assertions or descriptions. But as the user's role changes — as she becomes more comfortable with the task, maybe within the course of one encounter, maybe over a series of encounters — she needs less detail, and the dialogue acts shift to suggestions and hints. With progressive assistance, the situation is the reverse. Either because the task is thought to be intuitive or the user is presumed to be experienced, a given interaction will start with very little information, perhaps none, but the user's responses indicate that she is unclear about how to proceed and the interface progressively assists her with greater and greater detail — moving this time from hints to suggestions to instructions.

Register also influences the manner maxims very strongly (as well as aspects of manner that Grice doesn't concern himself with — such as politeness). In particular, notions of clarity (or its obverse, obscurity), directness, specificity, brevity, and orderliness are all (more or less) determined by the communicative resources of the register.

Let's look at something concrete.

An Example

> A dialogue is more than two monologues.
> — Max M. Kampelman

Here is a telephone exchange[13] between an executive and his assistant, the computer-generated agent Wildfire.

13: An extract of the demo, "Contacts," from the unfortunately defunct company Wildfire Communications, Inc.; http://www.wildfire.com/carrier_contacts.asp; accessed February 2003.

T_1	Jim	Create a contact.
T_2	Wildfire	For what?
T_3	Jim	This one [anaphorically, this is "Donna King"]
T_4	Wildfire	OK. Please say "Donna King."
T_5	Jim	Donna King.
T_6	Wildfire	They left a phone number: 617 960 9257. Would you like to file it?
T_7	Jim	Yes.
T_8	Wildfire	Which heading should I file it under?
T_9	Jim	Mobile.
T_{10}	Wildfire	Done. What else should I add?
T_{11}	Jim	That's all.
T_{12}	Wildfire	OK.

Jim and Wildfire are engaged in a series of actions. Jim issues a command (T_1), in as direct a format as if he was using an old command-line system. Wildfire wants more information, making a request for specification about how to set up the contact (T_2). Jim provides it with another command (T_3), satisfying her request. So far, they are engaged directly in task management, just trading obligatives. Now (T_4) Wildfire moves into dialogue management, acknowledging the command ("OK"); she also tacitly accepts it. In the same turn, she also makes a request of Jim. He immediately complies (T_5), his utterance constituting the compliance. (Since she is the projection of a system that is going to have to recognize the utterance "Donna King" when Jim utters it in the future, she solicits an utterance token from him.) Wildfire, in one turn (T_6), then makes two statements, followed by an offer that Jim promptly accepts (T_7). Wildfire asks for further direction about the task (T_8). Jim provides it, issuing another command (T_9). Wildfire, having silently complied with the command, confirms that compliance (and therefore entails that the task has been completed), and follows up with an offer for further assistance (T_{10}). Jim declines (T_{11}), and Wildfire acknowledges the declination (T_{12}). This analysis is also charted out in Table 4.2 (which includes dialogue management acts as well, for convenience and coverage).

All of the conversational maxims are followed perfectly in this exchange, evidenced by the fact that we hardly notice them at all. The amount of information is sufficient (quantity), except in a few cases, where Wildfire requests more; but the acts themselves ensure sufficiency. The turns are all relevant to the ones that precede them, and to the overall

	Turn	Utterance	Dialogue act	
			Task management	**Dialogue management**
T_1	Jim	Create a contact.	command	exchange initiation
T_2	Wildfire	For what?	acceptance (of command) constituent question	clarification request
T_3	Jim	This one [i.e., Donna King].	acceptance (of request) answer, directive	clarification
T_4	Wildfire	OK. Please say "Donna King."	acceptance (of directive) directive	confirmation turn assignment
T_5	Jim	Donna King.	compliance	acknowledgment
T_6	Wildfire	They left a phone number: 617 960 9257. Would you like to file it?	statement statement offer, yes/no question	exchange initiation turn holding turn assignment
T_7	Jim	Yes.	acceptance (of offer), answer	acknowledgment
T_8	Wildfire	Which heading should I file it under?	constituent question	clarification request
T_9	Jim	Mobile.	answer, directive	clarification
T_{10}	Wildfire	Done. What else should I add?	compliance offer, constituent question	confirmation of compliance turn assignment
T_{11}	Jim	That's all.	declination	exchange termination
T_{12}	Wildfire	OK.	acceptance (of declination)	acknowledgment

TABLE 4.2   Dialogue acts in an exchange with Wildfire

context; again, when relevance is in doubt, the acts ensure it. At T_6 Wildfire makes a couple of statements, following them up with a request which explains their relevance (the information in the second statement can be filed according to the category named in the first statement). They are clear, brief (but not too brief), direct, and orderly (the general category, phone number, for instance, is established before the specific category, mobile).

This exchange of dialogue acts takes place within the business register of telephone-call management. In terms of tenor, we have a phone user and a secretarial assistant. The field is phone use, which focuses the action (creating and filing contact information), and shapes the acts. The telephone user directs, the assistant complies, offers, and confirms;

when the assistant directs the user (at T_4), it is only pursuant to a task he has initiated (at T_1). (The phone is also the mode, of course, not just the field, so the exchange occurs in speech, rather than points-and-clicks, keypad pushing, or keyboard typing.)

As this exchange also shows, a dialogue is more than a pair of monologues. What is said in one turn depends on what is said in previous turns. But it goes further than the relevance of the information. The form and the function of one turn depends on the form and the function of previous turns — especially immediately previous turns. Questions are followed by answers, offers are followed by acceptances or declinations, the dialogue builds interactively. That is the topic of our next chapter.

Conversation

Understanding communication at the level of conversations is required for natural language processing, and it is very difficult indeed.
— Daryle Gardner-Bonneau

We have, to this point, sketched the phonetic and lexical atoms, the syntactic and semantic principles, for assembling the raw materials of conversations: utterances. We have spent some time charting the maxims for deploying those materials into contributions, and even more time drawing out the notion of dialogue act, the force those contributions have in the required contexts. But conversation is more than an assembly of deployed dialogic actions. The whole, as they say, is greater than the sum of its parts.

Conversation comes from the verb, *to converse*, shortened from the Middle English, *conversen*, "to associate with"; or, more literally, "to take turns with." It is, in short, fundamentally social. Talk is not just a form of action, it is a form of *collaborative* action (Clark and Wilkes-Gibbs, 1986[1]), and there are principles governing the collaboration.

Harvey Sacks, a founder of the sociological discipline we draw on heavily in this chapter, conversation analysis, calls these principles "the technology of conversation" (Sacks, 1992: 339). They govern who talks and who listens, how and when conversants swap roles, what they do when communication slips, how they assure each other when it doesn't, how they build meanings, effect understandings, and accomplish tasks. It's a subtle, if largely automatic, business, which needs to be made explicit and then automated, for voice interaction design. Not every gear and gate of the technology of conversation is directly rel-

---

1: For the fullest application of this perspective to speech systems, see the wonderful series of papers by Susan Brennan (especially, 1998a; also 1998b, 2000; Brennan and Hulteen, 1995; Brennan and Ohaeri, 1999; Cahn and Brennan, 1999).

evant for design work, but the general machinery is indispensable. Three notions are especially important from a design perspective — turn, dialogic pair, and groundskeeping.

Dialogue Management

The key to understanding the structure of conversations is to see that each [dialogue] act creates the possibility of a finite and usually quite limited set of appropriate . . . acts as replies.

— John Searle and Daniel Vanderveken

We saw Nancy and Hyla, in the last chapter, engage in a set of ritualistic procedures for beginning a telephone call. Here are Ilene and Charley ending one (she has made an inquiry about his driving to Syracuse with the hope that she can go along; he's not going, though, and has just told her):

T_1	Ilene	You know, that's all. Whenever you have the intentions of going, let me know.
T_2	Charley	Right.
T_3	Ilene	Okay?
T_4	Charley	Okay.
T_5	Ilene	Thanks anyway, Charlie.
T_6	Charley	Right.
T_7	Ilene	Okay?
T_8	Charley	Okay.
T_9	Ilene	Take care.
T_{10}	Charley	Speak to you.
T_{11}	Ilene	Bye-bye.
T_{12}	Charley	Bye.

(Hutchby, 2001: 71–2)

When humans end telephone calls there is usually a winding-down procedure because the participants are (more or less) equals, both of whom have invested in the call, and either of whom may wish to prolong it. So, they both give each other opportunities to do just that. Ilene states that she is done with the ride topic above (T_1's "that's all."), but wants Charley to know her interest has future relevance. Charley acknowledges that (T_2, "Right") and they formally close off the topic (T_3/T_4, "Okay?"/"Okay"). Ilene then initiates the closing of the overall discussion (T_5's "Thanks anyway"), which Charley acknowledges (T_6, "Right"), and they formally close off the discussion (T_7/T_8, "Okay?"/"Okay"). Finally, they terminate the call itself with pre-leave-takings (T_9/T_{10}, "Take care."/"Speak to you"), and actually exchange the ritual leave-taking formulas (T_{11}/T_{12}, "Bye-bye"/"Bye").

Yeesh. Don't people ever just say goodbye and hang up the phone? Well, yes, as a matter of fact, they do. Here is the conclusion of another information transaction about travel:

ARTIMIS   The provider of 36 68 00 00 is Météo-France. Do you want any other information?

Caller   No.

ARTIMIS   Goodbye.

Caller   [click]

Like Ilene, ARTIMIS closes off the previous topic and (albeit a bit crudely) leaves an opening for another topic; like Charley, the caller declines; ARTIMIS goes directly to the parting formula; and that's that. The distended verbal mambo that Ilene and Charley perform is typical of many classes of telephone calls, and it shows the subtleties of multi-turn interdependence that are inherent in conversational exchanges. But, especially in contrast with the ARTIMIS snippet, we can see that the elaborate maneuvers of human–human interaction are too delicate for the needs of voice interface design, at least for a long time to come.

But the general resources of human–human conversation are fundamental requisites, beginning with the crucial fact that dialogue acts very frequently come in pairs.

Dialogic Pairs

Most of the moves we make in conversation would not be made if we did not assume that they would be followed by responses whose nature we can predict. . . . a greeting calls for a greeting in exchange, a question for an answer, and an invitation for a response.
   — Ronald Wardhaugh

The most immediate relationship of cooperation in dialogue, and one that is of undoubted importance for voice interface design, is the dialogic pair.[2]

2: In the Conversation Analysis literature, from whence the basic research into this notion comes, this relationship is known widely as the *adjacency pair*, a label that is inadequate for a number of reasons. Like many of the terms

It is the dialogic pair that points toward an understanding of dialogue as sequential, not just serial. A dialogue is not just one utterance after another, by two or more speakers. It is a structured exchange of utterances, governed by function. Some dialogue acts, that is, call for other dialogue acts. Greetings call for greetings, questions call for answers, offers call for acceptances. (There are examples of dialogic pair match-ups in Tables 5.1 and 5.2. Please note that they are not in separate tables for any conceptual reasons, just for typographical convenience.) It is important to note that dialogic pairs — since they are combinations of dialogue acts — are *functional*, not *formal* categories. They are not always realized in the same way. Some are. If you say "Hi" to someone, they usually say "Hi" back (or a synonym — "Hello," "Yo," "Howdy"), but they might call you by name, or say something ostensibly very bizarre, like "Look who the cat dragged in." An offer might take many forms, and might be accepted (or rejected) in a variety of ways. If you welcome someone to your home, they usually say "Thank you"; if you thank someone, they usually say "You're welcome."

But — and this is what is especially important for voice interface design — while the ranges of forms are various, they are restricted; some are highly restricted.

Malcolm Coulthard offers a succinct account of dialogic pairs:

> *They are two utterances long; the utterances are produced successively by different speakers; the utterances are ordered — the first must belong to the class of [initiatives], the second to the class of [responses]; the utterances are related, not any [response] can follow any [initiative], but only an appropriate one.*
>
>   (Coulthard, 1985: 69)[3]

---

that come out of Sacks's (important and deservedly influential) work, it is awkward and a bit portentous. *Adjacency* suggests space, not time, perhaps reflecting the methodology of working primarily with written (therefore spatial) transcripts, not with raw conversations. And one of the most critical characteristics of conversational utterances, especially critical for interface design, is that they occur in time. You say them, and they're gone. Moreover, these pairs needn't occur in strict adjacency (or temporal contiguity); there can be "insertion sequences" separating the pair members. Of the various alternate terms (which collectively indicate a broad unhappiness with the original terminology), Deborah Schiffrin's *dialogic pair* seems best to me (Schiffrin, 1988: 268); other terms for these dual utterance packages include *exchange*, *exchange structure*, and (the appealing but unwieldy) *initiative-response unit*. The important point about dialogic pairs is not that they're (usually) adjacent/contiguous, but that they're reciprocally defining, each delineating the role of the other. (A question, for instance, may certainly occur in discourse without an answer, but its function can't be understood without reference to the notion of answer, and "answer" is incomprehensible without the notion of question.) Schegloff's definition of such pairs is based on "conditional relevance": "Given the first, the second is expectable; upon its occurrence it can be seen to be a second item to the first; upon its nonoccurrence it can be seen to be officially absent" (1968: 1083). They are mutually defining paired (and almost always sequenced) utterances.

3: As my interpolations betray, I have tinkered with Coulthard's terminology. He follows the legalistic conversation analysis phrasing of "first part" and "second part" that reifies sequence at the expense of function. I have adopted the terms used in dialogue-game theory, *initiative* and *response*, which maintain the sequential

Task-management dialogic pairs

First utterance (initiative)	Second utterance (response)
Apologize	Accept Reject
Assert	Agree Disagree/Correct
Assess	Agree Disagree/Correct
Compliment	Accept Reject/Downgrade Return
Yes/no-question	Affirm/Deny
Constituent-question	Answer
Greet	Greet
Identify	Recognize Reject
Invite	Accept Decline
Offer	Accept Reject
Offer-options	Select
Question	Answer
Recommend	Accept Decline
Request	Grant Decline
Summon	Answer
Take-leave-of	Take-leave-of
Thank	Accept-thanks

TABLE 5.1   Some natural dialogic pairings

Dialogue-management dialogic pairs

First utterance (initiative)	Second utterance (response)
Initiate dialogue	Accept Decline
Initiate exchange	Accept Decline
Take turn	Release turn Keep turn
Assign turn	Accept turn Decline turn
Request identification	Identify Decline
Identify	Recognize Reject
Request confirmation	Verify Decline
Self repair	Accept Reject
Other repair	Accept Reject/Self repair
Clarify	Acknowledge Reject/Correct
Repeat/Paraphrase	Acknowledge Reject/Correct
Echo	Confirm Reject/Repeat/Parphrase

TABLE 5.2 Some natural dialogic pairings

Saying the first utterance, that is, constrains the second utterance; as Bunt (1991) puts it, a speaker who produces one dialogue act exerts "reactive pressures" that predispose the hearer towards uttering some specific other dialogue act. More generally, the need for feedback is an instance of the broad psychological need that the gestalt theorists called *closure*. A request is not closed until it has a response; a question is not closed until it has an answer.

One act does not determine the next, just constrains it, but *constrain* is a word you should know and love as a voice interaction designer. The response-type becomes imminent; the system should be on alert, biasing the language model, and perhaps the recognizer toward the appropriate range of responses.

The most obvious, and most common, manifestation of this constraint is the immediate supplying of a specific response, as in this question/answer pair:

Caller   Who is its provider?

ARTIMIS   The provider of 36 68 00 00 is Météo-France.

Speech systems should be highly predictable in this way. But dialogic pairs do not always come as seamless units. Take this exchange

T_1 Caller   Do you have any flights to Miami on the 26th?

T_2 Agent   How many seats are you looking for?

T_3 Caller   One.

T_4 Agent   What time can you leave?

T_5 Caller   Some time in the afternoon.

T_6 Agent   Let me look . . . I'm not finding anything then . . . Can you leave earlier?

T_7 Caller   If I have to.

T_8 Agent   I've got a seat on an 11:00 a.m. flight on Treetop Airlines.

T_9 Caller   That'll be good.

(Geiss, 1995: 194)

---

indications but add functional indications. See, for instance, Hulstijn's characterization (2000: iv). There is, of course, an ambiguity between *initiative*, in the sense of controlling the dialogue flow, and *initiative* in this sense as the first member of a dialogic pair. The ambiguity is very systematic, however, and context generally clarifies the sense in given uses. And the ambiguity is also wholesome, because it highlights the connection between an utterance that functionally constrains a subsequent utterance and the notion of controlling the dialogue flow; the former is an instrument of the latter.

Turn T_1 is a yes/no interrogative, a question, initiating a dialogic pair, prospecting for an answer. But T_2 does not discharge this reactive pressure. T_2 is a constituent interrogative, another question, the initiative of another dialogic pair, an inquiry calling for an assertion. It is satisfied immediately (with T_3), but T_1 remains unrequited. In fact, there are three intervening dialogic pairs T_2/T_3, T_4/T_5, and T_6/T_7, before T_1 is eventually satisfied (with T_8). This exchange is a bit unusual in the length of the deferral, but deferrals are not at all uncommon.

The intervening dialogic pairs make up what is called an *insertion sequence*, a routine that is a necessary component in voice interface design. Repair, for instance, is impossible without inserting a problem-solving sequence into the dialogue; inevitably, this falls between members of a dialogic pair. And many transactions depend on assigning values to a number of slots — a departure-place, destination-place, departure-time, arrival-time, and seat-quantity in a travel booking operation, for instance; the from-account, the to-account, and the amount (and possibly time) in a bank-transfer operation; date and location for weather information. Some of the values will often be absent for the system, from a recognition failure or simply the user's only supplying partial information. Between the main initative and the main response (in the above example, T_1 and T_8), then, there will often be an insertion sequence that gathers the missing values.

For voice interaction purposes it is especially important to realize that the constraints exerted by the first utterance are not just functional, but conceptual. It's not just, for instance, that a question requires an answer. It requires a specific *sort* of answer, one that addresses the semantic needs of the question. It must "supply information that corresponds to meeting (or at least being relevant for reducing) the information need expressed by the question" (Bunt and Black, 2000: 18). Even highly formulaic utterances (like the greeting "How are you?" or "How's it going?", where the point is just to make contact or initiate an exchange), bias for coherent responses ("Fine" or "Not so bad," which address the superficial informational needs of the greeting, are more appropriate than "Hi!" or "Hot enough for ya?").

It is also important to notice that not all dialogue acts are joined as dialogic pairs, a matter (as all matters in speech) which frequently hangs on context. Take identifiers, for instance, informative acts which establish context by declaring who (or what) the speaking-agent is ("Acme Movie Information. How can I help you?"). Identifiers establish context, but do not call for a response. There may be a tendency to treat a variety of routine system-utterances in this manner. Compare these two exchanges:

Child/Mother (the parenthetical numbers indicate the time between utterances, in seconds)

Child Have to cut these Mummy.

(1.3)

Won't we Mummy?

(1.5)

Won't we?

Mother   Yes.

(Hutchby and Woofit, 1998: 42)

RailTel/User

System   Welcome to RailTel . . .

User   I want to go to Merseille tomorrow.

(Lamel, 1997: 72)

The examples are not, of course, the same. The first example is fairly typical of child/caregiver interactions, where the child is beginning to work out the rules of conversational interaction, and the caregiver is often just trying to get the spaghetti out of her hair. In this case, we have three initiatives (an assessment looking for agreement, followed by two questions looking for confirmation) before a response is finally triggered. These sorts of omissions happen sometimes in adult/adult exchanges, as well, and they're almost always noticeable. If you say hello to someone, or call to them, or ask them a question and they don't respond, you notice; if it is apparently deliberate, you notice and you're irritated.

RailTel may notice that its welcome is unrequited. It surely notices that it didn't get a chance to finish its utterance (it was set to ask "What information do you want?"). It is not, however, annoyed. There is ample — indeed, unmistakable — evidence that people orient themselves towards computers as if they were social agents. They treat them, in many ways, as if they were people. That is the most compelling argument for conversational interfaces, as well as a principal source of many of their problems. But people do not treat machines as they treat other people; they exploit the communicative strategies linked to information transfer and task completion, but much more rarely use the linguistic strategies of social-role management — the maintenance of face, the concern for feelings, that permeate our human–human interactions (at least with peers; some military officers, aristocrats, and the like, treat subordinates in a way that shows no concern for their face or feelings). The small courtesies and deferences are regularly scaled way back in human–computer interaction, and are often entirely absent. For instance, Brennan (1998) reports in a wizard experiment, that while the wizard greeted all users the same way, some of them knew "she" was a human and are others thought "she" was a computer, and that knowledge affected how users began their end of the conversations differently: "they always greeted human partners, while they greeted computer partners only half the time. [Further, they] almost always started out by directing complete-sentence, grammatical queries to human partners, and abbreviated, telegraphic strings to computer partners."

So, in the RailTel example, the user feels no obligation to requite the welcoming initiative, and goes directly to another, still unstated, initiative (the offer, "What information do you want?"). There are other reasons users might wish to avoid supplying their end of a dialogic pair. They might just feel like halting the interaction ("Goodbye" or [click]). They might be perplexed and want to issue a summons ("Help!"). Or they might want to switch dialogue contexts ("Traffic!"). On the system side, the "response" for a number of "initiatives" may well just be an action — for instance, when the user is navigating a message system with commands like Go Back, Pause, and Start Over, in command-and-control mode.

The strong possibility arises, then, that there will be unique dialogic pairs for human–computer speech interaction. Greetings, for instance, which are very highly constrained for return greetings in human–human interaction might correlate more highly with task directions of various kinds. These considerations link to the dialogic-pair notion of "preference" (Pomerantz, 1984): some responses are preferred with respect to initiatives, some aren't. Preference depends on context and on initiative-construction, but, roughly, when you invite someone to a party, you would prefer that they accept (or you wouldn't have invited them). There are additional implications here for relational coherence. Take this exchange, for instance:

Rose   If you'd care to come and visit a little while this morning, I'll give you a cup of coffee.

Bea   Well that's awfully sweet of you. I don't think I can make it this morning. I'm running an ad in the paper and I have to stay near the phone.

(Atkinson and Drew, 1979)

Rose issues an invitation, which prefers an acceptance; a simple "Sure, OK" or "I'd be delighted," and they could move on to another interaction, perhaps establishing a time. Bea, however, rejects the invitation. That calls for a reason. In the terminology of coherence relations, which we take up in the next chapter, she needs to *justify* her (dispreferred) response. The reactive pressures are substantial.

The important point, again, is that the first member of a dialogic pair constrains the second; or, conversely, that the second tends to satisfy the first. Either way, from the perspective of voice interaction design, this is a good thing. There are no guarantees, but there are strong cohesion and coherence predictabilities with dialogic pairs.

Turns

All utterances were supposed to start with the word "verbie" and end with the word "over"
to facilitate the recognition process (e.g., "verbie, the switch is up, over"). Subjects would
sometimes forget to start the utterance with "verbie" or forget to end it with "over."
— Ronnie W. Smith and D. Richard Hipp

Turn is a critically important concept for voice interaction design, because of the formal context in which it puts the utterance. But it is such an endemic notion to human social interaction that it requires very little explanation. It is a building block of cooperation, a primal element of interaction for a species that prizes both individuality and collectivity, that prizes being ourselves while working together. Two children who both want the same toy know (at least in principle) that fairness relies on each having a turn playing with it; when they grow up and marry, if they get along, they each have their turn picking movies and restaurants and using the TV remote. A game of Monopoly® is an intricately cooperative activity, with rules, tokens, fictional properties and buildings and institutions, an economy, and a really neat little wooden milk bottle; and it moves along because of the fundamental and untroublesome notion of "turn". But, in the entire 2,633-word official Parker Brothers Rules of Monopoly®, there is no definition for *turn*. There's no need.

There has been some unaccountable fretting among academics about what exactly a conversational turn is, but everyone except them — and perhaps those speech technologists and dialogue designers who insist on calling it a *state* in their system designs — already knows. Look at any one of the dialogue snippets studding this book. Look at the word(s) beside one name, occurring between the words associated with another name (or two other names). That's a turn. Look at a comic book, at the stuff in the speech balloon of a character, alternating with the speech balloons of other characters. That's a turn.

For the purposes of this book, I'll offer a definition: a turn consists of one agent's utterance(s) occurring between other agents' utterances (where agents can be human or computerized). There are, admittedly, complications. A first turn doesn't have any previous utterance(s); a last turn doesn't have any next utterance(s). Sometimes there is overlap — one turn starting before the other ends, two turns starting simultaneously, and the like. But *turn* is understood prototypically (as most words are), by its temporal-sequential relationship to other turns, not by special cases and clean boundaries. Let's move along.

Turn distribution in conversation is orderly, but not lock-step. The aptly named "simplest systematics" model (Sacks, Schegloff, and Jefferson, 1974), describes turn allocation by way of two rules (or four, depending on how you count):[4]

4: These rules, and the rules of Conversation Analysis in general (from which the simplest systematics model hails), are more like traffic regulations than mathematical formulas. It's wise to stop at stop signs, even at a rural crossroad in the middle of Canada's billiard-table province, Saskatchewan, when you can see to the horizon in every direction, and no vehicles are in sight. But not everyone does. The rules are sometimes bent, and dire consequences do not always follow. Still, Conversation Analysis has come under attack for using *rule* in this sense, to the point where Schegloff, exasperated, said "I am willing to adopt for now an alternate term, such as 'practice' or 'usage' (1992: 120).

1. a) If the concluding speaker selects a next speaker, then that speaker should take the next turn.

    b) If the concluding speaker doesn't make (or cue) a choice, then anybody can weigh in.

    c) If no one takes the floor, then the concluding speaker can start another turn.

2. Whatever option is taken, at the next relevant point, 1 a-c kick in again.

These systematics are simple enough, and surely conform to our intuitions of how we swap utterances among ourselves. In a two-party conversation (which is all that is relevant for current voice interfaces), when I finish, I can indicate that you're next; you can accept or decline, signaling your choice; and, depending on which of these options you take, one of us talks next (so long as turns are being allocated; when they're not, we're done); etc. But how does this reciprocal cuing occur?

Flow Regulating Dialogue Acts

> *This conversation is going on a little too fast: let's go back to the last remark but one.*
> — Humpty Dumpty
>   Lewis Carroll, *Alice Through the Looking Glass*

In Monopoly®, context usually suffices, but people occasionally resort to "your turn now" if the other player is counting her money, or staring out the window, or in another room rooting through the fridge. In conversation, context often does the job as well. If I'm pumping gas, and you pull in for a fill-up, we follow something of a script, such that when I say "How much?" the turn falls automatically to you; our utterances are functionally paired such that mine calls yours.

But context is not always sufficient for turn allocation, and, even during other-conversant fridge-rooting, the cues are not always so explicit as "Your turn now." Face-to-face turn management involves posture, eye contact, head movements and other physical signals, along with various verbal indications; on the phone, the verbal cues are all there is. Sometimes those cues are highly conspicuous, like the speaker's upturned intonation at the end of a question, which means "I'm done; you answer now." Sometimes they are more subtle, like the brief hearer utterance, "mm-hmm," which means "go on; I'm listening; I don't want a turn right now." Sometimes they aren't even there, like the silence served up at the end of a completed turn, which means "I'm done; anything to add?"

The dialogue acts relevant for turn allocation, the flow-regulating dialogue acts, have such labels as *initiate-dialogue*, *terminate-dialogue*, *take-turn*, *keep-turn*, *assign-turn*, and so on, including authentication sequences like *request-identification*, *identify*, and *recognize/reject* (acts which are important for establishing the right to participate in dialogues under a specific identity or a general role), and verification sequences (acts important for such things as ensuring agents are both clear about dates, inventory- or credit-card-numbers). (See Table 4.1 for a fuller list of flow-regulating acts.) There are two things we need to recall before proceeding. First, dialogue acts rarely travel alone. I don't mean that they participate in dialogic pairs — they do, but dialogic pairs are utterances by different agents. I mean that acts, in the same turn, by the same agent, are multiplex. They might be in clusters, and they are nearly always in amalgams. Second, dialogue acts have no specific verbal realizations; the same act can have many, many forms. So, the utterances in the ARTIMIS subdialogue we just looked at, for instance, play out something like the following (some redundancies and other details are elided):

Dialogue			**Dialogue acts**	
T_1	ARTIMIS	The provider of 36 68 00 00 is Météo-France.	TM	**answer, statement**
			DM	**keep turn**
		Do you want any other information?	TM	**offer**
			DM	**assign turn**
T_2	Caller	No.	TM	**decline**
			DM	**take turn, release turn**
T_3	ARTIMIS	Goodbye.	TM	**take-leave-of, release turn**
			DM	**terminate exchange**
T_4	Caller	[click]	TM	**terminate task**
			DM	**terminate dialogue**

All turns here have multiple dialogue acts, as many as four (T_1); any act might have been phrased differently (the declining act of T_2, for instance, might have been "Nope," "No, I don't," "Not right now," and so on); and some acts (T_4) are even nonlinguistically instantiated. T_1 is a dialogue-act cluster, with two utterances acting in different ways — "The provider..." is a response to a preceding act, while "Do you..." is an initiative for a forecasted act. A more familiar cluster-type would be, for instance, the opening, which often greets, identifies, and offers (as well as taking and assigning turns). And all of the utterances are dialogue-act amalgams; even the non-verbal "utterance," click, both terminates the task and the dialogue.

While there is variety, however, there are certainly also standard expressions and devices for specific dialogue-management acts — some universal, others more situation- or register-specific. Silence after an utterance (greater than a second or so), for instance, is a universal signal that the speaker is finished. Conversely, then, a familiar way of keeping the turn is just to fill all silences, even if it's just by "um" or "uh," or by stringing out the articulation of one or two words. Here's our old friend, not quite sure about what he's going to say next, but sure about one thing, that he doesn't want to surrender his turn yet:

> *"And so," said the Cat in the Hat,*
> *"So,*
>
>   *so,*
>
>   *so, . . .*
>
> *I will show you*
> *Another good game that I know.*
>   (Seuss, 1957: 27)

The string of *so*'s amount to the dialogue act of keeping turn.

The range of turn-management indicators for voice systems is, in principle, as wide and potentially delicate as the corresponding ones humans use when talking to each other. On the output side, the earliest dialogue-management acts in voice systems were beeps and tones to let the user know he could speak now. Such tones, sometimes called *earcons*, are the equivalent of a crude turn-assignment act, like "Speak!" Sometimes the input side had designated turn-assignment acts as well; in early prototypes of the Circuit Fix-it Shoppe, as we've seen, the user was required to utter *verbie* to signal "I'm taking my turn now" and *over* to mean "I'm done. Your turn." (Smith and Hipp, 1994: 183). These strategies, on both the output and the input side, are so counter-intuitive for language users (other than truck drivers or Ham radio aficionados) that they fail regularly, even for users who have been carefully trained. When someone gets involved in a task, artificial language behaviors tend to fall away.

Fortunately, recognizers are no longer so fragile that they require these kinds of practices, and voice interfaces use more natural cues. For letting the caller know it's her turn, speech systems tend to rely primarily on the functional characteristics of their utterances, and the prosodic cues that accompany those functions (still a challenge for synthetic voices). Relentlessly, this currently means questions ("How may I help you?"; "When do you want to leave?"; "Do you want any other information?"), though this is changing. Context and prosody are reinforced by silence: when the system finishes its turn, it clams up. Similarly, systems take silence-after-input as the primary signal that the user has completed her turn and is waiting for system output. And a completed system turn followed by silence from the user (of 1.5–3 seconds) is usually (and should usually) be interpreted as the user declining her turn, perhaps because she's not sure what to say; that is, as a call for explicit feedback.

Dialogue systems (usually) need to maintain turn while they are processing, or the user may feel compelled to speak. The best method for maintaining turn is the one humans use, avoiding silence, though "ums" and "uhs" (and dysfluencies of any sort) from a machine will give too much humanness to the systems, in a way that begs for debilitating input. Or, in a monotone synthetic voice, it might be the source of unintentional humour. The solution? Either a clearly delineating soundscape, one that indicates "I'm working on it; hang on," or judicious human-emulation, as in this example:

A   Let me look . . . I'm not finding anything then . . . Can you leave earlier?

(Geiss, 1995: 194)

Sometimes, of course, turn allocation goes wrong. Here are Nancy and Hyla again, with some added detail about their exchange:

Hyla: Oka ⌈y!

Nancy: ⌊Good. What's doing?

That little stretched bracket indicates that Nancy couldn't wait for Hyla to stop talking and her turn overlapped with Hyla's. Turn overlap is somewhat atypical in human–human conversation. The standard conversational protocol is "one party at a time" (Schegloff, 1968: 1076; see also Schegloff and Sacks, 1973: 293f), though we all know speakers and cohorts that deviate from this standard markedly. The lesson to take from Hyla and Nancy here, in any case, is just that turn overlap happens. In human–computer interaction, the computer overlaps the human:

User: two three one sixty-⌈one

System: ⌊Please speak the digits of the authorization number, pausing after each digit.

(Kamm, 1994: 435)

The human overlaps the computer, but isn't supposed to:

System: Do you want another transaction?

            Be ⌈ep!

User: ⌊Yes.

System: Remember to wait for the tone.

            Do you want another ⌈transaction?

User: ⌊Yes.

System: Beep!

(IBM, 2001: 165)

The human overlaps the computer, and gets away with it:

BeVocal: Name a city and state, or say⌈

Caller   ⌊Work!

BeVocal: Work! 982 Walsh Avenue, Santa Clara, California.

Is this correct?

(BeVocal.com, 2001)

Turn overlaps happen.[5] In human–human communication, there are various situation- and register-dependent etiquettes (as well as specific interpersonal relationships), ranging from fighting for the floor to joyfully talking in unison (to a couple rushing to finish each other's thoughts). In human–computer communication, the task-driven nature of the interactions, the requirements of the human, and the service functions of the computer all constrain those situations and registers. From a design perspective, especially, the major implication of turn overlaps is just that they need to be built in at the base.

Voice interfaces need to accommodate the hurry-up tendency of human agents in their dealings with computers. More than that, they need to respect this tendency better than humans generally do; which means, no competition for the floor except where truly necessary. Similarly, interfaces should not cut into the user's turn except when truly necessary. Only the last of the three scenarios we just saw, the human overlapping the computer's turn, should be a regular — indeed, inalienable — feature of speech systems. If there was a universal charter of speech-system user rights, it would be Article 2, right after "No advertisements."

Certain moments in a turn are more relevant for floor transitions than others — ends of phrases and clauses, in particular, and of natural breath units (which very often coincide with phrase and clause completions). Even when overlaps happen, at least in human–human interaction, they usually occur around these transition points; the hearer can see (or thinks she can see) where the speaker's turn is going, and just before he gets there she jumps in with the start of her turn.

These transition moments may well prove considerably less significant for voice-interface turn overlap than for human–human turn overlap. In keypad systems no one waits for a list of the options if they already know them (or, in any case, if they know the spe-

---

5: In the speech-system world, the user-on-system turn overlap is almost universally but unfortunately called *barge-in*, a term that has become so endemic to the industry that there is a VoiceXML attribute called *bargein*. Despite its ubiquity, I'll avoid it here (preferring just *turn overlap*), to bypass the moral overtones and general clumsiness of barge-in. (Such loutish terminology is characteristic in the telephony world, where, for instance, a hard-to-recognize individual is known as a *goat*, an easy-to-recognize individual, a *sheep*.) But, *barge-in*, *bargein*, *turn overlap*, or (as I've seen occasionally) *talk overlap* — whatever it's called — the functionality is extremely important.

cific option they want). Here is how my third exchange went one afternoon with a keypad call-screening front-end to a help line that I called twice earlier (and was cut off from both times midway through the service encounter):

BS   Welcome to Bell-Sympatico Internet service. Bienvenue au service d'internet Sympatico-Bell. Pour le service en Français, appuyer . . .

Me   Beep (= 3)

BS   We have received an invalid selection. Please try again. Nous avons reçu une sélection d'invalide. S'il vous plaît essayer encore.

Pour le service en Français, appuyer deux maintenant.

All Sympatico services are online and in full operation. You have five options . . .

Me   Beep (= 3)

BS   You have five . . .

Me   Beep (= 4)

BS   Thank you. Your call is being [etc.]

I was not waiting for the transition-relevant places; you wouldn't have been either, I trust. I was practicing what keypad-system designers call *dial-ahead*, one of the few mercies such systems afford (and then only after we have been forced to sit through enough menu-enumerations to figure out which numbers will get us where we want to go). Human–machine turn-overlap in voice systems, for familiar systems, becomes indistinguishable from talk-ahead (also sometimes called *talk through* or *cut-through*). Machine–human turn-overlap similarly does not attend particularly to the known transition points found — not that computers eagerly press the business ahead (which would aggravate callers endlessly), but that their reasons for overlap are always urgent (there are no reasons other than urgent ones to interrupt a caller — data loss, for instance, or legal obligations, or a hopelessly bewildered system that needs to go back and try over).

Groundskeeping

For people to contribute to discourse, a basic requirement is that they add to the common ground in an orderly way.
— Herbert Clark

The ground on which meaning is collaboratively constructed in conversation is of two general types. The first type is the *background*, which the conversational agents bring with them into the conversation. It is largely a function of the domain(s) in which the conversation takes place, along with whatever interactional history the agents share; roughly, these correspond to the field and tenor of the register. These shared elements can certainly be cut much more finely than this, but for the purposes of conversational interfaces, it can all be stuffed into the same sack, *background*. The second type is the *conversational ground*, which is built up over the course of a dialogue. They are highly interdependent; the background constrains what can be added to the conversational ground, while the conversational ground builds, in large part, by selectively evoking elements of the background.

Take this dialogic pair:

Ditko   Is he strong?

Lee   Listen, Bud, he's got radioactive blood.

The background here determines the meaning of the answer to Ditko's yes/no question. In the context of 1960s superhero cartoons, the answer is an emphatic affirmative ("yes, extraordinarily strong"). In the context of a medical exam, the answer is the reverse ("no, he's very sick; therefore, weak"). So, the background determines part of Lee's contribution to the conversational ground ("he is strong" or "he is weak"). But he also contributes specific information above and beyond his affirmative or negative response to Ditko's question (that is, the person they are both talking about has radioactive blood). Lee is building the conversational ground, but the way he builds it depends on the background.

Grounding needs to be calibrated for any kind of effective interaction. I might assume, for instance, that our textual ground, yours and mine, includes *fenks*. It's in this book, in earlier chapters. I put it there. But maybe you didn't read those sections, or you've forgotten them. I have a problem, as an author. So do you, as a reader. Socrates diagnosed your problem around 2,500 years ago, complaining about a new-fangled technology. "Writing," he said, "is unfortunately like painting." The problem is that paintings "have the attitude of life, and yet if you ask them a question they preserve a solemn silence." (*Phaedrus* 276ᵈ). You can ask a text all the questions you can think of, Socrates complains, and it just says the same damn thing over and over.

Authors have the reverse problem (also diagnosed by Socrates, as it happens — *Protagoras* 329ᵃ). I can write that damn thing down and not know that you're going to question it. There are things we can do. Authors can build in redundancies, provide resources (appendices, footnotes, glossaries, indexes), and the like, to try and make the reader's job easier. Readers can work with those resources, benefit from those redundancies, and do their best to puzzle out what the author is on about. But books are done deals. There is something missing from the immediate author/reader encounter, something big, that we could both use to help us build meaning if we were sitting across the table from each other now, or even at opposite ends of a telephone line.

Whatever the merits of books (which we both think are substantial, or we wouldn't be here together), what is missing is what gives dialogue its life's blood: feedback. Groundskeeping acts are about either supplying or soliciting feedback.

Feedback

Feedback is a method of controlling a system by reinserting into it the results of its past performance.
— Norbert Wiener

The notion of feedback comes to language studies from cybernetics (Yngve, 1970; Heritage, 1984), where it designates information about the results of a process which is then used to modify, or to maintain, that process. It comes in two flavors. Negative feedback indicates something has gone wrong, and needs to be adjusted. Positive feedback indicates that the process is successful. The analogy is so closely matched to interagent conversational management that it seems less like an analogy than evidence that conversation is cybernetic. Certainly human–computer speech interaction is. Any voice interface designer who thinks that human–computer speech interaction can't be characterized as "a system," in need of governance on the basis of performance, should start looking for another line of work.

Feedback is "part of a global controlling mechanism . . . which remains implicit when the dialogue proceeds successfully" (Derriks and Willem, 1998: 599; see also Clark and Schaefers, 1989; Brennan, 1998; and, for an application to computational dialogue, Cahn and Brennan, 1999). That is, you ask a question, you get an answer, and the very fact that you get an answer is confirmative feedback that you are getting somewhere. But feedback is also "made explicit" (Derriks and Willem 1998, 599); that explicit feedback, positive and negative, is where the groundskeeping dialogue acts come in. Here's an example:

T_1 Caller I want to leave on the first flight out of ORD, uh, Monday morning.

T_2 Amex Monday the first?

T_3 Caller The eighth! The eighth!

T_4 Amex OK.

(SRI/Amex, 1989; tape 1, call 1)

At T_1 the caller says he wants to leave from Chicago's O'Hare Airport (referenced by its code, ORD), but signals some hesitancy about the departure day. That may cue Amex (a human travel agent for American Express) to elaborate, or she may just be using a routine grounding strategy, but her dialogue-act response is a confirmation request as well as a question, which indicates (1) that she knows the day is Monday, and (2) that her tentative assumption is for Monday the first.

The caller quickly interrupts (there is a turn overlap here) with (1) an implicit rejection of Amex's assumption, and (2) an explicit clarification of his intention to travel on Monday the eighth. That is, the caller responds with negative feedback (prosodically signaled in part by the urgency, here shown via exclamation marks) — with an other-repair — and with the elided assertion, "[I want to leave on Monday] the eighth." Amex then responds with "OK," a confirmation of Monday the eighth, and the call carries on.

The whole exchange works, that is, because of explicit feedback: Amex's initial confirmation request (T_2) alerts the caller to her construal, triggering the repair (T_3), and her later confirmation (T_4) establishes they are now both on the same ground about the date.

Graphic interfaces are brilliant at feedback, consistently signaling the results of a user's behavior. Even if the designers don't wield their rich feedback potential effectively, the most abysmally responsive graphic interface is already light years ahead of voice interfaces: you feel the keys depress, you see the letters on the screen; you feel the pointing device in your hand, you watch the arrow move; you feel the physical button click, the virtual button darkens or depresses, and a new window appears. Graphic interfaces almost all use sound as an output modality as well, providing redundant auditory feedback to the tactile and visual feedback the user is getting. Their immediate responsiveness is hugely responsible for their success. And usefully implemented feedback features (progress bars, context-sensitive callouts, grayed-out items) can make an interface truly virtuous, bringing it to the point of genuine consideration for the user.

Human–human conversations are brilliant at feedback, which is hugely responsible for their success; the Amex example we just looked at is proof enough, but there is no shortage of evidence on this point.

Voice interfaces are not brilliant at feedback.

They are confined to the transient auditory groove, without the reciprocal tactile/visual channels available to graphic interfaces (nor the visual channel of face-to-face human conversation, in which even an eyebrow can contribute feedback). They operate only in time, not in space. So they have absolutely no hope of coming close to the comprehensive, distributed, redundant reassurances and calibrations of graphic interfaces. (Multimodal use of voice is another story, for another book.) But human–human telephone conversation has these limitations as well, and while you can't do drafting or word processing or statistical analysis over the phone with someone (at least not with very much elegance), people accomplish a heck of a lot by phone.

OK, computers aren't people. But they are fast, capacious, and in human–computer interaction they have two funds of human intelligence and creativity to draw on: the designers' and the user's. Those are formidable resources.

Cutting to the chase, there is nothing in the Amex exchange above that is even slightly beyond the reach of speech systems. This one, which you have already seen, parallels the Amex exchange quite closely (including the dialogue acts this time around):

	Dialogue	**Dialogue acts**	
Nicole	And on Wednesday?	**TM**	**constituent question**
		DM	**turn assignment**
SpeechActs	On Wednesday, November 17th at 2:15 pm Bob has "Return to Boston." At 5:15 pm, Bob has . . .	**TM**	**statements**
		DM	**turn-holding**
Nicole	Stop.	**TM**	**halt process**
		DM	**other-repair**
	Tell me what I have.	**TM**	**command**
		DM	**turn assignment**

(Yankelovich, 1994)

SpeechActs feeds back its assumption that Nicole wants to know about Bob's schedule (as well as its assumption about which Wednesday is in focus). Nicole rejects the assumption and directs SpeechActs to tell her about her own calendar. It does, retaining the assumption that the date-focus is still Wednesday, November 17th.

Like graphic interfaces, but highly impoverished in comparison, voice interfaces can provide system-state feedback. The soundscape can be used to advantage, in these circumstances, but if the processing is brief, the native strategies of human–human conversation are probably best. We've seen this example of conversational floor-holding earlier, but now look at it as the partial equivalent of a progress bar:

A   Let me look . . . I'm not finding anything then . . . Can you leave earlier?

(Geiss, 1995: 194)

Notice, in any case, that *something* needs to be done in terms of feedback if the system cannot yet respond directly to the input because it has further processing to do, but strategically still shouldn't relinquish the floor. The output channel cannot just fall into silence, because silence defaults to a turn-exchange cue. The user will start talking, and may push the dialogue into confusion.

In dialogue-act terms, positive feedback is an explicit signal that zeroes in on transmission success ("Got it"); negative feedback is an explicit signal that zeroes in on transmission failure ("I didn't get that"); but notice that every utterance in a dialogue (after the first) has some feedback component to it, and that the feedback must weight toward the positive or the negative. It could not be otherwise. If the utterance responds appropriately to the discourse that precedes it, the very appropriateness signals that the preceding utterance has been successfully processed. If the utterance responds inappropriately, that very inappropriateness signals that something has gone awry.

A specific subclass of positive feedback, backchanneling, is worth taking up briefly. A backchannel utterance is less a turn than support for another's turn.[6] 'It overlaps, but doesn't interrupt. As a dialogue act, a backchannel is one of the rare unidimensional acts, with no task-management function at all. It only concerns dialogue management, groundskeeping. It is the verbal equivalent of nodding (which it often accompanies in face-to-face conversation) and conveys roughly "go on/I'm attending/I got that":

T_1 Amex   Good afternoon. American Express. This is A.

T_2 Caller   A, this is B again, and I talked to C. D. again.

T_3 Amex   Mhm

T_4 Caller   About going from Hong Kong to Moscow. umm He made me feel like there had to be some other options besides going through Heathrow.

T_5 Amex   Ok uh

T_6 Caller   There had to be several different cities you can go through like Beijing, to Helsinki. There has to be some other options.

(SRI/Amex, 1989; tape 7, call 2)

Aside from the greeting (which goes unrequited here, notice; even in a human–human interaction the task can override normal conversational expectations), the Amex agent's utterances (that is, T_3 and T_5) are feeding back that she is paying attention but otherwise has no contribution to make yet; she is encouraging the caller to keep going. This is backchanneling. (For a detailed treatment, see Gardner, 2002.)

Eventually, when longish user narratives can be handled sufficiently well, system backchannels might prove important. At the moment, they really have only one job, but it is a significant one. Speech systems need to have sufficient sense of users' chunking tendencies to signal comprehension and encouragement in brief backchannel-like output. For instance, speakers often carve up telephone numbers, credit card information, addresses, and the like, at specific joints (such as area code, prefix, and identifier in phone numbers), and pause at those joints when they recite the information to someone, especially over the phone. The idea from human–human interaction, of course, is that the listener is writing down the numbers at the other end and needs a bit of processing time. If users leave gaps

---

6: Yngve (1970: 568) defines "the back channel" as something of a conduit from someone who doesn't have the floor (therefore, is not taking a turn) to someone who does. While some backchanneled utterances may meet the mechanical definition of turn I offered earlier, and some may not, I have no stake in the issue of whether backchanneled feedback is considered a turn or not.

(≥ 2 sec) at these joints, the system should issue small encouragements for them to continue, though a clear "yes" or "OK" would be strongly preferred over "mm-hmm." Likewise, the system should understand that the type of confirmations it requires in establishing the same information might be met by murmured or otherwise indistinct backchanneling and not feel obliged to query the input; what it really needs to listen for is a "no," which would not be backchanneled.

Voice interfaces are not brilliant at feedback, but humans are pretty darn good at it, which is fortunate because speech recognition is fragile enough that regular feedback that the system is still on track is a very important commodity. Early voice systems were so desperate for it that they taxed the user with constant demands, usually in the form of explicit confirmation requests. Remember the TES cross-examination from last chapter (and, in any case, a mercifully brief excerpt follows to remind you)?

> User   I want to leave from Torino.
>
> TES   Do you want to leave from Torino? Yes or No?
>
> User   Yes.

(Danieli and Gerbino, 1996)

Fortunately, there are more subtle methods of seeking confirmation that allow for a more natural interaction. (We will look at a taxonomy of feedback in Chapter 13.)

The importance of the conversational ground cannot be overestimated. In many ways, the only thing accomplished by a conversation is building and grooming the ground that agents can stand on together. To the extent that the agents can be said to have accomplished their goals, to that extent they have been successful at their collaborative grounding activities.

Repair

> *The reason that dialogue is such an effective means of communication is not because the thoughts of the participants are in such perfect harmony, but rather because the lack of harmony can be discovered and addressed when it is necessary.*
> — Martin Ringle and Bertram Bruce

One of the most important jobs of feedback, as in the Amex exchange ("The eighth! The eighth!"), or the SpeechActs exchange ("Stop. Tell me what I have."), is to uncover and repair communicative slippages. It is an axiom of interaction design, or should be, that users make no errors. (Humans do make plenty of errors, of course; some of us can't even walk and chew gum. But we're talking about *users* here, a special category of humans, and this axiom is the usability version of a familiar slogan in retail sales about another category of human, the slogan that claims the customer is always right. It's a matter of perspective, not ontology.)

An interface is after all only a subsystem that allows the user to communicate her intentions to the task-performing systems, and allows the task-performing system to communicate its functions and states to the user. When things go wrong in human–computer interaction, it is because of communication slippages; the interface has hindered or misdirected the reciprocal flow of intentions and feedback between human and computer. Something the system designers regarded as straightforwardly communicated by the interface, wasn't so straightforward to the user, or some action the user assumed would communicate her intent to perform some subtask communicated a completely different message to the system.

All human–machine interfaces should practice the no-user-errors religion. But voice interfaces, where there is all the more reason to believe that slippages are the fault of the system, cannot afford to worship in any other church. Moreover, the good thing about slippages is that they can be adjusted, calibrated, mended. Errors can only be corrected, and then only by wagging a finger at the error-making user and getting him to do what the system wants him to.

Slippages can be *repaired*, and they are, constantly, everyday, in conversation. Two examples:

Marty and Loes

Marty   Loes, do you have a calendar?

Loes   Yeah. (She reaches for her desk calendar.)

Marty   Do you have one that hangs on the wall?

Loes   Oh, you *want* one?

Marty   Yeah.

(Schegloff, 1992: 1321)

AT&T's "How May I Help You" System

HMIHY   What number would you like to call?

User   8 1 4 6 7 3 4 8 7 2

        [Recognition failure]

HMIHY   May I have that number again?

User   8 1 4 6 7 3 4 8 7 2

HMIHY   Thank you.

(Langkilde et al., 1999)

The first one is quite typical of human–human repairs: it is a misunderstanding, cleared up by adding specificity. Marty's intention is to *get* a calendar, but his request is hedged to the point that Loes just thinks he wants to *consult* a calendar; Marty becomes a bit more explicit, identifying the type of calendar he is asking about, and Loes makes the correct inference. The second example is quite typical of speech system repairs: it is a misrecognition, cleared up by repetition. HMIHY knows it got the number wrong, solicits a repeat of the utterance, and the user obliges.

On the page, the difference looks pretty minor (we read by sweeping over text), but it is substantial (just speak out the ten digits, wait a moment, and speak them out again verbatim, to get a sense of the temporal dimension represented of the HMIHY exchange particularly). Loes gets part of the request. She just isn't fully in synch with Marty's intention. Marty gets more specific, Loes gets it this time, and the request is satisfied. HMIHY, on the other hand, knows there is a problem, but can only ask for a repetition. Perhaps the failure has been total, but if the system caught part of the number — the area code, the prefix, or the line number (which the inference engine might be able to sus out with sequence or gaps or both) — it could construct a more helpful request, and a less taxing one for the user (repeating ourselves, especially repeating strings of digits, is an aggravating task for most of us): "Sorry, what was the area code again?"

Any time there is partial understanding, in short, it should be used. Even though Marty's request was ambiguous, Loes acts on her understanding right away, and Marty can tell which part of his intention she understands (that he wants a calendar) as well as which part she doesn't (the type of calendar). The Amex example is even clearer, since the agent's confirmation request includes her inference ("Monday the first?"), allowing the caller to see what got through successfully and what needs correcting.

Another example, one whose unnaturalness and potential for aggravation not even reading can obscure:

T_1 USER   Give me the price for AT&T.

T_2 SYSTEM   Hewlett-Packard is at 83 3/8, up 2 3/8 since yesterday.

T_3 USER   Give me the price for AT&T.

T_4 SYSTEM   Data General is at 63 1/4, up 1/3.

T_5 USER   Give me the price for AT&T.

T_6 SYSTEM   Hewlett-Packard is at 83 3/8, up 2 3/8.

T_7 USER   Give me the price for American Telephone and Telegraph.

T_8 SYSTEM   Excuse me?

(Levow, 1999)

The user tries valiantly to get the stock price for AT&T; the system doggedly refuses to notice that there has been a slippage (though it does make its recognition assumption clear, in a request-acknowledgement, allowing the user to diagnose the error; imagine if it had just come back with "83 3/8, up 2 3/8 since yesterday").

Communicative slippages are sometimes quite local — a background noise, a faulty signal, a moment of distraction — and the repairs can be just as local. The HMIHY example falls into this category. But sometimes they go on and on, as in the give-me-the-price-for-AT&T dialogue. Multiple-turn slippages, especially when the first error triggers later ones (the user may have increased his volume and decreased his tempo on T_3, for instance, which could have triggered the T_4 misrecognition), are known as error spiral (or error cascade, or error amplification). The very attributes that can trigger a spiral should be treated as diagnostics of trouble, and considerable design effort needs to be devoted to avoiding or cutting short cascading errors.

There is no single factor more important to the quality of the user's experience than the ease and naturalness with which repairs can be made. Effective repair strategies are the paramount voice interaction issue. And it is an issue that demonstrates how completely the speech-system product *is* the interaction: it implicates all levels of design, and hangs crucially on aspects of the recognition system. It's not a matter of building better discriminators, but of designing more receptive and intelligent discriminators, which can appeal to the background and the conversational ground, and make inferences based (in significant part at least) on human–human interaction patterns.

Slippages are breakdowns in coherence, and their repair is based on re-establishing coherence — of reference or of relation — which we will return to frequently in the book. But the repairs cannot be made without, first, detecting the breakdown, and second, by the system taking initiative to start the re-establishment process.

Initiative

> *[A mixed-initiative design] models the human–machine interaction after human collaborative problem solving. Rather than viewing the interaction as a series of commands, the interaction involves defining and discussing tasks, exploring ways to perform the task, and collaborating to get it done.*
> — James F. Allen, D. K. Byron, M. Dzikovsha, G. Ferguson, L. Galescu, A. Stent

A couple of millennia ago, Socrates was discoursing about discourse with an assortment of scholars. One of them was Polus, with whom he had the following exchange:

Socrates   Will you, who are so desirous to gratify others, afford a slight gratification to me?

Polus: I will.

Socrates	Will you ask me, what sort of an art is cookery?
Polus	What sort of an art is cookery?
Socrates	Not an art at all, Polus.
Polus:	What then?
Socrates	I should say an experience.
Polus	In what? I wish that you would explain to me.
Socrates	An experience in producing a sort of delight and gratification, Polus.

(Plato, *The Gorgias* 462)

It goes on. At first glance, this exchange instantiates the standard assumption about conversation, deeply engrained in the reciprocal notion of turn: that conversation is egalitarian. The turns here are pretty evenly distributed: Socrates says a few words, then Polus says a few, then it's back to Socrates. Socrates asks two questions and makes three assertions; for Polus, it's three questions and two assertions.

But many verbal interactions, like this one, where Socrates just leads Polus around by the nose, are far from balanced. There are asymmetries built into most relationships and most contexts, such that the flow of discourse is often largely in the control of one agent. The term for this in speech-system design is *initiative*; the one who has control of the dialogue flow has the initiative. Initiative is not a simple matter of who has the turn at any given moment, or how long the turns are, or even what form the utterances take, but of control over the flow of the discourse.

The first-level distinction to be made here is between *fixed initiative* (all the control by one agent) and *mixed initiative* (the control is shared). In Socrates's exchange with Polus, he has only a slight statistical edge in the overall number of words, but that does not come close to what is clearly a whopping asymmetry in initiative. The initiative is fixed at the setting: Socrates. It's *his* discourse, even to the point of literally putting words in Polus's mouth; Polus is only participating in the role of foil.

There are three logical initiative possibilities with a dialogue between two agents, call them *system* and *user*:

System initiative The system has all the control over the flow of the dialogue.

User initiative The user has all the control over the flow of the dialogue.

Mixed initiative Both agents share control over the flow of the dialogue, each able to assert (or relinquish) that control at any given point.

Since it is massively easier to know what to say and what to listen for if you have the initiative, dialogue systems have largely adopted the Socratic paradigm from their inception, taking all the initiative. Here is part of an exchange between a student and Jaime Carbonell's (text-based) SCHOLAR, perhaps the first intelligent tutoring system:

SCHOLAR   THE CAPITAL OF CHILE IS SANTIAGO. CORRECT OR INCORRECT?

Student: CORRECT.

SCHOLAR   THAT'S RIGHT. APPROX WHAT IS THE AREA OF BRAZIL?

Student   2,500,000.

SCHOLAR   WRONG. PLEASE INDICATE IF THE FOLLOWING STATEMENT IS CORRECT OR INCORRECT: THE AREA OF PARAGUAY IS APPROX. 47,432 SQUARE MILES.

(Carbonell, 1970: 192)

SCHOLAR was technically a mixed initiative system, as Carbonell proudly pointed out. There was even a mode called Mixinit! But it might be better termed a *bipolar initiative* system, since the student can take over the grilling for awhile, but there is no give-and-take exchange of initiative, as task or goal or agent mood might require. Here is a later interaction from the same session, when the student refuses to answer a question and abruptly takes the initiative with a system-directive:

SCHOLAR   NOW ANSWER THE QUESTION YOU DIDN'T ANSWER BEFORE.

Student   Q/A.

SCHOLAR   DO YOU WANT TO CHANGE THE GENERAL MODE OF THE INTERACTION TO Q/A?
PLEASE ANSWER Y ON N.

Student   Y.

SCHOLAR   OK, MR. JONES. LET'S GO AHEAD.

Student   PLEASE TELL ME MORE ABOUT PERU.

SCHOLAR   THE AREA OF PERU IS APPROX 480000 SQUARE MILES.
THE LANGUAGE IS SPANISH.
. . .

(Carbonell, 1970: 192)

SCHOLAR could respond as well as initiate, and had a range of dialogue acts that included asking, requesting, and even complaining ("YOU ARE TAKING TOO MUCH TIME . . ."). But it hung ruthlessly onto the initiative, aggressively pushing for answers, until and unless the user switched the mode to Q/A, whereupon it turned promptly meek and spoke only when spoken to. Aside from some very specific routines, concerning spelling and out-of-vocabulary words, there was no sense of a moment-to-moment negotiation in which the initiative might trade back and forth between the agents; a change in initiative required a change in mode. This initiative-swapping ability was impressive for a late 1960s real-time electronic tutor, but it falls very short of the ebb and flow of initiative in most human–human exchanges.

Ronnie Smith and D. Richard Hipp's Circuit Fix-it Shoppe had a more sophisticated initiative scheme, with four modes (1994, 12): *directive* (computer has all the control); *suggestive* (computer largely has control, but "is also willing to change the direction of the dialogue according to stated user preferences"); *declarative* (the user largely has control, "but the computer is free to mention relevant, though not required, facts as a response to the user's statements"); and *passive* (user has all the control).

What their quadruple scheme reveals (as does the inadequacy of the SCHOLAR bipolar scheme), is that truly mixed initiative is not just the capacity for one side or the other to dominate, but the opportunity for either to take the lead in contributing to the ground at any point. Conversational initiative, in fact, is best seen as a continuum, as in Figure 5.1, not as any combination of discrete categories — with system initiative at one end, user initiative at the other, and a range of mixed initiative conditions in the middle, variously favoring the system or the user as a function of field (a tutorial favoring the system, or a service encounter favoring the user) and tenor (an information seeker needs more initiative than an information provider, for instance). "Initiative is essentially a matter of degree," David Novick and his colleagues observe, "the degree to which the user is unconstrained" (1999: 166).

Mixed-initiative does not mean flipping a toggle like the SCHOLAR system, but neither does it mean "anything goes." What it means is voice-interface dialogues like the following, where the initiative shifts as the task-collaboration requires:

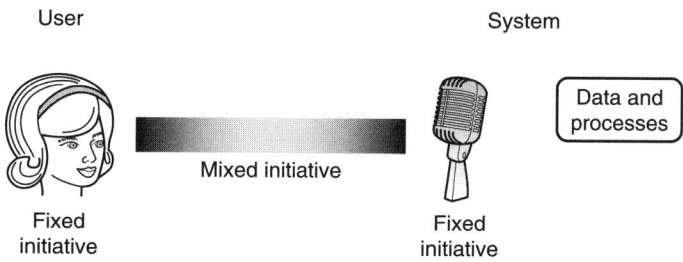

FIGURE 5.1   Initiative continuum

	Turn			Initiative
T_1	Caller	ring		**User**
T_2	RailTel	Welcome to RailTel . . .		**User**
T_3	Caller	I want to go to Merseille tomorrow.		**User**
T_4	RailTel	What is your departure city?		**System**
T_5	Caller	Lyon, around 8 in the morning.		**System**
T_6	RailTel	There is a train from Lyon to Merseille at 8:35 am tomorrow.		**System**
T_7	Caller	Thank you, goodbye.		**User**
T_8	RailTel	Goodbye.		**User**
T_9	Caller	click.		**User**

The caller initiates the exchange, and the dialogue proceeds under her initiative until T_4, when the RailTel needs some information to carry out the task, takes the initiative, and solicits a departure city. Initiative stays with RailTel until the task is over, when the user takes the initiative back to terminate the encounter. (And notice that while the dialogue initiative shifts back and forth, what Chu-Carroll and Brown, 1998, call the *task initiative* remains with the user. She phones up, gets her information, and that's that; the dialogue initiative is managed in a way to facilitate her task.)

We know from the keypad system experience that exclusively system-initiative designs are almost universally loathed. The occasional telephone geek who likes intricate keypad systems usually does so because, while the system output is designed to instruct the user turn-by-excruciating-turn which keys she is allowed to press, keypad interfaces almost universally allow dial-ahead — a way of providing the expert user a method to drive the interaction; in short, a type of initiative. But this sort of quasi-initiative does not work for voice interfaces, because even really, really good speech resolution is much slower than single-distinct-tone resolution.

Mixed-initiative design should be the default in voice design, with fixed-initiative design reserved for special tasks and contexts[7]. There are useful command-and-control

---

[7]: Several studies report results that make my claims about the importance of mixed-initiative design suspect. Kamm (1994), Oviat et al. (1994), Walker et al. (1997), and Walker et al. (1998) all found users to prefer interaction with system-initiative designs. I'm not at all sure, however, that any of these investigations extrapolate very far. All of them, for instance, were compromised by poorer recognition performance in the mixed-initiative conditions and I am skeptical that these results would stand up to longitudinal evaluation, since humans show a marked antipathy to being led around by the nose. If recognition performance was held stable across system- and mixed-initiative conditions, my guess is that these results would be reduced, eliminated, or reversed, and it is noteworthy

applications in which the initiative is fixed with the user (though these tend to be multi-modal, rather than voice-only), and largely system-fixed applications are clearly the way to go for certain form-filling and data-gathering operations, where the system effectively interviews the "user." But general-purpose voice interactions designed on a conversational model must be mixed-initiative.

"Certain designers," Jef Raskin notes, not disinterestedly, "consider forcing the user to stop and to work in lockstep with a planned sequence to be an interface advantage, in that the system is 'guiding' the user." (2000: 47). This design inclination is much, much stronger in voice systems than in graphic systems, because of engineering paranoia about recognition problems. As his disdainful quotes around *guiding* suggest, though, Raskin has very little time for this design sensibility, and he is one of the most tenacious user-advocates in the business. His advice should be followed, which comes down to: do not hinder user initiative.

Summary

Dialogues consist not only of elements that are directly motivated by the underlying task (like questions, answers, instructions), but also of elements motivated by the communicative task and controlling aspects that require attention in communication more generally, such as monitoring the attention, ensuring the correctness of understanding, taking turns, repairing communicative failures, etc.
   — Harry Bunt

While it may appear we are moving from one modular topic to another in our discussion of the language research useful for the design of voice interactions, it is important to notice that the modules overlap and interpenetrate in significant ways. For instance, repairs are ways to ensure that the quantity, quality, and clarity of the information is sufficient for task completion; that is, they are ways to ensure that Grice's maxims are followed. Indeed,

---

that the Walker papers found both ease of use and user satisfaction ratings improved for the mixed-initiative condition with experience. (Ronnie Smith, one of the pioneers of speech-system initiative research, argues that increased user-experience necessitates initiative sharing systems "as they gain expertise, provided the computer allows it. The ability to yield the initiative as users gain experience is essential if a dialog system is to be useful in practical applications involving repeat users.)" Walker et al. (1998) suggest that the poorer performance was at least in part the result of "user confusion about their available options" (which in turn fed the satisfaction ratings), and it is a fair bet that reducing the confusion, through experience or more intuitive design, or both, would inversely improve the satisfaction ratings. In any case, the results are certainly preliminary, and at best the question remains open. It may be that the results say nothing about different initiative styles for voice interaction design, but just about different receptions for the specific system- and mixed-initiative interfaces they tested. Chu-Carroll (2000) reports better satisfaction ratings with more "adaptive" mixed-initiative designs, which suggests that not all mixed initiative designs are created equal.

groundskeeping and flow regulating practices in general are motivated by the spirit of cooperation that Grice says governs conversation.

We cannot decide beforehand, however, which specific human–human conversational resources will be useful for human–computer interaction, and which won't, in any given voice interface. Designers always need to consider closely the particulars of the discourse domain, the tasks, and the agents. Cooperation manifests in different ways for different registers.

What we do know with confidence is which resource *categories* are useful for designing and analyzing voice interfaces. In this chapter we have discussed the categories related to **dialogue management** dialogue acts, which come in two flavors, **flow-regulating** acts and **groundskeeping** acts.

Flow-regulating dialogue acts are the ones that initiate and terminate dialogues, and exchanges within dialogues, which ensure orderly **turn** taking, and which generally govern the dialogic roles the agents play. The single most important conceptual notion for these management tasks is that dialogue acts exert reactive pressure on the agents to respond in limited ways — that is, that utterances come in **dialogic pairs** — such that a question calls for an answer, an offer calls for an acceptance or a declination, a turn assignment calls for a turn taking or turn declining dialogue act, and so on. In addition to traveling in pairs — matching acts by different agents — dialogue acts might also appear in clusters, and they nearly always amalgamate. A **cluster** might be a response to a previous initiative, followed by an initiative to a subsequent response, for instance ("The wife and kids are fine, thanks. And yours?"), whereas an **amalgam** is two acts united in the same utterance ("And yours?" is both an inquiry and a turn assignment, for example, since the hearer is called upon to take the floor in order to reply).

Groundskeeping dialogue acts enable conversational agents to build mutual understandings about the items of discourse (people, situations, parts, colors, sizes . . .), reassuring and correcting each other until they have calibrated their understandings sufficiently. Critically important are **feedback**, which concerns positive and negative indications about one agent's understanding of the other's utterances, and **repair**, which provides agents with the resources to overcome misunderstandings.

We also discussed **initiative** in this chapter, a concept related to which agent is most strongly influencing the flow of the dialogue, who is asking the questions, giving the orders, controlling the topic, and so on. In particular, we discussed an initiative continuum, noticing that while **fixed-initiative** systems (in particular, those in which it is fixed with the system) are much easier to manage and design, that **mixed-initiative** systems, in which the dialogue control shifts collaboratively as the task proceeds, are important for user satisfaction.

Glue

A dialogue is a coherent exchange of utterances.
— Joris Hulstijn

Dialogues, like all genres of discourse, are *about* something (or, often, multiple somethings). They have topics. The bits and pieces of conversations we've looked at are variously about such things as telephone numbers, employment centers, weather services, guns, addresses, email, cats, and mats. That is what makes them conversations, not babble. Conversations are coherent. If you see two people talking incoherently, even if they are following good turn management etiquette, and deploying dialogue acts appropriately, it would be difficult to call their activity a conversation. If each person's utterances are internally coherent, perhaps something like *interlacing monologues* would capture it; if there is not even that level of congruence, *gibberish*. Without topic coherence, they are not conversing.

The notion of topic is notoriously difficult to pin down, but we all have an intuitive sense of its applicability to discourse. Or, at any rate, we know that it makes sense on some level to consider the subjects, the events, the themes, the ideas, the people — the *contents* — of dialogues. Some dialogues, it's true, can be rather aimless, and participants may be hard pressed afterwards to answer the question, "What did you talk about?" But, to the extent that such a question is answerable ("baseball," "labor relations," "the ineffability of being") and perhaps refinable ("the Blue Jays' starting rotation," "the way City Hall treats maintenance workers," "Zen Buddhism"), lo, there is a topic.

In order for there to be an answer to "what did you talk about," the bits and pieces of the conversation have to consolidate and function collectively, not just on the syntactic level of a word string, or the dialogic-pair level of two matched utterances, but over the course of the entire dialogue. If they don't, communication breaks down, often fatally.

All communicative breakdowns in speech system interactions, whether they originate with the recognition engine, or the natural language processor, or the caller's lapses or misconstruals — *all* breakdowns — are fundamentally failures of the words to consolidate (see, e.g., Ardissono, 1998). In extreme breakdowns, the word-assemblage amounts to gibberish, a random heap of blather, a lexical pile:

> *that has sweat what you have a minus for the one year before that you you look have all along are right you feel that has . . .*
>   (Cooper er al., 2001: 3)

You may remember this incoherent snippet as a chunk of the bewildering speech-recognition output from Chapter 3; no matter, it's just a mess of words. There is no sense in which *sweat* and *minus* and *year* and *look* involve any mutual concepts here, no clear way in which we can say what this bizarre little text is about.

What's missing in a lexical pile like this is almost everything that helps words to function together. The words of a dialogue (or any other form of discourse) consolidate to function collectively when there are sufficient semantic, thematic, and perspectival overlaps between utterances, turns, and exchanges — that is to say, when they *cohere.*

Topic Management: Coherence and Cohesion

> *Coherence and cohesion are partial synonyms, but coherence is used chiefly in a figurative sense meaning "logically consistent, understandable," whereas cohesion is . . . simply "a sticking together."*
>   — Kenneth G. Wilson

When people traffic successfully in discourse, the words cohere, and a coherent dialogue is barely noticeable simply because it *is* coherent; *incoherence* is what sticks out like an embarrassing relative. Even crazy stretches of language like the ones found in *Alice Through the Looking Glass*, for all their apparent anarchy, are largely coherent:

> *But I was thinking of a plan*
> *To dye one's whiskers green,*
> *And always use so large a fan*
> *That they could not be seen.*
>   (Carroll, 1962: 104)

Dye and *green* cohere because they both implicate the notion of color; *plans* fits nicely with *thinking*; among the properties fans can have is largeness; and so on. If the poem, "Haddock's Eyes," didn't have this level of coherence, it wouldn't even be amusing, just a baffling heap of blather like "that has sweat what you have a minus . . ."

Coherence operates in alliance with another, very similar but differently angled, notion in discourse, *cohesion*. Through one of those fairly common perversities of linguistic history, *coherence* and *cohesion* in fact both derive from the same Latin term, *cohaero*, "to adhere," but they differ in what they glue.

Coherence is a conceptual adhesive. It glues words and utterances together into discourses on the basis of mutually implicated meanings. Cohesion is a structural adhesive. It glues together not the ideas, but the physical words and phrases; it operates at the level that Halliday and Hasan (1976) call the "texture" of discourse. Coherence glues content, cohesion form. The first is a semantico-pragmatic notion, the second lexico-syntactic. Coherence is manifest when words (therefore concepts) like *dye* and *green* show up together; cohesion is manifest when any two discourse segments are linked by a word like *but* or *and* or *that*.

Both notions are continuous, rather than binary. Discourses are more-or-less coherent, more-or-less cohesive. So, for instance, example 1a is coherent, 1b less so; neither of them is particularly cohesive, though examples 1c and 1d are.

1a   I love to collect classic automobiles. My favorite car is my 1899 Duryea.

1b   I love to collect classic automobiles. My favorite car is my 1993 Toyota.

1c   I love to collect classic automobiles, and my favorite car is my 1993 Toyota.

1d   I love to collect classic automobiles, but my favorite car is my 1993 Toyota.

(Mann and Thompson, 1987: 57; actually,
they used "1973 Toyota," but that probably *is* a classic now)

The sentences of 1a are coherent because *1899 Duryea* seems to fit the conceptual bill for a "classic automobile;" even if you don't know what a Duryea is (I don't), you're prepared to accept it (perhaps temporarily, until you can consult an authoritative source about it) as satisfying the designation of "classic automobile." The accompanying date supports this acceptance, of course, by falling into the appropriate range for *classic* in the company of *automobile*. But the sentences of 1b cohere much less well, because *1993 Toyota* does not satisfy that designation (it is relatively noncoherent, but the word <u>in</u>coherent does not apply).

Neither sentence pair — not 1a nor 1b — displays much cohesiveness, however, because there are only minimal lexical cues (only the referential overlap of *car* and *automobiles*), and no syntactic cues, that link the paired sentences.

In contrast, sentences 1c and 1d *are* cohesive, because of the lexical and syntactic linking effected by the punctuation and the conjunctions, although the conjunctions do different work. Sentence 1c is no more coherent than the sentence-pair of 1b, because *1993 Toyota* still does not fit the conceptual bill for a "classic automobile," which the use of *and*

Sentences	Coherent		Cohesive	
	yes/no	**Explanation**	**yes/no**	**Explanation**
1a	yes	1899 Duryea satisfies "classic automobile"	no	No linking term
1b	no	1993 Toyota does not satisfy "classic automobile"	no	No linking term
1c	no	1993 Toyota does not satisfy "classic automobile;" *and* implies satisfaction	yes	One linking term (*and*)
1d	yes	1993 Toyota does not satisfy "classic automobile;" *but* implies a contrast, and therefore, *non*-satisfaction	yes	One linking term (*but*)

TABLE 6.1   Discourse coherence and cohesion

suggests it does. Sentence 1d, on the other hand, is coherent exactly because *1993 Toyota* does not fit the bill, just as the use of *but* suggests.[1]

The examples of 1, then, represent the logically possible combinations of coherence and cohesion: 1a is coherent, but not cohesive; 1b is neither coherent nor cohesive; 1c is noncoherent but cohesive; 1d is both coherent and cohesive. Keeping in mind that these are relative and not binary notions (something that the yes/no designation misrepresents), this sketch of coherence and cohesion relations is served up in Table 6.1.

It is important to realize that while coherence is necessary, and cohesion is significant, for effective discourse construction, both are a long way from sufficient for that task. Take this dialogue, for instance:

1: Humans are very good at imposing coherence, which makes it difficult to find a text that is truly incoherent. So, for instance, a reader of 1b or 1c (perhaps you) might impose a coherence that brought "1993 Toyota" somehow into the sphere of "classic automobile" by way of some notion such as speaker credibility — for instance, that the writer of those sentences *really believes* the 1993 Toyota is a classic automobile (and therefore doesn't know much about the subject area); or that the 1993 Toyota *really is* a classic automobile (and the reader now has to revise his domain-assignments); or perhaps that the writer is trying to trick, or just spoof, the hearer with that assignment. Whatever. Coherence takes place in the hearer/reader's interpretation of a text; the best the speaker/writer can do is forecast possible interpretations and steer toward (projected) coherent ones, by juxtaposition or cue. Coherence relations are really just labels for interpretive moves; as Moeschler (2001) puts it, "discourse coherence is basically an interpretation problem."

Referential	The same entities and/or actions recur in the discourse
Relational	Two discourse constituents stand in specific conceptual relations to each other

TABLE 6.2 Dimensions of discourse coherence

T_1 System Customer number 4. Is that correct?

T_2 User No, customer number 3.

T_3 System Customer number 10 does not exist. Please state your customer number.

T_4 User Customer number 3.

T_5 System Customer number 3.

T_6 User Yes.

(Bernsen et al., 1996: 13)

It would be hard to find a more coherent and cohesive exchange, but if I were the caller I would almost rather have gibberish. It's not the recognition errors that compromise this exchange so badly. Certainly they don't compromise the coherence or the cohesion; by concentrating the exchange so closely on one topic, they actually enhance coherence, and two identical contiguous turns (T_4 and T_5) are the epitome of cohesion. It's the triviality of the coherence and cohesion that damns this exchange. There is only one technique of cohesion in the customer-number-3 exchange, and pretty much the same technique of coherence, and repetition is very wearying. What voice interaction design requires is access to the full range of cohesion and coherence options, which I outline in this section.

Effective discourses, then, are networks of coherent and cohesive text elements (for dialogues, we call those elements *utterances*) that consolidate into coherent structures. But, of course, that's not quite all there is to the story; there are several dimensions of cohesion and coherence that play roles in how discourses hang together.

There are two dimensions on which to peg degrees of coherence: *referential* and *relational*. Cohesion not only has two subdivisions as well — *associative* and *connective* — those two work hand-in-glove with the two forms of coherence. These dimensions, along with rough accounts of their functioning (to be fleshed out below), are arrayed in Tables 6.2 and 6.3.[2]

2: The coherence/cohesion scheme represented in Tables 6.2 and 6.3, and discussed throughout this section, is very substantially influenced by a number of scholars, but the configuration, along with several of the labels, are mine.

	Methods	Coherence support
Associative	repetition, anaphora, ellipsis	referential
Connective	coordinators, subordinators	relational

TABLE 6.3 Dimensions of discourse cohesion

The reason that you may be a bit lost right now, and perhaps annoyed with me, is that I haven't given you very much in this section beyond a big clump of abstract and unrealized terms. But they will serve you well, I hope, as the handles for important concepts, which we will now take up serially. As the handles attach to the concepts, we will be building coherence. I have labeled all these handle-and-concept divisions earnestly below, to build them cohesively. Our target is a consolidated discussion of the ways in which words consolidate.

Coherence

> *Under the category of Relation, I place a single maxim, namely, "Be relevant."*
> — H. Paul Grice

Coherence comes in two varieties, referential and relational.
Take these two pieces, a pair of well-formed sentences (adapted from Hobbs, 1979):

2   Oriana took a train from Paris to Istanbul. She loves spinach.

They are coherent, sort of; they are coherent to the extent that they are both about Oriana. Specifically, they are *referentially* coherent, because both sentences (at least under normal assumptions for *she*) refer to the same entity in the world, Oriana. But coherence is a graded notion, and example 2 is a less immediately coherent pair of sentences than example 3:

3   Oriana took a train from Paris to Istanbul. She loves Byzantine architecture.

The two sentences in example 3 function more coherently on first pass than the two in example 2, because "Byzantine architecture" serves a much clearer role as an object of Oriana's love with respect to the going-to-Istanbul prior sentence than "spinach." That is, the sentences of 3 have a clear relation holding between them — they are *relationally* more coherent — because the second sentence reveals a clear purpose for the trip (to experience Byzantine architecture). Now, example 2 is not *in*coherent, and the right context could make it at least equally coherent with 3 (say, a rare variety of spinach grown in the greenhouses of Antalya is only available in the markets of Istanbul; voila, the concept in the second sentence can now serve as a purpose for activity described in the first). But you have to work a little harder to see a connection, if one isn't provided.

Discourse is coherent to the extent that readers and hearers can make sense out of it on the basis of what they already know; it relies on the background of the utterances. Referential coherence, the variety we will now take up very shortly, is the easiest to manage because it depends on a small and rigid set of correspondences. For instance, English pronouns agree in person, number, and sometimes gender, with their corresponding nouns. Change pronoun number, for instance, and our pair of sentences is jarred immediately out of its coherence groove:

4   Oriana took a train from Paris to Istanbul. They love Byzantine architecture.

Relational coherence, the other variety of coherence, which we will postpone for a while, is considerably trickier to manage, because it depends not only on a larger and spongier set of correspondences but also on a fair amount of domain knowledge.

Referential Coherence

Costello	*Well then who's on first?*
Abbott	*Yes.*
Costello	*I mean the fellow's name.*
Abbott	*Who.*
Costello	*The guy on first.*
Abbott	*Who.*
Costello	*The first baseman.*
Abbott	*Who.*
Costello	*The guy playing . . .*
Abbott	*Who is on first!*

— Bud Abbott and Lou Costello

A dialogue is referentially coherent to the extent that expressions refer to the same entities. The Who's-on-first routine violates referential coherence because Abbott uses *Who* as a proper noun, referring to a specific individual, while Costello uses *who* as a constituent-question pronoun, with an indeterminate reference he is trying unhappily to have satisfied. (And, if you ever want to illustrate how pedantry can kill a joke, you have my permission to quote that last sentence.)

Here's a more mundane breakdown in referential coherence:

User   Give me the price for AT&T.

System   Hewlett-Packard is at 83 3/8, up 2 3/8 since yesterday.

(Levow, 1999)

The user wants to reference the stock of the company American Telephone and Telegraph with the phonetic pattern that is transcribed by Levow as *AT&T*; the system thinks that pattern references the name of another stock altogether (perhaps in its acoustic pattern bank as "HP"). Coherence is not absent from this pair of utterances, otherwise they couldn't be juxtaposed sensibly. The shared deployment of the concept of price, the governing theme of the stock market, and the linked perspectives of information-seeker and information-provider all give the exchange coherence. But it breaks down on the matter of reference for one acoustic pattern (something, in fact, which is only diagnosable because the other elements of coherence are present). The exchange runs into trouble because of a recognition failure.

Recall (or look back at) the longer dialogue this exchange comes from, discussed in the last chapter. It falls apart completely — from a logic failure, from not doing an inter-utterance comparison, from not attending carefully enough to grounding, and from not carrying out the appropriate inferences (all of which deafens the system to the fact that the dialogue is going desperately wrong for the user). *All* of these deficiencies might have been solved by attending to a notion of referential coherence that reached further back than the last utterance. The system gives the price for Hewlett-Packard twice, without cluing in to the fact that it is rare for someone to ask for the same stock price twice in three turns, and, even more obliviously, by failing to notice that three concurrent requests employ the "same" (that is, highly similar) phonetic patterns. (And, the real insult-to-injury comes with not even allowing a disambiguation strategy to the user, who offers a much longer, and there-fore less ambiguous, pattern to reference the same stock).

Here's a similar referential ambiguity, much better handled:

T$_1$ MailCall   Welcome to MailCall. Who's this?

T$_2$ User   Matt Marx.

T$_3$ MailCall   Nat Parker, correct?

T$_4$ User   No, Matt Marx.

T$_5$ MailCall   I thought I heard "Nat Parker" again, but you just said that. Is that right?

(Marx and Schmandt, 1994)

MailCall works to ensure that the dialogue enjoys genuine referential coherence by signaling that it knows there is something strange about two identical utterances in such circumstances. The stock system, on the other hand, obliviously chugs along.

One more example of successful attention to referential coherence, which also illus-trates how it can work in multiple, overlapping ways:

T$_1$ Caller   Do I have any emails from Anna about elephants?

T$_2$ MailSec   No, you have no emails from Anna Cordon about elephants, but you have one message from Keith Preston about elephants.

(Williams, 1996, cited in Churcher et al., 1997)

The first clause of MailSec's response (T_2) references four entities evoked by T_1. The second clause goes on to reference the caller, email (with a partial synonym), and elephants each a second time, introducing a new referent in the bargain, Keith Preston, all in the process of creating a distinct new entity into the discourse (one message from Keith Preston about elephants).

The exchange also shows an intelligent use of search criteria, because MailSec is stretching itself to ensure a maximally satisfactory, coherent answer. The answer "No" would be coherent in this context, as would be "You have no emails from Anna Cordon about elephants." But MailSec pursues referential coherence further, by searching for any email at all about elephants (maybe the caller doesn't care, and maybe there are thirty messages from various people about elephants — in which case MailSec should draw other inferences about the query — but this is a reasonable guess at constructing a turn that builds the dialogue on referential coherence).

Ensuring referential coherence is a critical aspect of groundskeeping for speech systems. For instance, if someone is shopping for a digital camera and wants to review several models with the agent, he needs to be confident that when he asks for the price or the specs on "the Sony," it's the same model the system actually tells him about. What's crucial, in fact, about dialogic referential coherence, is just verified agreement. It doesn't so much matter what something is called, as long as both parties use that term.

Susan Brennan's (1998) work on how speakers develop conceptual pacts for referring expressions is especially illustrative in this regard. Taking Rorschach-like graphics, she elicited descriptions from a group of speakers getting a multiplicity of responses. The one in Figure 6.1, for instance, drew the following referring phrases:

5   the bird diving straight down

    the candle

    the rocket ship

    the angel upside down with sleeves

    the Olympic torch

    the Canada symbol

    shapes on top of shapes

    the airplane flying straight down

    the anchor

    a bat

    the man jumping in the air with bell bottoms on

FIGURE 6.1   Abstract shape from Brennan (1998), based on research from Stellman and
Brennan (1993)

In conversation, however, it doesn't much matter whether you call some abstract piece
of geometry a *candle*, a *bat*, or a *man jumping in the air with bell bottoms on*, so long as
both parties form a pact and agree to the reference. Referential coherence in conversations,
in short, is often a product of negotiation.

Relational Coherence

> *A week has seven days. Every day I feed my cat. Cats have four legs. The cat is on the mat.*
> Mat *has three letters.*
> — Nils Erik Enkvist

Effecting, then maintaining, referential coherence requires solid design; it especially
requires good groundskeeping. But it is a relatively uniform notion. However tricky it might
be to build systems that engage in referentially coherent dialogues, the notion itself is conceptually straightforward: talking together about the same entity for a given duration.

Relational coherence is more subtle. It involves a grab bag of notions concerning the
notional linkages among utterances (depending on the researcher, counts of these linkage
types range between 2 and over 100 (Knott and Sanders, 1979; Mann and Thompson, 1988,
catalogue 23).[3] There is an appendix to this chapter that defines and exemplifies some of

---

3: This is an opportune time to point out that my treatment of relational coherence here follows closely, if opportunistically, the Rhetorical Structure Theory program of Mann and Thompson (especially 1987, 1988; in concert

the more significant relations, but let's take up a few relations now, starting with this brief exchange:

T₁ Fix-it This is the circuit fix-it shop. How may I help you?

T₂ User I want to fix this rs111.

T₃ Fix-it I am familiar with that circuit. The LED is supposed to be displaying alternately-flashing one and seven.

T₄ User The LED is displaying alternately-flashing one and the top corner of a seven.

(Smith and Hipp, 1994: 6)

The exchange is coherent, both within-turn and between-turns. Any English speaker can see that right away. It has referential coherence (Fix-it uses *I* twice of itself, Fix-it's *you* and User's *I* are coreferential, as are *rs111* and *that circuit*, *The LED* is used congruently by both agents, and so on).

But the coherence of most sensible discourse goes beyond harmonies of reference. People also understand language by seeing networks of motive and development and elaboration among utterances. The epigram at the head of this section by Enkvist illustrates this network clearly, by absence; it is referentially coherent, but, beyond that, it's completely screwy. The words and the sentences have meaning, and there may even be a way in which we can say the whole text has meaning. But, if so, it is an additive meaning: five independent propositions, none building on or linked to any other. There is no collective meaning. It is incoherent, the sort of thing you might hear muttered at you by an unwashed and unkempt stranger on a city street. The Fix-it/User exchange, on the other hand, has collective meaning; the sentences are *related* to one another. The two sentences of T₁ are coherent because the first one provides the background with which to interpret the offer of help expressed by the second. The two sentences of T₃ are coherent for a similar reason: the first sentence supplies a justification for crediting the second sentence.

---

with the articles and supplements on Mann's web site, http://www.sil.org/~mannb/rst/). The term *coherence relation* comes from Hobbs (1983), however, as does some of the research; points which Mann and Thompson apparently prefer not to notice. I fiddle with Mann and Thompson's terminology slightly, mostly for reasons of cohesion. For instance, most of their labels are nouns, but *Justify* is a verb; I change it to (the noun) *Justification*. Strictly speaking, too, Mann and Thompson investigate the relations between "text spans" and confine their work to "written monologue," noting their framework "has not yet been expanded to describe dialogue or multilogue" (1988: 244). Their work has been expanded in those directions by others, however, and has been quite influential in computational linguistics generally. See, in particular, Stent and Allen (2000) for the application of relational coherence to dialogue systems.

The relevant coherence relations are called, not coincidentally, *Background* and *Justification*.

As with many coherence relations, these two are asymmetrical. One element is more central than the other (the one for which the background or the justification is being supplied): it is the *nucleus* of the relation. The other element is less central, functioning largely (but not necessarily exclusively) to support the nucleus in some way (in the way signaled by the name of the relation): it is the *satellite*.

Consider another exchange, from a dialogue with Johanna Moore's PEA, exemplifying the coherence relation, Purpose, one the most important linkages between pieces of discourse:

Purpose

T_1 PEA   You should replace (SETQ X 1) with (SETF X 1).

T_2 User   Why?

T_3 PEA   I'm trying to enhance the maintainability of the program by applying transformations that enhance maintainability. SETQ-TO-SETF is a transformation that enhances maintainability.

(Moore, 1995: 193)

PEA recommends a course of action to the user in the nucleus, T_1; it's a commissive dialogue act, a piece of advice. But users — they're people after all — don't always like to do things mindlessly and this one asks for a reason. PEA supplies it in the satellite, T_3, giving the Purpose behind its directive. (The appendix provides a more detailed analysis of this exchange in relational coherence terms.)

Contrast the relation of Enablement, in which the information in the satellite provides the ability to perform the action encoded by the nucleus:

Enablement

T_1 Danluft   Hello, this is the Danluft reservation service for domestic flights. Do you know how to use this system?

T_2 Caller   Nope, I don't. No.

T_3 Danluft   The system can reserve tickets for Danish domestic flights. You use it by answering the system's questions. In addition, you may use the two special commands "repeat" and "change" to have the most recent information repeated or changed.

(Bernsen et al., 1998: 34)

This example is slightly different than the previous ones, in that there are *two* satellites cohering with the nucleus. All of the elements are in T₃. The nucleus is "The system can reserve tickets for Danish domestic flights." One satellite gives general enabling information ("You use it by answering the system's questions"), the other gives more specific enabling instructions, for manipulating the system directly (and, briefly, assuming the initiative), if the general instructions go awry.

Notice that Danluft might, in fact, have started with T₃ (perhaps with a "hello" thrown in for sociability, and a change of "The" to "This" for better referential coherence). So, what's T₁ doing there? Part of it is sociability, and part of it is a screening mechanism, to allow people who are already familiar with Danluft the option of getting right along with the task. But it is also serving satellite functions for the informational nucleus of the exchange. Danluft's T₁, that is, serves the general orienting role we saw a page or so earlier, *Background*.

Some coherence relations are so strongly required in dialogue (largely by way of the Cooperative Principle) that their absence is boorish. Contrast the alternate responses (T₂ₐ, T₂ᵦ) to the request in T₁:

T₁   Caller   Could I please have my balance?

T₂ₐ   Bank   No.

T₂ᵦ   Bank   No, I'm sorry. That part of our system seems to be down at the moment.

The response in T₂ᵦ is the better one, clearly, because of its adoption of the coherence relation, Justification. Perhaps more detail might be provided ("the lines are down"), but *some* justification needs to be provided for the system's refusal or it will be interpreted as flat-out rude.

Relational coherence, like referential coherence, can work within a single utterance or turn, and it can work between turns, to unify stretches of dialogue. Also like referential coherence, which can evoke multiple referents in overlapping ways, several coherence relations can be in play collectively. For instance, the coherence of this greeting and response exchange depends, in varying degrees, on three relations:

T₁   United   Welcome to United Airlines flight information system. I'll be able to help you get information on all United, United Express, and United Shuttle flights. Enter or say the flight number, or say "I don't know" if you're unsure.

T₂   Caller   Flight 455.

(Kotelly, 2003: 154)

We can separate out the coherence units in this exchange as outlined in Table 6.4. Now, T₁ᵦ is an Elaboration of T₁ₐ, the fact that the system can provide information on three types

T$_1$	a		Welcome to United Airlines flight information system
	b		I'll be able to help you get information on all United, United Express, and United Shuttle flights
	c		Enter or say the flight number
	d	i	say "I don't know"
		ii	if you're unsure
T$_2$			Flight 455

TABLE 6.4   A breakdown of the coherence units for the United Airlines exchange (on page X)

of United flights adding specifics to the functioning of United Airlines information system. Those two work together as satellites in the Background relation, providing the context in which to interpret T$_{1c}$ and T$_{1d}$. (More generally, of course, T$_{1a}$ provides the defining background for the entire dialogue, with elaboration from T$_{1b}$, but it is their local coherence relation with T$_{1c}$ and T$_{1d}$ that is relevant for this dialogue snippet.) Units T$_{1c}$ and T$_{1d}$, for their part, are what Enable T$_2$. But T$_{1d}$ includes another relation, Condition: the clause T$_{1d(ii)}$ is the condition under which the caller would perform the action described in T$_{1d(i)}$. Explicating multiple interactions of this sort is always tough to follow in prose (and I, for one, wouldn't have a hope in hell of following it by speech). Graphically, however, you can see how the elements of this little exchange relate to each other and build a coherent set of utterances: Figure 6.2 represents these relations visually (using the formalism of Daniel Marcu; see, e.g., his 2000).

Coherence is the way in which bits of significance link with other bits of significance to form a meaningful discourse. It is usually sufficient on its own to let readers and hearers follow the contours of that discourse. But coherence on its own is not always elegant or efficient, and completely on its own, it renders the discourse stark, if not downright alien. Language has a surface structure as well as a conceptual understructure, and the surface needs to reinforce and complete the frame of significance. Discourse needs cohesion.

Cohesion

The discourse structure of a conversation is . . . reinforced by the cohesion, which explicitly ties together the related parts.
— Michael Halliday and Ruqaiya Hasan

Cohesion works very closely with coherence. It has two subdivisions: *associative* and *connective*.

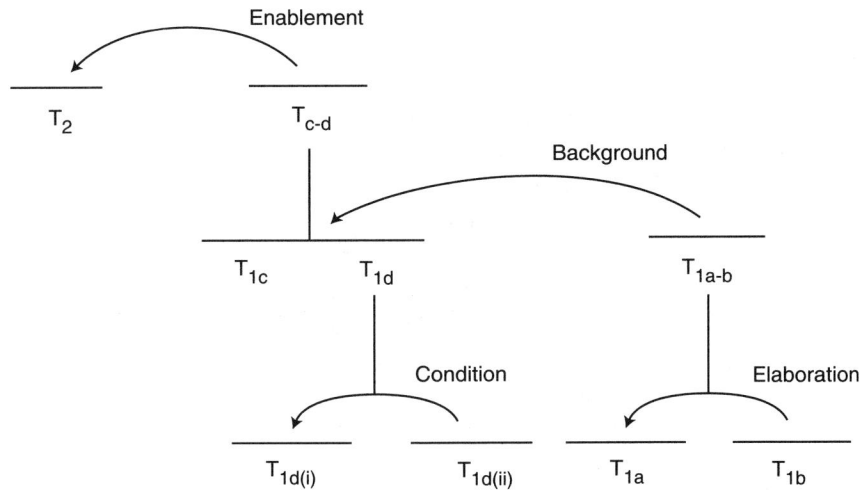

FIGURE 6.2   Graphic representation of the coherence relations in a dialogue from Kotelly (2003: 154), as charted in Table 6.4

The first reinforces referential coherence; the second reinforces relational coherence. Associative cohesion, more precisely, concerns the *forms* of reference, and connective cohesion provides the lexical signposts for coherence relations.

Associative Cohesion

> *Two guys are fishing along the coast one morning, with miserable luck. They see another guy with a boat full of fish and ask him where they're biting. He says "just up the mouth of that river, as soon as the water turns fresh." So off they go. A little distance up the river, one guy asks the other to dip a bucket overboard, bring it back up, and taste the water. He does. "Salty," he says. A little bit later the first guy says, "Try it again." "Still salty" is the verdict.*
>
> *They continue on that way, stopping and testing every mile or so all day long — "try it again," "still salty;" "try it again," "still salty" — until the day is getting dark and they find themselves in a weedy little swamp. "Try it again," says the first guy, wearily. The second guy tastes the water again, and says "still salty." The first guy says, "This is ridiculous. We've been heading up this river all day, and we must be almost at the end. We're never going to find that spot."*
>
> *"Ya," says the second guy, "and the bucket's almost empty now, too."*
>
> — Anonymous

There are three core mechanisms of associative cohesion, all of which we've seen in various forms already — repetition, anaphora, and ellipsis. The bucket's-almost-empty joke turns on the second one, anaphora, going wrong.

Jokes are usually highly coherent little chunks of discourse, but they often turn on a moment of incoherence, where someone's actions or words prove to be out of synch with someone else's (and/or the listener's), just in time for the punch line. In this case, the first guy and the second guy have distinct, and therefore mutually incoherent, ideas of what "it" refers to in this context. They exhibit a breakdown in referential coherence.

Referential coherence is aided and abetted by associative cohesion, by the use of words (or in the case of ellipsis, word-gaps) that chain to an initial reference. As the joke shows, *in*coherence can be aided and abetted by cohesion, as well; or, perhaps, a sly form of incohesion. We throw our anaphoric lot in with the first guy in this joke. Something is wrong, we begin to suspect, but surely not the association of *do it* with the action, "get a new sample and taste that sample." We coast along until the end, way-the-hell-and-gone up the river, and we are ready to curse fate or the other fisherman along with our two heroes, when the second guy is revealed as a bonehead. But, in fact, the second guy is behaving pretty logically, very much like a computer might, and his association of *do it* with "test the (original and only) water sample" is both more coherent and more cohesive than ours (or, at least than mine, when I first heard the joke). Maybe we should lay the blame on the first guy, for the low quality assurance standards he has for information (for incredible laxness in what we will later be calling *grounding criteria*). A simple "is that a new sample?", if not a visual check, could have saved a way-the-hell-and-gone trip up the river, and possibly filled the boat in the bargain.

The simplest (and most monotonous) way to reinforce reference is just to use exactly the same patterns to reference the same entities or concepts, using the dialogue acts of repeating (for one's own utterances) or echo (for one's partner's utterances). This procedure is frequently necessary in voice systems, especially along with flow-regulating dialogue acts, like identification or verification:

T_1 Caller   Hello. Could I speak to Mrs. Salmagundi?

T_2 Operator   Mrs.?

T_3 Caller   Salmagundi.

T_4 Operator   Salmagundi?

T_5 Caller   Yes

T_6 Operator   Please hold on . . .

(Balkanski and Huraultl-Plantet, 2000: 921)

In T_2 there is a partial-echo of a sort quite common in dialogue: the part that is heard (or as much of it as is relevant) gets repeated, with a trailing, rising intonation to request that the missing part be supplied. In T_4 there is another echo (with the same intonation pattern, this time seeking confirmation or correction). This kind of prosodic control is problematic with synthetic voices, and predicting which elements might need this treatment in a recorded-voice system can also be problematic. But in both cases it is worth the effort: verbatim echoing can often be the best way to groundskeep reference when the stakes are high.

Repetition can also be a good way to keep structures parallel, and parallel structures are generally easier to process. For instance, the difference between sentences 6a and 6b is lexically negligible (the repetition of one monosyllabic, unstressed function word), but that small repetition affects prosody and effects parallelism and reduces the user's processing burden:

6a   Do you want me to wait or end the call?

   b   Do you want me to wait or to end the call?

Outside of literal repetition, the road to associative cohesion is through anaphora (the use of proforms) and ellipsis (the use of ommision) — which is actually a bit weird. Anaphora and ellipsis are universally seen by discourse scholars as strategies of cohesion, but in fact they *reduce* cohesion and introduce uncertainty. Two identical noun phrases are completely cohesive; a noun phrase and a pronoun, or a noun phrase and a gap, are less cohesive. Here is a classic example:

7   Ten theoretical linguists and eight language engineers were present. They were constantly heckling them.

Cohesion is reduced to the point where (referential) coherence is affected, since it isn't at all clear who was constantly heckling whom. Contrast full or partial repetition, which brings cohesion and coherence along with it:

8   Ten theoretical linguists and eight language engineers were present. The linguists were constantly heckling the engineers.

On the other hand, a sacrifice in cohesion (so long as coherence can be maintained) is well worth the price, because of the gains in fluency and naturalness. Compare sentences 9b and 9c in the context of 9a:

9a   Ten theoretical linguists attended the news conference given by eight language engineers.

9b   They were constantly heckling them.

9c   The linguists were constantly heckling the engineers.

In the context of sentence 9a, 9c becomes tedious, because of the background, our cultural understanding of who are generally the hecklers, and who are the hecklees, at a news conference.

Utterances of the 9c sort are deadly, but not uncommon, in voice interfaces. Michael Cohen and his colleagues offer a nice example of how pronouns can rescue a brutal dialogue. Check out this dialogue snippet:

System   You have five bookmarks. Here's the first bookmark. . . . Next
         bookmark. . . . That was the last bookmark.

User   Delete a bookmark.

System   Which bookmark would you like to delete?

Now, compare this version:

System   You have five bookmarks. Here's the first one. . . . Next one. . . . That was the
         last one.

User   Delete a bookmark.

System   Which one would you like to delete?

(Cohen, Giangola, and Blalogh, 2004: 138–139).

Anaphora and ellipsis reduce cohesion, but they do so in a controlled way that brings fluency and naturalness to the interface. They also pick up the pace of a dialogue, since the fuller phrases take longer to utter. Take this tapering example, for instance:

FedEx   What's the approximate length of your package, in inches? Please round off
         to the nearest inch.

Caller   Eleven inches.

FedEx   OK, what's the approximate width, in inches?

Caller   Twelve inches.

FedEx   What's the height?

(Kotelly, 2003: 167)

Blade Kotelly calls a sequence of system utterances with this sort of escalating ellipsis, *scaffolding utterances* (or *prompts*), because the utterances build on each other; in doing so, they make for a much more rapid, easier exchange.

As the most fluid mechanisms of cohesion, anaphora and ellipsis are the primary methods of reinforcing referential coherence. They help build chains of reference. Compare the real TES (Train Enquiry Service) example, 10a, with the hypothetical version that follows it, 10b:

10a   There is a train from Torino Porta Nuova to Milano Centrale. It leaves at 7:10 p.m. and arrives at 8:55 p.m. Do you want more detailed information about it?

(Danieli and Gerbino, 1996)

10b   There is a train from Torino Porta Nuova to Milano Centrale. Departure is at 7:10 p.m. and arrival is at 8:55 p.m. Do you want more detailed information?

Example 10b is still coherent, but without the anaphora ("It leaves" and "about it") and the ellipsis ("Ø arrives") it is more abrupt, more vague, and somewhat stuffy. In little doses like 10b, this sort of bureaucratese can be tolerable, but over the course of a dialogue it can become oppressive and wearying; the hearer has to work just a little bit harder inferentially threading the information together. With 10a, the *train-it-Ø* chain of reference makes TES's turn more cohesive for the hearer.

Connective Cohesion

The continuity that is provided by cohesion consists, in the most general terms, in expressing at each stage in the discourse the points of contact with what has gone before.
— Michael Halliday and Ruqaiya Hasan

Associative cohesion is the lexico-semantic reinforcement of referential coherence. Connective cohesion has a similar role with respect to relational coherence. Connective words achieve cohesion by pointing the hearer towards the affiliations the speaker wants her to make between and among utterances. They work closely with relational coherence, but are logically independent of it. Recall two of our classic-automobile sentences:

1c   I love to collect classic automobiles, and my favorite car is my 1993 Toyota.

1d   I love to collect classic automobiles, but my favorite car is my 1993 Toyota.

The connective *and* seems weird in sentence 1c because the expected coherence relation between the two clauses (given the normal construals of "classic automobiles" and "1993 Toyota") implicates incompatibility in some way, but *and* more naturally supports conceptual compatibility. The hearer/reader of 1c is going to work hard to optimize the coherence of 1c, because that's what hearers/readers usually do (hence, the insights of Grice). In sentence 1d, however, the connective *but* seems right at home: it suggests incompatibility, so it falls in line with (in fact, reinforces) the expected coherence relation.

The strongest determinant of coherence relations is the combination of context and understood semantic attributes (what sort of cars qualify as classic, for instance), but connective cohesion can nudge coherence relations in one direction or another. Consider sentence 1e now, for which the most natural coherence relation is not contrast (as it is for 1d), but concession:

1e   Although I love to collect classic automobiles, my favorite car is my 1993
Toyota.

There is still an inherent contrast between the two clauses (in particular, between "classic
automobile" and "1993 Toyota," but the focus of sentence 1e has shifted somewhat toward
the speaker's attitude, which acknowledges the incompatibility and its implications. What
connective cohesion does, then, is to make more explicit the coherence relations intrinsic
to the context and the propositions — drawing out the shared understanding of speaker and
hearer. That clearly-shared understanding helps to make the discourse more natural, the
most obvious benefit to voice interaction design.

Mike Cohen and his colleagues offer a nice contrast between the sort of prompting
we're familiar with from primitive systems and a set of alternatives that includes connective cohesion. Here's the primitive version:

Please say the date.

Please say the start time.

Please say the duration.

Please say the subject.

And here's the version incorporating *connective* (and <u>associative</u>) cohesion:

First tell me the date.

Next, I'll need the time <u>it</u> starts.

Thanks. ⟨pause⟩ *Now* how long is <u>it</u> supposed to last?

Last of <u>all</u>, I just need a brief description . . .

(Cohen, Giangola, and Balogh, 2004: 140–141)

The second set of utterances is clearly more natural, more fluent, more conversational. The
reason these utterances hit the conversational target so much better than the primitive set
of prompts is the coherence-cohesion match up. The coherence relation for both sets of
utterances is one that Mann and Thompson (1988) just call *Sequence*: one item follows
another. The primitive version, among its other faults, doesn't signal that relation at all.
The connectively cohesive example signals it literally at every turn, building a shared
understanding of the discourse.

The Cohen et al. example shows the importance of connective cohesion in a system-
initiative, form-filling routine (effectively, an interview). But, as the very word *turn*
indicates, the fundamental coherence of a conversation generally, no matter what other relations may be implicated, depends on the notion of sequence: one turn after another. These

turns — more particularly, the contributions they encode — need to be kept straight. The common ground at any given point is a matter of knowing what contributions have been made, so that the current contribution can follow them coherently. And cohesively: the cohesion devices for signaling awareness of previous turns involve words like *now*, *another*, *again*, *also*, *similarly*, *too*, and so on — words that convey awareness of the track of the exchange, the conversational ground.

For instance, if a caller has declined hearing a list of details earlier, but it is appropriate to make the offer again, it makes a huge difference to say it this way "Would you like to hear that list of details now?" If the caller has asked about one movie that is a critic's pick, and then he asks about a second that falls into the same (or similar) category, a human would say "This movie is also an Ebert pick," or "This one is a Roeper favorite, too." This sort of record keeping, of course, is absolutely trivial for a computer.

In addition to its job of reinforcing coherence relations, which it serves in any variety of discourse, from poetry to engineering reports to knock-knock jokes, connective cohesion has a specific function in conversation, coordinating turn management. Overwhelmingly, it is used to hold the floor — by grafting *and* onto the ends of clauses, for instance, to signal that there is more coming. But occasionally they are used to decline the floor as well — by uttering a single question-intoned connective, like "And . . . ?" or "So . . . ?", to signal that more is desired.

Closely related to those turn-declining connectives are the little head nods, assenting murmurs, and yeses that backchannel one agent's attention to the other. Emanuel Schegloff (1982: 80) calls them "continuers," because they serve to inform the speaking agent that she should continue speaking. As Gavin Churcher and his colleagues point out, although they don't "fulfill an act in themselves" as fully lexical, stressed connectives do, backchanneling gestures and vocables "add cohesion to a dialogue and reassure the speaker that his or her partner is paying attention." (Churcher et al., 1997: 4).

Summary

Only connect!
— E. M. Forster

Dialogues hang together like all discourses, if they do, because of networks of reference and relations. The words, utterances, turns, and exchanges consolidate into dialogues because they invoke the same entities and they support each other, on the two elemental levels of language: content and form. Those networks affect the **coherence** and the **cohesion** of the dialogue. Coherence is a function of the conceptual networks, cohesion of the formal networks.

Coherence has two aspects, **referential** and **relational**. Referential coherence follows from utterances that evoke the same objects or ideas, just the way this one and the previous one both make reference to the notion of coherence generally and referential coherence specifically. Relational coherence follows from utterances that operate together in some established way, just as this clause (the one starting with *just*) provides an example of the claim in the previous clause (starting with *Relational*).

Cohesion likewise has two aspects, both of which complement their respective coherence aspects: **associative**, which complements referential coherence, and **connective**, which complements relational coherence.

Referential coherence concerns the conceptual networks of reference, so associative cohesion provides the words and other formal strategies for expressing those networks. It works by **repetition**, the way *associative* is repeated in this paragraph, and *cohesion*, and *coherence*, and so on. It works by **anaphor**, the use of pronouns, the way "which" and "their" function in the previous paragraph. And it works by **ellipsis**, the strategic omission of words, the way we understand in this paragraph that *associative*, when it occurs, really means *associative cohesion*, but *cohesion* is left out to increase the tempo of the prose a bit.

Connective cohesion concerns the linking words that make coherence relations explicit, the way a word like *because* signals a relation of causation between clauses ("Itchy tripped because Scratchy stuck out his foot."), and the way *but* signals a relation of contrast ("Elmer was slow but he was elegant.").

Appendix: Coherence Relations

Speech or writing that has coherence is logical, consistent, and clear. It hangs together.
— Kenneth G. Wilson

For ease of reference, what follows is a list of coherence relations, with definitions and examples, including the relations outlined in the main text.[4] The list is drawn from Mann and Thompson (1988), with definitions adapted from Mann's website, and influenced by Dirven and Verspoor (1998).

A coherence relation holds between two speech units, at least one of which is a *nucleus*, the other usually being a *satellite*. The nucleus carries the brunt of the message, and the satellite is subordinate conceptually (and often syntactically, but that is a matter of cohesion). Some coherence relations hold between two nuclei, when the message depends equally on both speech units.

4: Many of the analyses in this appendix are too shallow. Coherence relations usually come in clusters, and the examples in this appendix are not exceptions to that tendency. If you see multiple relations in a given dialogue snippet, you're probably right. I provide the examples just to illustrate the individual coherence relation that is most prominent in the snippet, but it is not always the only one present.

In what follows, <u><u>nuclei</u></u> are double-underscored, <u>satellites</u> single-underscored. Coherence relations often co-occur with specific connective cohesion devices, but think of these only as diagnostics (if "but" is present, for instance, or if you can sensibly add it, then chances are good that you have an instance of the coherence relation, *Contrast*). I list some of the more common cohesive terms with their associated coherence relations. The list is alphabetical.

Background
The satellite increases the ability of hearer to understand the nucleus.

Example

AOL   <u><u>You're at the AOL main menu</u></u>. If you know what you want, just <u>say the keyword now</u>.

Caller   Email.

(AOL demo, 2002)

Correlated cohesion devices
well, oh, did you know, . . . ; *welcome to* X, *thanks for calling* X, . . .

Cause
The satellite causes the nucleus; this relation is the inverse of Result (the difference being a matter of which speech unit the focus falls upon).

Example

User   <u>Delete it.</u>

Chatter   <u><u>Message deleted.</u></u>

(Marx, 1995: 62)

Correlated cohesion devices
Subordination; *because, as a result of, therefore,* . . .

Circumstance
The satellite sets a framework in the subject matter within which the hearer interprets the nucleus.

Example

Thrifty   You are flying in, correct?

David   Uh-huh.

Thrifty   Sorry, I didn't understand. For more information say "Help." <u>I have the location where you're picking up the car</u>, now I just want to know: <u>Are you flying into that airport, yes or no</u>?

(Kestenbaum, 2002)

Correlated cohesion devices
No cohesion devices correlate strongly with Circumstance.

Concession
The speaker acknowledges a potential or apparent incompatability between the nucleus and the satellite (in a situation where the compatability of the nucleus and satellite is otherwise desirable).

Example

Office Manager   I don't know when Nicole will be back. She left the following message at 2:34. "Hi! I had to step out, but I will be back by 4:00 at the latest." Do you want to leave a message?

User   Yeh. Nicole, I have the demo ready, so stop by when you get a chance.

Office Manager   <u>I'm clueless as to what you said</u>, but <u>I recorded it all</u>. What do you want done with the recording? Say save it, forget it, or record.

(Yankelovich, 1997)

Correlated cohesion devices
though, *although*, *while*, *but*, . . . ; also *sorry* and related apologetic terminology

---

Condition
The realization of the nucleus depends on the realization of the satellite.

Example

AOL   <u>If you know what you want</u>, just <u>say the keyword now</u>.

Caller   Movies.

AOL   Hello and welcome to AOL Movie Phone. <u>If you know the name of the movie you'd like to see</u>, <u>say it now</u>, or to . . .

(AOL, 2002)

Correlated cohesion devices
if, . . . *then*

Contrast

Two nuclei are the same in many respects but different in at least one key way, and they are compared with reference to the difference(s).

Example

AirTran   Great, what's the arrival city?

Caller   Norfolk, Virginia.

AirTran   I'm sorry, but <u>AirTran doesn't currently serve Norfolk, Virginia</u>. However, <u>AirTran does fly to the Newport News-Williamsburg International Airport</u> in Newport News, Virginia — which is relatively close by. Do you want me to check for flights at that city?

(Kotelly, 2003: 170)

Correlated cohesion devices

but, yet, however, conversely, on the other hand, . . .

Elaboration

The satellite presents additional detail about the nucleus in one or more of the following ways:

NUCLEUS	SATELLITE
set	member
abstraction	instance
whole	part
process	step
object	attribute
generalization	specific

Examples

User   <u>Tell me about my bills</u>.

CM   Which bills do you want to know about?

User   <u>Mastercard and phone</u>.

(Karat et al., 1999: 27)

User   What are <u>the daily departure times of flights to Munich</u>?

System   <u>That's 7:45, 8:30, 9:50, 14:30, 18:25, and 20:30.</u>

(Bunt and Black, 2000: 29)

Caller   <u>Call 555 465 2468</u>

System   *Calling 555 465 2468.* By the way, <u>when you're done with this call, you don't have to hang up.</u> ⟨pause⟩ Just say. "Come back."

(Cohen, Giangola, and Balogh, 2004: 142)

Correlated cohesion devices
also, as well as, another, . . .

Enablement
The satellite increases the hearer's ability to perform the action in nucleus.

Example

Danluft   <u>The system can reserve tickets for Danish domestic flights.</u> <u>You use it by answering the system's questions.</u> In addition <u>you may use the two special commands "repeat" and "change" to have the most recent information repeated or changed</u>.

(Bernsen et al., 1998: 34)

Correlated cohesion devices
Modal verbs.

Evaluation
The satellite identifies the degree of the speaker's positive regard for nucleus.

Example

Jen   Hi, John! How's it goin? I know you're into rock, so <u>check out our cool new featured CD, *Out of the Ashes*, by Exit.</u> It's on sale this week, for $9.95. Here's the first track:

♫ ♫ ♫ [partial track plays] ♫ ♫ ♫

Caller   Add it to my shopping cart.

Jen   OK. I've added it. <u>I *love* that CD</u>, too. Would you like to listen to something else?

(General Magic demo, 2003)

Correlated cohesion devices
Verbs of support and endorsement; approbatory adjectives.

Evidence
The satellite increases the hearer's belief in the truth or effectiveness of the nucleus.

Example

A   <u>Helen did not come to the party</u>.

B   How do you know that?

A   <u>Her car wasn't there</u>.

(Kreutel and Mathesun, 1996: 367)

Correlated cohesion devices
since, because, therefore, . . .

Justification
The satellite increases the hearer's willingness to accept the speaker's right to present the nucleus.

Examples

Thrifty   <u>At what airport or city are you picking up the car?</u>

User   ⟨timeout⟩

Thrifty   <u>Sometimes rates and availability can depend on the location where you're picking up the car</u>.

(Kotelly, 2003: 82)

Correlated cohesion devices
No cohesion devices correlate strongly with Justification.

Motivation
The satellite increases the hearer's desire to perform the action in the nucleus.

Example

Student   Can I take Computability instead of Fundamental Algorithms?

ADVISOR   Yes, you can take it. But <u>Fundamental Algorithms would be more helpful</u> because <u>it is a prerequisite for all software courses</u>.

Student	But I plan on concentrating on theory.
ADVISOR	<u>You still need to take the required software courses</u>.
Student	But I would much rather take Computability.
ADVISOR	<u>It would be a mistake</u>.
Student	Ok, ok.

(Elhadad, 1990)

Correlated cohesion devices
No cohesion devices correlate strongly with Motivation.

Purpose
The satellite can be realized through the activity in the nucleus.

Example

System	<u>You should replace (SETQ X 1) with (SETF X 1)</u>.
User	Why?
System	<u>I'm trying to enhance the maintainability of the program by applying transformations that enhance maintainability. SETQ-TO-SETF is a transformation that enhances maintainability</u>.

(Moore, 1995: 240)

Correlated cohesion devices
Subordination; *in order to*

Result
The nucleus causes the satellite; this relation is the inverse of Cause (the difference being a matter of which speech unit the focus falls upon).

Examples

UA	OK, <u>I'll look for flights that have that itinerary</u>, hold on. ⟨Database look-up⟩ <u>I found a few flights that match that itinerary — three to be exact</u>.

(Kotelly, 2003: 158)

User	Please <u>book the flights</u>.
System	<u>I have booked the flights</u>.

(Walker et al., 2000)

Correlated cohesion devices

Subordination; *because, as a result of*

Restatement

This coherence relation is paraphrase, useful for varieties of explanation, for confirmations and clarifications, as well as corrections: the satellite restates the nucleus. (Mann and Thompson say the satellite and nucleus are of comparable bulk, but the nucleus is more central to speaker's purposes.)

Example

Sundial   Paris, Brest, when would you like to leave?

Caller   Next Thursday.

System: Where do you want to go?

User: From here to Cologne.

System: When do you want to go?

User: At 10:00 in the morning.

System: You want to go from Munich to Cologne at 10:00 a.m.?

(Kolzer, 1999)

Correlated cohesion devices

No cohesion devices correlate strongly with Restatement in confirmations; in corrections, they can come with a prefatory phrase like "I said . . . "

Sequence

There is a succession relationship between the nuclei.

Example

ELVIS   In the messages from Kim, there's one message about "Interviewing Antonio" and one message about "A meeting today." The first message is titled "Interviewing Antonio." It says, "I'd like to interview him. I could also go on to lunch. Kim."

(Walker, 2000)

ELVIS (EmaiL Voice Interactive System) says there are two messages from Kim and introduces them in a specific sequence; then it begins reading out the message that it listed first. The phrasing is not especially optimal, and the order is barely noticeable (exactly because it exhibits the expected coherence), but ELVIS employs a relational coherence pattern based on contiguity: an established order is followed.

Example

System   <u>Now what is the result of substituting this value</u>?

Student   (gives correct answer)

System   OK. Next problem. <u>Mr. Jones has bought</u> . . .

(Bunt, 1995: 20)

Correlated cohesion devices

Next, then, after, before

Solution

The nucleus is a solution to the problem presented in the satellite.

Example

System: There are ambulances in Pittsford and Webster.

User: OK. Use one from Pittsford.

System: Do you know that <u>Route 96 is blocked due to construction</u>?

User: Oh. <u>Let's use the interstate instead</u>.

(Allen et al., 2001)

Correlated cohesion devices

No cohesion devices correlate strongly with Solution.

Diction

*A new monk arrives at the monastery and is assigned to help copy the old texts by hand. He notices, however, that they are copying from copies, not the original manuscripts. So, he goes to the head monk and points out that if there were an error in the first copy, it would be continued in all subsequent copies. The head monk says, "We have been copying from the copies for centuries, my son, but you make a good point." So, the head monk goes down into the cellar with one of the recent copies to check it against the original manuscript. Hours go by and nobody sees him. So the new monk goes down into the cellar to look for him. He hears sobbing coming from the back of the cellar and finds the old monk leaning over one of the original manuscripts, in tears. He asks the old monk what is wrong. In a choked voice came the reply, "The word is **celebrate**."*
— Anonymous

Designing an effective voice interface is, more than anything, a matter of what ancient rhetoricians called *diction*, choosing the optimal word for a given audience, purpose, and context.

Diction has sponsored other words, of course, most prominently the word we use to label one of our most powerful reference tools, the dictionary. Dictionaries were originally invented as instruments for improving social interaction, and marketed to the upwardly mobile, giving them the lexical keys they needed to do business, participate in culture, adopt prestige dialects, acquire power, and so on. They included phonological and orthographical guidance, to help rid speakers of nonstandard pronunciations and to help writers compose a properly polished letter. But they have never really been very good at supporting word choice, diction. You have to know something about the relevant word already before you can find an entry and use it; the choice comes before the entry.

Peter Mark Roget's guiding mission, he of *Roget's Thesaurus*, was to solve the diction problem, and his general principles are the reverse of those behind a dictionary:

> *The purpose of an ordinary dictionary is simply to explain the meaning of words; and the problem of which it professes to furnish the solution may be stated thus — The word being given, to find its signification, or the idea it is intended to convey. The object aimed at in the present undertaking is exactly the converse of this: namely, — The idea being given, to find the word, or words, by which that idea may be most fitly and aptly expressed.*
>   (1925: xiii)

This last line offers as precise a definition of the purpose behind lexical selection in voice interaction design as there is: given an idea of the functions the user wants to perform, one needs to choose the most apt, the fittest, words to allow her to perform them.

The interaction must be crafted on the basis of what the *user* is *mostly likely* to say, at any given point, and of what the *system should* say; in short, on the basis of diction.

The only way to know what to say and what to listen for is to know the patterns of usage in the register at hand, and the best way to investigate those patterns is to chart them out in a representative collection of utterances from that register; that is, in a corpus. One investigative tool that is particularly powerful is simply aligning all the uses of a given word or vocable in the corpus — assembling a concordance — and snooping through those uses. But this method has an inherent liability. If the word is truly characteristic of the speakers and their purposes, there will be too many examples to snoop through; if the number of examples is highly manageable, then the usages may not be especially functional in the register, and therefore only of marginal interest to designing the interface. So, two other strategies are necessary to plumb diction sufficiently in a register: collocation and colligation.

Both notions concern the ways in which words cluster together, the ways in which they keep "habitual company," in Firth's useful phrasing (Firth, 1968: 182). Collocation, as we have seen, is simple proximal co-occurrence — for instance, that *silly* tends to show up a fair amount proximal to *ass* in mid-twentieth century, upper-crusty British registers. Colligation is *structured* proximal co-occurrence — the fact that the phrase *a silly ass* tends to show up a lot in such registers (that is, not just that *silly* is close to *ass* in such registers, but precedes it and modifies it adjectivally).

The vocabulary problem represents the biggest challenge in voice interaction design. Diction — that is, careful attention to lexical choices — represents the clearest solution; indeed, the only solution. Concordances, collocations, and colligations — all with respect to an appropriate corpus — are the best methods to investigate usage, and inform the choices of diction in building an effective voice interface.

Usage

One cannot guess how a word functions. One has to look at its use and learn from that.
— Ludwig Wittgenstein

Voice interaction design requires thorough investigation of the characteristic diction of the register in which the speech system participates, which means assembling and analyzing a corpus of utterances.

The most immediately useful analysis of a corpus is simply to assemble every example in the collection of a given word or phrase. Every example, that is, can be *concordanced*. A *concordance* (switching now to the noun) is the output of a search routine which combs the corpus for multiple tokens of a specific sequence of linguistic units (phones, syllables, letters, words).

Take the vocable, *deposit*, a partial concordance of which (drawn from a large corpus of mundane American English) you can find in Figure 7.1.[1] *Webster's* regards the verb form of *deposit* to be primary, probably for historical reasons (it comes to English from the Latin verb *deponere*, to "lay aside" or "put down"), but the concordance in Figure 7.1 makes it clear that the noun is far more basic in terms of usage. Only one of the 40 examples is used as a verb (36). But a substantial portion of those noun tokens (44%) function like adjectives (that is, modifying other nouns), as in "deposit account," "deposit slip," and "deposit bottle" — a tendency that goes unnoticed in dictionaries. The dominant usage of *deposit*, this concordance tells us, is overwhelmingly financial (a tendency that *is* noticed by dictionaries): 95% of the examples evoke finance in some way (only 6 and 38 do not). The financial proclivities of *deposit*, however, show up a good deal more subtly in this list than most dictionaries reveal. For instance, we recognize that one use of *deposit* means that the money involved may come back at some point — two indicate this specifically (26 and 27), while a few others just invoke it, because of facts we know about the world (specifically what we know about security deposits, as mentioned in 5 and 32, and what we know about bottle deposits, as in 7).

Now, the examples in this (partial) concordance all come from a broad cross-section of American registers, a noteworthy proportion of which are frankly commercial.[2] This sort of general-purpose corpus data is important for voice-interface work, and the team lexicographer should certainly have a subscription to one of the major corpus services, or a

---

1: The concordance is of the <u>vocable</u>, to get at broad trends. That is, the search was done on the relevant seven consecutive letters, *deposit* — not on either of the words *deposit* (i.e., the noun, as in "leave a *deposit*" or the verb, as in "*deposit* an envelope"), either of which would have produced narrower results.

2: The sources are books of all varieties, radio programming, and a range of short-lived information texts (pamphlets, brochures, operating instructions, and the like).

1	00 per day with advance reservation	deposit.	KAYAKING AND SOLO CANOES &
2	that if you invest in a money market	deposit	account rather than a money fund,
3	products like CDs and money market	deposit	accounts. You may, of course, be
4	of savings accounts, money market	deposit	accounts, and money funds as
5	for rent money, the security	deposit	and the fee she paid to the
6	taste buds are all formed; some fat	deposit	beneath skin. The infant is
7	it too hard. To hell with it. When a	deposit	bottle is broken you don't get
8	glass. In a corner, there's a safe	deposit	box that has been blasted open.
9	Please Note: A safe	deposit	box is available for the
10	be sent to you when you send in your	deposit.	Bring your own fishing gear, and
11	of golf. Plus, if you send us your	deposit	by September 29, 1995, you can
12	installation, billing, payment,	deposit	complaint and service records,
13	her balance. "We could put down a	deposit,"	he said, "hold it for a while,
14	All checks must be drawn on funds on	deposit	in the U.S. If your account is a
15	return to the original intention of	deposit	insurance, protecting the small
16	Seidman is chair of the Federal	deposit	Insurance Corporation. Regulators
17	reforms that are called for in the	deposit	insurance system. Clark: The
18	nation's largest banks. The Federal	deposit	Insurance Corporation Fund, which
19	and poor management of the Federal	deposit	Insurance Corporation and the
20	the acting chairman of the Federal	deposit	Insurance Corporation, said that
21	his money. This loss is offset by	deposit	interest required by local rules,
22	All prices are subject to change. A	deposit	is required. Ask about the
23	the Normandy also accept Eurocard. A	deposit	is required. The tourist
24	will be completed. A $5,000	deposit	is one form of consideration.
25	is sold by a real estate broker, the	deposit	is usually placed in the broker's
26	reasons. The $75 tuition	deposit	is non-refundable. 1. If the
27	the deal is off and the purchaser's	deposit	must be refunded in full. What
28	Providian Home Loans First	deposit	National Bank First deposit
29	an account. 1 (We) authorize First	deposit	National Bank to contact me and to
30	insurance carrier. If you leave a	deposit	of $100 for a future visit
31	envelope All you pay is a	deposit	of $211.55 then $75 a
32	GUARANTEED APPROVAL. NO security	deposit	required. [c] telephone
33	to pay a $5,000 deposit. The	deposit	shall be credited to the buyers at
34	enclosed. Include a voided check or	deposit	slip for convenient monthly
35	cover, register, and one-part	deposit	slips with every order. Call
36	it, make it into jewelry, whatever,	deposit	the money they got for it in their
37	buy a brokered CD. Certificates of	deposit	usually pay a fixed rate of
38	feet lower than the old calcite	deposit	which formed the cavern floor. He
39	The Free Access Certificate of	deposit	will renew into the same type for
40	purchaser you want to offer a small	deposit.	You also want to assure that you

FIGURE 7.1   A 40-line concordance of *deposit* From the Collins COBUILD Bank of English, nine-million-word corpus of "American books, ephemera, and radio," available from http://titania.cobuild.collins.co.uk/

license to work with some large and general corpus. But the primary domain of interest will always implicate a more refined corpus, for what computational linguists are in the habit of calling a *sublanguage*, but what we will continue to call a *register*. (Among other problems, the word *sublanguage* has a misleading faux precision to it that is best eliminated in voice-system design discussions.)

Other, more refined, corpora than the one behind Figure 7.1 would no doubt reveal other tendencies. A corpus which included mostly health discourse, for instance, would likely include far more usages of the sort in line 6; a concordance of material from geological or mining registers would lean towards usages like line 38; a collection of texts from a biological register (in which animals deposit feces; birds, reptiles, and insects deposit their eggs; and so on) would lean towards uses not even represented in Figure 7.1. The lexicographer must shape and build such a corpus, depending on the field(s) for which the system is designed, and this specifically constructed corpus then becomes the foundation of the interface lexicon.

Take another example, *withdrawal*, concordanced in Figure 7.2, from the same large, general-purpose, American English corpus. The *deposit* example has probably prepared you for an onslaught of financial examples (that is, you've likely undergone a species of lexical entrainment); I know I was ready for such examples, when I chose the closest thing to an antonym of *deposit* I could think of, and ran the program. But 77% of the examples are military, with the remaining examples scattered among several other categories (drugs, sex, emotional involvement); only one (line 6) is clearly financial. The results here undoubtedly reflect the period in which the data was harvested, and the registers from which they were harvested, but it is hard to escape the conclusion that *withdrawal* has decidedly other primary allegiances than *deposit*.

Withdrawal is exclusively a noun (though it frequently functions adjectivally, as in lines 34 and 36). Unlike *deposit*, the corresponding verb form is distinct, *withdraw*, which has interesting implications for recognition. *Withdrawal* offers a notably bigger acoustic pattern for the recognizer to bite, and would therefore be preferred from a simple recognition perspective. A financial voice system, of course, should be prepared for both. But, from an interface standpoint, the caller might be encouraged to use the noun. It's not just that the noun has an extra syllable, but that it implicates a highly predictable verb to boot (*make*) and a preposition (*of*), which moves the recognition problem from a lexical matter to a phrasal matter. Compare the relevant amount of the signal a recognizer would have to work with in 1a and 1b:

1a   I would like to *withdraw* $100.

  b   I would like to *make a withdrawal of* $100.

A decision about whether to prime for 1a or 1b in a given interface can't be made solely on the basis of general-purpose data, of course, nor on the needs of the recognizer alone. It needs to rest on a range of far more specific considerations, not the least of which are the particular patterns revealed by the appropriate register corpus.

Concordancing is also a very reliable way to probe synonyms. Take just the few examples in Figures 7.3 and 7.4. When you set synonymous terms side by side in concordances,

1	a force to monitor the Iraqi	withdrawal	and establish a buffer zone
2	retreating, they don't call it a	withdrawal,	and the fighting continues.
3	[p] On Monday, we got wind of the	withdrawal,	and after our attempts to reach
4	January 15th deadline for Iraqi	withdrawal	bears down on the nations in the
5	Hussein personally order an Iraqi	withdrawal,	but Saddam today gave no
6	many annuities have substantial	withdrawal	charges if you take your money
7	would last as long as the	withdrawal	continued in good order. Paul
8	punished that can be expressed as	withdrawal,	depression, suicidal ideation,
9	taken from Kuwait before the Iraqi	withdrawal.	Every day, they say, the Iraqis
10	is marked by the infant's further	withdrawal	from participation in the
11	He could not argue for America's	withdrawal	from Berlin and the 'consequent
12	demand for an unconditional Iraqi	withdrawal	from Kuwait. Corey Flintoff,
13	it calls for Iraq's unconditional	withdrawal	from Kuwait, guarantees Iraq's
14	is in the Soviet proposal—Iraqi	withdrawal	from Kuwait—and it's not being
15	state until last night, when the	withdrawal	from Kuwait started and is
16	24 hours, the time has come for a	withdrawal	from a part of our country—just
17	and 338, which call for an Israeli	withdrawal	from occupied territories in
18	has a proposal for a three-year US	withdrawal	from Subic Bay Naval Base, but it
19	initiative. They also see that	withdrawal	is likely. There's a couple of
20	behind Iraqi lines, so any Iraqi	withdrawal	is likely to have to actually
21	observing what could be an Iraqi	withdawal.	More after headlines from Carl
22	condom should be held throughout	withdrawal.	No matter how nice and cozy it
23	a different tack, proposing the	withdrawal	of the brigades from Luttwitz's
24	life, and it must not perish. The	withdrawal	of blood from the surface of his
25	offer calls for the simultaneous	withdrawal	of Iraqi forces from Kuwait and
26	and Shevardnadze engineered the	withdrawal	of Soviet troops. Where Brezhnev
27	He says, for example, that the	withdrawal	of the combat brigade in Cuba is
28	the north of the country and the	withdrawal	of the government troops in the
29	to the talks was compounded by the	withdrawal	of the leader of the Croat
30	the Iraqi nation announcing such a	withdrawal,	officers with airborne troops
31	finish. If Saddam were to feign a	withdrawal	or actually begin to pull out of
32	The Iraqis were not without a	withdrawal	plan, of sorts. Unfortunately for
33	the section entitled "Systematic	withdrawal	Plan" is replaced with the
34	the drug, complete the physical	withdrawal	process, and the addiction was
35	way home. He said the promised	withdrawal	should satisfy the UN Security
36	with cravings - all the time. But	withdrawal	symptoms are temporary. They
37	cut off and reduced to a fighting	withdrawal,	the military and political
38	as I see it, is now to ensure the	withdrawal	—the unconditional withdrawal—
39	a withdrawal and a re—retreat. A	withdrawal,	they say, is a sort of planned
40	so the 1960s saw their massive	withdrawal.	This withdrawal began in the

FIGURE 7.2   A 40-line concordance of *withdrawal* From the Collins COBUILD Bank of English, nine-million-word corpus of "American books, ephemera, and radio," available from http://titania.cobuild.collins.co.uk/

it becomes much easier to compare their functions, senses, and connotations. In particular, a simple swap test (coupled with native intuitions) can get quickly at the range of applicability for given words. It is relatively natural to use the verb *educate* for some of the contexts in which *train* is used, but others sound bizarre, perhaps even unacceptable (in

1	it can do what it does best,	educate. Excerpt from 'All Things
2	stories that both entertain and	educate. What are these storybooks
3	wealth itself and the need to	educate all citizens to its dangers.
4	now he wasn't prepared to	educate an FBI agent. Anyway,"
5	, the writer meant to alarm,	educate, and mobilize his readers all
6	in by the Christian Coalition to	educate and to make sure the people
7	point in spending much money to	educate blacks because they needed only
8	factory owners and parents to	educate children, even if it's in
9	for paper recycling, and help	educate employees about recycling and
10	business rather than to merely	educate first-time buyers, Fourth,

FIGURE 7.3   A 10-line concordance of *educate*

1	announced that it intended to	train 100 reserve officers in the
2	Gerard says. 'I knew how to	train a boxer, hit here, hit there.
3	and soundly a rider can also	train a horse to use his body like a
4	dollars to recruit, arm and	train a first contingent of 500
5	by fear, sort of, "We'd better	train all these unwashed ignorant kids
6	plan). Instead of helping to	train and educate homeless and poor
7	the therapist was able to	train him to answer questions and to
8	and her husband, Mike, who	train horses up in the Yakimaw Valley.
9	What a money machine it is to	train hundreds and hundreds of
10	away from the money-changers and	train it instead on powerful

FIGURE 7.4   A 10-line concordance of *train*

lines 3, 8, and 10, for instance; and 2 would result in ambiguity about the skills and/or information at issue). On the other hand, *train* substitutes for *educate* more widely, though it alters the sense somewhat, even to the point of potentially giving offense (7).

Concordancing is the most immediate tool for plumbing usage, both broadly and deeply. It provides a wide, cross-sectional survey of words in context, letting you get a rapid sense of usage patterns. And it allows quick navigation, at least in digital form (paper concordances allow more laborious navigation), into a corpus, each line linking to the exact point in the text from which it was drawn, providing a more thorough probing of specific individual usages. But it has the inherent limitation of a size/representativeness trade off. If the examples are plentiful enough to be representative of the register, there are too many to examine closely; if the examples are few enough to examine closely, they may not be representative enough of the register.

Understanding diction requires additional methods — chiefly, exploring lexical associations.

Lexical Friends and Relatives

It is clear that words do not occur at random.
— John Sinclair

Words seek each other's company. They form habitual associations (*salary* and *career* and *professional* are in one such club; *paycheck* and *job* and *trade* are in another); that is, they collocate. And they gather into recurrent phrases (*jump the gun*; *hold the phone*; *a wing and a prayer*); they colligate.

Collocation

[Collocation investigates] key words, pivotal words, leading words, by presenting them in the company they usually keep.
— John Rupert Firth

Collocation, the functional pressures in a register that place words together, shows up in a corpus as the statistical tendency of one word to occur in the presence of another. It condenses the significant patterns of a concordance.

If we run a collocation analysis of *educate* on the Bank of English corpus, we find that, of the 278 occurrences, 47 show up within four words of *people*. Eyeballing it, that's a noteworthy pattern, but we can probe that impression further with a simple computational routine to produce a co-occurrence measure of the statistical significance for *people* to show up near *educate*. Cobuild's Mutual Information statistic, for instance, yields the relatively high value of 6.3 for this co-occurrence.[3] *Children*, an important class of people, also patterns significantly with *educate* (5.2), as does *about* (5.6).

Including *train*, and a couple of other terms from the same semantic domain, we get the spread of collocational results reported in Table 7.1. None of the other words, we note, collocate significantly with *people*, though three of them do (at much lower levels of statistical significance) with *children*.

3: I've adopted the Mutual Information (MI) measure for this illustration largely out of convenience; there are several such measures, and all of them yield very comparable results. Another common statistic is the association ratio, and T-tests are also frequently employed. Effectively, they all just provide an indication of the tendency for word X to show up proximal to word Y (where *proximal* is standardly within four words before or after Y) at a nonchance level. For instance, *tree* will show up a reasonable amount in a general-purpose corpus. So will *Christmas*. Let's say the probability of any randomly chosen word in the corpus being *tree* is p(t). What these statistics do, then, is use p(t) to control for the chance co-occurrence of *tree* with *Christmas*. Mutual information, in particular, uses the probability of *tree* occurring in any random eight contiguous words (that is, p(t) * 8) as a coefficient to calibrate the frequency with which it shows up when those eight words are not random, but centered on *Christmas*. By the way, *tree* occurs in the 450-million-word Bank of English corpus 2839 times, 165 of which are within four words to either side of *Christmas*, making for a mutual information statistic of 6.1.

Node Words	Collocates
educate (278)	people (47; 6.3), about (41; 5.6), children (29; 5.2)
train (4,296)	help, helped (24; 6.3), harder (10; 5.8), teachers (10; 4.5)
teach (1,746)	lesson (49; 7.2), instructors (5; 6.4), learn (25; 4.2), children (4; 1.7)
instruct (119)	bank (7; 2.6), children (4; 1.7), lawyer (3; 1.7)

TABLE 7.1   Some collocational patterns from the Bank of English corpus
The first parenthetical number is the number of occurrences in the corpus. The second number (for the collocates) is the Mutual Information measure of the significance of collocate's co-occurrence with the primary word.

You can see why recognition engineers care so much about collocation: if the recognizer resolves *people* but is having trouble with a pattern that might be *educate* but might also be *simulate*, *crew rate*, *fire grate*, or *whateverate*, a simple disambiguation strategy based on collocation can help return the correct reading.

But collocation is equally valuable for interaction design. We know from results like those in Table 7.1 that *people* and *children* can help prime for *educate*, that *lesson* can help prime for *teach*, that *train* and *children* do not mix as easily as *teach* and *children*. And so on. We know more about what words to expect from the user, and about how to induce certain words from the user. We know more about the very immediate fields of usage.

Colligation is similarly useful for both the recognition engine and the parser, on one side, and interaction design on the other.

Colligation

*A word's colligations describe what it **typically** does grammatically.*
   — Michael Hoey

There are two aspects to colligation — one, sometimes called "the idiom principle," concerns the ligatures among specific words; the other, without a snappy label, concerns the ligatures between specific words and grammatical structures.

Idioms are phrases, like "kick the bucket," that carry a meaning distinct from their constituents and therefore function pretty much like individual words (the phrase is synonymous, for instance, with *die*). John Sinclair, the one who coined "the idiom principle," generalizes this notion, observing that "a language user has available to him or her a large number of semi-preconstructed phrases that constitute single choices, even though they might appear to be analysable into segments" (1991: 110). We considered such a phrase earlier, *make a withdrawal* in a banking register, which suggests how useful it is for priming (and the recognition engine) to take account of colligations.

Collocation is good for tuning the recognition routines. If you know that *deposit* and *dollar* strongly collocate with each other, then you can use each word (with calculated probabilities) to support the resolution of the other. Colligation is different. If you know that *deposit* colligates into the phrases *make a deposit* and *money market deposit account*, then you can tune the system to listen for those specific phrases, those big acoustic clumps. For instance, choosing between the two candidates in sample 2 might be much easier given colligation facts about the discourse:

2a   blah blah **calling card** blah blah

b   blah blah **calling Carl** blah blah

Colligation, too, has substantial implications for system output, for talking the talk of a register. Language use is much more routine than most people assume (if it wasn't, we'd be out of a job), and investigating colligation is investigating the routines in a corpus.

The lack of command over a register's routines is one of the markers of someone who is out of place, who doesn't know the lingo. Think of standard-issue idioms in English. For some reason,[4] we describe a heavy rainfall as "raining cats and dogs," not *dogs and cats*, not *cats and canaries*, not *popsicles and pizza*, but *cats and dogs*. It's a staple joke with second-language speakers on sitcoms that they get these colligations wrong, and there's a million of 'em: *ebb and flow*, *nook and cranny*, *life and limb*, . . . Some of these groupings (recall Grice's maxim of manner — order), have cognitive/communicative motivations, but what's important for understanding the tone of a register is just that they *have* lots of established colligations.

At a more general level — the level without the snappy label — colligation tells you what structural affinities words have; for instance, in which grammatical slots one might find certain words. Some nouns, for instance, have a tendency to be subjects or objects (if objects, they may be direct, indirect, or oblique) or locatives, and so on; certain verbs tend to follow modals, adjectives to be predicate or nominal. Take *deposit*. As a noun, *deposit* is most likely to be a direct object ("I would like to make a deposit"), less likely to be a subject ("My deposit will be $300"), and quite unlikely to be an indirect object ("The bank assigned the deposit a good interest rate"). These sorts of colligations can be very valuable in diagnosing (and consequently, for responding to) utterance and interaction patterns — for instance, the word *no* colligating at the head of an assertion or request correlates strongly with changes or corrections to the conversational ground.

4: Best guesses: a permutation of the Greek *catadupe* or "waterfall," or of the Latin *cata doxas* ("contrary to experience", i.e., an unusually heavy rainfall), perhaps spliced with various mythologies or folklores (Lockhart, 1988).

Summary

Words are not as satisfactory as we should like them to be, but, like our neighbors, we have got to live with them and must make the best and not the worst of them.
— Samuel Butler

The cardinal focus of voice interaction design, **diction**, is a matter of finding the words that accomplish the job most fitly and aptly, which means knowing the words — how they sound and, as we've investigated in this chapter, how they function. The idea is to stock the system vocabularies with words that are acoustically distinct enough from each other to minimize confusion, and acoustically robust enough in themselves to give the recognizer something to bite into. Oh, and they have to represent the register, too. But that's not so easy, so we need to calibrate both our choices and the resolution strategies with a very thorough knowledge of lexical usage patterns, prominently including networks of associated words.

The traditional device for plumbing word usage has been the dictionary, but dictionaries provide little more than rough sketches of a given word's functional range. Those sketches can certainly be useful, but they are far too preliminary for voice interface needs. Coming at diction from the side of meaning, thesauri are also generally useful instruments. They are more than adequate for the diction concerns of writing and speaking most of the time. But for designing speech-system behavior, they provide little more than early-phase brainstorming material, or suggestions to follow up with more sensitive tools.

In recent decades, however, these traditional instruments have been supplemented by an incredible new digital tool, one indispensable for voice interaction design, just as it is indispensable for speech recognition and natural-language understanding, the **corpus**. Corpora quickly reveal broad swatches of two critical types of information about words, **collocation** and **colligation**. Both types tell you about the habitual company words keep: the first tells you the general company, the other words they tend to travel with; the second tells you about specific company, the other words they line up along side of.

Collocation analyses provide priming data, which can help us predispose users to say what our recognizers are best prepared for them to say. They also provide lexical-association data, so we can decide what the system vocabularies should contain — not just on the basis of the most common terms, but also on the basis of the role they might play in skeins of affiliation with those common terms. Colligation analyses provide lexical-conglomeration data, so we can ferret out multiword acoustic clumps to populate the vocabulary, and for the agent to speak, helping both to prime for those clumps, and to sound native.

Speech recognition, everyone realizes, cannot proceed without massive computational support, not only in terms of a pure recognizer that can store vast catalogues of acoustic patterns, analyze incoming signals into candidate word patterns, and rapidly compare those candidates to the relevant catalogued patterns. It cannot proceed without an array of tools and strategies that come from the close, ingenious examination of big-to-huge collections of language — corpora. Voice interaction design has very similar requirements.

Design

Design from the human, out.
— Wesley E. Woodson

Once the product's task is known, design the interface first.
— Jef Raskin

The design challenge is really to figure out some way to constrain what people say.
— Kate Dobroth

We can build integrated, natural-language dialog systems, even when working with error-prone recognition engines and imperfect grammars, by designing the dialog flow to reduce the likelihood of errors and to enable quick error recovery.
— Wlodek Zadrozny, Catherine Wolf, Nanda Kambhatla, and Yiming Ye

Crafting Voice Interfaces

. . . a reasoned habit of mind in making.
  — Aristotle

Voice interaction design? The Greeks had a word for it; well, for the whole field of inter-action design, in which voice is playing an increasingly important role. Or they would have had a word for it — if they had the processors, circuit boards, and information economies to drive a concern for such an enterprise. Interaction design is what they would have called a *techne*.

Techne is usually translated as "art," but that doesn't quite get at it. (Or we could just say "Interaction design is an art," and leave the Greeks out of it.) The main problem with *art* is that it has accrued romantic connotations of individual creative genius: Beethoven or Pollack or Whitman throwing away all rules, and all interest in communication, and expressing their souls directly in notes or paint or vocables.

Techne is the word at the root of the ultimate term of the twenty-first century, *technology*, which, if we bring the notion of art along with us, takes us in the right direction, toward the governing importance of functionality. But "art + technology" doesn't get directly at the full-blooded heart of a *techne* either, because *technology* has accrued an opposite set of connotations, of soulless electrons pulsing through silicone chips.

But *techne* has also left its imprint on another English word, *technique*, which suffi-ciently rounds out an account of the word in contemporary terms. *Technique* isn't enough on its own; a repertoire of good techniques is indispensable in any design work, but the word *technique* can suggest a mere knack or trick. There was nothing "mere" about a *techne*, at least not the way Aristotle tells it; he described it as "a reasoned habit of mind in making" (*Nichomachean Ethics*, 1140a). Thought, routine, and creativity, says Aristotle, are integral to a *techne*.

Voice interaction design, as a special case of interaction design, draws on the implications of all three of our key terms: the inventiveness suggested by *art*, the ordered methods suggested by *technology*, and the skill suggested by *technique*.[1] The first part of this book lays out the raw materials and the tools of the art of speech and its communicative strategies. This section sets down the ordered methods and provides the techniques.

There are some special affordances in the process for developing a voice interface that are not as readily available in other development cycles, and the specifics of the process are unique to the medium, but the overall development cycle is no different from that of any complex system for human users: analyze, design, test, release-and-monitor in one overall process, and iteratively throughout that process. There are three distinct stages where the focus is on each of these processes, but within each all three repeat cyclically. It's the familiar wheels-within-wheels of user-centered design. If this cycle is taken seriously, it is not a product-development cycle, however, it is an interface-design cycle.

What you absolutely cannot let the product folks do is leave the interface for later on down the line. Picking over the corpses of "good" products with "bad" interfaces has long been a favored pastime among design professionals. There is a veritable boneyard of such examples.

The question to ask is not the one that development engineers sometimes throw at interaction specialists, "Which came first, the technology or the interface?" This question is crazy on two key counts: (1) there are no technologies without interfaces, not even something as simple and familiar as a claw hammer; and (2) the interface *is* technology. But if someone should ask you such a question, the one to ask back is, "Which came first, the technology or the user?"

Many software companies see interfaces as application-support systems — as collections of doodads that make it easier for users to operate the application, a set of features that is the real story. This view is blockheaded. Software is an enabled set of functions, and the enabling strategies (the sum of which is the interface) are at least as important as the functions. Often the enabling strategies are more important than the functions. Indeed, the strategies and the functions should be so seamlessly intertwined that even the technophiles will not be able to draw a line in the silicone: everything on this side is "application," everything on that side is "interface." There will just be a product, which works well or doesn't.

What this means for software development is a product-development cycle that is an interface-development cycle — a process in which the interface is not an add-on, after the

---

1: As I am sure you suspect, we could go on almost indefinitely tracing out the tentacles of *techne*. Its deep importance to the ancient world, and inevitably to us, can be gauged fairly well by the residue it has left in our language (*tectonic, technician, technocracy, Technicolor*™). Through Latin cognates, it is also implicated in a range of other words like *text, textile, context,* and *tissue*; it's even lurking around behind terms like *TKO* in boxing.

features-functionality-marketing gizmos are in place, and is not even an early stage in the cycle. It *is* the cycle.

This position certainly strikes me as right for products sporting graphic interfaces, but for speech systems it is absolutely unavoidable. There is really no alternative. Any other position is — how to put this delicately? — delusional. The product *is* the interface. The speech-recognition and natural-language-processing engines are interface support mechanisms, nothing more, and surely nothing less. The product is the interface. The product is the interface. Believe it. Users do.

This section of the book outlines the ways in which the interface can realize the seamless integration with data and processes that users expect and need. This chapter, in particular, charts some of the fundamental concerns in that integration. First, we look at the issues behind implementing a voice interface to begin with — is the task, and the information flow-rate of that task, such that voicing it can satisfy the user's related goals? We also take up issues related to the voicing of the web, and attend two morality lectures, one about overcoming the design intuitions instilled in us from graphic interface interaction, the other about the perniciousness of menus.

To Voice or Not to Voice

The virtue of a thing is relative to its proper work.
   — Aristotle

The first question to ask, at the very outset of the design process, is whether or not the artifact/application/process is appropriate for voicing. Some clear guidelines, such as the ones in Tables 8.1 and 8.2 (adapted from Sun MicroSystems recommendations), are helpful here. These are not hard and fast laws, and the considerations should be balanced against each other.

As the tables indicate, the primary factors in deciding whether to voice an artifact, application, task, or process are the contextual implications for speech input and speech output. In general, speech input is especially suited to circumstances in which the user's hands are busy; speech output is especially suited to circumstances in which the user's eyes are busy.

But other factors are relevant as well, largely having to do with information flow and the nature of speech. Spoken language is highly variable, which constrains the input; spoken language is highly transient, which constrains the output (or, more accurately, constrains the ability of the user to absorb that output). High-information speech input is tough on the system, because of the recognition and natural-language processing demands it imposes; high-information speech output, especially in parallel presentations (for instance, in a comparison task), is tough on the user.

Voice input	
Appropriate when . . .	**Inappropriate when . . .**
• there is no other input mode available • no other input mode is practical in the device context (e.g., a key entry or pointing-device system doesn't suit the density of the information that must be input) • no other input mode is practical in the task context (e.g., the task requires the user's hands to be occupied, such as in driving or maintenance and repair) • no other input mode is practical in the user's knowledge context (e.g., users cannot type with sufficient speed, or are illiterate) • no other input mode is practical in the user's physical context (e.g., user's hands or arms are physically disabled)	• the task requires users to talk with others while engaging in it • the environment is very noisy • tasks are easier with other input modes (e.g., choosing from lists)

TABLE 8.1   To voice or not to voice: Input concerns (adapted from Sun MicroSystems, 1998: 20)

Speech (and audio) output	
Appropriate when . . .	**Inappropriate when . . .**
• there is no other output mode available • no other output mode is practical in the device context (e.g., there is no visual display) • no other output mode is practical in the user's knowledge context (e.g., users are illiterate) • no other output mode is practical in the user's physical context (e.g., user's eyes are physically disabled) • the task requires the user's eyes to be looking at something other than a visual display (e.g., driving, maintenance and repair)	• large quantities of information must be presented to the user • confidentiality is important (e.g., for privacy or security) • the environment is very noisy • tasks are easier with other output modes (e.g., comparing data items is a significant aspect of the task)

TABLE 8.2   To voice or not to voice: Output concerns (adapted from Sun MicroSystems, 1998: 21)

An ideal candidate for voicing is a travel-booking task, which is why it was the focus of the both the massive DARPA initiative in the United States and the RailTel initiative in Europe; and why, in turn, so much of the speech system literature concerns itself with flight and train information and booking tasks. Travel booking is ideal because it involves manageable amounts of information; because that information chunks naturally into specific functional units (dates, times, locations); and because it has a history (and, not coincidentally, an available discourse base to study) of voice-only interaction.

A moderately poor candidate for voicing is an email application: a task that has an exclusively graphic-output, direct-manipulation input legacy that involves large quantities of information (e.g., scanning an inbox-list for source, subject-line, and urgency information, among other categories); and a task that often involves a species of comparison (keeping a received message open, for instance, or quoting-and-responding to specific parts of a received message). However, email applications were in fact the focus of early and interesting work, especially at Sun MicroSystems and MIT. Despite the lack of appropriateness of the task to voice modality, the growing importance of email in the nineties, for an increasingly mobile user population, led to research into voice-enabling email, which in turn helped spark the turn to conversational models.

With the evolution of protocols and devices for mobile email that retain visual output and provided for (marginally) adequate direct-manipulation input — that is, with the development of PDAs, tablets, beefed-up cell phones, and other portable digital appliances — voice-only email applications are understandably gone. (There are still rich multimodal possibilities for portable email development, especially combining visual output with vocal input.)

Or take a domain like banking. Automatic bank machines are poor candidates for voicing, because of security or privacy concerns. It is easier to keep access codes and bank balances secret with manipulation and visual display than with speaking. While voice-identification holds promise for a security replacement of keyed-in access codes, retinal or finger-print scans have the same bio-uniqueness with more confidentiality (they can't be tape recorded, for instance). A kiosk with old-fashioned, closed-door phone-booth characteristics might be workable for protecting information in a voice-enabled banking kiosk, but if you're primarily using voice and ears for the transaction, there are few reasons to go to a determinate location and talk to a machine.

But these concerns all have to do with the lack of privacy in the physical context of a banking machine in a public space. Most of the *operations* a bank machine performs are not only available for voice-enabling, but desirable in other contexts — chiefly, the privacy of a home phone, or the portability (and potential privacy) of a mobile phone.

Paying bills, transferring funds, checking balances, and so on, are all prime candidates for voice interaction. Again, it is the information flow that is most significant. Aside from deposits and withdrawals (which, for the foreseeable future anyway, still require physical exchanges of checks or cash, and consequently call for a determinate location), simple

	Low output	**Moderate output**	**High output**
Low input	telephone polling	weather report	credit history
Moderate input	pizza order	vehicle navigation	online encyclopedia
High input	questionnaire	auto mechanic	advisory service

TABLE 8.3   Chart of information flow in typical tasks (Adapted from Novick et al., 1999: 174)

banking operations are ideal candidates for a voice interface. The tasks involve manageable amounts of information; the information chunks naturally into specific functional units (amounts, companies, accounts); and it has a history of telephone interaction. Indeed, for banking operations there is a history of tortuous telephone interactions, via keypad interfaces, so the addition of voicing to the transactions will be potentially welcomed enthusiastically by the customer base.

David Novick and his colleagues offer a useful chart, keyed to information flow, for plotting the demands placed on voicing a task, reproduced here as Table 8.3.

Low-input (that is, a low information flow from user to system), low-output (system to user) situations are the easiest to automate; high-input, high-output situations are the hardest. For instance, a telephone polling application, with relatively brief and specific pieces of information flowing back and forth, lends itself to a largely fixed-initiative interface with clearly determined vocabulary at every turn. "Age?" the system says, and "Sex?" and "Please rate your current television viewing as High, Moderate, or Low," and so on; to responses like "37," "Male," "Low," and so on. At the other extreme — high input and output — an advisory service has so many contingencies at almost any point of the interaction that building an effective speech system is difficult in the extreme.

Of course, all the situations in Table 8.3 are variable by register and task. An advisory service for matching wine to food would be quite manageable — fairly constrained for task, and involving a narrow register. It wouldn't be trivial. The variables are still extensive. But scripting a structured interaction which controls the way those variables enter the interaction, and therefore controls the vocabulary, is still quite manageable. On the other hand, an advisory system for family medicine would be an incredibly difficult undertaking. The possible symptoms and the available descriptors are legion, the potential combinations more so; and the system might need to probe for unreported (and possibly unnoticed) symptoms, particularly on the basis of earlier input.

And that presupposes a well-structured database. What if the database is already in existence and poorly structured, or structured for another medium, or both? What if we're talking about the Web?

Voicing the Web

Within a year or so, people will be shopping by talking to their Web pages, and the Web pages will be talking back.
    — Anne Eisenberg (2000)

In the late 1990s through the turn of the century, speech technology companies were blooming all over the gadget-strewn, high-tech landscape, with promises that the dazzlingly gigantic World Wide Web was just a phone call away. "You don't have to have a browser on the mobile phone to access personal or Internet information," said Ben Linden, the VP-Marketing of Phone.com. "You just dial into a WAP-based system and talk" (Schwartz, 1999). In the very thick of it, Daniel Eisenberg characterized this frenzy as "the latest, hysterical high-tech land grab" (2001).

After the 2001 dot-com bust, the story changed. Enthusiasm waned with the economy, cynicism filled its vacated attitudinal niche, and articles like "Voice Recognition: Another Dead End" (Dvorak, 2002) became the order of the day. "Who really wants to listen to a computer read a web page over the phone?" complained one author (Tweney, 2001). Well, nobody. That's stupid. It's like asking who wants to see a video in a magazine, looking at one frame after another, page after page after page. Videos are a medium that require motion, something that magazines can't provide. The web is a medium that requires space, something telephones can't provide.

Web pages are graphical through and through, and the text — the source for any spoken output a voiced version would utilize — is often the biggest liability. Don't take my word for it. Here's a web designer, exposing what he calls the "dirty little secret" of web designers: that they ignore or slight content because of their graphical proclivities. We "crave eye-candy," he said:

> *We can't get enough beautifully designed, Flash-enhanced, supercharged pixels-per-inch into our cerebral cortexes. We're junkies of all that is ephemeral and we tend to eschew the big ideas — content that engages us, motivates, inspires, and informs us.*
>
> (Powell, 2001)

This poverty of content generally, text specifically, has recently motivated Jakob Nielsen, among others, to turn his web-design attentions forcefully to information richness, and the poverty of semantic structure is largely behind the growing migration to XML.

But it is not a simple matter, even if the content is rich and richly structured, to build useful interactions. It is not even simple most of the time to generate useful speech output. How do you read bullets? Rollovers? Animations? What do you do about pop-ups? Your site includes email reviews: How do you read abbreviations (IMHO)? Quoted text? Headers? Sigfiles? Emoticons?

Who wants to listen to the Web over the phone? Nobody. The Web is just an endless collection of graphic interfaces to data. But tweak the question a bit until we get something sensible, and we can proceed: Who wants to *talk* to the Web? Who wants to use their phone to bypass the graphic interface and access all that massive, pulsing data behind it? Probably everyone with a phone, the need or desire for a quick answer, and reasons for avoiding graphic displays and/or direct manipulation.

Voice enabling the Web for the sake of voice enabling the Web is pointless, but there are web sites galore with promise for useful speech interaction. The trick — provided there is a service with a potential customer base — is not to treat the web site itself as primary, but as the graphic interface to data the customer wants to access. It's the data, not the site, that is key. The particular challenges of the Web are the challenges of how the data is formatted.

Some kinds of visual data, and even such phenomena as links, can be voiced, but only if the perspective is changed. We need to envision not a graphic interface read over the phone but an interface that is designed expressly for vocal interaction; a *parallel* interface that might exploit elements of the visual interface, but is primarily another channel to the same data, not a *parasitic* interface, grafted opportunistically onto the graphic interface. Don't take it from me. Take it from a pioneer:

> *Speech interface design needs to be a separate design effort. If you just tack it onto the graphical interface, your application is doomed to mediocrity, if not failure.*

(Yankelovich, 2000: 319)

When framed in these terms, it's not a question of who wants the Web read to them over their phone, but hundreds of far more specific questions. Who wants to get some information and evaluations on a new DVD player before making the purchase? Who wants to know how Research in Motion is trading the morning after the merger? Who wants to know what the traffic is like on the I-90, before taking the onramp? Who wants the score of the Jays/Red Sox game last night, because he had a bet with his neighbor that he would mow her lawn if the Sox won? Who wants to know the score of every Jays/Red Sox game this year, *before* he bets with his neighbor?

In 1990, one could easily have asked "Who wants to see a computer display a newspaper?" and lots of people probably did ask it dismissively. In retrospect, it's easy enough to see that only a lack of imagination could have sponsored that question. Nobody would want to, if it meant scrolling horizontally and vertically around the full-size image of a multicolumn broadsheet. But, break it up into story modules, rearrange the "page," preserve the newspaper tradition of partial stories on the front page, but in smaller doses now, put in links, provide the compression and bandwidth for graphics, even toss in some video, and newspapers became one of the success stories of the web — a success wholly based on the reconfiguration of the content to suit the new medium.

Designing in this new medium requires learning about the principles of spoken language. Yet it is almost equally important to unlearn the principles of graphic language — not just in the spelling-and-punctuation sense of prose conventions, but in the margins-and-bolding sense of page layout, and in the icons-and-buttons sense of graphic interfaces, which evolved from principles of page layout. The next section takes up two particularly significant aspects of unlearning from graphic design.

Bad Habits

... She seems to have prepared her witchcraft in such a way that a spoken word is necessary to accomplish her designs, and these spoken words are known only to herself.
  — L. Frank Baum

The *graphic* in *graphic(al) interface* originates from the Greek *graphein* "to write." By way of typography, graphic interfaces are layouts, with text and pictures and abstract geometrical objects deployed in various proportions. Our encounters with graphic interfaces are governed at least as much by the written language in button labels, menu items, dialogue boxes, file names, field definitions, tool descriptions, and help text, as they are by pictures and shapes (and, of course, vastly more than they are by the few impoverished sounds incorporated into various nooks and crannies of some interfaces).

All of that, every bit of it, gets in the way seeing a voice interface as the temporally determined auditory experience it is, but an especially troublesome infection from graphic thinking, which also comes out of keypad (and, thereby, voice-response) interfaces, is the hierarchical menu. Firstly, we need to layout carefully the differences between voice and graphic interfaces.

A Voice Interface is Not an Auditory GUI

Talking and listening are activities that are intimately bound to social conventions of interpersonal communication acquired throughout our development. Pressing buttons and viewing displays trigger a rather different set of associations and behaviors.
  — J. A. Waterworth

Many of the early graphic interfaces were dismal experiences because they were translated directly from command-line systems. Graphic interfaces got better as they began to be designed expressly for visual display and direct input, not kluged into those modes. The lesson for voice interfaces is unmistakable.

Almost all the training, literature, and implementation of interactive design has been overwhelmingly based in graphic principles for over a decade now. And almost all of us, designers and users, are disposed to see computers by how we get at their functionality, through their interfaces. Since we are so thoroughly GUIfied, no matter how virtuous we hope to be about keeping voice interfaces and graphic interfaces distinct in our minds, there is a huge amount of inertia to overcome.

So, repeat after me: "A voice interface is not an auditory GUI." One cannot be translated into the other. Voice and graphic interfaces can be brought into certain functional parallels, much as a movie and a novel can share characters, themes, and plotlines. But also like movies and books, they are fundamentally different media, entailing fundamentally different experiences, and calling for very different design considerations.

Not only are voice and graphic interfaces not the same, they are almost completely inverse. Their strengths and weaknesses, in particular, are largely antithetical.

Input

Speech input can be employed when the hands are busy, and/or when the user is not near the relevant device. It requires speech (and hearing), but not sight or muscular control. Direct-manipulation input occupies the hands and eyes, tying the user to the device. It requires sight and muscular control, but not speech or hearing (though auditory alerts and feedback are often used redundantly). See Table 8.1.

Output

Auditory output can be employed when the eyes are busy, and/or not near the relevant device. It requires hearing, but not sight (or, for that matter, literacy). Visual output occupies the eyes (and also requires literacy, for menus, system messages, tool labels, . . .), tying the user to the device. See Table 8.2.

Stability

Voice interfaces are primarily temporal. Graphic interfaces are primarily spatial, though interaction with them happens over time. They very effectively deploy temporal devices, such as progress bars, and they depend on motion — the change of spatial relations over time — for much of their functionality. This difference has a profound impact on their interactive characteristics. Speech (along with sound, generally) is transient; graphics are stable.

Naturalness

Natural is as natural does, of course, but speech is a much more pervasive and deeply ingrained phenomenon than hand-centric input devices, which must be learned (though the expertise can be quite portable). My seven-year-old son can talk your ear off, and come back for the other one after lunch. But he can't type. My three-year-old daughter can mold her world very efficiently with speech — get what she wants, express how she feels, create wonderful songs — but she needs a great big arrow, thick-edged screen shapes, and a slow track ball to use the computer.

Information display

Because they are inherently stable, and because they have writing, pictures, and motion, graphic interfaces can display incredible quantities of information. Voice interfaces cannot "display" anything.

What's more, people can utilize those incredible quantities of information very efficiently. Sight has a very broad bandwidth; it can take in lots of information, and substantially control its flow, allowing people to zero in on what they care about and ignore the rest. Hearing has narrow bandwidth: the information has to come in serially, and although some filtering takes place (you don't hear the hum of the computer or the patter of the rain unless you listen for it), it is much less efficient — a bunch of simultaneous displayed images can be visually sorted through, one at a time; a bunch of simultaneously broadcast voices are a babble.

General mechanisms of attention operate in both domains, of course, which allow us to sort through stimuli, but the focal, directional, and peripheral characteristics of vision give it a massive advantage.

Information flow rate

Speech is not the ultimate method for information transfer. It's damn good, but it's not ultimate. If it was, we would never have invented writing, pictures, oscilloscopes. (It's also not independent of activity, of course; see Table 8.3 for the information-flow-rate effects of specific voice tasks.) Even in terms of language, graphic display is more efficient for getting information into people than speech is. We can read faster than we can hear.[2] On the other hand (sticking with language), we can speak faster than we can type.[3]

Of course we don't, however, stick with language when we are inputting information. It's very difficult to quantify something like information flow rate in a comparison of speech and direct manipulation, but Bruce Tognazinni (2001) sees it this way:

The clutter of words, icons, and buttons that obscure our screens today are the result of the severely limited vocabulary of the mouse. The only word it knows is "click," so you have to find an instance of the word you want to convey to the computer, then say "click" while you hover over it.

He exaggerates to make his point. Pointing devices have a syntax, too, not just a vocabulary, and that leads to a wider repertoire of actions than just point-click. Double-clicking

---

2: The average adult reading speed in the United States is 250 to 300 words per minute, which can actually increase with serial presentation, up to 800 words per minute (Bailey and Bailey, 1999). On the other hand, we hear words most comfortably at around the 150 to 160 words per minute rate. That's the rate at which "books on tape" are recorded (Williams, 1998). This, too, can go higher — up over 200 words per minute, with compression (Omoigui, He, Gupta, Grudin, and Sanocki, 1999).

3: In one study of dictation, for instance, people spoke to computers at about 105 words per minute, while typing rate for transcription was around 33 words per minute, for composition 19 words per minute (Karat, Halverson, Horn, and Karat, 1999).

and half-clicking (depressing and holding) afford launching, dragging, and sliding actions; with multibutton devices, right-clicking and chording open up other functions. But Tognazinni's defining observation holds: the input vocabulary (and syntax) is vastly richer with natural language; indeed, even a severely constrained artificial language has much denser informational input possibilities than direct manipulation (which is why command-language interfaces still have their champions).

Saying "bold, italic, small caps, 14-point Futura" takes me about two seconds at normal rates, for instance; performing the same action on captured text with a (very efficient) formatting palette takes about eight seconds.

Repertoire

Graphic interfaces have a much wider repertoire of actions: pointing, clicking (double-clicking, right-clicking, chording, . . .) dragging, and sliding. Voice interfaces don't have the evolutionary history of graphic interfaces, and no doubt specific catalogues of actions — for more rapid decisions, undoing actions, automatic data-entry, and the like — will evolve as the interfaces mature. But at the heart, there is really only speaking and listening.

Uses in the future development of voice interfaces for volume and pitch, for exploiting the cocktail-party effect, for tonal "objects," for whatever innovations and practices develop to optimize speaking and listening as an interactive medium with computers, will come far more from an understanding of vocal and auditory modalities than from analogies to the current graphic systems.

The most dangerous aspect of a graphic-interface design sensibility for voice system design is the omnipresence of the menu metaphor, made even worse by its contamination of the design literature for keypad and voice-response systems. For voice interaction design, menus are a curse.

Why Menus Are Very, Very Bad for Voice Interaction Design

System   *State the type of cuisine served by the restaurant which you are attempting to locate. Some examples would be Thai, Italian, Indian, or Chinese.*

Caller   *Mexican.*

System   *You chose "Mexican." If this choice is correct, say "yes" or press the pound key now. If you did not intend to choose Mexican cuisine, say "no" or press the star key now.*
— Thom Stark

"And it went on and on and ON like that," Stark (2001) continues, reporting on his encounter with a speech system for finding a restaurant in New York. "By my count, it took fifteen individual steps from the initial query to get at last to the application's ultimate recommendation — a moderately priced Mexican place on the Upper West Side — and the whole process was so excruciating that I just wanted to bite somebody by the time it was over."

Thom Stark was up against the sort of auditory interaction design component called "a menu." The name is not appropriate, and obscures its dreadfulness by borrowing a label from a domain in which it denotes a very successful (though very different) design component. What is a menu?

"The so-called 'menu' in one form or another," Bruce Balentine tells us, "has been a commonplace device in almost every type and style of interface for at least two decades" (1999: 205). Before that, it was a pretty handy document for choosing a BLT, a service it still provides very ably in those restaurants where the waiters don't memorize everything and greet you with an oration.

A menu is, literally, a (detailed) list; we borrowed the word from the French in the 19th century, by way of *menu de repas* ("list of the details of what is served at a meal"). Before that, it goes back to the Latin, *mintus* ("small in size, amount, or degree" as well as "possessing or involving minute knowledge," similar to its English descendant, *minutiae*). Hence, a detailed list.

A menu is visual. It supports grouping (appetizers, salads, deserts, . . .), and fosters a comparison (usually on several variables such as price, foodstuff, and preparation). On a page, or a blackboard, you can take in all the items in a glance, search through them, look back at ones you've forgotten, skip over several to look at the fourth one in the list, and so on, eventually settling on a BLT. Ignoring the additional toggling, cascading, pop-down, tear-off functionalities, all of which change the dynamic a bit, this describes pretty well what you can do with a graphic-interface menu as well, though without the gustatory rewards.

You can do none of those things with the so-called speech menu. You can just listen.

Through various functionalities, it's true, you can go back to preset points and hear the list start up from the beginning again, you can skip up or down a hierarchy, and even choose-before-you-hear if you know what's coming. But it takes stamina, concentration, and a very good memory to get around a hierarchical auditory list structure in even a remote approximation of how you can get around a graphic menu, on your computer or in a restaurant.

On a graphic interface, using a menu is usually a four-step process:

1. Move the pointer to the menu title, to select the menu.

2. Depress the mouse button, to invoke the menu.

3. Keeping the button depressed, move the mouse, to select the option.

4. Release the button, to activate the selection (and return to prior state).

Putting it this way — putting it serially, into language, that is — makes it seem like a complicated process. It's not complicated, of course, on a graphic interface. We have a motor-script that gives this cluster of actions the illusion of a single action, and it comes with stepwise feedback. You go to the menu title, depress the button, navigate to the menu item, and release the button (all the while seeing the menu appear, the items highlight, the menu disappear, and getting the motor feedback of tension against the pressed finger, beginning, sustaining, ending). But voice interfaces make this single-action illusion impossible to maintain. You listen, listen, listen, choose.

The truth is inescapable: there are not now, nor have there ever been, keypad menus or voice-system menus. There are auditory interaction schemes using telephones *called* "menus," and to that extent, they are menus. What you call something is what you call something. But the term *menu* was not chosen for accuracy, but because "of contemporary biases [in the 1980s] that made the term 'menu' synonymous with 'user friendly' and therefore lowered certain obstacles to product development" (Balentine, 1999: 206). The name for this interaction scheme might as well have been "Shadrack" or "Paris Garters" or "BLT"; the original metaphor breaks down so completely when you move to a sound modality, without stability, vision, feedback, or space.

The metaphor is wrong. The label is therefore misleading. And the interaction scheme coded by that label is generally gruesome. But it gets worse. The menu metaphor has an offshoot that hopelessly contaminates design; users are said to "navigate" these menus. Instructions are offered to users about "going back up to the main menu," for instance, and designers envision users who know "where they are" in a menu, and "where item X is" in the menu. This talk is all spatial talk, not temporal talk.

Let's take an example of "menu navigation." You've chosen an option — say, "Mexican" — and you await further instructions. Let's say, too, that the system you're talking to doesn't require the constant verification that the one Thom Stark used did, and that you can go directly ahead. Now what? Well, now you get another list, something like:

AcmeMenu   State the price range of the restaurant which you are attempting to locate. Some examples would be Inexpensive, Reasonable, Moderate, or Expensive.

Then, if the system's goals are satisfied here, perhaps you get something like:

AcmeMenu   State the location of the restaurant which you are attempting to locate. Some examples would be Uptown, Midtown, Downtown, or Brooklyn.

Etcetera, etcetera, etcetera.

And what happens if you get to the last option, and it just doesn't apply? There's an inexpensive Mexican restaurant in Midtown Manhattan, with a family atmosphere, but it doesn't take credit cards, or maybe just not your card type. Then you've got to reassemble the whole list of attributes again, perhaps starting with Indian this time, and hope for the best. It is exactly this sort of mindless, interminable list of alternatives, coming to dead ends or taking wrong forks, that gives voice systems a bad name (and keypad systems before/along-side them).

None of this means that designers can escape hierarchies, nor even that option-presentation/selection interaction schemes have to be avoided at all costs. From a design perspective, it is virtually inevitable to think of interaction points as nodes in a hierarchy. And option-presentation/selection schemes are quite robust for recognition; when the user can only say three or four distinct words, and you tell her what they are, recognition is vastly easier. But, as Victor Zue (1997) laconically puts it, that success comes "at the cost of user annoyance due to its inflexibility."

The presentation of, and enforced selection from, a list of options should be a fall-back position, when the interface gets in trouble, or a knowingly awkward compromise for circumstances when the database is so large and unwieldy that you can't make it support a more graceful interaction. And a list-after-list-after-list interaction design, without allowing the user more natural input — something other than recitation — is tantamount to design failure.

The ruling metaphor should be one of conversation and cooperative exchange, not of menu navigation; of building common understandings, not of traveling through hierarchical conceptual terrain; of sharing information, not of directing movements. Menu selection in a graphic interface, remember, is a four-step procedure that seems like one; we need to develop the same sort of perceptual ease with voice-interface interaction. A conversational metaphor has exactly that effect. There may be four or five user utterances, including repairs, to get some local task accomplished (finding a suitable restaurant), but it will seem like one event to the user if it is carried out cooperatively over time, rather than dictated under the illusion of space. The "steps" disappear. (See, for some specifics, "Broadening the Interaction" in Chapter 14, which ontlines the technique often called "flattening the menu.")

Habitability

> *Habitability, by definition, is about achieving a balance between system and user behavior.*
>   — Kate Hone and Chris Baber

In human–computer interaction, the members of a small cluster of terms regularly fall into temporary vogues or antipathies, become buzzwords for a while, then foils, and undergo mutual substitutions; *natural*, *compatible*, and *intuitive* are among the most common. These words all point in slightly different directions, but one or more of them is always in active use (while others are in disrepute) because they share a core notion that never goes away: accommodating machines to human dispositions. While vague and occasionally misleading (what, ultimately, could truly be "natural" about interacting with technology?), those terms retain their usefulness by pointing to the heart of the job we do.

We use them because we need a vocabulary to express the importance of matching user expectations to system behaviors. Users expect, for instance, that vertical scales will signal an increase by upward movement, and a decrease by downward movement. Inherent dispositions and cultural reinforcement correlate up with more and down with less, and a computer system that employs scales conforming to those expectations can legitimately be called compatible, intuitive, even natural; so can a thermometer.

All of these words are common in speech-system design articles, but *natural* is by far the most widely used, probably because we are a naturally talkative species. But as generally handy as these words are, the attacks have a point. They *are* vague, and their use in graphic-interface design increases their vagueness for voice work, and their anchorage in ordinary language can make them misleading, Fortunately, there is an able and important word which captures the critical sense of accommodating technology to human dispositions, and is localized to language-based interaction to boot: *habitability*.

The term comes from a very important 1968 paper by William C. Watt (see also Oettinger, 1965). Here's how he describes it:

> *Habitability is a property of a sublanguage (a proper subset of the set of, for example, English sentences), of a universe of discourse (like that of the* Airline Guide*), and of human speakers (ordinary speakers of English); a sublanguage is "fully" habitable for a given universe [of discourse; i.e., a discourse domain] and for a given group of speakers if those speakers can freely use, for example, English sentences appropriate to the universe of discourse without overstepping the bounds of the sublanguage.*
>
> (Watt, 1968:340)

The word is ugly, Watt's definition is unfortunately tied to a Carnapian/Chomskyian view of a language as a set of sentences, and the notion of sublanguage is correspondingly suspect.[4] But for all that baggage, the notion is right at the very heart of voice interaction design. Its virtues are legion. It foregrounds

- speakers
- language behavior
- conceptual constraints

And it

4: *Sublanguage* appears to have been coined by Zellig Harris (1968, 1982), an important figure who was influenced by Rudolph Carnap, who in turn strongly influenced Noam Chomsky, and also markedly influenced the development of computational linguistics. The definition of a language as a set of sentences, made famous in Chomsky's *Syntactic Structures* (1957: 13), while it helps to get at some problems, is a hopelessly optimistic and idealized fiction, which suggests language is a closed system, amenable to full specification with mathematical precision. It's not, and it's not. Of all the relevant topics broached in this book, the overwhelming prevalence of context and the nonsentential character of many utterances are enough to make the point several times over. This definition and its implications are not generally an issue in interface design, which understands both the critical importance of context and the primacy of utterances, except when the contaminated term, *sublanguage*, comes up.

- encompasses a range of mappings among speakers, concepts, and language behaviors.

Plus it

- instantiates the beautiful metaphor of a discourse model as a domain in which users can dwell; in which "users can express themselves without straying over the [model's] boundaries" (Watt, 1968: 338).

This book generally, and this section specifically, is about the design and development of speech-system habitability.

Habitability is not enough on its own, of course, because the interface needs to be designed, scripted, and realized in a way that allows the foundation of habitability to support effective interactions. But it is a bedroch necessity.

Summary

Intellect itself, however, moves nothing, but only the intellect which aims at an end and is practical; for this rules the productive intellect, as well, since every one who makes makes for an end.
  — Aristotle

In this chapter we looked at the appropriate philosophical approach to general interaction design, and more specifically to voice interaction design: it is a creative, rule-guided, skill-based enterprise of the sort the Ancient Greeks called a **techne**. The closest word we have in modern English to this cluster of implications is *craft*, but it is too encumbered with notions of macramé and glass-blowing to do the job. Voice interaction design is a techne.

We considered the appropriate relationship between the speech technology and the interface, to the extent that they can be separated, and stressed that the technology is there to support the interaction, not the other way a round.

We outlined the **considerations behind choosing a voice interface** for a service or database. In particular, we considered that speech input is especially suited to circumstances where the user's hands are busy or otherwise incapable of interacting via object manipulation, and that speech output is especially suited to circumstances where the user's eyes are busy or otherwise incapable of interacting via visual display. These circumstances might be physical, or they might be hardware driven — the digital appliance may not support rapid, accurate object manipulation or sufficiently effective visual display.

We also took an **information-flow** perspective with respect to the degree of difficulty that a range of tasks present to voice enablement. Low information exchange rates, between user and system, lend themselves quite naturally to voice interaction design; high information flow — user to system, system to user, or both ways — presents more challenges to design.

We considered the gargantuan data pool known as the **World Wide Web** for its voicing possibilities. All of the considerations about hands and eyes and information flow apply, of course, for the decision about voicing a given corner of the Web or not.

We took up the necessity of unlearning some of our habits of mind developed from graphic interface interaction and the legacy of keypad interaction. Voice interfaces are not auditory GUIs. They are **temporal** and **transient**, not spatial and persistent. And speech-system **"menus,"** so called, are especially problematic in the way they enforce **hierarchical** (spatial) interactions on a naturally batch-processing medium, spoken language.

The final topic we took up was **habitability**, the accommodation a speech system makes to the user's register that allows him or her to interact with it without straying beyond its resources (vocabulary, syntactic structure, dialogue acts, exchanges).

In This Section

Ironically, the bane of speech-driven interfaces is the very tool which makes them possible: the speech recognizer. One can never be completely sure that the recognizer has understood correctly.
—Nicole Yankelovich

There are a great many significant issues in the design of voice interfaces, but the single most compelling one is the indeterminacy of the input; one can never be completely sure, as Yankelovich says, that the recognizer has understood correctly. This section is overwhelmingly about (1) designing interfaces that increase the probabilities of correct understanding while (2) ensuring a satisfactory user experience. These joint goals — the balance of system capabilities with user and task needs — are effectively the definition of habitability.

An outline of the chapters in this section follows, but I want first to point out, emphatically, that there is nothing in this chapter sequence that is meant to suggest a development sequence. There is a loose chronological arrangement to the extent that the user-and-task-analysis and the discourse-modeling chapters come early in the section as well as early in the development process, but otherwise the chapter sequence is wholly incidental. There is nothing, for instance, in this arrangement that should lead you to believe that you must decide on the agent before you can start scripting (in fact, scripting concerns might well influence agent design), or that testing comes only at the very end of the development cycle (in fact, it comes throughout).

The Team and the Process

This chapter outlines the structure of a voice-interface team, and sketches the development process. The team leader, an Interaction Architect, coordinates the talents and specialties of lexicography experts, interactive dialogue writers, and soundscape designers, with guidance from technology and subject-matter experts. This team works best in the spiral software development model. As with the design and deployment of any significant arti-fact, this development process needs to proceed with the participation and direction of quality assurance and usability experts.

Users, Tasks

Here we detail the data-gathering techniques and goals of building user profiles and task analyses, emphasizing issues of context and purpose, while attending to variables such as initiative, complexity, and quality. The two stages are *observation*, which requires an approach stressing empathy, richness, and openness, and *interview*, which requires asking a set of structured and task-centered questions.

Building the Discourse Model

In close parallel to the previous chapter, this chapter relys on natural dialogue studies and illustrates modeling the discourse that users deploy to perform their tasks. In it, we focus on the habitability goals of getting the vocabulary, utterance structure, and interaction pat-terns right. The primary vehicles to this modeling are two register-specific instruments, a corpus and a digital lexicon.

Agents

Here we outline the criteria behind agent design, with a somewhat extended excursion into the personification debate — on whether voice interfaces should mirror the behaviors of persons. They should. Once the personification issues are examined, the chapter lays out the selection criteria behind such choices as recorded or synthetic voicing, male or female agent gender, and standard or nonstandard dialect, as well as specifying the primary attributes of character, personality, and emotion, and outlining the procedures behind casting voice talent.

Dialogue Matters

Things go wrong with speech systems, no matter how meticulously the vocabulary and interactive patterns are crafted, and this chapter is largely focused on how to minimize those slippages, through preventative measures like effective prompting, stocking the vocabulary, and certain resolution strategies; and on how to fix the slippages that do occur, through strategies that guide the user in carrying out the repair process. It also contains a taxonomy of errors and slippages, and treatments of several other significant dialogue matters related to prevention and repair — managing initiative, tapering and expanding feedback or prompts, and concerns that arise from working with pre-existing text and the human or keypad legacies often inherited by speech systems.

Scripting

Crafting an interactive, goal-directed conversational speech application is effectively the design of an expert system, one that supports the specific purposive verbal behaviors we call *dialogues*. In this chapter we develop the well known, expert-system notion of a knowledge script as a format for developing the dialogue, planning the call flow, and specifying the overall design of a voice interface.

Iterative Evaluation

This chapter concerns Wizard of Oz testing and other methods of usability engineering — full fledged usability tests, heuristic evaluations, pluralistic talk-throughs, beta-tests, and field studies.

Conclusion — Pursuing Habitability

This chapter wraps up the themes and perspectives of the book.

The Team and the Process

[Speech] systems require a lengthy development phase which is data and labor intensive, and heavy involvement by experts who meticulously craft the vocabulary, grammar, and semantics for the specific domain.
— Roni Rosenfeld

Building a complex interactive artifact takes a team, especially an artifact built on unstable input drawn from that whopping-big, highly contingent, personal/social phenomenon: natural language. In this chapter, we outline the functions necessary for such a team, treating them largely as individual job descriptions, though suggesting ways they might be combined. We also sketch out the iterative development process this team needs to work within.

The Team

If the finished parts are going to work together, they must be developed by groups that share a common picture of what each part must accomplish.
— K. Eric Drexler

Voice interface development needs a specific, dedicated team of specialists, which is both independent of, and equal to, the team working more directly on the recognition engine, the language-understanding components, and the database. As Alan Cooper tells it, programmers want the development process "to be smooth and easy". But users want "the interaction . . . to be smooth and easy. These two objectives almost never result in the same program. In the computer industry today, the programmers are given the responsibility to create interaction that makes the user happy, but in the unrelenting grip of this conflict of interest, they simply cannot do so" (1999). Graphic interface development now understands this conflict, in principle if not always in practice, but voice interface development still puts far too many of its eggs in the programmers' basket.

This section advances an interaction-design team structure that goes beyond current practices, and way beyond recent practices, for speech system development. A team structure of this extent will not always be necessary. Some personnel may be less essential for the success of a voice project than others — depending on the product itself (especially the information-flow variables outlined in the previous chapter) and the legacy products (especially usability research and task analyses associated with them), as well as on deadline and budget concerns. It is conceivable that one energetic, dedicated, creative person can design and implement an interface for a reasonably tidy system with a well-known user base — a call-routing system, say, for a small office with a clearly defined clientele — with an off-the-shelf recognizer. For a retail system, on the other hand, with multiple tasks, high grounding criteria, and heterogeneous users, the requirement is for scores of people (some of whom may have other overlapping assignments, some of whom may be contracted). But body-count issues aside, for even moderate success, the team does need to perform all the *functions* represented by the job titles in this list:

- Interaction Architect
- Lexicographer
- Interactive-Dialogue Writer
- Soundscape Designer
- Quality Assurance Prime
- Usability Prime
- Research Prime
- Speech Technology Expert
- Subject-Matter Expert

Team Organization

> *As it is the power of exchanging that gives occasion to the division of labor, so the extent of this division must always be limited by the extent of that power, or, in other words, by the extent of the market. When the market is very small, no person can have any encouragement to dedicate himself entirely to one employment.*
> — Adam Smith

All of these functions are critical for the success of a voice interface — in various combinations, depending on the project, its goals, and the outsourcing. The ideal arrangement is to have all of these functions represented by distinct individuals, so they can work synergistically with each other and the other stakeholders, under one (virtual or real) roof. But some roles might be naturally combined under the job description of one individual, and some roles might be served by people seconded from other groups.

The speech technology expert, for instance, might be adjunct to the team, in the sense of being directly involved in building the recognition and language tools, but made available for regular consultation about the interface and participating in the planning and the reviews (though, ideally, the speech technologist on the interface design team should be a dedicated member of the group). Several of the other roles, too, might also work in an adjunct sort of way. For instance, the Quality Assurance and Usability Primes will often be responsible for several overlapping projects (not all of them necessarily voice projects), with deeper involvement in each project at different times, and explicit sign-off obligations at established gates; only rarely, perhaps with a company just entering the voice market, will they be dedicated exclusively to one voice interface project. The Subject-Matter Expert is the role most likely served by an adjunct, provided by the client.

Interaction Architect

Unless . . . either philosophers become kings in our states or those whom we now call our kings and rulers take to the pursuit of philosophy seriously and adequately, . . . there can be no cessation of troubles.
— Plato

Most current speech-system development teams call the position that corresponds most closely to the Interaction Architect by other labels — *Dialogue Team Leader*, or *Speech Application Project Manager*, or the like, usually emphasizing technical credentials, especially coding experience, and/or time served in the telephony world. This view is mistaken, deeply so; the emphasis should be on design qualifications and knowledge of conversational interaction. It's not that programmers do not have the interface sensibilities and usability awareness to guide the design effectively. Some do, but not many. The reverse situation, a design specialist with software savvy, is far more common. And in general it is usually easier, by a country mile, to bring a designer sufficiently up to speed on the technology than it is to bring technologists sufficiently up to speed on design considerations, the quite-different sense of creativity, and the user-centered intuitions (let alone damping the technology-centered intuitions).

The Interaction Architect should be an amalgam of guru, artist, and manager; the language magnate of the team, for whom the notions of Gricean cooperativity and register and logical form and conversational protocols are second nature; she should also be the chief artisan of habitability, responsible for the overall structure and tone of the interface; and she should be the leader, the one who directs the team that develops the system. The term *architect* here is meant somewhat more literally than it generally is in computer development: someone who has the vision, who guides the team that draws up plans to execute that vision, and who oversees the execution.

Plato has attracted a large amount of flack over the millennia for his argument on the superiority of his ilk over the rest of us: for being a philosopher who said the world would only get better when philosophers were kings, or kings were philosophers. I am open to similar amounts of flack. I am an interaction designer saying the products will only get better if designers guide the project, or the project guiders are designers. Guilty. There are three responses:

1. I am far from the only one saying that interaction design should be in the driver's seat of product development. Jef Raskin, Alan Cooper, and Bruce Tognazzini, for three prominent voices, recurrently argue pretty much the same thing (e.g., Raskin, 2000; Cooper, 1995, 1999; Tognazinni, 1996). And here's an injunction from Donald Norman to much the same end, that design comes first, technology only in the service of design:

   > *Designers of all sorts — industrial design, graphics design, and interaction design —*
   > *[should be] working as a team from the very beginning of the concept of a product. First*
   > *of all, to decide what the product should be . . . Second, to decide what its function*
   > *should be, and third, how it behaves.*
   >   (Norman, 2000: 25)

2. This need for interaction-design-from-the-git-go is even more acute for speech products than other types of interactive objects. A speech-recognition product is, for the users, *only* an interface, nothing else.

3. Plato was right.

Credentials

The Architect should have a strong combination of management, language, and design training/experience. This person's credentials should include:

- a degree in human factors, with several years experience in interactive language technology (remembering that language technology in the sense I'm using it in this book is as much about conversational analysis and pragmatics as it is about speech recognition or synthesis), and strong indications of both leadership and project management skills

  OR

- a degree in language technology, with several years experience in interactive design, and strong indications of both leadership and project management skills

OR

- a degree in almost any field, with extensive experience in interactive language technology, at least some of which is in project management

Functions

"Abstractly, software architecture involves the description of elements from which systems are built, interactions among these elements, patterns that guide their composition, and constraints on these patterns. . . . In addition to specifying the structure and topology of the system, the architecture shows the correspondence between the system requirements and elements of the constructed system, thereby providing some rationale for the design decisions" (Shaw and Garlan, 1996: 1, 3).

The Architect's role is to:

- coordinate the user and task analyses

- develop the product specification

- shape the design of the product

- drive the design, by matching its language behavior to its users' language behavior

- drive the design, by giving the team direction, focus, and leadership

- develop project plans, deadlines, and completion criteria

- ensure objectives are met

- develop, guide, and sign off on the design specification

Lexicographer

> *The ideal candidate will have a love of both words and computers.*
> — from a job ad for "lexicographer"

Someone on the team must have primary responsibility for the construction of the discourse model, the nuts-and-bolts specifics of its habilitability. Someone has to coordinate the study of the register(s) that the system needs to speak and understand. Different disciplines have different names for this sort of expertise — *terminologist*, and *domain analyst*, as well as *lexicographer*, and a good many people fitting the label *corpus linguist* would also be appropriate — but the team needs someone whose primary job is to conduct and coordinate the research into the language of service domains. In the early days of graphic-interface development the need for graphic experts rapidly became acute, and recruitment from the ranks of graphic artists and typographers soon followed — to consult on, and shortly after to contribute directly to, the design of the interface. The analogous situation in current voice interface development is the need for lexicographers.

Credentials

The ideal qualifications for the lexicographer is experience in exactly this work, but it is very rare. In its absence, possible credentials include degrees in any of the following disciplines, especially in combination with experience in user research, field work, and/or corpus studies:

- Lexicography

- Lexicology

- Terminography

- Anthropology (with linguistic training)

- Sociology (with sociolinguistic training)

- Social psychology (with discourse-analytic training)

- Philology (a pursuit that involves intensive lexical study; such degrees usually come out of modern or ancient language departments)

- Knowledge engineering (with a natural-language focus)

- Linguistics (especially with a lexical or morphological focus, field-study training, and/or corpus research)

This list is loosely ranked, in the sense that people with credentials in the first-bulleted field (lexicography) are somewhat more likely to be suitable than people with credentials in the last (linguistics), but there are manifold complications. Firstly, lexicographical training is rare. There are few degree programs in lexicography in the Anglo-American academic systems. But there are, increasingly, classes in lexicography in departments of linguistics, computer science, and various modern languages.

Nor will just anyone do who fits criteria in that list. A voice interface lexicographer has to have the right sensibilities. Even a trained and experienced lexicographer, for instance, could be a serious liability to the design team if he is the wrong type of lexicographer. There are basically two wrong types (though they often inhabit the same body). "Every language," wrote Samuel Johnson, the patron saint of English dictionary makers, has "its improprieties and absurdities, which it is the duty of the lexicographer to correct or proscribe" (Willinsky, 1994: 231 n2). No, Dr. Johnson, it isn't — at least not if you're involved in speech system design. There certainly may be words one would want to discourage in system interaction, and others one would want to encourage, but the decisions to these ends must be made on the basis of utility, not linguistic morality. And, don't kid yourself, *morality* is the appropriate term; words bring out fiercely Puritan streaks in many people.[1] The first type of lexicographer to avoid, then, is one who sees his work as partially corrective or purifying. The second type is equally common, one with an allegiance to print generally and fine print particularly, who would rather hunt in the pages of *The New Yorker* than listen to talk radio. Lexicography has long had a commitment to print authorization; words rarely get in dictionaries unless they appear in print.

The lexicographer either has to be intimate with computational linguistics, or people who are intimate with that field need to be put at his disposal. The job cannot be done without working closely with corpora, and working with corpora without computational resources is as primitive as typesetting by hand.

Functions

The Lexicographer needs to:

- seek out text corpora (there are lots of scattered, voluminous text collections, many based on spoken tokens)

- interview and observe discourse users in the relevant domains (catalogue shoppers and sales people, for instance)

- deploy digital collection techniques

- create and manage domain-specific corpora

- analyze discourse domains, via

  - text corpora, speech corpora, interviews, and observation

  - web research (a peculiar kind of text corpus)

  - media (especially verbally driven media, such as talk radio)

  - whatever (this is a new field)

- create and manage lexical databases, especially the interface lexicon

---

1: Look, for evidence aplenty, at the language columns of newspapers and magazines; the overwhelming bulk of them are about what people *should* say, and only what they *do* say to the extent it can be lectured against or ridiculed; often, such columns include as a leitmotif the complaint that civilization is going to hell on a handcart along with, say, the misuse of verbs. Steven Pinker calls such people *language mavens* (1994), and he might have added an adjunct category of maven wanabees. A great number of people who study and love language have corrosive notions of correctness (corrosive for voice interface design in any case). Terminographers, while in general excellent candidates for voice-interface lexicography, are often especially prone to prescribe and proscribe, since standardization is an important component of their profession.

Interactive-Dialogue Writers

The first draft of everything is shit.
— Ernest Hemingway

The difference between adequate speech systems and great systems, Blade Kotelly says, is in the words and the phrasing of the system output. "Great systems," he says, "do it with an elegance worthy of a haiku; their meaning and impact are clear and immediate, and not a single word is wasted" (2003).Writing the system side of the dialogue is not only among the most critical in speech system development, but also among the most misunderstood. Current ads for "Dialogue Developer" or "Script Writer" or "Prompt Designer" are almost always more concerned about proficiency in Java or C++ or VoiceXML than in a natural language, like English. For reasons that are easier to explain (technomyopia) than to understand, this situation is not a scandal in the industry. "In my experience, effective prompt and message design is the most difficult aspect of [a voice] application development for software developers," Mike Farley (2001) has written. "Yet in many cases developers must write 'spoken' content because there's no one else to do it." An expression from the 1950s comes to mind: "There oughta be a law." It should be an indictable offense for the most critical element of the interface to be crafted by people with no experience, training, or interest in it. This is a job for language professionals, for people who are not satisfied with the first or second version or third version of a system utterance, but who will work and craft and refine it until they have something they can be as proud of as they would be of a haiku.

Thirty years ago, most user documentation was dreary, system-centered, and forbidding to all but the most masochistic users. The reason was easy to explain: technomyopia. The company allegiances were to the technology, not to the users, and writers were not even allowed in the door unless they had a bachelor of science in some branch of engineering that proved they could generate code or fire up an oscilloscope, but said nothing of their facility with verbs. Pocket protectors were mandatory. That meant the writers were either failed engineers who couldn't get a better gig, or, far more rarely, engineers who had an interest in language and communication, and who had sought the position out. The latter were among the vanguard that vastly improved the standards for documentation to the (not always unproblematic but) more sophisticated, context-sensitive, user-centered suite of information support instruments now available for any successful computer products, from brochures and manuals to roll-over labels. This change came with the universal recognition that information developers must be experts in communication first, user advocates a close second, and technological proficients a very distant third (in practice, most are technosavvy, which is what attracts them to this communicative arena; in principle, technoliterate is all that is required).

The meaning of this historical parable should be clear. The words spoken by the system bear a huge amount of the interactive freight: along with the soundscape, they completely represent the system to the user, delivering the necessary information, coaxing the appropriate responses, making the requisite repairs, and providing the personality of the system. Leaving them to someone with little interest and less training in communication — especially as the technology itself licenses a move beyond the "Me system, you caller" stage into genuinely conversational exchanges — is industrial suicide.

Credentials

These people need training and/or experience in professional communication, along with creativity, imagination, and excellent interpersonal communication instincts. Their credentials should satisfy the following criteria:

Necessary requirements

- creative talent

- degree in Professional Communication or in a field like Rhetoric with workplace experience (that is, the degree should include a significant co-op or intern component)

  OR

- Several years experience in professional communication

Desirable requirements

- Experience with nonprint media, especially those which utilize voice and sound

- Knowledge of linguistics/pragmatics/conversational analysis

Functions

Interactive-Dialogue Writers script half of a wide array of projected dialogues, a subtle and creative job. Specifically, these folk need to:

- write dialogue

- anticipate and prepare for voice interactive scenarios

- maintain personae consistency

- put the interface design into words by, *inter alia*:

  - following conversational maxims

  - understanding dialogue-act and dialogic-pair requirements

  - shaping interaction via expansion, tapering, and entrainment

- fostering naturalness via devices like ellipsis, anaphora, and constrained variability

- cuing and recognizing conversational turns

- scripting feedback

- building and profiting from conversational ground

- building, changing, and responding to topic elaborations or shifts

- managing coherence

- incorporating cohesion

Soundscape Designer

Surely it is part of the meaning of an American to sound like one.
— John Rupert Firth

The soundscape of a voice interface corresponds very closely with what graphic designers used to call the "look and feel" of the interface, a somewhat amorphous phrase that means "everything which contributes to the aesthetic functionality of the interface." In a voice interface, these components are primarily music and auditory feedback cues for orientation, navigation, and branding, as well as the vocal characteristics of the agents.

The job of developing a soundscape is frequently left until very late in the process, and often farmed out. Ideally, it should be developed hand-in-glove with the rest of the interface, with exactly the same opportunities for prototyping and iterative development.

Credentials

I'll suggest a set of credentials shortly, but this position is the most difficult to create a shopping list for. Roughly, soundscape design calls for a musician with a penchant for ambient music, a creative sampling disposition, and a thorough knowledge of industrial design; experience with film soundtracks could be beneficial, or in the auditory aspects of multimedia, or in various aspects of radio engineering (drama, advertising, news magazines). A degree, or even experience, is not as important here as inspired creativity, but only of the right sort. In particular, there is no room for the sort of expressive creativity that lies behind much music (and art generally): "this is in my soul, and I have to get it out." What is needed is a deeply communicative creativity: "this is the message, and I have to get it across."

But a degree and a record of experience serve two important purposes in helping to fill this position. They signal a disposition to work in the communicative realms of sound.

And they suggest a body of informing theory and a repertoire of techniques for putting that theory into practice.

Requirements

- creative talent

- degree in Communication, with a concentration or a thesis in sound design

  OR

- several years experience in radio engineering

  OR

- degree or several years experience in multimedia design, with an emphasis on sound

Functions

The soundscape designer will have to

- design the non-speech audio of the voice interface

- develop and maintain a library of audio cues

- compose or commission music

- manage (or find and contract) a recording studio

- supervise (or conduct) the recording and engineering of sound elements

He may also have to (and, if not him, then the Architect, in consultation with him):

- specify the vocal characteristics of agents (recorded and/or synthetic)

- hire (or contract) the vocal talent

Quality Assurance Prime

Quality is Job 1
   — Ford

The responsibility for ensuring the quality and integrity of the dialogues should be concentrated in one individual. For all the familiar quality-assurance reasons, this person should be autonomous of the dialogue writers, and certainly of the Interaction Architect. But a natural place to concentrate multiple roles into one position would be to combine the quality assurance job function with lexicographical job function, or possibly with usability.

Credentials

This person needs training/experience in editing in a professional communication context. Their credentials should satisfy the following criteria:

Necessary requirements

- experience with non-print media which utilize voice and sound

- critical talent

- solid interpersonal communication skills

- degree in Professional Communication, and experience in professional communication, at least some of which should be in an editing or quality assurance role

  OR

- several years experience in professional communication, at least some of which should be in an editing or quality assurance role

Desirable requirements

- knowledge of linguistics/pragmatics/conversational analysis

Functions

This role corresponds to a traditional editor: a gatekeeper who maintains the dialogue standards, and has final approval on all scripts before they are recorded. The Quality Assurance Prime must

- develop and enforce dialogue standards

- maintain, and ensure adherence to, scripting guidelines with respect to constrained variability, convergence, and the like

- proof and edit dialogue scripts

- ensure script integrity, with respect to issues of naturalness, personae consistency, and register

- ensure adherence to general and domain-specific lexical databases

- sign off on scripts before they are recorded

- sign off on design specification

Usability Prime

If our quest is to actually design more usable computer artifacts, then a better knowledge of the "users" is required as a part of our analysis — one that sees people acting in a situation, with motives, and intentions, in interaction with others and the environment.
   — Liam J. Bannon and Susanne Bødker

Usability has grown an infinite variety of extensions since its incorporation into software development in the 1980s; it has become what the deconstructionists call "an empty signifier," meaning nothing more in some contexts than "we had a passing whim about the users." But usability was originally anchored very solidly in user testing, an anchor for this job description as well. The Usability Prime needs to be a general-purpose user advocate in the design and development of a voice interface, but *everybody* involved in the design and development of an interface needs to be a general-purpose user advocate. What distinguishes the Usability Prime is that she must be a *specific-purpose* user advocate, with the specificity coming from recurrent testing (of concepts, prototypes, and products) against samples from the (real or projected) user population.

Credentials

The Usability Prime must have training and/or experience in human-computer interaction testing, and should have very good linguistic sensibilities. The credentials should satisfy the following requirements:

Necessary Requirements

- degree in Human-Computer Interaction, or Cognitive Psychology, or any of the Behavioral Sciences, with direct experience in usability testing

   OR

- several years experience in interface or documentation testing

Desirable Requirements

- experience with voice interface and/or speech technologies

- knowledge of linguistics/pragmatics/conversational analysis

Functions

This person will be responsible chiefly for usability assessment throughout the development cycle, beginning with Wizard of Oz testing, through prototype tests, to late-iteration usability testing. The Usability Prime must:

- analyze user characteristics and needs

- analyze tasks

- design, conduct, and analyze expert reviews, heuristic evaluations, and other usability inspection methods

- design, conduct, and analyze Wizard of Oz tests

- design, conduct, and analyze usability tests

- design, conduct, and analyze beta tests

- collect and analyze postrelease user feedback

- sign off at prototype-to-model and model-to-service stages

Research Prime

Research is formalized curiosity. It is poking and prying with a purpose.
— Zora Neale Hurston

This role is more clearly a function than a specific position. Only the largest and most interface-dedicated corporations will have the luxury of a full-time Research Prime, but this role easily fits into the possible job descriptions of almost any of the interface positions (with the Quality Assurance Prime and the Lexicographer perhaps the most natural). It might even be distributed among several other positions; indeed, to some degree all of the design folk need to keep informed, and all should participate in the process of following developments, investigating new products, and keeping each other current.

But somebody on the development team has to be responsible for gathering and disseminating voice-interface research. Somebody has to stay current with the journals, attend the conferences, and analyze and learn from other voice interfaces (including, especially, any interfaces competing in the same market).

Original empirical research is less critical to product development, but if opportunities arise in that area for, say, the Usability Prime or the Lexicographer, they should be encouraged to take them up, for the payoff in improved current and future product development, but also for conference presentation, to raise the profile and prestige of the company.

Technology Expert

The user-interface designer needs to work closely with a "speech technologist" who has an intimate knowledge of the recognizer and its capabilities.
— Amir Mané, Susan Boyce, Demetrios Karis, and Nicole Yankelovich

The team needs a technology expert, not just one who knows the field well (the theories, the products, the history), but one with close contact to the technology development team, so she knows exactly what the specific system is capable of. I have been promoting human factors at the expense of programming throughout this section and this book. And I'm not about to stop here. The interface should be in the driver's seat. But the recognition engine drives the car.

More generally, there is an inescapable corollary to my regular slogan, "the interface is the product." The design of the interface can only proceed by way of a deep responsibility to the affordances and constraints of all components of the product. The recognizers both afford the entire enterprise and constrain it tremendously.

Subject-Matter Expert

Our discussion will be adequate if it has as much clearness as the subject-matter admits of.
— Aristotle

There is one further role, which is almost always played by someone seconded to the team, that of Subject-Matter Expert.

Interface development needs regular contact with a Subject-matter Expert, to varying degrees, depending on the service the system is providing. In the past, voice (and earlier, graphic) interface development has consorted somewhat too thickly with domain experts, at the expense of consulting with users. But it is equally unwise to go too far in the other direction, and forget the importance of specific subject-matter expertise.

The lexicographer, in particular, frequently needs to work closely with someone who knows the tasks, the procedures, and the vocabulary of the product's domain, from the position of an expert (rather than the position of a user): for instance, a banker, a stock broker, or a meteorologist. Subject-Matter Experts should rarely have the final word. The primary allegiance should always be to users. But experts are frequently a source of very valuable insight, especially into issues of structure and coverage.

The Process

Rapid prototyping, constant user testing, and incorporating user feedback into subsequent versions is . . . essential in arriving at interfaces that successfully deploy auditory interaction.
— T.V. Raman

There are three critical components to the design process of a voice interface:

- **Analyze the register**, including its users and the tasks they use the register to perform

- **Design the interaction model**; that is, model tasks for the users within the register

- **Instantiate and test the model**, prominently including tests with users and tasks

The first component boots the design process up, launches it, gets it off the ground; some people call this stage "predesign" work. It's not. But it is the beginning, and it is a relatively discrete phase. The second two components are heavily interdependent and iterative: design, instantiate, and test; design, instantiate, and test; design, instantiate, and test; . . .

Iterative design has some drawbacks; in particular, it brings a level of uncertainty and instability into the development schedule, a level that makes some engineers and managers uncomfortable. But the benefits outweigh those drawbacks so dramatically as to dissolve them — bringing not just user considerations, but users themselves, into the design process. User behavior is complex and not easily predicted, especially with respect to new technologies and/or new products. Iterative design is perhaps unnecessary for developing a new claw hammer; the tasks and human physiology are well understood, and centuries of Darwinian market pressures have sorted out the critical variables (size, shape, materials). But what this means for claw-hammer design is just that the iterations have largely run their course. Darwinian evolution, in the marketplace just as in nature, *is* an iterative development cycle.

We have comparatively none of this sorting information for human–machine vocal interaction. We have a huge amount of it for human–human vocal interaction, which gives us a design foundation, and a fair amount of it for human–machine nonvocal interaction, which gives us indications (as well as many misleading suggestions) for building on that foundation. But a foundation and some methods is not enough to bypass an iterative process that involves people and tasks, and uses them to draft and refine the blueprints and build the structure.

A useful picture of the process is the familiar helical growth line, popularized by Barry Boehm as the "spiral development model" in a variety of articles and talks since the late eighties, but available in many flavors from many vendors; a highly attenuated version of which I offer in Figure 9.1.[2]

What is most important about the spiral model is the metaphor, the centrifugally expanding growth, which illustrates beautifully three aspects of an effective development cycle for complex, interactive artifacts:

(1) The design gets "bigger" as it moves along (more investment behind it, more detail, closer to market reality).

(2) Development requires systematically revisiting the same design questions from those increasingly bigger perspectives.

2: See, in particular, Boehm (2000), and the references therein, for the developed model. Boehm calls the unauthorized clones of his model, "hazardous spiral-lookalikes" (2000: 5 et passim) — a category my version here may be destined to join, although I am concerned only with the geometrical analogy, not with the institutional details, and my version may therefore be too sketchy to merit the insult.

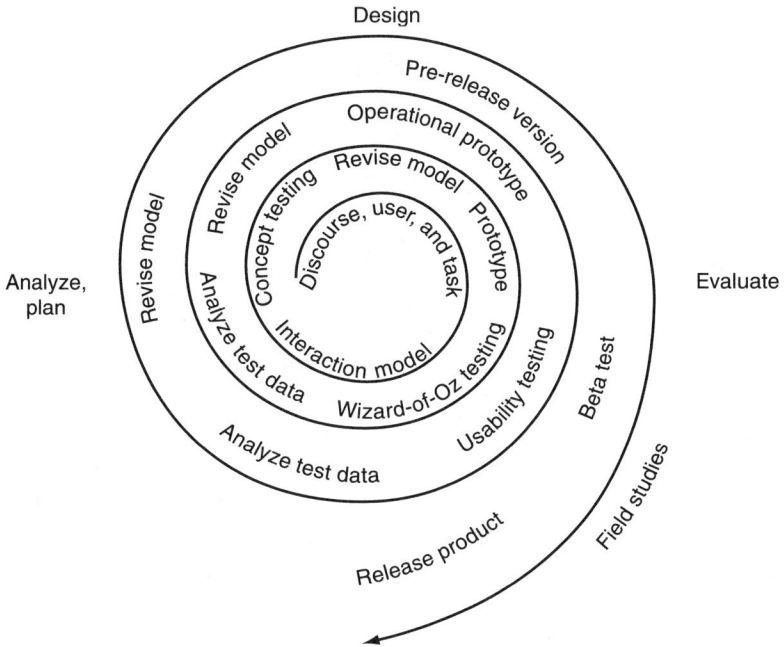

Design

Pre-release version

Operational prototype

Revise model

Revise model

Concept testing

Discourse, user, and task

Analyze, plan

Revise model

Prototype

Analyze test data

Interaction model

Evaluate

Wizard-of-Oz testing

Usability testing

Beta test

Analyze test data

Field studies

Release product

FIGURE 9.1   A spiral model for voice interface development

(3) The developmental process is continuous, rather than discrete, with one stage blurring into the next.

There are certainly administrative boundaries (quality-assurance checkpoints, sign-offs) in interactive-product development, and there are certainly places at which primary methodologies shift. But in design practice those boundaries are more like membranes than walls; they are very permeable.

The spiral metaphor misses one critical element of the design process, however — the essence of iteration, the epicycles of creativity that define each major developmental phase. Wizard of Oz testing, for instance, might require three or four iterations — a pilot study followed by somewhat fuller testing phases — with the results of one set of tests fed into the development of the interaction model for the next set of tests. And so on.

Summary

Specialization is a feature of every complex organization, be it social or natural, a school system, garden, book, or mammalian body.
— Catharine R. Stimpson

It takes a team, working over time, with shared, motivated criteria; established methods; and clear goals, to design and deliver an interactive product as sophisticated as a voice interface. This chapter outlined the functions that must be served within such a team. The leader is the **Interaction Architect**, responsible not only for personnel management of the team, but for its vision, as well, and for the overall tone and habitability of the interface. She crafts the design specification. The **Lexicographer** is responsible for studying the register and crafting an **interface lexicon**, its vocabulary. The **Interactive-Dialogue Writers** are the scripters of the system, writing its speech output. The **Soundscape Designer**, on the other hand, composes the system's nonspeech audio, as well as contributing to the design of the agents and managing any vocal talent. The **Quality Assurance Prime** is responsible for the quality and integrity of the system's output, ensuring that the dialogue writers maintain the appropriate range of tones, in the appropriate interaction points, and that they stay within vocabulary. The **Usability Prime** studies the users and their tasks upfront, designs and carries out Wizard of Oz and usability tests, and stays in productive contact with the end-users through field studies. The team also needs a **Technology Expert**, who guides the design from a close awareness of the affordances and constraints of the recognition and language components, and a **Subject-Matter Expert**, who serves a similar purpose from a close awareness of the tasks and purposes of the service domain.

We also considered briefly the **development process**, chiefly that it is metaphorically a spiral, which cycles iteratively through phases of analysis and planning, designing, and evaluating the product as it grows in effectiveness. I call this process a "voice-interface development process," where others might say "product development process," because I take very seriously the notion that a speech system *is* a voice interface. The product is the interface; the interface is the product. Most emphatically, the recognition and language-understanding components of the speech system are not "the product", to which an interface is appended, late in the cycle, as the industry has too long assumed. They are the interface-enabling components.

Users, Tasks

The key to the development of conversational/agent interfaces lies in the development of detailed models of the task and the user.
— Philip J. Hayes

A speech system has to be highly tuned to specifics: specific people, speaking within specific discourse domains, for specific purposes, to specific devices. Further, like human agents (though not entirely, of course), speech systems have the resources to probe the sequences of words they hear for confirmation and correction. The job remains daunting, but it is eminently tractable, and amounts to knowing (researching) the same three things you need to know for any interactive design work:

- who

- what

- how

For *voice* interaction design work, the how is discourse.

It is the last of these three areas — the opportunity and the necessity of researching the *medium* of the interaction, not just the actors and their actions — that most distinguishes the early design phases of speech-system development from designing and building other interactive artifacts. Discourse analysis is by far the most important element of the design of voice interaction — virtually the whole point of the design.

These three research areas — users, tasks, discourse — are distinct only in the abstract; in the flux and buzz of activity they commingle inextricably. Pursue them simultaneously, not in sequence. This phase of the development cycle corresponds closely to a participant task analysis in graphic interface design, and we can adopt that label without much fuss, so long as it remains clear that it includes as a central component something which is minor or nonexistent in the graphic domain: the close study of discourse. It is at this stage that the construction of the language model begins.

The entire design team should be involved in the user/task/discourse analysis phase — most prominently, the Usability Prime and the Lexicographer.

This chapter concerns users and tasks: in Halliday's terms, the tenor and field of the discourse. The following chapter concerns the elements of the registers that they define.

User Profiles

To achieve user-friendliness, as with any other kind of friendliness, you first have to know who you have got to be friends with. We know who we are, but who are they?
— Janet Whitcut

John Gould and Clayton Lewis call the human component of an interactive system "a coprocessor of largely unpredictable behavior" (1985: 305). True. That's why we need to test, and test again.

Gould and Lewis say further that "there is no data sheet on this coprocessor" (1985: 305). One can see their point here; the available and gatherable information lacks the precision of a cpu spec sheet (though we have highly accurate data about characteristics like auditory and vocal ranges). But there usually are data sheets of various sorts, from various sources, which you need to compile, collate, winnow, and augment. You can't get very far without a decent data sheet on half the final system. The first step in any design is compiling that data sheet: figuring out as much as you can about who is going to use the product.

You need to start with whatever data you can glean that has already been gathered — by the marketing department, by vendors, customers, competitors, web sites, by the research-and-development people who had the idea for this product in the first place, . . . anyone and everyone.

But any data that you get from sources other than your own direct contact with users needs to be treated with caution, filtering out as much of the sources' agendas as possible. Often the context of collection gives you enough information for the filtering: marketers care more about income and ephemeral trends than about experience and professional circumstances; web data is selfreported by people who sometimes willfully distort it because they've been required to enter it on the way to a free download; customers have often collected it for reasons and in ways that are no longer apparent. Take the situation where a bank wants you to provide a service for its customers. It will hand over reams of data it has collected. (Or maybe not. Sometimes you have to go looking; in big organizations, data gets misplaced, but they have lots of it somewhere.) Big-corporation data, though, is often so decontextualized that you can no longer see the politics behind its collection (to support someone's project? to sink someone else's? to answer a competitor's data?). Use it as a starting point, but use it cautiously.

Wherever the initial user research comes from, you will have to augment it by direct contact with users. This need is so critical that you might as well fold up your tent if you think primary user research is too much bother, or not really necessary, or wasteful, because the bank did it already. Remember: this is a *user* interface you're building; avoiding contact with the users is arrogant and foolish. Contact with users will tell you about them; it will tell you about their language, the medium through which they will perform their tasks and it will tell you about the tasks your system will support. All of these elements shape the design directly and feed into instruments (especially user testing instruments) that will further shape the design.

David Attwater and his colleagues, for instance, note the importance of two very important, immediate characteristics of users for general-purpose, telephone-based services (voice or keypad):

Victim or volunteer

> Is the user expecting automation or is he an unsuspecting victim just trying to accomplish a task?

Frequent or infrequent

> Is the user well primed and experienced, or naïve, with little or no previous interaction with the service?

These two characteristics exert a very powerful influence on the interaction style, and therefore on the design. "It is also extremely common," they note, "for these two dimensions to pair up into frequent volunteers and infrequent victims. By definition, frequent callers to a service will quickly come to expect automation and become volunteers if they continue to call" (Attwater et al., 2000: 280).

More specific services call for more specific user descriptions. Building user descriptions involves two levels of profiling — the general population of users, and any relevant subcategories from that population — along parameters that might include the following:

- demographics
  - age
  - gender
  - socioeconomics

- attitude
  - what they like
  - what they don't like
- personality
  - outgoing — reserved
  - cerebral — emotional
  - aggressive — passive
- experience/training
  - general
  - domain
  - system
- linguistic characteristics
  - first- or second-language users
  - dialects
  - professional registers
- phone use[1]
  - business
  - personal
    - business (dental appointments and the like)
    - social
  - consumer
  - information

Some of these categories may not be useful, and most of them require their own fine tunings. For professional registers, it's probably enough to have labels (*legal*, *financial*, and the like), especially since the specific register characteristics need to be plotted out very

---

1: Even if the product is not accessed by phone, this data is highly relevant.

carefully through the entire development cycle; for phone use, a graduated scale (with landmarks like high, moderate, and low) would be more appropriate. To the extent that the product involves specialized activities (installation, repair, translation), specialized hardware (headset, directional speaker/microphone), or specialized purposes (personal financial services, comparison shopping for electronics), the user details might be quite specific.

The best way to gather this information is with a questionnaire-observation-interview strategy, involving the design team fully. Certainly the lexicographer, the writers, and the usability prime should all be involved. But, since gathering this data combines with gathering task data, we can defer methodology for a few pages.

One more thing, however, specifically about user profiling: do not close the book on it, ever, as long as your product exists. You may be, for instance, quite wrong about who is going to use the product, and even why — a potential that is especially plausible for novel products and services. The telephone, for instance, was conceived culturally and economically (and marketed on the basis of this conception), as a business tool for men at work. But among its earliest and most influential adopters were "isolated midwestern farmers' wives who used it to support their social networks" (Lacohee and Anderson, 2001; De Sola Pool, 1977). It may also be that your users change over the course of the product's life.

In either case (error or change), once the product is out, you need to pay attention to who is using it (and how, and for what), remaining alert to design (and marketing) changes implied by what you learn. Mayhew (1999: 44) recommends revisiting user profiles every two years, which is reasonable, but given the novelty of speech systems, the *first* revisit should happen (1) if the user population calls attention to itself in some unexpected way; or (2) in six months; whichever, as they say in the warrantee business, comes first.

Task Analysis

A task analysis should be conducted before developing the system requirements to guide the choice and design of the system functionality; the ultimate usability of the product is actually determined at this stage.
— David Kieras

The "who" and the "what" are often so closely connected that it's not always easy to tell which category to slot data into. For instance, *why* people use the phone (for social or business reasons) is "user data," but *where* they use the phone (home, car, office, train) is "task data." *How often* they use the phone may be task-related, but it might also be the central variable in their experience, traditionally a user category. These bodies of data also function together, side by side, in the design process. In developing user tests, for instance, the user profiles shape participant selection, while the task analysis feeds scenario development for those tests.

In the same places, and at the same time, as you gather user data, you need to harvest task data — with whatever information you can gather from previous research, by the marketing department, customers, research-and-development folk, and so on, always remembering to apply the filters. Documentation for current tasks (a quick reference card for a keypad system, for instance) is potentially a useful source, and there is almost always lots of text on web sites (task-related ones, if they exist, or the focal one if you're voice-enabling a pre-existing site). But, as with building user profiles, the primary method of task analysis comes through direct contact with the people who are performing the tasks. User research puts more of a premium on talking with them, task analysis on observing them, but it is the combination of observation and interview that reaps the information.

Task is not a particularly well-defined term in human–computer interaction. Is it, for instance, getting two return tickets, Copenhagen to Aalborg, on Tuesday, 27th October, departing at 7:00 or 7:30, and coming back by 17:25? Or is finding out about availability one task, booking the tickets a second task, and paying for them a third task? But we have bigger fish to fry than what *a* task is, so long as we identify *tasks* as those activities (however subdivided) that constitute goal-directed interactive behavior. With voice systems especially, the capacity to collapse "steps" or "tasks" into one turn, in a range of combinations, is endemic.

The bigger fish are those activities and behaviors themselves, not their labels. With graphic interface development, working out the behaviors to call tasks and analyzing them is relatively straightforward: the activities are in all likelihood computerized already, so the task analysis largely involves charting the interactive behavior. With voice interfaces, the activities might involve human–human interaction (for instance, booking flights), they might involve graphic interaction (e-planners), they might involve phone-based keypad interaction (voice mail system), or they might involve all three (personal banking). What these activities are, and where they are, condition how much observation you need to do.

The observation is crucial. Kieras (1997: 1402) tells a crazy but not uncommon story of an early digital datebook, which effectively translated a paper day-planner into electronics. But the translation was based on form only, not function. It "included no clock, no alarms, and no awareness of the current date," all of which would have been trivial to implement. There was also no way to schedule repeated meetings (except by a tedious cut-and-paste kluge). As Kieras points out, task analysis would have looked not at the structure (or not only at the structure) of the day book, but at what people do with it and, therefore, what they would have liked an electronic version to do for them. The example (it's only one of many Kieras documents) may seem only to indict boneheaded engineers, but it's far from isolated.

I worked on a phone-based voice interface for a weather information service, coming into the project late, only to discover a fully-coded and operational product that didn't answer the sorts of questions people want to ask about weather. Travelers want to know about flight conditions, sailors want to know about boating conditions, skiers want to know about snow conditions. And the system could talk about most of these conditions, just not in any kind of responsive way. Visibility could not be queried, for instance; it just came as part of a list of "more about weather" (after high and low temperatures and precipitation). Barometric pressure and wind speed came as part of that list, too, but couldn't be requested specifically. Travel advisories and small-craft warnings were not volunteered, or even made available upon request, but again only came as part of the "more about weather" option. Ski reports, which the system didn't have in its database, could have been acquired easily and made to answer the interests of the callers. And so on.

More boneheaded engineers? No, not here, and not with the day-planner developers. The day-planner team was simply translating a paper product into an electronic one. The weather team was just turning a graphic web site into a talking web site. They got the elements right; what they missed was the interaction design, by not starting with the task and making the product answer to that task, rather than providing mere fidelity to a product based in another mode or medium. It's important to emphasize this fact. Getting the fidelity right is hard work. Getting the hardware and software to function in efficient synergy is hard work. It's creative work. It's indispensable work. And it's what the engineers are good at. But it's not enough. The uses a product will be put to must be understood from the outset, or no amount of hard, creative work can save it. The tasks must be understood, and understood early.

One behavior at a time, you document the procedure. There are a variety of variously rigorous methods, from systems engineering, human factors, and interaction design. What they all have in common is the decomposition of a suite of behaviors toward a goal into discrete steps or actions, yielding a rich description of how users accomplish their goals. The core task description may be a series of (possibly nested) actions, or a flow chart, but however you document the task you need to embed the analysis of that procedure in an account that includes a variety of other metrics. And since our target is not an abstract understanding of the existing procedure, but an opportunistic understanding of that procedure with an eye toward translating it into speech interactivity, those dimensions need to include grounding, initiative, and terminology.[2] The categories of analysis will shift with the task and the system, but they should include:

- goal(s)

- grounding

- procedure

- terminology

---

[2]: One of the side effects of a task analysis, of course, is an understanding of how well the current system functions, and if your goal is not to replace that system but augment it in some way (as in the following example, with email), you can make the results of the analysis available to the designers of that system, possibly horsetrading them for resources (interviewers, usability participants, earlier data that team might have collected, and so on).

- system actions
- initiative
  - user
  - system
- complexity
  - overall rating
  - number and type of actions
  - number and type of subtasks
  - number and type of actions per subtask
  - number and type of information units involved
- volume
- problems
  - breakdowns
  - bottlenecks
- environment
- continuity
- criticality
  - overall
  - individual actions
  - presence of economic commitments
- duration
- frequency
- quality

Taking a very simple and familiar task (replying to an email using MS Entourage), writing up a task analysis takes the format partially outlined in Table 10.1 (or one that approximates it). The full write-up is given in the Appendix to this chapter.

You might need additional categories. For instance, the situation might call for more specificity with respect to subtask, or more detail with respect to system vocabulary (such

Goal	To reply to an email
Ground	From received-message window, or main mail window; in either case, the received message, the icons, and the menu titles are all visible. All menus are drop-down, to reveal active menu text, with shortcuts; all icons have pop-up labels. Main mail window and/or received-message window remains visible behind reply-message window. The reply-message window includes main tool bar, information fields, and editing toolbar.
Procedure (=user actions)	Choose reply icon from toolbar, OR Reply from Message menu, OR CMD^R Enter/Edit text Option: add attachment Option: insert (picture, sound, movie, etc.) Option: add/change/remove/replace signature Choose Send now OR Send later OR CMD^RETURN OR OPT^ CMD^RETURN
Terminology	Reply From To Cc Send now, Send later Attachment (Add, Remove) Insert
System actions	Presents reply-message window (recipient, copy, and subject fields completed). Depending on application settings quoted message and/or signature may be in the edit pane. Removes reply-message window upon sending. Presents "Message sending" dialog with progress bar. Potential action: presenting system-status message; e.g., "Mail could not be sent using the account 'RaHa.' Explanation: A connection failure has occurred. Error: -23016"

TABLE 10.1 Sample Task Analysis (Continued in the appendix to this chapter)

as informal labels or pop-up descriptions for icons). And you need to cross-reference each write-up with ones for any implicated or required subtask. Be careful, and be true to the analysis, in adding categories, however. Some people like to distinguish between *observable* system actions (effectively, feedback) and *unobservable* system actions, for instance. But from the perspective of task analysis, there are no unobservable system actions; this requires inside knowledge that only confounds task analysis. On the other hand, noting potential system actions is consistent with task analysis, at least the ones apparent from observation of current or previous observations. Be equally careful in eliminating categories; you can't know beforehand what will prove useful, so it is best to err on the side of comprehensiveness.

Notice that the *goal* determines everything that follows, not the application; for instance the user can quit the application at any time, delete the reply, save it to draft, and so on. But these are functions the interface offers, not actions that satisfy the goal of replying to a message, so they are left out of the task analysis. General interface characteristics will usually be available from system documentation, so you don't need to duplicate the data (if, however, you are tracking a task that is more fluid — like a call to a catalogue service — you'll need to chart the range of potential general user behaviors separately).

Be especially alert to data that might implicate voice interaction, like terminology and initiative, though you also need to resist the urge to start designing the interface at the data-gathering stage, which could begin to limit your observations.

If your target task is a phone-based interaction already, then the task analysis is more immediately translatable into system design. I will defer a fuller discussion of this aspect until the next chapter (on discourse analysis), since task analysis and discourse analysis merge at the level of goal-directed conversations. But Figure 10.1 illustrates this overlap nicely, a depiction of the task-management structure of a request for an operator-assisted call, by David Attwater and his colleagues.

Attwater (with Mike Edgington, Peter Durston, and Steve Whittaker) breaks the dialogue into four exchanges — the task identification, where the operator and caller work out the problem; the task specification, where they identify the specific task they will have to complete; the task itself, an information gathering phase where the caller provides, the operator receives, and they both ground, the number needed to execute the call; and the final exchange, in which the task is completed.[3] There is, of course, no guarantee that a speech-system interaction would follow this format, some of the phatic niceties ("please," "thank you," "even for us") would surely drop away, and finer-grained analysis is required to flesh out the interaction — cataloguing dialogue acts, looking at the encounter more broadly, and the like — in order to build a foundation of data on which to develop the register model. But their work beautifully shows the way that task analysis and discourse analysis meld when the task already depends on conversational interaction.

Figure 10.1 represents the first-cut structural analysis of the discourse. The subtasks and the interaction patterns need to be investigated more fully, and a range of such dialogues needs to be distilled down to a basic knowledge representation of the system-side performance, before we can start building towards implementable voice-interface scripts. (See Figure 11.6, in Chapter 11, for a fuller analysis of a similar interaction.)

Gathering the Data

> *To develop a usable product, you have to know, understand, and work with people who represent the actual or potential users of the product. No one can substitute for them.*
> — Joseph S. Dumas and Janice C. Redish.

Most complex behaviors have automatic, even ritual, components to them. I have seen people assembling sophisticated missile guidance systems with methods that

---

3: I have altered their categories somewhat. Their original terms are, respectively, *problem specification*, *task identification*, *information gathering*, and *task completion* (Attwater et al., 2000).

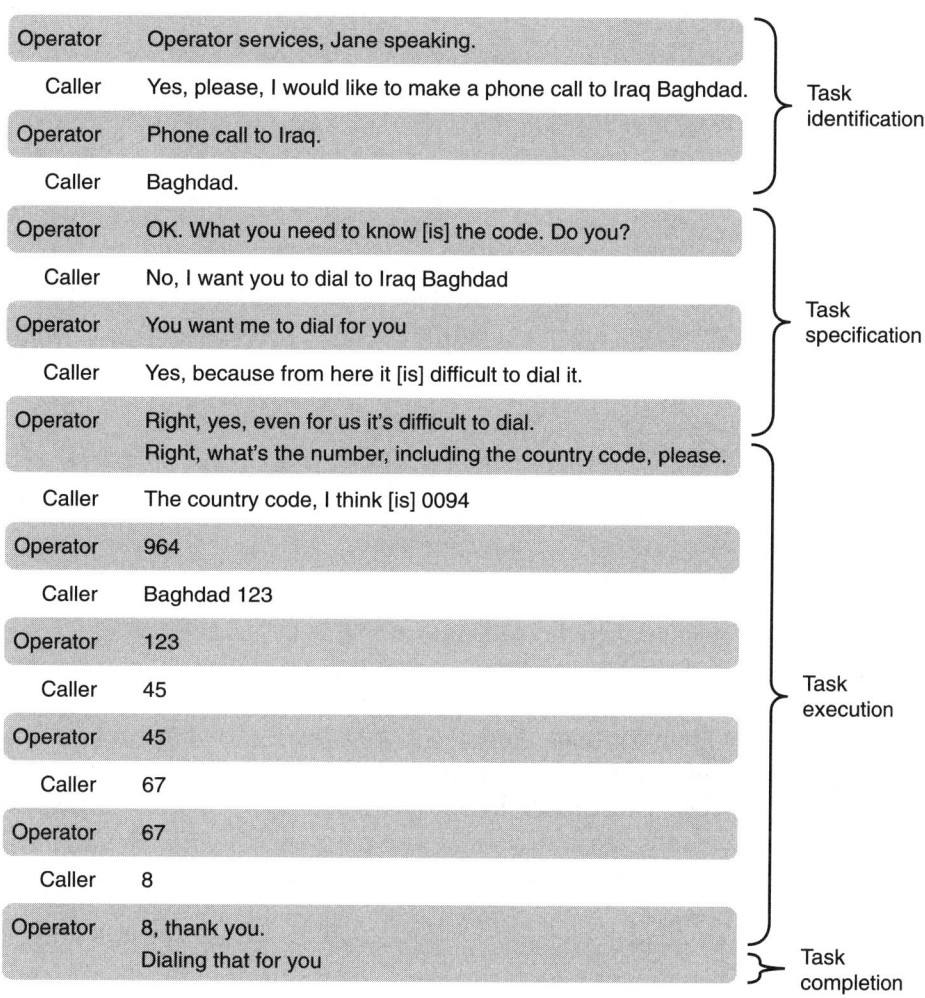

FIGURE 10.1   A structural analysis of an operator assisted call (adapted form Attwater et al., 2000:281)

involved a high degree of superstition ("It worked when I did it this way before"). And there are uncounted stories in the Naked City about usability inspections that uncovered activities so fundamental to the culture of the workers they were invisible to all but the inspector. In short, just asking them what they do is not sufficient; you have to observe. But you can't get at their attitudes and motivations by just observing; you have to talk to them too.

User analyses and task analyses go hand in hand. Both require interviewing. Both require observing. Both depend on a range of quantitative, semiquantitative, and qualitative data. Quantitative data includes simple binary values (which can be counted), like male/female; hard numbers, like age; and measurements, like task frequency. Semiquantitative data includes scalar answers to questions. Qualitative is pretty much everything else — impressions, interpretations, and hunches about why people say and do what they say and do in the performance of tasks.

In any contact with users, at any point in the development and maintenance of a product, the one thing you have to make absolutely clear to them is that *they* are not being evaluated in any way. They are collaborators in the design process, and you should make sure they feel that way.

Observation Phase

To truly understand the work of a set of users . . . you must go to their workplace, learn the user jargon, and observe and talk with a representative set of users of all key types.
— Deborah Mayhew

Start off by letting them know who you are (even if there has been prior contact about this), and how you are hoping they can help your project. Tell them about how long it will take and what it will involve. (Make sure you talk to anyone else relevant in the local chain of command about these matters, too.) And shoot the breeze a bit; mutual verbal comfort is important — *phatic communion*, if you remember the jargon of the trade.

Then, on their timetable, just settle in and observe. Always leave the interview for later, until after you've had a chance to watch the user for a while (usually this means after the observation, but I've had success breaking it up as well: watch, interview, watch again).

You can ask for explanations, and even ask them to describe what they're doing and thinking,[4] but the most important part of the observation phase is just that the users feel comfortable enough to perform fairly naturally. Standard counsel is to be unobtrusive, and that's a good first-order policy. But some people are quite uncomfortable with an earnest spy in the room, and are more at ease chatting with the observer about what they're doing than simply doing it under scrutiny. Some participants even fall naturally into a play-by-play or think-aloud protocol. And anything to make them comfortable is reasonable. Dray

---

4: Verbal protocols (getting people to talk about their tasks as they perform them) are virtually useless further down the voice-interface development pipeline. Talking about talking while you're talking is hard enough to figure out; it's impossible to do. But at the user/task-analysis stage, these protocols can be a very rich source of data (depending on the task and media, of course; if the task is buying plane tickets over the phone, verbal protocols are useless here too).

and Mrazek (1996), in an ethnographic study of home computer use, brought dinner with them.

Audio data will present itself in any case, though, and it is best to have a recording device running (with the participant's permission). Videorecording is not generally a good idea; it is somewhat more obtrusive, always more expensive, tends to stretch the analysis out forever, and is rarely necessary for a project that only implicates sound. But a digital audio record can be very valuable, something you can scan to help with the integrity of your notes and reporting, and which you can incorporate into interview records if appropriate.

While you may end up distributing some of your "interview" questions throughout the observation phase, you will also very likely have more questions suggested to you by what you see; if it is inappropriate to get them answered right away, note them and integrate them into the interview.

Have a stopwatch, for durational data. Take detailed notes, including running tallies for frequency information, which you should work into a more permanent record soon after the session. Use the audio record to help rework your notes.

Observing people to plumb the meaning behind their activities is, as many interaction specialists have noted, a rudimentary species of anthropology. There are scores of methodological and ideological differences between (on the one hand) the real ethnography of cultural anthropologists, who develop and refine an interpretive description of specific cultures (a description profoundly rooted in the culture's own epistemological and ontological categories, through principled immersion in its practices) and (on the other hand) figuring out how to design or optimize a tool by watching people at work. But the defining difference is in the motive. Anthropology is motivated only by the production of knowledge about what it means to be human; observational task analysis is motivated only by figuring out how to design or optimize a tool. The former is philosophical, in the grand original sense of loving (*philo*) knowledge (*sophia*). The latter is technical, in the grand original sense of craft (*techne*). It is purposive. It is for making things.

Three notions are especially important to borrow from the conceptual kit bag of cultural ethnography to inform interactive design, to lean the philosophical toward the technical:

- **Empathy**

Put yourself in the user's shoes as much as possible, adopting her point of view; Kieras (1997) calls the point of this disposition *intuition building*. You learn to think responsively about the task in the categories and terminology of the user. Indeed, to the extent possible, *be* a user. For highly specific applications (installations, for instance), that may not be possible on anything but a very artificial level. But for general-purpose activities (banking, shopping, travel arranging), you should roll your sleeves up and participate directly. Open an account. Buy a sweater. Arrange a trip to Vail. (You can always return the sweater or cancel the booking.)

- **Richness**

Be a promiscuous data gatherer: observation, hallway chatter, structured interviews, user guides, spec sheets, . . . anything is fair game. Don't turn down any task-related information a user offers you, and ask for copies of anything that looks relevant. The difference between thorough task analysis and true ethnography is the sanctity the material is accorded. Think of the archive you collect not as documents and artifacts to be examined in detail, but as reference materials: you don't scrutinize it all, but it might come in handy for consultation, or just snooping around in, for ideas and insights.

It may also be that the service you are voice-enabling is instantiated in a number of current formats. Personal banking, for instance, may have in-person, automated-teller, web, human-to-human phone, and keypad phone venues, all with widely overlapping functionality. You might think that just the phone interactions call for observation and data gathering, and they should certainly be the focus, but there may be much you can learn from web and automated-teller systems as well, and you can always profit from watching human–human interactions. People can get sports scores from the paper, the TV, the Web, and the radio. Which among these offer user/task functionality that a speech-system sports service could best accommodate?

Talk to the indirect users, as well. In a travel-booking service, it may well be that the secretaries do the booking, but others do the traveling.

- **Openness**

Remain available to novel or unexpected strategies or insights that might present themselves; avoid preconceptions. This attitude is particularly important because conversational voice interfaces are new, and because most of our intuitions are trained in other directions.

For instance, a typical graphic-interface intuition is that relentless linguistic consistency is a good thing, especially in system messages and forced choice points. "Strive for consistency" is Shneiderman's Golden Rule #1 for interface design (1998: 74). Jakob Nielsen has a book dedicated to it (1989). A slight difference in phrasing causes the user some uncertainty, some cognitive friction, reducing both comfort and efficiency. If nothing else, it forces her to read something that she would be happier responding to with a familiar scan. She has received a string of "memory error" messages, why is it now a "memory allocation error?" She has been asked to respond "OK" in confirmation in all her other interactions, why is the button now labeled "yes"? But whatever virtues relentless linguistic consistency has for graphic display — where people automatically scan for familiarity, rather than read for redundant assurance — it is full of vices for auditory presentation, where there is no scanning option and the messages must unfold completely each time. Consistent repetition is as quick a route to aggravation as there is. Every two-year-old knows that the fastest way to drive someone buggy is to repeat yourself over and over.

An engineering-trained intuition venerates speed and accuracy. But one of the most robust findings of voice interface research is that neither of these qualities matters nearly so much to users as an overall sense of comfort, satisfactory task completion, and a sense that the system puts a premium on cooperation.

An intuition trained in technical communication values specificity and directness. Users of voice interfaces, however, would rather get a good hint (to be explicated in a later turn if it doesn't do the trick) than listen to instructions listing the ten words correlated to the ten actions now feasible.

(All of the preceding examples are subject to qualification — chiefly by user and by task — which only reiterates the value of openness. Don't assume there is a standard operating procedure except by witnessing it, and even then remember the limits of the data.)

There are three things to watch for (and, if necessary, to probe), particularly with respect to goal satisfaction:

- successes

- failures

- strategies

None of these can be taken at face value, of course, since you are not just working on an update of the current modus operandi, but on a translation of the task(s) to another modus altogether. But all of them can be suggestive of ways to proceed, directions to avoid, and metaphors that can guide the voice interaction. You need to be in a state of constant opportunistic vigilance, for ways that a speech system might be steered toward the same goal satisfaction.

The range of candidate applications and services is much narrower for voice interaction than for graphic interaction; they have to be amenable to temporal, sequential, linear interactivity, and relatively independent of both visual display and object-manipulative input. They must be detached from (or already free of) space, vision, and movement; and they must be translatable into (or already instantiated as) turn-trading vocal interaction. These conditions mean that as much of your observation may result in filtering off functions and characteristics from the observed system as may result in migrating functions between the modalities. There is much potential, for instance, in voice-enabling web sites (or, rather, the databases behind many of them), but one look at a standard-issue capacious hub-site, like Yahoo or Netscape, with tabs, headlines, ads, text-input fields, tables, icons, buttons, thumbnails, and lists-upon-lists, and you know something's gotta give.

Voice interaction design will, of course, result in new functions and new characteristics as well. One thing the visual displays can't give you, even those bloated hub-sites, is serial information. You can find it, and even array it yourself in a quasi-serial fashion (a visuo-spatial display can't be truly serial). But that is quite different from asking for directions, or a recipe, or installation instructions, and getting them, one step at a time, as you need them, serially.

The nature of voice interfaces, too, and of the services for which they are appropriate, means that the task analysis is not always limited to observing and talking to only the users.

You can't often interrogate the system behind the task, if that system is as fully digitized as electronic mail, but many candidate systems for voice interfaces involve humans, whom you *can* talk to: call centers, for instance, and catalogue operations, and, most famously, travel planning. Voice interfaces for such tasks amount to building an expert system (see Chapter 14), and the humans are the experts. Talking to people whom everyone can see your speech system will eventually displace in principle (if not them specifically, then certainly their roles) calls for an extraordinary amount of tact and often calls on a goodly amount of patience as well. The main rule is to assure them that the point of the analysis is to find ways to offload the more trivial and routinized tasks from human operators.

The possibilities for statistical analysis of many phone-based activities are also very rich, and wherever possible the information should be gathered. For instance, how often do people ask to transfer funds and pay bills during the same calls? In what order? How often do they pay utility bills versus credit card bills (when they have both) in the same encounter? What time of the month do transfer activities and payment activities correlate? How often do callers inquire about interest rates on their mortgages versus inquiring about their transaction history? All of these pieces of information can generate the probabilities of dialogue acts and their contents at particular points in exchanges, which informs the design of the system: what to listen for, when, and what to say, when.

Interview Phase

> *Well, Waddaya know?*
> — Groucho Marx (among others)

The interview should follow the observational phase quite closely — perhaps after a coffee or lunch break, or even right away — to capitalize on both your and the participant's freshness about the interaction, and to develop naturally from the relationship you've built up with the participant.

In the observational phase of the task analysis, you may gather lots of direct-response data, some of it by interesplicing "interview" questions with the observation, some of it by asking questions prompted by the interaction you're watching (and which may then be added to your stock of interview questions), some of it by the participant falling into a spontaneous or an induced verbal protocol. Observation will also present questions to you, or whole investigative lines, that didn't occur to you when you were working out the structure of the formal interview. Now is the time to round both of these factors out, editing on the fly.

Structure is important, in the sense of having the bulk of your questions worked out in advance. Structuring the interview is especially important for consistency and reliability of findings (if Fred and Betty say something, it may be valuable or not, but its value is more certain if you have Barney and Wilma's opinions on the matter too). But repeating a question the participant has already answered (unless the answer was partial or otherwise unsatisfactory), just because it is Question #7 and you have just asked Question #6, is pointless and annoying. Similarly, if a good new question presents itself, don't avoid it just because it's the last interview and you haven't asked anyone else the same thing (one data point is of limited use, but it is usually better than zero data points on a suggestive issue). Be structured, but be flexible.

Keep the interview relaxed and informal, but keep it on track; don't waste their time. Listen to design ideas they suggest, and actively solicit design ideas at some point, but don't fall into a design meeting with the participant. Your primary job is to gather data, not build a model.

Audio record the interviews. Never videotape them, unless you need gesture data for multimodal reasons, or a record for reporting purposes. Video won't help, and can hinder, the analysis, and generally adds an unnecessary layer of equipment and other resources to the project. Analytically, only use the electronic recording as a backup reference tool. Don't depend on it. Take comprehensive notes, and rework them as soon after the encounter as possible; the quick shorthand reminders to yourself that you scribble down during the observation and interview will rapidly become vaguely allusive puzzles.

Prompt users regularly for elaborations. Compare their activities to other related activities and get their reactions. Ask about other modalities (prominently, of course, voice) for the same activities.

Don't suggest answers, or finish sentences, for the interviewee.

We'll get to a set of task-centered question templates shortly, for anchoring the interview, but the interview should start with questions that build the user profile. You'll have their name, professional role, general experience level, and gender information already (the first three should be part of the recruiting procedure; the last is usually self-evident), but a few more categorical data points will round out your profile and get the interview rolling with some automatic responses. One way to get people talking is just to get them talking.

Age, length of experience, relevant background (education, training, previous occupations), and guesstimates of various related voice activities (duration and frequency of overall phone usage, for instance, as well as social and business dimensions of phone use).

It is also worth collecting linguistic information: first language, other languages, years speaking English (or the language of the target application), and country of origin (in the United States, perhaps region of origin). None of this will be of direct use to you in the design work, but it is easy to collect, can help you establish good routines of participant classification, makes tracking the interface development longitudinally a little more efficient, and is the sort of data that, at various points in the development, especially once recognizers are deployed, may help the recognition team interpret the representativeness of the results.

With age data, some people are squeamish about specificity. The two most common interviewer work-arounds are to ask for a birthdate rather than an age, or to ask for placement in a range. The latter gives the participants more wiggle room, and people tend to regard themselves more as falling into the "twenty-something" range anyway, than as "being twenty-eight." And range data is perfectly suited to the analysis in any case. If they fudge, they fudge.

Any attidudinal questions are best answered initially on a scalar basis, and then elaborated as required. It is better to ask a question that allows the user to position herself relatively, than a question which forces a sharp decision. Rather than "Do you like using email" (implicating a *yes* or *no* reply), solicit responses to statements like the following:[5]

I like email

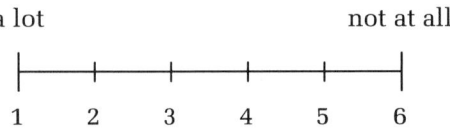

5: As the evaluation scales indicate, I advocate the use of even-point scales (specifically, six-point scales) — in this instance, and in the Oz/usability context discussed in Chapter 16. The almost religious norm in testing is for odd-point scales generally, and the famous five-or-seven-point Likert scale in particular. As far as I can tell, this preference is just deference to the field of social psychology, whence their inventor, Rensis Likert, hailed, and where odd-point scales are preferred because (among other things) they allow the respondent to orient his evaluation with respect to a neutral midpoint. My own preference is for even-point scales precisely because they have no midpoint. They force respondents to commit themselves in either a negative or a positive direction (actually, if combined, as I combine them, with segmented continua, it allows a dogged centrist to sit on the fence, but it biases against fence-sitting). Specific design testing does not aspire to the rigor of social scientists' research in any other way; using a scale that licenses neutrality, just because social scientists do, when we are interested in opinion and attitude, seems misguided to me. (Note that I am not talking about genuine social science research that concerns interfaces or technology, just the specific research that goes into the design of specific products, where sample sizes almost always preclude statistical rigor anyway.)

In part, this procedure allows you to peg the relative strength of the attitudes, but it also gets users thinking about their responses in sufficient detail to sponsor an elaboration ("I like using it because it is a better record of details than voice mail, but not as messy as paper. I hate the endless trivia that ends up in my mailbox, though.")

A good slate of task-interrogating questions to start with should be based on the following templates[6]:

- **When do you do X?**

  - generates timing data (when in relation to overall time)

  - generates frequency data (when as in "how often")

  - generates sequence data (when in relation to other actions)

- **How often do you do X?**

  - generates frequency data (ask if you don't get frequency information from "when")

- **Where do you do X?**

  - generates environmental data

  - generates sequence data (if taken as where as in relation to other actions)

- **Why do you do X?**

  - explicates goals

  - supplies motives

  - illustrates cognitive strategies

- **How do you do X?**

  - illustrates solution strategies

  - illustrates procedures

And, wherever relevant, interspersed through these questions, or during the observation, or collected for one part of the interview:

6: See Mayhew (2000: 88–89) for a similar list, and for additional suggestions. My own list develops from Greco-Roman stasis theory, which has made its way into contemporary research methodologies as the famous 5-Ws (or, sometimes, 5-Ws and an H). It's remarkable how resilient that approach is for investigating human behavior. See Crowley and Hawhee (1999) for a thorough, modern account of stasis theory.

- **What is Y?**

    - generates definitions

    - generates terminology

These questions get directly at the tasks: what they are and how, when, where, and why they are done. Questions on this scheme also, as various activities fill in the variable, X, and various objects and instances fill in the variable, Y, get at problems, errors, particularized strategies, work-arounds, and the like.

Towards the end of the interview, you also need to ask the three standard-issue collaborative-design questions, opening the responses up, tapping attitudes fairly directly, and encouraging participant suggestions for modification:

- What do you like most about the activity/service/system?

- What do you dislike most about the activity/service/system?

- What would you change about the activity/service/system?

And two questions should up the participant-design quotient even further, turning the interview explicitly toward voice interface concerns (though not if you are observing voice interaction already, in which case these questions will already have been answered):

- What sorts of things would you like to ask the system?

- What sorts of things would you like the system to tell you?

These questions might get effectively the same answers ("If I get an announcement about a conference, I don't want to listen to the whole thing. Why can't I just ask where and when it is and if Donald Norman will be one of the presenters?" and "I'd like it to be able to tell me what's in a message if I ask it.").

But you might also get some very different responses to these two parallel questions, depending on how the users envision their relationship with the system (in particular, concerning initiative: whether they see the system as having, or being capable of taking, initiative and whether they see that as desirable). For instance, users might want a system that functions more as a personal assistant than as a text-reader ("Well, if it could see whether the sender was in my contact list, and ask about adding her, that would be handy.").

Other Methods

> *You can observe a lot just by watching.*
> — Yogi Berra

At this stage in the development cycle, there are no other methods as richly generative of data, or (therefore) as cost effective, as participant task analysis. Questionnaires, for instance, can be helpful as follow-up instruments, and deploying a few scalar questions during the participant task analysis (particularly during the interview) can give you some insight (so long as the responses to them are elaborated through subsequent questions). But a questionnaire on its own won't tell you enough about users or their tasks to be worth the energy and expense, and will tell you even less about their discourse. (And, in any case, questionnaires are not worth very much at all unless you are a professional in their construction and analysis, or you hire professionals.) Focus groups, too, have their uses; in particular, they can help marketers gauge the commercial terminology and imagery that might accompany product development, and can occasionally feed back good suggestions about functionality, agent personality, and the like. But focus groups are too amorphous and too unreliable to inform product development at the outset.

I would never discourage these or other techniques as augmentations to user-task-discourse analysis, and all methods which return information about the users' goal-directed behaviors. Focus grouping, in particular — or, actually, a kind of inverse focus grouping, a debrief grouping — is often a good way to conclude the participant task-analysis phase, bringing a subset of the participants (the ones who generated the best data) together for a meeting to cap the early data-collection phase.

But there is no substitute for participant task analysis at the beginning of a project, and it is flat-out the best method for investigating the material a voice interface is built from: the users' discourse. It also has the inestimable value of giving the designers an early opportunity to begin building intuition about the task.

A significant component of intuition is linguistic, knowing the language of the natives, and a goodly amount of time in any participant task analysis should be devoted to harvesting words. As Mayhew (2000) points out recurrently, you have to learn the users' vocabulary in order to make the interviews flow anyway; nothing puts a participant more on edge than having something they call a kadiddle repeatedly referenced as the "valve-like-thingamajig-what's-it-called-the-kadiddle?" And even the people it doesn't annoy, by convergence and courtesy, fall out of their natural discourse and say things like "the-valve-like-thingamajig-we-call-a-*kadiddle*." The last thing you want is to induce the user to speak *your* language.

With voice interaction design, the need to learn the vocabulary, and the other elements of users' task-driven discourse, is urgent to the point of being definitional.

Summary

In many ways the saying "Know thyself" is not well said. It were more practical to say "Know other people!"
    — Menander

In this chapter, we have looked at the first two elements of the first-stage triumvirate for voice interaction design, building user profiles and analyzing their tasks (the third element, taken up in the next chapter, is charting the discourse).

Building **user profiles** begins by capitalizing on any data that may already have been gathered on the users, and then elaborating and/or refining that data with direct research of the population. You need to gather several categories of information about the users, including demographic, attitudinal, personality, experience, linguistic, and phone use. (Notice that this last category, because of the methdological parallels with voice-only interfaces, is important to chart even if the system you're developing is not accessed by phone.)

The users are the "who" of the analysis phase; equally important is the analysis of the **tasks** they perform, and which the system will assist them with. The overriding consideration here is the **goal** of the task. The analysis must frame everything with respect to it, carefully distinguishing it from the tools (applications, appliances, instruments) used to accomplish that goal. It must also outline the significant elements that **ground** the task, the **procedures** followed, the **terminology** used by both the user and the tool(s), and the **system actions** (effectively, feedback, since unobservable system actions are not part of task analysis). A range of other variables are also useful to chart — especially important for voice interaction design is **initiative**, but also factors such as **complexity**, **problems** encountered, task **environment**, **continuity**, **criticality** of task itself or the data it may involve, **duration**, **frequency**, and the **quality** of the task as it relates to the tool.

The final sections of the chapter outline **data-gathering** best practices and rules of thumb, outlining its two phases, observation and interviewing. The **observation** phase requires mostly attentiveness and the right disposition, towards **empathy** (putting yourself in the users' shoes), **richness** (keeping your eyes open for as much as possible, whether it seems immediately relevant or not), and **openness** (in particular, dampening the "correct way" to do things that you may be importing from prior knowledge about the task, the application, or from prior sensibilities related to enterprises like graphic interface design). The **interview** phase involves asking a set of **structured** and **task-centered** questions. Although statistics are often of limited rigor in task analysis, and need to be interpreted cautiously, questions which seek opinions should be **scalar**, which can provide rough indicators of satisfaction/dissatisfaction, and should always be followed up by requests for **elaboration**. The general scope of the interview concerns **when** the tasks are done, **how often**, **where**, **why**, and **how** they are performed. Particularly important for voice interaction design, too, is the **what**, which can lead quickly to terminological issues — words and their meanings, which feeds nicely into the discourse analysis that is the final element of the initial stages of the design process, and which occupies the next chapter.

Appendix: Sample Task Analysis

People engage in tasks and activities. They make discoveries and encounter difficulties. They experience insight and satisfaction, frustration, and failure. At length, their tasks and experiences help to define requirements for future technology.
    — John M. Carroll

What follows is a sample analysis for a simple email task, replying to an email using MS Entourage, the first part of which is given in the main text as Table 10.1. If you are new to task analysis, or somewhat rusty, the best approach is just to trace out this analysis alongside your own performance of the task, then move to other tasks in the same application, and then to other applications, and on to tasks in other domains altogether, perhaps augmenting or even dropping some of these analytic categories. (Only drop very cautiously; it may be that volume, for instance, proves unhelpful, or that environment is so redundant it serves no purpose, but it may also be that you just don't see the relevance yet.)

There's nothing sacred about this particular method of analysis; almost everybody develops their own procedures, refines their own categories, and configures their own instruments. Ledger books in hard copy, for instance, or spreadsheets in digital form, are very useful for physically charting the tasks, but choosing one of those forms over the other (and, if you go digital, using one display over another) significantly affects the way you can enter this data. Ledger sheets allow for more simultaneous columns; digital entry allows for automating the entry in ways that can free up more time for observation (but also in ways that can make you lazy).

Goal	To reply to an email
Ground	From received-message window, or main mail window; in either case, the received message, the icons, and the menu titles are all visible. All menus are drop-down, to reveal active menu text, with shortcuts; all icons have pop-up labels. Main mail window and/or received-message window remains visible behind reply-message window. The reply-message window includes main tool bar, information fields, and editing toolbar.

Procedure (=user actions)	Choose reply icon from toolbar, OR Reply from Message menu, OR CMD^R
	Enter/Edit text
	   Option: add attachment
	   Option: insert (picture, sound, movie, etc.)
	   Option: add/change/remove/replace signature
	Choose Send now OR Send later OR CMD^RETURN OR
	   OPT^ CMD^RETURN

Terminology	Reply From To Cc Send now, Send later Attachment (Add, Remove) Insert
System actions	Presents reply-message window (recipient, copy, and subject fields completed). Depending on application settings, quoted message and/or signature may be in the edit pane. Removes reply-message window upon sending. Presents "Message sending" dialog with progress bar. Potential action: presenting system-status message; e.g., "Mail could not be sent using the account 'RaHa'.' Explanation: A connection failure has occurred. Error: -23016"

Initiative	Mixed, but dominantly user
User	Invokes reply function Edits message Terminates reply function (by sending)
System	Only in case of critical system-status message (that is, alerting the user something is impeding her goal), or from some system function unrelated, or indirectly related to the task (e.g., a new message alert, or a calendar alert)
Complexity	Low, given adequate domain and system knowledge
Number of user actions	3 (+3 possible options)
Number of subtasks	1 (editing reply message text[7])
Number of actions per subtask	Variable, ranging from one (inserting text) to a dozen or more (deleting, formatting, inserting, attaching . . . many of which are more properly subtasks than actions)
Volume	Message mean per day = 27; 80% trigger replies
Problems	Few and minor
Breakdowns	Trying to reply from wrong account; working offline
Bottlenecks	None
Goal	**To reply to an email**
Environment	Office, home
Criticality	Variable (depending on the triggering message)
Overall	Variable (depending on the triggering message)
Individual actions	Variable (depending on the triggering message), but every action can obstruct the goal if it goes wrong
Duration	Rapid (mean reply time 110 seconds)
Continuity	Intermittent
Frequency	22/day
Quality	Variable; there is no detectable quality impedance from the interface; quality depends on message volume, content, relevance, and so on

---

7: A separate task analysis is required for editing.

11

Building the Discourse Model

Build complete domain languages.
  — Alexander Rudnicky

Once upon a time, in the Speech Applications Group at SunMicrosystems, the designers tackled an email application. They knew the user population well — in fact, they and their fellow Sun techies *were* the user population. They knew the task well. They had been using the application for years. And they had a base of good ideas for voicing that task. "We had decided that the application had to support reading, skipping, deleting, and replying to new messages," Nicole Yankelovich recalls. "We also knew that our users would be familiar with the graphical interface for performing these tasks." That, it seemed, was that: "We thought we knew enough to proceed."

You can see where this is headed, I'm sure. From the very first participant, in the very first user test, of the first prototype,

> . . . it [was] obvious that there were major design flaws. Callers over the telephone were overwhelmed by the same volume and organization of mail headers that worked so effectively in the graphical interface.
>   (Yankelovich, 1997)

The moral of the story: they didn't know the discourse.

Yankelovich tells this parable to illustrate the critical importance of early-cycle discourse work for the development of voice interaction, work that she calls *natural dialogue studies*.[1] These studies "capture human–human interactions in the domain of the target

---

1: Actually, she calls them *predesign studies*, an egregious term for several reasons, but a few years later, in a paper with Jennifer Lai, she calls them *natural dialogue studies*, a term so much more satisfactory that it is worth committing anachronism over (Lai and Yankelovich, 2002).

application . . . [in order to use it] as the basis for the speech interface design." Among the methodological principles Yankelovich enumerates for natural dialogue studies — a variant of participant task analysis that focuses closely on discourse — three are directly related to building the discourse model:

- Collecting appropriate vocabulary

- Determining commonly used grammatical constructs

- Discovering effective interaction patterns

These are data-collection objectives, for building the discourse model underlying an effective speech system, but they have clear, point-by-point design implications, for building the voice interface:

- Get the vocabulary right.

- Get the utterance structures right.

- Get the interaction patterns right.

These principles I now hereby dub "the habitability goals," recalling William Watt's definition of a habitable speech system as "one in which its users can express themselves without straying over the [model's] boundaries" (1968: 338). Full and absolute habitability — a system that anyone can talk to about anything they want in any manner they like — is, we all know, science fiction. But Watt does not define habitability in absolute terms, and no one in computer speech technologies should think in such terms either, at least not in this century.

Watt's notion of habitability is a condition of some *specific speech system* relative to some *specific group of speakers* (1968: 340). To achieve an interface that satisfies this condition, for users to talk with comfort, we need to know their vocabulary, their constructions, their patterns of interaction. If they are to stay within the system's boundaries, we have to build systems that deploy those words, structures, and interactive routines with grace and clarity. This chapter is about satisfying the habitability goals, chiefly through building and managing two highly interdependent tools for setting the boundaries, a lexicon and a corpus, both developed specifically for design work.

Everyone knows that speech *recognition* work cannot proceed without massive computational support, not just in terms of a pure recognizer that can store a huge catalogue of acoustic models, analyze incoming signals for their spectral properties, and rapidly compare those properties to the relevant catalogued models in order to generate a ranked list of word candidates. It cannot proceed without an array of tools and strategies that come from the close, ingenious examination of big-to-huge collections of language — corpora —

to help resolve which of those candidates (in light of other candidates from the same input) is most likely.

However, what not everyone in the industry knows is that voice *interaction* design — the human-factors partner of speech recognition engineering — has precisely the same requirements. The speech recognition (and natural language processing) side and the inter-action side can certainly share resources, working with corpora that overlap in significant ways (that is, which draw on some of the same sources), but both sides also have unique requirements. In particular, speech recognition has a need for large, general-purpose acoustic corpora (for extracting the relevant phonetic patterns) and smaller, user-domain corpora (for collocation data and register-specific phonetic patterns). Voice interaction design really only needs the smaller, more specific register corpora, for building the lexicon that maps the discourse model.

In this chapter, we outline the methods for gathering and structuring the data that populates the design corpus. This work substantially overlaps the user and task studies outlined in Chapter 10. Further, it must proceed, hand-in-glove, with building the design lexicon, the second major topic in this chapter. The corpus and the lexicon are what allow us to satisfy the first habitability goal, to get the vocabulary right. An absolutely critical component of the design corpus is a body of task-driven, spoken dialogues in the target domain of the speech system. Close analysis of that subcorpus helps to achieve the other two habitability goals. We also look in this chapter at ways to get the utterance structures right, and get the interaction patterns right.

Gathering the corpus data requires a range of opportunistic methods, from armchair cogitation to interviews, naturalistic observation, machine-readable texts, even web-harvesting, but two selection principles are essential. First, the discourse must be judiciously selected, despite the necessarily somewhat loose grab-bag of sources. Look wherever you can think of for the data, but don't use it indiscriminately. The sources must be identified and segmentable, so that you can draw your analyses wisely. And, whatever other sources you draw on, it is critical that you include a range of spoken, task-driven dia-logues, even if you have to manufacture them. As the corpus is gathered, it must also be tagged — marked minimally for word class, syntactic role, and dialogue acts — so that it can support a range of analyses and feed the design lexicon for the interface project.

Everything follows from the words in a voice interface. A speech system with even moderate sophistication cannot succeed without a thorough understanding of the domain vocabulary. The lexicon is where that understanding is encapsulated — what the words are and what their variations, travel-companions, alternates, and associates are. We will look at the longstanding instruments for understanding words and their use, dictionaries and thesauri, extracting lessons for the composition of a voice interface lexicon.

Words, however, don't just stack up. They add up. They fall into predictable constructions with specific meanings and uses. Different fields favor different constructions — unique colligations, characteristic turn units, chosen dialogue acts, preferred syntactic forms and cohesion strategies. Understanding the discourse means knowing these utterance constructions, and we will look at discourse-analytic ways of coming to know them.

The greatest virtue of voice interfaces is that they allow people to interact with a body of computational processes and data with little-to-no learning curve, drawing directly on a natural repertoire of conversational strategies and intuitions. But that repertoire realizes its optimal articulations in register-specific ways. The field of the register encourages specific interaction patterns, favoring certain dialogic pairs and coherence relations, grouped into specific exchange structures and task substructures. Studying captured and/or induced spoken dialogues is the only way to discover the effective interaction patterns users deploy, so that the voice interaction design can emulate them and pursue habitability.

Collecting Appropriate Vocabulary

The words! I collected them in all shapes and sizes and hung them like bangles in my mind.
    — Hortense Calisher

The corpus feeds the lexicon; the lexicon maps the corpus. From a hard-research standpoint, the corpus takes priority; from a design perspective, they develop together. The design team will have bookshelf and online dictionaries, some of them general, some very specific, depending on the field and what is available. They will also have a large, general purpose corpus, and perhaps more specific corpora, depending again on what is available in the domain of the speech system. But a voice interaction project of any complexity needs to build two specifically tailored, mutually vitalizing instruments — a design lexicon and a design corpus — starting very early in the development cycle, and refining them throughout. They grow together.

The project lexicon begins with what Furnas et al. (1987) call the Armchair method: the lexicographer cogitates on the discourse and her own personal word hoard, and assembles lists of relevant words, interspersed with an impressionistic investigation of any domain-related media (periodicals, radio, video, web materials). Other people should be drawn quickly into the process — other interface personnel, recognition and processing engineers, a focus group of users — which brings more words into the mix, raises the level of control against idiosyncrasy, and has a variety of beneficial effects that make it a reasonable early methodology. As Uriel Weinreich said, it is "wasteful to put the whole burden [of defining] on the lexicographer, or any other lone semantic descriptivist." The job is too big for one person, and usually too register-dependent to exclude users from the definitional work: "Why not enlist," Weinreich added, "the help of a sample of speakers of the language?" (1962: 42).

Sound, usable advice, but in the end more armchairs are just more armchairs. Corpus building has to start early as well, overlapping the armchair lexicography.

Corpus analysis trumps intuition every time, even rooms full of collective intuition. Intuition can calibrate corpus findings in certain applied instances (that is, in design work), but it can never displace or overrule those findings, so long as the corpus in question is representative.

Building the Corpus

The unaided human mind simply cannot discover all the significant patterns, let alone group them and rank them in order of importance.
— Kenneth W. Church and Patrick Hanks

Constructing a design lexicon for a voice interface requires the attention of a corpus-knowledgeable lexicographer, but there is rarely a corpus available that is sufficiently specific to work from. The team lexicographer usually needs to create and manage one, if not from scratch then from close to it. (If there is such a register-specific corpus available — if, for instance, you're working with a call center that has been recording calls — the interface design process can be streamlined considerably. As voice interfaces become better, more widespread, and more serious about understanding their relevant discourse domains, these corpora will begin to grow, and later iterations of speech systems will benefit immeasurably.)

Gathering the Data

We must begin by investigating the nature of discourse.
— Socrates

Once domain specialists and user groups are brought into the process, the corpus can be fed. The first step in taking the necessary step beyond armchair intuitions is to record and analyze — and you should start capturing the discourse among users, domain specialists, and lexicographers/designers from the outset.

In these interviews, discussions, focus-gabs, John Sinclair's elicitation slogan "difficult to recall, . . . easy to recognize" (1987: xviii) becomes the ruling principle. Keep asking questions like "Can you say X in this situation?" and "What does it mean if you say X here?" and "How would you interpret X if someone said it at this point?" and, especially, "Can you give me an example of X used in context here, please?" That is, keep putting the language into context. It's a lot like a multiple-choice test about their discourse. Your informants may not be able to generate all the appropriate, inappropriate, and squishy usages on demand, but they can invariably tell, with very little reflection, what's acceptable and what's not (and they can also supply and refine definitions for the lexicon). And any corpus texts generated in this stage have to be treated much differently from truly environmentally-embedded speech.

Nothing can match naturalistic observation and unobtrusive recording for reliability.

Interviews, think-aloud protocols, and other elicitation techniques are all valuable, but they all involve what William Labov (1972) called "the observer's paradox." The more you draw an informant's attention to the language she is using, the more you alter that language, often pushing it towards exaggerated pronunciations and prestige diction. You need to gather as much data as you can when the language users are only minimally aware they are being observed. (See Bénjoint, 1983 on this question, especially pages 70–73.)

If the service you are voicing is phone-based already, gathering the corpus is vastly easier. If it is not, the corpus becomes more difficult to collect, and will in all likelihood (unless, say, electronic eavesdropping laws get changed) end up at least somewhat impoverished. But interviewing users by phone or setting up tasks for people divided by a phone line are low-fi substitutes for natural data. For voicing an appointments calendar, for instance, Yankelovich and her colleagues used the ingenious technique of simply having users call someone who had a calendar in front of her, asking her schedule-related questions, and they quickly manufactured very useful data like the following:

Ben   I would like to figure out what Tom's calendar looks like for next
            week . . . Specifically, the late afternoon of the 23rd and the 24th.

Assistant   OK. 23rd. He has a meeting at 9 o'clock.

Ben   No, how about later that afternoon?

Assistant   Ok, it's open.

Ben   How about the 24th?

Assistant   24th. What time?

                                                        (Yankelovich, 1997)

This technique — filtering tasks through phone encounters — is indispensable when there is no indigenous source of task/domain spoken dialogues. As a prototype takes shape, phone-based Wizard encounters are also useful ways to gather data early in the development cycle. As the product cycle advances, spoken data collected from pilot studies and Wizard and usability tests can augment the corpus in limited ways; harvesting from field trials, and even full market-release services, can help build a powerful and comprehensive corpus.[2] But such data must be used very judiciously. In particular (as Chapter 15 takes up)

2: Many researchers have found that making their systems widely available and providing real payoffs to callers (that is, free information about topics of interest, like movie times and locations, weather conditions, and the like) priming issues seriously compromise the use of such data for building the system's vocabulary.

The sooner you can collect real-world spoken interactions related to the task(s) you're voicing, the better. Depending on the product you're developing, you might harvest from talk radio (much of which is genre driven: sports, politics, finances, gardening), or specific TV programs found in the bezillion-channel universe. Popular media has substantial limitations — in particular, the hosts (and professional guests) tend to have quite nonstandard articulations — but they can have compensatory virtues of various sorts as well. For instance, background music, station-IDs, pet locutions, topic segmentation strategies, and the like can all help feed other aspects of the interface design. The watchword here is, as always, *judicious*. Use this material opportunistically, should you need it, but always with an eye to its liabilities; be sure to keep the corpus data easily segmentable and clearly marked as to source. Always provide for the isolation of the data by source (filtering off media-source data from interview-source data, for instance), to define subcorpora.

Minimally, the corpus will need to divide easily into two principal subcorpora, one based on spoken materials, one on textual sources. If you can harvest phone-based interactions, or have other ways of gathering lots of field- and task-specific vocal data, then the textual corpus can round things out. The spoken corpus should always be weighted more heavily, but it is usually much easier to get texts.

A text corpus is not so much necessary because of the design work — which is ideally confined as much as possible to the verbal modality — but because of the convenience and availability of textual sources. Both subcorpora must have textual representations (even with the spoken source material, a large amount of the work has to be visual because it involves contrasting and comparing, which our brains are much better at doing spatially than temporally).

The design team and the recognition team have to work closely in all respects for the development of a successful speech system, but one of the most important areas of collaboration involves their respective corpora. The recognition team calls their spoken corpus "training data." Its purpose is different, but it's a big storehouse of potentially relevant data all the same.

But more importantly, much more importantly, your data — the spoken subcorpus harvested from user interviews, focus groups, naturalistic recording, and assorted media —

---

has been incredibly productive for them, in terms of refining and strategizing the interface (as well as recognition robustness). See, for instance, Polifroni (1998); Glass and Hazen (1998). This technique has not been used commercially however, where it could be very profitable in all senses of the term. The trick is finding the right payoff. Voice-interface telephone banking systems, for instance, might involve something like reduced fee structures for the first year, or reduced interest rates after hitting a threshold of use. While such a suggestion could strike most flinty-hearted bankers as akin to throwing money off the roofs of their financial towers, it can have distinct rewards in customer satisfaction, more effective systems, reduced maintenance, and personnel costs.

should be made available to the recognition side for training purposes. As above, the articulations in some genres, like talk radio, can comprise parts of the design corpus for use as training data, and the corpus needs to be readily segmentable, but much of it will still be useful: in-domain training data is the gold standard (Rudnicky et al., 2000). Some recognition folk (or, more frequently, their bosses) tend to think one token is as good as another, as long as it's "North American" or "British," as required; if there are enough tokens fed into the hopper, they figure, the recognizer is fully prepared. It's all about quantity and statistical procedure. That line of thinking ignores the sociolinguistic effect of register on pronunciation.

Quality counts too.

Although I cringe to recall it, there was an in-register pronunciation of the word *really* required by the pseudo-hippie registers I participated in as a callow youth. You can still see the word spelled in a way that reflects this pronunciation, in old *Fabulous Furry Freak Brothers* comix: *rilly*. I never used that pronunciation with my parents or teachers, but it was the only one I used with my peers. My only consolation is that I now get to see others propagating similar distortions on language in the Pursuit of Cool. Among the most common as I write this (but probably so five-minutes-ago by the time you read it) are the bisyllabic pronunciation of *cool* (coo-uhl!), and the really long vowel of *sweet* (as a term of approbation, not as a description of sugar — sweeeet!).

These are the extremes — the bell-bottom pants of articulation — and therefore easily spotted. But they are not isolated. In subtler ways, such acoustic traits are definitive of registers.

One of the strongest effects of convergence is that people in communities (professional just as much as generational, geographical, or socioeconomic) gravitate towards specific pronunciations. Catching these pronunciations can make the recognizer more robust for those communities.

Different companies have different financial and administrative (and political) structures, so how the spoken-register data benefits both the interface and the recognition teams is highly variable. The best arrangement, however, is usually for the design team to collect it, because they need a far more intimate understanding of the register than the recognition team. The design team should collect it, but the recognition team should share heavily in funding the collection.

The design team also needs a text-based component to the corpus, however, which the recognition team has little need for (though the natural-language understanding group might have some interest in it). For the text subcorpus, any source of domain-related text is potentially usable. Machine-readable source is obviously the most convenient, and documentation directly related to the product or service you are voicing is especially handy — pamphlets for financial services, user guides for computer applications, promotional material for information services. But this must be balanced by data from more neutral, or competing, sources — pamphlets from other institutions, user guides for related applications, promotional materials from other information sources.

One very ripe source of machine-readable text for the corpus is the Web. Bots that hunt for terms and constructions already known (or even, at the outset, suspected) to characterize the relevant domain can help locate sites from which scrubbers can then harvest data. The Web is not without its problems as a source of data,[3] we all know, but these can be overestimated. Chiefly, there are reliability concerns, about the age, literacy, first-language background, nationality, and various predilections of the authors, which are consequently reflected in the texts they generate. These are concerns at some level for all corpus data sources, however, and their implications are primarily for sampling strategies. There are also potential complications from the rampant data replication of the Web, but this can be turned to advantage. For instance, if the same or nearly identical texts show up on multiple sites, the tendency may be to discard them (or discard all but one) so that they don't skew the statistics, and some controls certainly need to be built in. But there is also an audience effect to consider, not just for the Web, but for all public documents. Popularity of discourse corresponds to both the influence and the representativeness of the discourse, both of which are significant for speech-system design. That is, you only need one copy of a given text in the corpus, but having information about its popularity can help in building the project lexicon and inform diction decisions for the speech system.

Instances of *use* are clearly the overriding corpus criterion, but remember that we are not building corpora and lexica to represent abstractly how people talk — what pure linguists and mainstream lexicographers are doing. We're building them also to reflect what people might know, recognize, or comprehend in an applied context. So instances of *exposure* can be relevant as well. How many times someone is exposed to a usage, and/or how

---

3: I'm talking here about considerations associated with the representativeness and overall reliability of the data. But there are social considerations as well. The site owners may not want you to take the data. The information could be proprietary (a subscriber chat room, for instance), or copyrighted (news sources), or secure simply because of the nature of site (financial transaction sites). There may be various levels of protection associated with this sort of data, or not. Even when you only want material for statistical purposes — never directly using what you harvest, but making decisions based on the analysis of what you harvest — you need to be careful where you go and what you take. Certainly, you need to secure the necessary permissions before harvesting. And you should follow the Robots Exclusion Protocol (or any standards-backed protocol or regulations that replace it; the current specification is available at *http://www.robotstxt.org/wc/robots.html*). The protocol falls more into the category of etiquette than of standards or regulations, but it is an important one, the equivalent of someone who invites you into her home, expecting you not to go into certain rooms or drawers. There are other etiquettes one should follow as well. In particular, you should not do any harvesting at peak times (and check the server location: your quiet times may not be the same as its quiet times); you should take small amounts of data at a time, rather than great big whacks of it; and you should not continually harvest the same data (that is, rather than taking everything on every visit and then discarding what you don't need, you should refrain from harvesting what you already have records for). Additionally, there are technical concerns to web harvesting — the strong likelihood of unwanted clutter in the data, for instance, that requires filtering (javascript, graphics, ads); and the notorious transience of web sites means one might disappear suddenly, or that a harvesting script which works today might be disabled tomorrow, precluding return visits.

many people are exposed to "one" usage can be significant for how recognizable a term might be, or will become; hit counters (and, for print media, circulation data) might be used as a way to gauge the audience effect for individual sites.

The Web is a wide, deep ocean of text (and, increasingly, audio and video, which is currently more difficult to harvest, but has considerable potential for the future), and it is therefore virtually guaranteed to have sources you can use. It becomes especially important if the service you are developing voices a web site.

Tagging the Corpus

The process of encoding or tagging a corpus is best regarded as the process of making explicit a set of more or less interpretive judgments about the material of which it is composed.
— Lou Burnard

Conversation applications are systems whose purpose is only to engage in conversation with a user, not to interface him with data and tasks; the most famous of them is Colby's ELIZA. In talking about their design and implementation, Ken Colby says "there is no escape from large amounts of sheer drudgery and dog work" (1999: 7). He might have been talking about voice interaction work generally, and compiling the corpus and the lexicon that drive the design provide a very big whack of that dog work. The corpus, in particular, needs not only to be collected, but structured — which means tagged (or annotated). A corpus is not a great long string of utterances. It is a great long string of marked-up utterances. There are tools that automate corpus tagging, but their success rates are not so high that they can be used without close supervision and numerous hands-on calibrations.

Here, as throughout the book, we won't concern ourselves with specific tools — or in this case even theories and methods — but concentrate only on objectives: what the resulting corpus should be capable of. The technology implicated in voice interaction design is highly changeable in general. But with corpus research and computational lexicography the situation is even more acute (see, for instance, Bird and Harrington, 2001: 1).

A register corpus for voice interface design should be tagged fairly simply — for parts of speech, syntactic roles, and discourse functions. The corpus does not need to probe theories or highlight methodologies, so the tagging can also be somewhat partial. Dialogue-act tagging, for instance, is tremendously important, but can be confined just to the most exploitable texts (the spoken dialogic texts).

Word class needs to be tagged. The general lexicographic practice is for highly specific mark-up here, not just NOUN, but SINGULAR COMMON NOUN and PLURAL TEMPORAL NOUN, and the like. There's no reason to curb this practice for design work, if the tools or (especially) the team lexicographer's work habits support it. Who knows what might prove useful at some point? But nor is there any reason to pursue that level of analysis if it demands much in the way of extra labor or other resources. It is essential to know what part of speech a word is serving, but the basic categories are largely sufficient for design work.

Syntactic elements should be tagged as well. Again, the grain need not be too fine. Syntactically, the basic phrase and sentence patterns are sufficient, along with Subject, Object, Indirect Object, and Complement. Semantic annotation might also be extremely useful, particularly for semantic field and metonymic relationships, but that will almost always prove to be a luxury. Unless good automation comes along in this area, the lexicographer should, if she uses semantic tagging at all, confine it to a small subset of the most representative texts in the corpus.

The most important subcorpus will of course be the set of spoken dialogic texts, and they should be annotated for discourse function, especially for dialogue acts.

The spoken subcorpus, too, should be tagged for marked dysfluencies. They correlate very highly with communicative slippages and repairs, which you need to analyze for planning repair strategies.

Building the Lexicon

Having a great store of the right words that one can employ is the basis, so to speak, the foundation of the whole thing.
— Marcus Tullius Cicero

By far the most important of the habitability goals is the first — getting the vocabulary right, achieving what Ogden and Bernick (1997) call "lexical habitability." And assembling a corpus is only half the job. The other half is constructing a reservoir of the basic lexical building blocks of the interface, the design lexicon. The two instruments are really two aspects of a single functioning whole. The corpus populates the lexicon, but the lexicon charts the corpus; indeed, it is fundamentally an interface to the corpus.

The lexicon must be well-built, carefully maintained, and richly detailed, a unique repository of the domain register's vocabulary, specifically crafted for interface design and development.

It has important similarities to those traditional word maps — bookshelf dictionaries and thesauri — but it is at once more limited and more comprehensive than either. It is more limited because the number of headwords is fewer, including only words used in the domain register of the speech system.[4] It is more comprehensive because the number of

---

4: I'm using *headword* in this book (also occasionally just *word*, where the context is clear), for what lexicographers most commonly call a *lemma*. It is not really a word, but a principled abstraction. The noun headword *bank*, for instance, stands not just for the word *bank* but also for the words *banks*, *bank's*, and *banks'*, though not for words like *banked* and *banking*, which would be covered in the dictionary by the verb headword, *bank*. A headword, in short, represents a principled range of variants.

datafields for each headword is greater, including concepts such as collocation, colligation, and metonyms that are largely absent from bookshelf dictionaries.

It is also, necessarily, what Kenneth Church calls a *virtual lexicon*, one that is primarily a digital database, with fluid configuration and output possibilities (Ooi, 1998: 70ff), including multimedia — most significantly, audio. There is no single, privileged entry in the sense of a traditional dictionary, just various datafields associated with each headword.

Dictionaries

> *Dictionaries are a certain kind of expert system.*
> — W. Lender

A dictionary, the prototypical reference book we consult (or, too often, fail to consult) when we need to know something about a word, is a collection of entries, all of which include at least three types of information: orthography, pronunciation, and sense. A standard-issue dictionary entry looks pretty much like this one, adapted from the 1913 *Webster's Revised Unabridged Dictionary*:[5]

> . . . methods of the Fenians.

Fenks \fengks\, n.

> The refuse whale blubber, used as a manure, and in the manufacture of Prussian blue.

Fennel \fe·nel\, n. . . .

This entry is, first of all and oppressively, in alphabetic thrall. The entry is keyed by a headword: a privileged representation of the word in English orthography. It includes an account of the pronunciation (between backslashes), an identification of the part of speech (n., for noun), and a definition, representing the sense. If there are orthographic variants, either in the word itself or in its morphology, then the variants are included and tagged (AmerE *color*, BrE *colour*; *buddy*, *buddies*). If there is gross phonological variation, that is also included (*buoy* can be either /boi/ or /bü·e/). These are the bones of a prototypical, off-the-shelf dictionary entry: an alphabetically privileged, orthographically specified headword, with pronunciation, part of speech, and definition.

The project lexicon needs a good deal more information than these few datafields represent — orthography, pronunciation, part of speech, and definition — and much of this chapter details those needs. But let's start with the skinny, Figure 11.1, a representation of

---

5: I am just discussing book-dictionaries here, because they are prototypical and, except for navigation, they have migrated pretty much intact to online formats.

deposit (637)   /dɛ • pa • zət/   🗣 IV

NOUN; there is also a VERB variant.
1. Money put into an account, a KIND-OF sum.
2. The act of putting money into an account, a TYPE-OF transaction.

FIGURE 11.1   The basic datafields for *deposit* in a design lexicon for an American English banking register. 🗣 is an icon that triggers a meun of pronunciations; **IV** stands for "in-vocabulary," meaning that it is among the recognizer's acoustic models. The other option for this binary datafield is OOV, standing "out-of-vocabulary," meaning that it is not among the recognizer's models; **VERB** is a link which calls an entry for the verb variant of *deposit*.

the basic information an interface design lexicon would have for the banking-register noun *deposit*.

There are two elements in even this narrow glimpse of the design lexicon that do not correspond to a bookshelf dictionary, one related to the speech system's discourse model, the other to the design corpus. The system element is the abbreviation *IV*, representing the binary system-vocabulary datafield (that is, the datafield which indicates whether the headword is represented among the recognizer's acoustic models). *IV* means that it is in the recognizer's vocabulary; the inverse term, *OOV*, would mean that the relevant word is out-of-vocabulary for the recognizer. These designations, of course, remain very pliable throughout the design cycle, words moving into and out of the register model as it develops. But its role in the speech system is among the most critical pieces of information about a word. (There is, by the way, nothing sacred about this coding. Abbreviations like *IV* and *OOV* are simply convenient for black-and-white print display. A graphic coding, especially color, is a much more efficient convention for signaling whether a word in the design lexicon is in the recognizer or not.)

The corpus element in Figure 11.1 is the parenthetical number after the headword, 637, which represents the number of number of tokens corresponding to the headword that occur in the design corpus — an especially useful piece of knowledge for calculating what the most likely words and phrases are for your user to utter, and for planning the ones with which to populate and structure the system's vocabulary. This particular number (overall number of occurrences) can only change as the corpus grows (or, conceivably, shrinks), but the project lexicon also needs to be able to generate more specific numbers — for instance, the number of occurrences in the spoken dialogue subcorpus — all of which should be labeled.

But, as Figure 11.1 shows, an abbreviated "entry" looks pretty much like a conventional dictionary entry: the headword itself, which anchors and orients the datafields; its pronunciation, in both transcription and (multiple) audio files, because sound is the currency of a speech system; its part of speech, for deployment; its definition, for function;

and its spelling, which you really need only for comfort most of the time (though it is sometimes useful to remain aware of variants, for text-to-speech engines which may need to read more than one ortholect). The definition, of course, is more specific than a general-purpose dictionary entry for *deposit* would be (nothing about geological or biological or retail deposits, for instance), because it is constrained by the register.

Pronunciation

Teacher	*And what is this animal called, Jimmy? [Pointing to a picture of a typical canis domesticus]*
Wee Jimmy	*Please, Miss, that's a dug, Miss.*
Teacher	*[frowning] No, no, Jimmy. It's not a dug. It's a dog.*
	Pause
Wee Jimmy	*That's funny. It looks like a dug.*

     — Tom McArthur

Pronunciation, too, is constrained by the register, though the constraints are somewhat more subtle, and much less easily revealed in Figure 11.1 than if we had audio files in this book. The inclusion of audio files is one the most immediate advantages of a digital lexicon over a paper-based product like a bookshelf dictionary or a book about voice interaction design.

At the outset, perhaps before much of the corpus is gathered or tagged, there may be only one audio file, from a standard database; some words, peculiar to the register, or otherwise not found in a standard database of pronunciations, may have no audio file at all. But several should be made available for each headword, as the lexicon, the corpus, and the interface develop. There may be multiple agents fronting the interface, for example, and each of them should be represented in the relevant audio files (unless, of course, the specific word is out of *their* vocabulary — some systems, especially those with both natural- and synthetic-voiced agents, may have different agent vocabularies). And the corpus will have multiple tokens to offer, which have the great virtue of coming in a context (words have different acoustic features when spoken within an utterance, of course, than when they are spoken in isolation). Emphatically, the audio files should not be confined to the sorts of isolated and idealized target pronunciations common with online dictionaries. Figure 11.2 illustrates these options as a drop-down menu triggered by the audio file icon.

But there may be substantially distinct pronunciations as well — regionally, professionally, generationally — and there may also be multiple variations within the same register. Some pronunciations are so distinct that they affect recognition in highly specific ways. Take geographical place names. They can be incredibly difficult to chart. There is a town in Indiana that was named after the grand palace of the French Sun King, Louis Quatorze. But the name of the palace is pronounced /vɛrsɑj/, rhyming with *mare sigh*. The

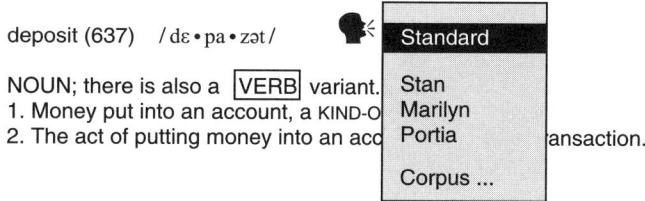

FIGURE 11.2   A drop-down menu of audio files for *deposit*

town in Indiana (at least to the locals) is pronounced /vʌrselz/, rhyming with *fur sails*. Personal names, street names, and business names can be similarly problematic.[6] Professional registers have various idiosyncratic pronunciations for their jargons. The acronym *GUI* is almost always /gu·i/ (goo-ey), for instance, rarely /dʒi·ju·ɑj/ (G.U.I.), while *URL* is almost always /ju·ar·ɛl/ (U.R.L.), rarely /ʌrl/ (like the proper name, *Earl*).

Depending on the service, context, and size of the recognition vocabulary, among other variables, the interface will have to treat these differently. With personal names in a corporate reception system, alternate pronunciations (indeed, *mis*pronunciations) need to be accommodated, as well as resolution strategies such as spelling. With geographical names in a national weather or travel service, only the most prominent alternatives, if any, could be accommodated; with a regional service, the local pronunciations must be accommodated. For professional jargons, the dominant pronunciation should always be spoken, but significant variants should be in the recognizer.

But whatever the system accommodates, or fails to accommodate, the design lexicon requires both transcriptions and acoustic files for the significant variants of all key terms; one job of the lexicon is to keep comprehensive track of the system vocabulary, but an equally important job is to guide the decisions designers make about that vocabulary. Figure 11.3 shows multiple pronunciation datafields for the headword, *GUI*.

Bookshelf dictionaries nearly always privilege a single pronunciation, usually from the prestige dialect of a specific national variety. Speech systems do not often have that luxury. Indeed, conventional dictionaries, often compiled by people with rather strict standards of "correctness," tend to focus most of their efforts on the exemplary, rather than the mundane. This tendency of conventional dictionaries toward virtuosity, rather than typicality, certainly encompasses usage, our next topic.

6: See Jannedy and Mobius (1997) for a succinct list of some speech-system complications with proper nouns, and a list of useful references. Attwater et al. (2000: 283) discusses regionalisms in the United Kingdom in the context of speech recognition.

GUI (811)   / gu • i / (783)   IV

/ dʒi • ju • aj / (28)   IV

FIGURE 11.3 A representation of a double-pronunciation entry for *GUI*

Usage Data

You shall know a word by the company it keeps!
— J. R. Firth

Like the best book-shelf dictionaries, the design lexicon needs to illustrate usage, too, but illustrate it far more fully, by several orders of magnitude, because even the best definitions are insufficient to get at meaning and function with enough precision to support design work. The definition is a wonderful invention, which abstracts from the flux and sheer bulk of innumerable occurrences a "meaning," a descriptive paraphrase pointing to which concepts are invoked by typical uses of headwords. But definitions are necessarily idealized fictions, which even a casual look at actual usage betrays. That's understandable — equations are idealized fictions, too — but it is precisely why good dictionaries supplement definitions with representative *in situ* instances of usage, and why your sixth-grade teacher didn't believe you knew what a word meant until you could use it in a sentence. Nothing substitutes for context.

Those conventional dictionaries that include usage examples, however, overwhelmingly go for the gold — something from Shakespeare or John F. Kennedy, rather than the daily linguistic iron of examples from a small town newspaper, a pamphlet, or a business letter. They are also confined by medium to a very few examples, usually one. For voice interaction design, we need the iron, and lots of it: mundane usages, in the quantity that only a corpus can provide.

The lexicon's datafields should include a random assortment of *in situ* corpus usages, perhaps three to five — and should link to a concordance extracted from the design corpus. And, because we are constructing a system with verbal output — projected usages, if you like — a sampling of those should be available as well.

The voice interaction design lexicon, however, is not just a beefed-up and digitized conventional dictionary — or, if it is, the beefing up and digitization is not confined to putting the elements of a conventional dictionary on steroids. Other elements are necessary as well to satisfy the system's habitability needs and guide its diction. Recall that Peter Mark Roget's complaint about dictionaries is you have to start with the word, and then find its signification (to which we can add pronunciation, usage, and so on), but that the job of diction is to start with the signification — the *task*, in our terms — and from there "to find the word, or words, by which [it] may be most fitly and aptly expressed" (Roget, 1925: xiii).

We can't fully disentangle signification from expression, tasks from terms, of course, but we can take a lesson from Roget, who approached the problem from the opposite direction of the dictionary, and build our lexicon in a way that allows us to come at design from a purpose-driven way, not just a word-driven way.

Thesauri

> *Every workman in the exercise of his art should be provided with proper implements.*
> — Peter Mark Roget

The most dramatic difference between a dictionary and a thesaurus is that an entry in a thesaurus has fewer bones, and a lot more flesh. It has a headword as well, but that word doesn't have the same authority, and sometimes looks almost arbitrarily chosen. It often comes with a descriptive phrase, and is subordinated to both that description (if present) and a unique numeric identifier. The "headword" is really just the label for a sack of other words. Here is entry 371 of *Roget's Classic American edition*:

371. [The economy or management of plants.] Agriculture. — N. agriculture, cultivation, husbandry, farming; georgics, geoponics; tillage, tilth, agronomy, gardening, spade husbandry, vintage; hort-, arbor-, silv-, citr-, vit-, flor-iculture; intensive culture; landscape gardening; forestry; afforestation.

husbandman, horticulturist, citriculturist, gardener, florist; agricult-or, -urist; yeoman, farmer, cultivator, tiller of the soil, ploughman, sower, reaper; woodcutter, backwoodsman, forester; vine-grower, vintager; Boer; Triptolemus.

field, meadow, garden; botanic —, winter —, ornamental —, flower —, kitchen —, truck —, market —, hop-garden; nursery; green-, hot-, glass-house; conservatory, cucumber frame, cloche, bed, border, seed-plot; grass-plat, lawn; park &c. (pleasure ground) 840; parterre, shrubbery, plantation, avenue, arboretum, pinery, pinetum, orchard; vineyard, vinery; orangery; farm &c. (abode) 189.

V. cultivate; till, — the soil; farm, garden; sow, plant; reap, mow, cut; manure, dress the ground, dig, delve, dibble, hoe, plough, plow, harrow, rake, weed, lop and top; force, transplant, thin out, bed out, prune, graft.

Adj. Agr-icultural, -arian, -estic.

arable; predial, rural, rustic, country, bucolic; Bœtian; horticultural.

Roget's organization is decidedly not alphabetic.[7] Its genius, in fact, is that it has nothing to do with *form* at all. To understand words, as Thomas Hobbes put it, one must "observe the distinction betwixt the soul of Words and the body; betwixt that in them which is corporeal, and that in them which is spiritual." (Harwood, 1986: 181). Lexicographers, unfortunately, didn't adopt Hobbes's terms, and use two really ugly, bulky words instead: *semasiological* (for the *body*) and *onomasiological* (for the *soul*). The first concerns all lexical matters determined by the form of the word; the second, all matters determined by the meaning of the word. (For instance, homonymy is a semasiological issue, because it is determined by the acoustic form; synonymy and antonymy are onomasiological because they are determined by the sense[s].) Dictionaries are traditionally structured semasiologically, thesauri onomasiologically. I'll largely avoid these words, using rough synonymns like *formal* and *conceptual* most of the time, but they are important terminological keys for reading the literature of lexicography.

Dictionaries are organized by the body; Peter Mark Roget wanted the soul. The specifics of his organizational structure aren't especially important for our purposes, but his philosophical tendency is: *Roget's Thesaurus* is entirely conceptual.

A thesaurus is often thought to be a laundry list of synonyms. But a quick glance at virtually any entry, like 371, makes it clear that the groupings are more capacious than that: several pairs, for instance, are antonyms, like *reaper* and *sower, prune* and *graft,* perhaps *meadow* and *garden; pinery* and *orchard* have clear overlap in terms of structure, but are far too distinct in function to be confused for synonyms, as are *florist* and *farmer; geoponics* and *Triptolemus* and *parterre* are . . . well, I'm not sure what they are, other than arcane, but if we can find a source on them, and we look them up, we can bet they will have something to do with human-tended flora.

Roget describes the plan for his entries as "a copious store of words and phrases, adapted to express all the recognizable shades and modifications of the general idea under which those words and phrases are arranged." (1925: xiv). The plan calls for (1) quantity and (2) conceptual correlation, two essential characteristics of the design lexicon. Both characteristics need to be more restricted than they are in Roget's masterpiece, in a judicious, register-governed way; the chances of needing *Triptolemus*, even in an agricultural advice system, are remote. But the design lexicon needs lots of options, and those options critically include synonyms and metonyms.

7: There *are* alphabetic thesauri, it's true, or at least there are misbegotten alphabetic collections of words that purport to be thesauri, including the miserable *Roget's II*, but they are a different species (dictionaries of synonyms), born because paper-based thesauri are not especially easy to navigate by hurry-up users.

Synonyms and Metonyms

The mapping of terms to referents is many-to-one.
— Susan E. Brennan

Synonyms are a given. A user-interface design lexicon needs to include synonyms. As I've been nagging all along, they have to be deployed judiciously, but you need to have rapid access to terms with sufficient semantic overlap in the register to be uttered in comparable circumstances. How you structure the vocabulary and the utterances of a voiced travel system, for instance, is subject to all sorts of contingencies, but if it involves flying into and/or out of Chicago, you can't make the right decisions unless you know that *ORD* and *O'Hare* and *Chicago International* are all terms mapping to the same referent. You can't make the right decisions about a confirmation exchange unless you know that *yes*, *sure*, *OK*, and even *righty-oh* overlap sufficiently that they all amount to the same response for a question like "Should I book that for you now?" Synonyms are a given, and the design lexicon should record them.

But metonyms? What the heck do we need metonyms for? Let's start with a dialogue. My son, at twenty-seven months, was trying to accomplish that age-old task of getting someone to pass him the salt:[8]

T_1 Galen   I want some salt, please, Mommie.

T_2 Mother   Just a minute, Darlo.

T_3 Galen   Some salt please, Mommie.

T_4 Mother   [silence; I have no timing data, but it was a standard short duration of benign neglect, a few seconds; she was reading the paper]

T_5 Galen   Some of that white stuff there.

He is paraphrasing *salt* here; more specifically, at T_5 he is using a lexical relation known as metonymy; that is, using a word (*white*) that stands in an attributive relationship to the target word he wants to evoke (*salt*).[9] In this case, our hero's mother was just preoccupied,

---

8: I know what you're thinking — "Asking for salt!? This is too perfect: it uses the staple pragmatics example, and it illustrates a communication repair, and it's *his* kid!" But I'm not inventing this. For one thing, I could never make up the word "Darlo."

9: *Metonym* is a term associated with long-standing confusions. It is related to the figure of speech, metonymy, which literally means "name change," but which conventionally means calling something by the name of something that it's related to in a part-to-whole way. The confusion comes from whether it can only serve the part role in the relationship (as in "thirty head of cattle" or "my new set of wheels is a Ford"), or whether it can serve

but it was a reasonable guess that her silence meant she didn't understand *salt* — he had already given her two shots at it — and a sound repair strategy to employ metonymy.

People use metonyms all the time; for instance, when they have lexical search difficulties ("Oh what's-his-name, you know, the comedian. He had a ski-jump nose. He was in those road-movies"), or when they just want to be eloquent ("Coca-Cola stocks lose their fizz," winks the headline writer), or, sometimes, when they really are eloquent ("Friends, Romans, Countrymen, Lend me your ears," wrote Shakespeare for Mark Antony).

We don't need our machines to be eloquent, of course, nor should we build them to expect a caller's attempts to be eloquent, should one stray in that direction. But metonyms are instruments of the mundane too, as the *white/salt* example shows. Here is a speech system, combining synonymy and metonymy for a clarification routine:

T$_1$ User   Talk faster.

T$_2$ System   Did you say to increase the speech output rate?

T$_3$ User   Yes.

T$_4$ System   Increasing the speech output rate.

(Marx, 1995: 85)

Metonyms, like collocates, are very useful for both priming and resolving, though in somewhat different ways, and, like synonyms, are prime instruments of habitability because they are tools of paraphrase.

More generally, an understanding of metonymic relationships goes a long way to achieving what Ogden and Bernick (1997) term "conceptual habitability," since metonymy (along with synonymy and antonymy) expresses the fundamental conceptual relationships among terms: part-of, composite-of, element-of, and set-of. The inference engine associated with a speech system needs to know such things as a "storm is type of weather," a "storm might be a blizzard," and "snow comes with a blizzard." At the level of the lexicon, these relationships are metonymic. And they are often the fall-back terms when trying to resolve errors. Here is a typical user trying to get past a misrecognition, for instance:

---

as the whole (as in "Canada won the gold"), and what its connection is to another figure of speech, synecdoche. As they say in New Jersey, fuhgedabowdit. Rhetoricians can't get their act together, and I'm using the word here to identify any attributive relation, especially in paraphrase. Several technical terms from lexicography that you might come across in this connection are *meronym* (also *partonym*), which means a part-relationship (the child in a PART-OF relation, using object-oriented terminology); *holonym*, its opposite, which means a whole-relationship (the parent in a PART-OF relation; in the pair of words, *finger/hand*, *finger* is the meronym, *hand* is the holonym); *hyponym*, which indicates a subset relationship (the child in a KIND-OF or IS-A relation); and *hypernym*, its opposite, which indicates a superset relationship (the parent in a KIND-OF or IS-A relation; in the pair of words *schnauzer/dog*, *schnauzer* is the hyponym, *dog* is the hypernym). I'm using *metonym* for all of these lexical (and, in fact, phrasal) relations.

User	What meal is served on this flight?
System	Here is the airline for the flight from Atlanta to Baltimore.
User	Will there be food on the flight?

(Stifelman, 1993)

Absolute freedom to paraphrase, for Watt (1968), is the defining characteristic of a "fully habitable" speech system. Absolute freedom is not, of course, a realistic goal for a dialogue system in general use, and metonymy reveals this limitation as well as anything might. The potential metonyms for a given term comprise a vast semantic web (as we see from any *Roget* entry). The design lexicon could not begin to accommodate them all. But any metonyms that show up in the corpus, especially ordinary ones, like *white* for *salt*, or *increase + rate* for *faster*, should be noted, and every headword should link to a datafield of metonyms. That datafield may be null, but in effect it is something of a sack for notable conceptually related terms that aren't synonyms.

The lexicon needs to accommodate semantic variations, not with any great precision, but with enough coverage that designers can make informed choices about what words the system says, and what words it hears, both in primary situations and in repair. Voice interfaces are expert systems for conversation, and people often use semantic alterations in their conversations, especially for aspects of grounding or repair work, through paraphrase.

The significant relations of critical words need to be structured in the specific language model that the speech system employs, and these relations should be registered in the lexicon. It is not enough, for instance, in a travel domain to know that the word *Chicago* is a KIND-OF the word *city*, unless you also know that *city* is a KIND-OF *destination* and a KIND-OF *departure* point.

The "Entry"

The lexicographer is making an inventory; that is his business.
— Richard C. Trench

The voice interface lexicon, again, is not a collection of entries in the bookshelf sense — or, for that matter, the usual electronic sense. It is a specifically designed tool for choosing the fundamental building blocks of the interaction: words. It has to be searchable, and sortable, by a range of criteria, including:

* theme

* frequency

* pronunciation

- syllable structure
- orthography

There are no entries, per se, since the output depends on the search and sort criteria, but the lexicon should generate a body of information with respect to a given word, or words, which has much in common with traditional entries. Figure 11.1 represents a bare-bones look at the word *deposit* in a financial register from a design lexicon developed for a voice-banking system. A maximal "entry" in the same circumstances would include the following information.

- Headword

  (keyed to the term/concept searched)
- Overall frequency
- Orthography
  - Standard
  - Alternates
- Pronunciation
  - Standard
    - Transcription
    - Audio files
  - Alternate
    - Transcription
    - Audio files
- Morphological variants
  - Inflectional
  - Derivational
- Part of speech
- Definition(s)
- Usage
  - Multiple examples from the corpus
  - Multiple examples from the speech system output

- Collocations

- Colligations

  - Phrases

  - Idioms

  - Compounds

- Synonyms

- Metonyms

- Notes

We've looked at most of these elements, some in great detail, in this chapter and elsewhere in the book — all but the last item. Words are notoriously idiosyncratic, and some registers can have quite specific and unexpected usage attributes for many of their words; the notes datafield is just an extended text field for the lexicographer's observations about these idiosyncracies and any assorted relevant facts or notions.

A representation of a maximally specified "entry" for *deposit* in a voice-interface lexicon is given as Figure 11.4.

Getting the vocabulary right is the most critical habitability goal. And a richly crafted design lexicon, drawn from a representative corpus, is the most critical design tool for achieving lexical habitability. But — have all the words you want — if the interface does not also utilize the appropriate utterance structures and interaction patterns, the system will fail. The design corpus is essential here, too.

Determining Commonly Used Grammatical Constructs

Madame Merle gives her excellent advice, but it's a good deal like giving a child a dictionary to learn a language with. He can look out the words, but he can't put them together.
— Henry James

A well-tagged corpus will automatically generate displays of the sort given as Figure 11.5. It is a dialogue representation of a by-now familiar sort — this one between a customer and an agent collaborating on a travel-arrangements task — charted by utterances, turns, and dialogue acts. If you're lucky, your corpus will naturally contain many specimens like this dialogue for you to dissect in similar ways; less lucky and your corpus will unnaturally contain such specimens, after you have filtered the target tasks through the

deposit (637)   /dɛ • pa • zə t/   ))) IV

NOUN; there is also a VERB variant
1. Money put into an account, a KIND-OF sum
2. The act of putting money into an account, a KIND-OF transaction.

Variants

NOUN FORMS: deposits, deposit's
RELATED FORMS: deposit (verb), depositor, depository

Usage

that if you invest in a money market	deposit account rather than a money fund,
products like CDs and money market	deposit accounts. You may, of course, be
Please note: A safe	deposit box is available for the
All checks must be drawn on funds on	deposit in the U.S. if your account is a
enclosed. Include a voided check or	deposit slip for convenient monthly

concordance

There were five deposits in [month].
Your last deposit was on [date], ...[amount].
You cannot make a deposit by telephone.

more

Collocations

additional	(2633; 6.5)	bank	(11926; 5.1)
money	(25274; 5.5)	Into	(81207; 2.3)

more

Colligations

deposit insurance, deposit base, deposit protection, certificate of deposit, safe deposit box, safety deposit box, make a deposit of [amount]

Synonyms	Metonyms
NONE	account, checking, saving
	balance, interest rate

more   more

Notes

Deposit (noun) has the virtue that there are no real synonyms in personal and commercial banking, though there are several paraphrases. The chief paraphrase involves the verb form of d*eposit* ("to deposit" = "to make a deposit".)

Outside of banking, both the noun and the verb denote the use of money to secure obligations (a partial payment, to ensure the vendor doesn't sell the goods to someone else, for instance; or or an amount held in trust for a leesee against the possibility of damage to the leased space).

FIGURE 11.4   A maximally specified design-lexicon "entry" for the noun *deposit* in an American English banking register

telephone; even less lucky, and you may have to wait for a prototype (coded or counter-feited) to generate useful enough dialogic data. But you can't really get a voicing project off the ground in any reliable way until you have dialogues to analyze and learn from.

Figure 11.5 is not an academic research tool, which would call for many careful distinctions and the adherence to clear methodological principles. For instance, most of the turns (all but perhaps T_{15}) end with turn-assignment protocols, but only one is labeled here for observing the overt dialogue act, turn-assignment. On the other hand, there is only one "true" yes/no question among the acts (T_{13}). The rest are exclusively requests ("Could you give me the name of the guest, please?" Ann says in T_8; the caller doesn't affirm or negate, but says "Tom Wilkinson," bypassing the question to satisfy the request). Yet, every utterance that takes the syntactic form of a yes/no question is identified (sloppily, a pragmatic theorist might complain) as a yes/no dialogue act. Other complaints are admissible as well.

But turn-assignment is redundant throughout the dialogue, and is only worth marking where there was an explicit, active assignment of turn (as in T_4, where the greetings have already been exchanged, and this "Hello" basically means "OK, I'm here, what's up? Talk to me"). And, for design purposes, the form of a request is at least as important as the pragmatic fact of that request. We are, after all, pursuing the habitability goal of implementing the appropriate grammatical constructs (effectively, what Ogden calls "syntactic habitability" — 1988). The charting laid out Figure 11.5, the sort of analysis we need to do to lay the foundations of a voice interface, is frankly opportunistic — driven by design intentions, not theoretical concerns.

Charts like Figure 11.5 are the first step in understanding the structure of dialogues. On the one hand, we need to uncover relevant colligations from the corpus. The people who speak, and collectively develop, a register favor certain expressions. On the other hand, we need to tabulate the specifics of such dialogues in greater detail, to get a sense of the microstructure of the register. To participate effectively in a register, our machines have to deploy those colligations and characteristic structures regularly and appropriately, too.

The dialogue represented in Figure 11.5 does allow us to see some colligations in action, however, which can point us to ways we need to investigate the corpus. One would have to test these patterns against the corpus as a whole, for instance, to see if they are characteristic of the register or just peculiar to Ann, but it is surely significant that her T_8 and T_{14} requests both begin "Could you give me. . . ."

Colligation has another side, of course, the one without a snappy label, which helps identify in what grammatical slots one might find certain words or certain categories of words. Again, we need much more general data than we have in Figure 11.5, but there are some illustrations in that dialogue of what we need to look for. Take Ann's pronoun use, for instance. When she solicits information — in directive dialogue acts — she uses a first-

T_1	Caller	Ring	Summons, dialogue initiation
T_2	Ann	Good morning, this is the reservation desk, Ann speaking.	Response, greeting Identification (functional), statement Identification (personal), statement
T_3	Caller	Good morning.	Greeting, acknowledgment
T_4	Ann	Hello.	Greeting
T_5	Caller	I would like to make a reservation for March 2, please.	Request, exchange initiation, statement Mitigation
T_6	Ann	Yes, for one night only?	Acceptance-to, confirmation Yes/no question, clarification request
T_7	Caller	For one night.	Clarification, statement
T_8	Ann	Could you give me the name of the guest, please?	Request, yes/no question Mitigation
T_9	Caller	Tom Wilkinson.	Acceptance-to, compliance, statement
T_{10}	Ann	Wilkinson.	Echo, acknowledgment, statement
T_{11}	Caller	Wilkinson.	Echo, confirmation, statement
T_{12}	Ann	Okay, will the guest pay the bill in our hotel?	Acknowledgment Yes/no question
T_{13}	Caller	Just one moment..... No, could you send the bill to REM Slochteren, please?	Turn-holding Negation Request, yes/no question Mitigation
T_{14}	Ann	Okay, could you give me a fax number?	Acceptance-to Request, yes/no question
T_{15}	Caller	78.	Acceptance, compliance, statement
T_{16}	Ann	Yes.	Acknowledgment
T_{17}	Caller	9208	Compliance, statement
T_{18}	Ann	Okay thanks, we will send a fax for affirmation.	Acknowledgment Thanking Statement
T_{19}	Caller	Yes?	Acknowledgment request
T_{20}	Ann	Thanks.	Thanking, acknowldgment
T_{21}	Caller	Thanks, bye.	Thanking, exchange termination Leave-taking, dialogue termination
T_{22}	Ann	Bye.	Leave-taking

FIGURE 11.5 A human–human travel-arrangements dialogue, charted by utterances, turns, and dialogue acts. The data is adapted from Steuten (1997)

Dialogue Act		Caller			Agent			Dialogue		
		Freq.	Words	Tokens	Freq.	Words	Tokens	Freq.	Words	Tokens
Identification	Personal				1	2	2	1	2	2
	Functional				1	5	5	1	5	5
Thanking		1	1	1	2	1	1	3	1	1
Mitigation		2	1	1	1	1	1	3	1	1
Greeting		1	2	2	2	1.5	1.5	3	1	1.6
Leave-taking		1	1	1	1	1	1	2	1	1
Clarification		1	1	1				1	1	1
Clarification-request					1	3	3	1	3	3
Echo		2	2	2	1	1	1	3	1	1.3
Question	Yes/no	1	11	11	4	5	6.5	5	5	7.4
	Constituent									
Statement		6	3.3	4.4	3	2.7	2.7	9	2.9	3.3
Answer		2	1.5	1.5				2	1.5	1.5
Confirmation		1	1	1	1	1	1	1	1	1
Acceptance-that										
Acceptance-of										
Acceptance-to		2	1.5	1.5	2	1	1	4	1.25	1.5
Negation		1	1	1				1	1	1
Turn	assignment									
	holding	1	3	3				1	3	3
Exchange	initiation	1	10	10				1	10	10
	termination	1	1	1				1	1	1
Dialogue	initiation	1	ø	ø				1	ø	ø
	termination	1	1	1				1	1	1
Summons		1	ø	ø				1	ø	ø
Response					1	2	2	1	2	2
Request		2	9.5	9.5	2	5.5	8	4	6.25	8.75
Command										
Directive										
Offer										
Acknowledgment		1	1	1	5	1	1	6	1	1
Acknowledgment-request		1	1	1				1	1	1
Compliance		3	2.3	2.6				3	2.3	2.6
Average per turn		2.7	3.5	4	2.4	3.5	5.1	2.6	2.8	4.5
Average per act			1.2	1.4		1.4	2		1.1	1.7
Total (for acts, type I token)		21	29	39	44	15	23	39	56	25	52	62	100
Average acts per turn			2.9			2			2.4	

TABLE 11.1 Dialogue act, average word, and average taken counts for a 22-turn human–human travel-arrangements dialogue (as charted in Figure 11.5)

person singular in object position ("Could you give me . . ." — T_8, T_{14}). But when the obligations flow the other way — in commissive dialogue acts, she uses a first person plural in subject position ("We will send . . ." — T_{18}).

Associative cohesion is also in notable evidence in the dialogue, with a significant tendency for the agents to use repetition, though neither anaphors nor ellipses are used in any uncommon way. Connective cohesion is in very low evidence, both within and between the turns.

We need also to tabulate the micromechanics of such dialogues: the mean number of tokens per turn and per dialogue act, the mean number of words per turn and per act, the number of acts per turn, the words and tokens per agent, the correlation between dialogue acts and agent roles (especially to the extent the existing roles can be forecast into system and user roles), and other variables that might be of interest. Table 11.1 serves up such a tabulation, for the dialogue of Figure 11.5.

Such tables really only serve their purpose when they collate data from multiple dialogues, with a diversity of agents, which helps to control for idiosyncracy, personality, and other particularities, and gets at broader patterns of the register generally. But Table 11.1 serves well as an illustration. We notice right away, for instance, that the dialogue has a modest and efficient vocabulary — 62 words, which get the job done, 100 tokens, over 22 turns. Syntactically, statements and questions dominate, with brief interjections (*hello*, *bye*, *please*, *thanks*, *OK*) distributed broadly. The turns overall are also modest and efficient, the longest one clocking in at 11 tokens, many only a token or two, and a mean token-to-turn rate of 4.5.

Table 11.1, however, goes beyond grammatical constructs to interaction patterns, the last of the habitability goals, to which we now turn.

Discovering Effective Interaction Patterns

It is a common observation, and a common-sense one, that talk in interaction comes in what might be called clumps.
    — Emanuel A. Schegloff

Dialogue data calibrates the vocabulary and syntax of the system by revealing the basic elements and structures of the register in action. But the words and structures of the register can be explored in a variety of ways, from a variety of sources in the corpus — pamphlets, interviews, web materials, even email. What corpus dialogues are absolutely essential for is to reveal interaction patterns the interface can build from, shaping architectural decisions about the system callflow. Ogden and Bernick, (1997) call this (or something very much like it) "functional habitability."

Table 11.1 (more specifically, similar tables of data, extracted from a collection of dialogues) is a first cut at understanding the patterns underlying functional habitability.

The data show that while the caller has the overall initiative, Ann carries the lower-order initiative, negotiating the specifics of the booking. In the terms of Jennifer Chu-Carroll and Michael Brown (1998), the caller has the "task initiative," while Ann has the "dialogue initiative." The caller not only initiates and terminates the dialogue, he also initiates and terminates the task exchange, and the token count of the exchange-initiation act signals that he begins it with minimal preliminaries, getting right to the purpose. Further, he issues both more dialogue acts overall (29 to 23) and more distinct types of dialogue acts (21 to 15). On the other hand, Ann asks the bulk of the questions; the caller makes most of the statements. She talks somewhat more overall (measured in token count) but he contributes somewhat more overall (measured in dialogue acts).

There is a notable asymmetry in turn and act length. Most acts are quite brief, but the heaviest task-management acts (obligatives and informatives), are comparatively long. The informatives tend not only to be long, but frequent; requests, acknowledgments, acceptances, and echoes are also fairly frequent. Without knowing anything beyond this data, we could safely diagnose the dialogue as information-driven with moderate to high grounding criteria.

While Table 11.1 does not comprehensively list all possible dialogue acts, it is noteworthy that a few obligatives we might expect to be involved in such a task — particularly directives and offers — are absent. Along with the high number of phatic acts (mitigators, thankings, greetings, leave-takings), this data suggests a very cooperative encounter. The low turn (22 each) and act (25 distinct types, 52 overall) counts indicate a very business-like encounter, of the sort that is especially amenable to voicing. The prevalence of echoes strongly suggests a listening strategy that cues for repetitions and partial repetitions (in addition to confirmation/disconfirmation, for instance, with yes/no questions).

We have to return directly to the dialogues, however, to get more fully at the interaction patterns. Figure 11.6 is another view of the hotel-booking dialogue, this time with the added mark-up of a task analysis (similar to the one we saw last chapter, in Figure 10.1).

Starting with dialogic pairs, we see many of the expected match-ups. The summons gets a response, greetings and leave-takings are traded, clarification-requests are followed by clarifications. All the requests are accepted. But the pairings are not quite as systematic as we might predict. In particular, there are five questions (T_6, T_8, T_{12}, T_{13}, T_{14}), but only two of them get direct answers (T_6/T_7, T_{12}/T_{13}), and a specific request for acknowledgment (T_{19}) is effectively ignored (T_{20}). Interactively, these are a byproduct of the efficiency we noticed characterizing the dialogue. The failure of Ann to give the caller the acknowledgment is perhaps just an oversight, but all of the "missing" answers are a function of higher-order reactive pressures. T_6 and T_{12} end in direct questions, and strongly pressure for direct answers (affirmative or negative), and they get them. But, for instance, T_{14} is more strongly a request (effectively "Please tell me the fax number") and pressures for an acceptance/compliance, which it gets promptly; T_8 is much the same, in pressure and in satisfaction,

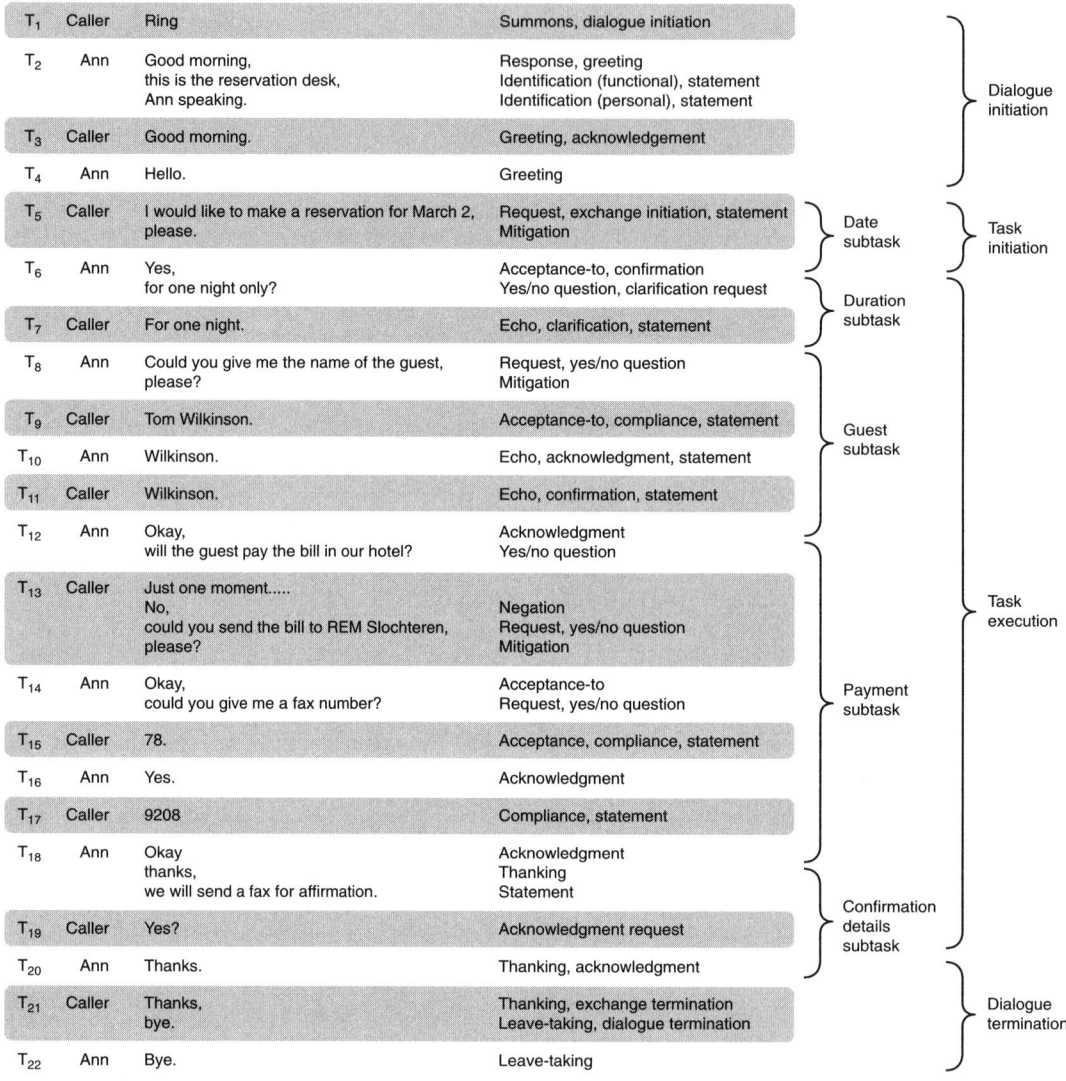

T_1	Caller	Ring	Summons, dialogue initiation
T_2	Ann	Good morning, this is the reservation desk, Ann speaking.	Response, greeting Identification (functional), statement Identification (personal), statement
T_3	Caller	Good morning.	Greeting, acknowledgement
T_4	Ann	Hello.	Greeting
T_5	Caller	I would like to make a reservation for March 2, please.	Request, exchange initiation, statement Mitigation
T_6	Ann	Yes, for one night only?	Acceptance-to, confirmation Yes/no question, clarification request
T_7	Caller	For one night.	Echo, clarification, statement
T_8	Ann	Could you give me the name of the guest, please?	Request, yes/no question Mitigation
T_9	Caller	Tom Wilkinson.	Acceptance-to, compliance, statement
T_{10}	Ann	Wilkinson.	Echo, acknowledgment, statement
T_{11}	Caller	Wilkinson.	Echo, confirmation, statement
T_{12}	Ann	Okay, will the guest pay the bill in our hotel?	Acknowledgment Yes/no question
T_{13}	Caller	Just one moment..... No, could you send the bill to REM Slochteren, please?	Negation Request, yes/no question Mitigation
T_{14}	Ann	Okay, could you give me a fax number?	Acceptance-to Request, yes/no question
T_{15}	Caller	78.	Acceptance, compliance, statement
T_{16}	Ann	Yes.	Acknowledgment
T_{17}	Caller	9208	Compliance, statement
T_{18}	Ann	Okay thanks, we will send a fax for affirmation.	Acknowledgment Thanking Statement
T_{19}	Caller	Yes?	Acknowledgment request
T_{20}	Ann	Thanks.	Thanking, acknowledgment
T_{21}	Caller	Thanks, bye.	Thanking, exchange termination Leave-taking, dialogue termination
T_{22}	Ann	Bye.	Leave-taking

Task structure (right-hand annotations): Dialogue initiation; Task initiation; Date subtask; Duration subtask; Guest subtask; Task execution; Payment subtask; Confirmation details subtask; Dialogue termination.

FIGURE 11.6 A human–human travel-arrangements dialogue, segmented by task. The data is adapted from Steuten (1997)

while T_{13} can only get the acceptance, because the compliance has to be nonverbal (that is, Ann, or somebody, has to go physically to the fax machine and send the document). Task-driven dialogues, in particular, tend toward the efficient discharge of reactive pressures, with responses to only the most goal-oriented elements of the initiatives, and the dialogue charted in Figures 11.1 and 11.6 is certainly a case in point.

Turning to the structure of the dialogue, we see an initiation and termination frame that is moderately involved (four turns and three turns respectively, comprising over 30% of the dialogue, measured in turns). But once the task gets underway, the dialogue hums through five stages very efficiently. The task is identified and specified in the same task (compare Figure 10.1, for instance, from the previous chapter), and still in the same turn (T_5), the date(s)-of-stay subtask is initiated. Most of the subtasks, in fact, overlap the next subtask, one wrapping up as the other one commences, in the same (booking-agent) turn. At T_6, for instance, Ann acknowledges the initiation of the reservation task, simultaneously acknowledging the date ("Yes"), and starts the duration-of-stay subtask ("for one night only?"); at T_{12}, she closes off the guest-identity subtask ("Okay") and begins the payment subtask ("Will the guest . . ."). Ann actually works very much as one might project the interface working. She has four routines to carry out. She needs to get the date(s), confirming the duration as an error-check on the date(s). She needs to get the name of the guest(s). She needs to get payment details. And she needs to get contact information from the caller to ensure the confirmation can be sent. Done.

Referential coherence is established and maintained with a similar mechanical efficiency, one reference after another. It's not fully clear to Ann what "reservation for March 2" refers to (at T_5), so she checks her likeliest surmise ("for one night only?" — T_6), and the caller confirms it ("for one night only" — T_7). That reference is now locked and grounded for the duration of the dialogue. The guest is referenced in five consecutive turns (T_8–T_{12}), securing shared reference, and never mentioned again.

Coherence relations are largely localized to elaboration (the particulars of the booking) and sequence (date, duration, guest, payment, and confirmation). We don't know if there is anything sacrosanct or optimal about this sequence, something we could only tell by consulting a collection of dialogues centering on the same task, but there is no doubt some preferred order, or some range of preferred orders.

Some words and phrases tend very strongly to show up in specific locations of specific discourses ("Once upon a time" at the beginning, "Happily ever after" at the end, of certain narratives, for instance; Methods sections at the beginning, Results at the end, of experimental reports; "Martin, Bob" in the middle of phone books). Michael Hoey calls this tendency "textual colligation," and dialogues exhibit it in varying degrees. Greetings and leave-takings are the most obvious examples, but they can be fairly subtle and unexpected as well. Bernsen et al., for instance, found in field studies of travel-booking tasks that traveler identities tended strongly to be fixed before dates and times were worked out (1998: 166).

In the reservation task of Figures 11.1 and 11.6, there are at least a couple of these textual colligations that seem inevitable — the confirmation-details subtask at the end, in particular, and the initiating request at the beginning — but there may be other preferences, perhaps even a rigid pre-established order, to the task structure. Maybe the name-of-guest(s) task always precedes the billing task, and the duration-of-stay task always precedes the

Take call.

Get dates.

(Get length of stay.)

Get name of guest.

(Give price.)

Get payment details.

Confirm payment details.

Confirm travel details.

(Provide additional details.)

End call.

FIGURE 11.7 A distillation of hotel-booking dialogues into a system-side knowledge representation. Parentheses indicate optional components (schemas)

billing task, but the guest and duration tasks do not have a preferred order with respect to each other. We can only tell by looking; discovering effective interaction patterns depends on searching for such tendencies through all the related dialogues in the corpus.

Task duration, too, is important. In this dialogue, the billing subtask takes the longest (measured in turns, acts, words, and tokens) — which may be coincidental to this encounter, may be quite typical, or may be a specific function of only those billing subtasks where addresses or fax numbers have to be collected. It's always hard to say (though not to guess) without a group of dialogues to generalize from, but turns/acts/words/tokens-per-task and per-subtask are important measures for interaction patterns.

Ultimately, to understand the task, we need to distill the data gathered from the analysis of many specific dialogues, like the one in Figure 11.6, into a bare-bones representation of the overall pattern, something that very much resembles a knowledge-representation script of the sort proposed by Roger Schank and Robert Abelson (1977) and now widely populates the artificial-intelligence landscape. Figure 11.7 is such a distillation.

We take up the elaboration of such representations into implementable scripts in Chapter 14.

Summary

It is a characteristic of what the human brings, the variability in vocabulary usage, that dominates the problem.

— George W. Furnas, Thomas K. Landauer, Louis M. Gomez, Susan T. Dumais

In this chapter we have explored methods for achieving these three **habitability goals**:

- Get the vocabulary right.

- Get the utterance structures right.

- Get the interaction patterns right.

Habitability is a multiplex notion, because ease and comfort in using a language (sublanguage, register) depends on multiple elements, chiefly words (for **lexical habitability**), constructions (for **syntactic habitability**), and interactive rules (for **functional habitability**).

A keyboard-based interactive language system can concern itself more with syntactic and functional (and conceptual) habitability, but because a speech system depends so heavily on acoustic models, lexical habitability is by far the most important concern. Getting the vocabulary right depends on two essential instruments, a corpus and a lexicon.

A **voice-interface corpus** is a register-specific collection of discourse from the field that the interface will service, opportunistically gathered from various textual sources, but crucially including **task-driven spoken dialogues**. The sources in the corpus, because of the necessarily loose collection procedures, need to be easily separable, so that subcorpora can be used independently, and tagged for part of speech, syntactic role, and dialogue act. The job of the corpus is to provide data supporting the pursuit of all three habitability goals, but most particularly to populate the project lexicon.

The **voice-interface lexicon** is a register-specific combination of traditional dictionaries and traditional thesauri, in a **digital format**. It contains word-based information to inform the diction of the interface, including frequency data (for occurrences within the corpus), pronunciation, meaning, usage (concordanced), lexical variants, collocates, colligates, synonyms, and metonyms. The examples in this chapter of what such a lexicon should look like and how it should behave are my own design. Your lexicon will be your design, too — built chiefly by the team lexicographer — but there is no shortage of models to follow and adapt, many of them freely available. Most of them are general-purpose virtual lexicons, none of which are as extensive as the one I have sketched here (few include pronunciation or concordancing, for instance), but they are illustrative all the same. Figure 11.8, for instance, is Apple Computer's Sherlock, with a navigation bar of variant and associated terms, and a second pane for synonyms and antonyms. Figure 11.9 is a typical window of another ingenious lexical database, WordNet. Developed by the Cognitive Science Laboratory at Princeton University, under the direction of George Miller, WordNet is modeled on psycholinguistic theories of word storage and retrieval. The initial headword

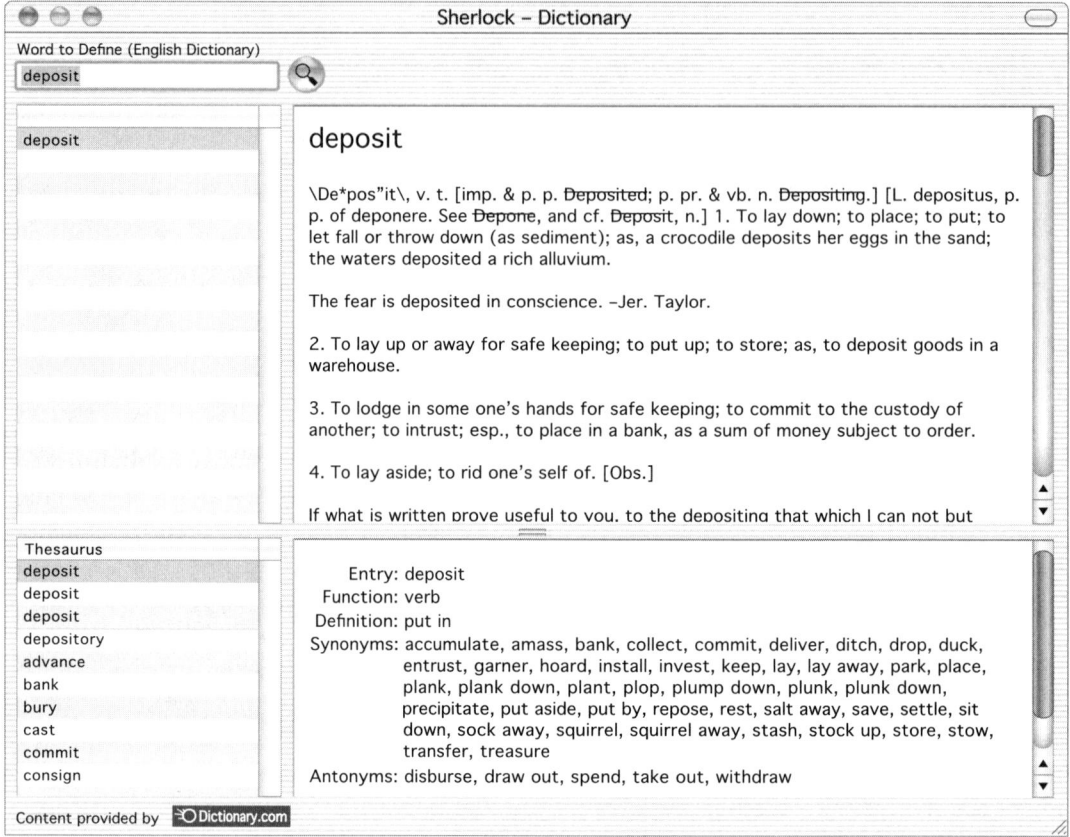

FIGURE 11.8   A typical window of Apple Computer's Sherlock dictionary

data is only a range of definitions, but keying on one specific sense of the term allows one to generate lists of synonyms, metonyms (= coordinate terms, hypernyms, hyponyms, and meronyms), morphological variants (= derivationally related forms), and usage statistics (= familiarity).

Syntactic and functional habitability are also essential to the success of a voice interface. The first, **getting the utterance construction right**, also depends on corpus research, into the colligations favored in the register, turn length variability, preferred dialogue acts, common syntactic forms, and cohesion strategies. The second, **getting the interactive patterns right**, requires drawing task-driven spoken dialogues from the corpus and charting them out to discover referential and relational coherence practices, dialogic pairings, dia-

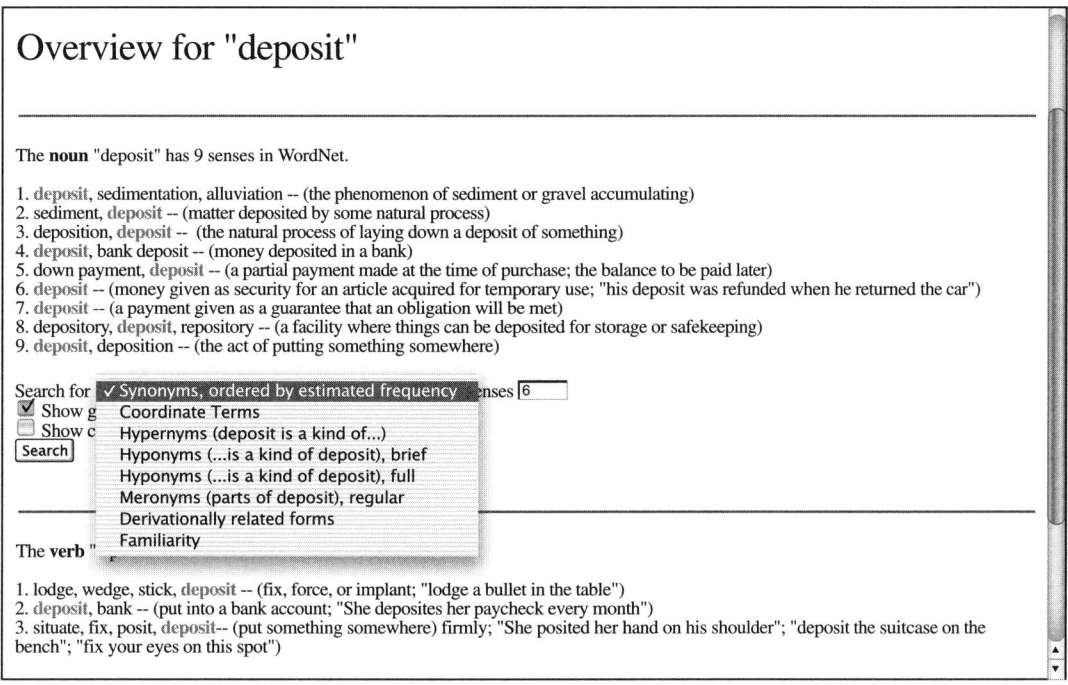

Overview for "deposit"

The **noun** "deposit" has 9 senses in WordNet.

1. deposit, sedimentation, alluviation -- (the phenomenon of sediment or gravel accumulating)
2. sediment, deposit -- (matter deposited by some natural process)
3. deposition, deposit -- (the natural process of laying down a deposit of something)
4. deposit, bank deposit -- (money deposited in a bank)
5. down payment, deposit -- (a partial payment made at the time of purchase; the balance to be paid later)
6. deposit -- (money given as security for an article acquired for temporary use; "his deposit was refunded when he returned the car")
7. deposit -- (a payment given as a guarantee that an obligation will be met)
8. depository, deposit, repository -- (a facility where things can be deposited for storage or safekeeping)
9. deposit, deposition -- (the act of putting something somewhere)

Search for ✓ Synonyms, ordered by estimated frequency   enses 6
  ☑ Show g   Coordinate Terms
  ☐ Show c   Hypernyms (deposit is a kind of...)
  [Search]   Hyponyms (...is a kind of deposit), brief
              Hyponyms (...is a kind of deposit), full
              Meronyms (parts of deposit), regular
              Derivationally related forms
The **verb** "   Familiarity

1. lodge, wedge, stick, deposit -- (fix, force, or implant; "lodge a bullet in the table")
2. deposit, bank -- (put into a bank account; "She deposites her paycheck every month")
3. situate, fix, posit, deposit-- (put something somewhere) firmly; "She posited her hand on his shoulder"; "deposit the suitcase on the bench"; "fix your eyes on this spot")

FIGURE 11.9 A typical window of George Miller's WordNet

logue flow, and task segmenting. This work simply cannot be done without field-specific, task-driven spoken dialogues in the corpus. Your job will always be much easier, therefore, if you have such dialogues from naturalistically collected observational data, but if you have to jerry-rig it — for instance, by filtering the tasks through a telephone line — that is far better than guesswork.

Contrary to popular impression — but exactly what Yankelovich and her colleagues discovered — speakers, even specialist speakers in specialist domains "are not necessarily accurate reporters of usage, even their own" (Sinclair, 1987: xviii). The SunMicrosystems voice interface team *was* a collection of specialist speakers in a specialist domain. And the first thing they recognized as soon as they had a prototype to test is that they didn't know the discourse well enough to proceed and had to go back to the dialogic drawing board, and this time start with linguistic data gathering.

It's not that the Sun group invented voice-interface data collection (see, for instance, Delomier, 1989; Guyomard, 1987; Fraser, 1991; Day and Boyce, 1993); in fact, Yankelovich concedes the group was remarkably naïve in thinking they could get by without early discourse research. But, as a tidy little parable on the voice-interface requirements for (1) getting to users/speakers early on, and (2) doing close discourse analysis, their story is worth telling, and telling again, until the lesson takes. What the Sun Speech Applications Group discovered, the hard way, is that you have to conduct extensive **natural dialogue studies** at the outset of the project, which, feeding into an iterative development process, gives your speech system the most promising early-cycle start to eventually living happily ever after.

Agents

Embarked on a conversation, you must fairly quickly start making judgements about what you can accomplish as you begin to "read" the other person. . . . You must make tentative judgements about the kind of person you are dealing with. Your views are likely to be heavily influenced by the kind of language he or she is using. Is he or she assertive, quiet, talkative, polite, rude, or what?
— Ronald Wardhaugh

Designing the agent(s) who will represent your system is among the most critical decisions you need to make.[1] The design can be revised, and should be revisited throughout the development cycle, but agents are the color, shapes, typefaces, alignments, groupings — the overall style — of the interaction. Agents represent the interface. They set the tone. They are its character. They *personify* the interface, in a way that is completely nonmetaphorical.

1: Just to be clear about what this chapter concerns, the word *agent* here is used in the same general sense as I use it in the rest of the book — as the entity one can most immediately identify as the source of some piece of language, an entity like me writing this, or you muttering "What the hell is he using a Fancy-Dan word like *entity* for?" or HMIHY asking "How may I help you" or SpeechActs saying "You currently have 'Lunch with Kate Ehrlich' until 2 pm" or, for that matter, HAL-9000 saying "Dave?" That is, the entity can be "real" or fictional or the front-end of a speech system; it's just the thing with the name that is associated with the language. This chapter does not, however, concern the ubiquitous "software agents" or "intelligent agents" that are beginning to permeate complex computer systems. There are some very interesting ways that these notions intersect with voice interaction; see, e.g., Turunen and Hakulinen (2001). This work has fascinating implications for agents at the level I am discussing here, even human agents (their implied metaphysics recall Marvin Minsky's [1986] *Society of Mind* speculations), and immense promise for more elegant conversational agents. But those discussions are all beyond the human-factors focus of this book.

In this chapter we will take up the primary areas of consideration in agent design, the ones that shape decisions about how many agents you need, whether they should be real or synthetic or an integrated team of both, whether they should be male or female, adopt a mainstream or nonstandard dialect, what elements of character are critical, and what sort of personality and emotions they should manifest.

There is also, unfortunately a substantial baggage of controversy that we need to dispel before we can move on to any of these specific considerations, the issue that is often labeled *anthropomorphism*. I prefer the term *personification*. Technically, personification is the trope whereby abstract notions (love, greed, wisdom) are fashioned into persons for rhetorical effect. The prototypical domain of personification is allegory (think *Pilgrim's Progress*). Technically, anthropomorphism is the trope whereby concrete nonhuman entities (deer, rabbits, skunks) are given human attributes, almost invariably including speech, for rhetorical or aesthetic effect. The prototypical domain of anthropomorphism is the cartoon (think Disney's *Bambi*). Neither term is directly applicable to interaction design — which is really a commingling of the abstract (software) and the concrete (hardware) — and both terms have often been used to encompass the other.

Nothing much hangs in the balance of this abstract-versus-real-but-still-not-human distinction, but I am happier with the label *personification*, the less ugly of the two, and the one that reveals the motives beneath both tropes: to make *things* (concrete or abstract) seem like *persons*. And the anti-anthropomorphism objection, after all, is to making interfaces like persons.

Using *personification* also has two useful byproducts. It gets rid of the sexism; *anthropos* means "man." And it gets rid of the non sequiturs; *morphism* means "shaped as." Never mind even Bambi's mother, how can we make *anthropomorphism* accommodate the exaggeratedly not-*man*-shaped Jessica Rabbit? Say, wait a minute, not even Bambi is man-*shaped*. It's Bambi's behavior, far more than his appearance, that effects the personification.

Personification

It is by no means unusual to find subjects saying "please" and "thank you" in their exchanges with what they thought was a machine [in a wizard study]. In one sense this is comparable to thanking a kettle for boiling.
  — Robin Woofit, Norman M. Fraser, Nigel Gilbert, and Scott McGlashan

There is a vocal constituency in the design community that opposes personification in interaction design generally, and in voice-interface design specifically. The constituency is misguided, but substantial. We need to address its arguments and concerns. Our first move is to confess.

Admit it, *you* do it. I'll confess, *I* do it. Look around, we *all* do it. We talk to machines. There is a genre of stupid-user stories that circulate on the Internet, and a considerable number of them hinge on personification, like this one:

> *A confused caller to IBM was having troubles printing documents. He told the technician that the computer had said it "couldn't find printer." The user had tried turning the computer screen to face the printer, but that his computer still couldn't "see" the printer. (http://dot.com.do/techchallenged.htm)*

These stories circulate so that techies can laugh at novices, and so that novices can laugh at people even further down the expertise food chain than them. But computer design not only encourages personification; it virtually mandates personification. Notice, for instance, that our stupid user here was following the system language: *find* and *see* are perceptual terms. He was led in the direction of his personification by the messages he got from his computer.

It's not just desktop computers, or such devices as those ubiquitous automatic-banking machines, that use bits of language, deploy dialogue acts, respond, and solicit responses. *All* devices of interactive technology stand toward us in a way that makes talking to them inevitable. Interaction — because it is turn-driven, predicated on cooperation, and topic-focused — always resembles conversation, a resemblance that approaches identity when language is added.

But it's not just the design of this machinery; in fact, it's not even just machinery. It's *our* design, and it's virtually everything. As far back as our species goes, we've been talking to things — sometimes worshipping, sometimes fearing, sometimes just shooting the breeze; it's the way we're wired. When kids are learning to talk, when elderly people reach the point where they forget there are other humans observing, or just don't care, when psychotics enter their own world, this penchant is more noticeable by the circumspect rest of us. But we all do it.

Here is our old friend, Aristotle, weighing in on the topic, on Homer's systematic "practice of giving metaphorical life to lifeless things":

> *all such [personifications] are distinguished by the effect of activity they convey. Thus,*

- *Downward anon to the valley rebounded the boulder <u>remorseless</u>; and*

- *The <u>[bitter]</u> arrow flew; and*

- *Flying on <u>eagerly</u>; and*

- *Stuck in the earth, still <u>panting</u> to feed on the flesh of the heroes; and*

- *The point of the spear <u>in its fury</u> drove full through his breastbone.*

In all these examples the things have the effect of being active because they are made into living beings; shameless behavior and fury and so on are all forms of [human] activity.
(*Rhetoric* 1411[b])

It is significant that Homeric verse is the residue of an extensive oral tradition, a pre-literate mode of encoding and transmitting knowledge, a mode which taps more directly into cognitive dispositions than, say, books on interactive design. The poetry is constructed so that the bard can recall it easily, and so that it lodges with the audience as well. Metaphor, metonymy, repetitio, antithesis — the figurative blueprints that Homeric language is built to — all correspond directly to well-known cognitive principles (and, as Aristotle points out, personification is a species of metaphor).

Personification is primitive, pervasive, and taps into cognitive structures. It also works very well as a technique to help decipher the world around us. (Which is surely related to its primitive, pervasive, and cognitive character; evolution has a way of encouraging successful strategies, and punishing unsuccessful ones.) Daniel Dennett has a long and illuminating discussion of personification as a strategy for understanding the world (although, being a philosopher, he is obliged to call it something really portentous; his label is "the intentional stance"). The discussion is worth quoting at length, so I will:

The strategy . . . works on most other mammals most of the time. For instance, you can use it to design better traps to catch those mammals, by reasoning about what the creature knows or believes about various things, what it prefers, what it wants to avoid. The strategy works on birds, and on fish, and on reptiles, and on insects and spiders, and even on such lowly and unenterprising creatures as clams (once a clam <u>believes</u> there is danger about, it will not relax its grip on its closed shell until it is <u>convinced</u> that the danger has passed). It also works on some artifacts: the chess-playing computer will not take your knight because it <u>knows</u> that there is a line of ensuing play that would lead to losing its rook, and it does not <u>want</u> that to happen. More modestly, the thermostat will turn off the boiler as soon as it comes to <u>believe</u> the room has reached the desired temperature.

The strategy even works for plants. In a locale with late spring storms, you should plant apple varieties that are particularly <u>cautious</u> about <u>concluding</u> that it is spring — which is when they <u>want</u> to blossom, of course. It even works for such inanimate and apparently undesigned phenomena as lightning. An electrician once explained to me how he worked out how to protect my underground water pump from lightning damage: lightning, he said, always <u>wants</u> to find the best way to ground, but sometimes it <u>gets tricked</u> into taking the second-best paths. You can protect the pump by making another, better path more <u>obvious</u> to the lightning.

(1997: 65; the emphasis is both mine and Dennets, intermixed)

So, personification is rampant, cognitively entrenched, useful for catching mice and protecting water pumps. It is also useful, Dennett notes, for figuring out artifacts, like computers. Why are so many interaction designers saying bad things about it?

The Case against Interface Personification

Unfortunately, cartoon characters were not successful in the heavily promoted, but short-lived, home-computing product from Microsoft called BOB. Users could choose from a variety of on-screen characters who spoke in cartoon bubbles with phrases such as: What a team we are, What shall we do next, Ben? And Good job so far, Ben.
— Ben Shneiderman

The case against personifying computer systems is championed most thoroughly by interface pundit Ben Shneiderman. He is deeply antipathetic to personification, calling it "deceptive, counterproductive, and morally offensive to me" (Brennan et al., 1992: 69).

His best-known assault is in his *Designing the User Interface*, now into its third edition and widely regarded as The Orthodoxy of interface design. The case is more innuendo than argument, but it goes like this:

There is a great temptation to have computers "talk" as though they were people. It is a primitive urge that designers often follow, and that children and many adults accept without hesitation. . . . Children accept human-like references and qualities for almost any object, from Humpty-Dumpty to Tootle the Train. Adults reserve the anthropomorphic references for objects of special attraction, such as cars, ships, or computers.

This primitive, childish urge is bad for user interfaces because, well, because computers are not people:

Attributions of intelligence, autonomy, free will, or knowledge to computers can deceive, confuse, and mislead users. The suggestion that computers can think, know, or understand may give users an erroneous model of how computers work and what the machines' capacities are. Ultimately, the deception becomes apparent, and users may feel poorly treated.
  (1998: 380)

I will resist the urge to analyze the conceptual repetitions, the strategic modal verbs, the guilt-by-association-with-children theme, and the bad form of drawing Tootle into it, and move directly to an epitome of the argument, which resembles a *modus ponens*:

People have a tendency to treat nonhuman things as human-like.

One of the class of things people perpetrate this tendency on is computers.

Computers are not human.

Therefore, to support this tendency in computer design is to set people up for disappointment, and ill-feeling toward the computer.

Shneiderman, as we have already seen, is certainly right about two things: personification is primitive, and it is pervasive. And there's no disputing that people practice personification on their computers.

Backing this argument, Shneiderman's evidence is slim — little more than a catalogue of false hopes dashed, all the way back to ENIAC, and a few ambiguous lab studies. There have been manifold failures linked to personification, and almost everyone who has used a computer has some smug parrot or talking paper clip they've been deeply aggravated with. It's curious, then, that people would persist in using the personification strategy on their computers, as we most certainly continue to do, parrots and paper-clips notwithstanding. We orient ourselves toward interactive technologies in particular as if they were social actors (Reeves and Nass, 1996). We are, let's not forget, social animals. We live in collectives, we form and populate institutions, we cooperate to build and design software, we gossip. There is even a moderately compelling theory that language began chiefly to afford gossiping (Dunbar, 1997). And one of the reasons that informational and interactive technologies — television and computers especially — promote addictive behaviors is surely that they are satisfying neurobiological needs or desires.

BOB, we all know, did not serve those needs very well. That's not because BOB was rife with personifications. That's because BOB wasn't well designed — the creatures were annoying, their speech balloons were full of inanities, they distracted from tasks more than they supported them, and they didn't have any decent gossip. And, in case no one was noticing, vast numbers of software products without annoying cartoons have also failed.

Still, the antipersonification argument has been a serious worry for a notable faction of the speech-system community. Personification, they hold, is especially crippling for speech systems because of recognition fragility; human-like agents, they fear, "create unrealistic user expectations that lead to errors and then disappointment with the system" (Boyce, 1999: 40). Bruce Balentine, for instance, regards the central cause of error spirals, the most debilitating condition in voice-interface interaction, to be the creation of "false impressions" which encourage "users to produce highly intelligent, socially aware, and unbounded input" (1999: 210). And Noyes (2001) goes so far as to call speech (and writing) unnatural for human–computer interaction, because they are modes developed for human–human interaction, and machines aren't humans. The problem is a real one, though the objection has perhaps been overstated.[2]

2: Arguments to this end, for instance, usually don't go much further than citing studies a decade or more old (e.g., Jones et al., 1990), or wagging their fingers at *Star Trek* for generating, in the mantra of this contingent, "unrealistic expectations."

What is abundantly clear, however, is that your users will orient themselves socially toward your system. It's not only a design philosophy that has long been serving their technology interactions, it's a cognitive disposition. Both the design and the disposition are ratcheted up to the nth degree by the use of language as the near-exclusive medium of interaction, speech as the channel. We've always talked to machines; now they're talking back.

The only question, then, is how you deal with it — how, as designers of talking machines, you shape that talk, how you talk to users. Your choices are, put flatly, between (1) ways that consciously hinder personification, in the hope that you can thereby preclude chat mode or undue expectations; and (2) ways that capitalize on personification, in the belief that there are more fruitful ways to constrain users' language behaviors.

One approach to hindering personification in voice-interface design has been to avoid the use of "I," one of Shneidermann's "Non-anthropomorphic Guidelines" (1998: 385). Some voice-interface designers have regarded the "not-I" injunction as something of an inoculation against overly familiar behavior, a shield that blocks users from falling out of task mode and into chat mode. The technique, though, is less than fruitful. In fact, users barely notice the presence of "I," and their performance proves unimpeded by its presence (Boyce, 1999: 42; Huang et al., 2001). Further, users classify systems that use "I" higher in satisfaction ratings. Further yet, avoiding "I" has grammatical repercussions that may result in awkwardness and cognitive friction. (The notoriously tortuous writing of many scientists and engineers, for instance, is largely motivated by trying to side-step personal reference.) Users personify the interface irrespective of the pronouns it uses or avoids; aversion has no benefits, adds friction, and correlates with lower satisfaction ratings.

Similar prohibitions have been expressed against contractions and other speech elements characteristic of informality. On the surface, these are register concerns. In human–human interactions, the use of phonological reductions, familiar forms of address, and "common" lexical choices ("Yo, dude, how'd you lose the lamer?") are appropriate or not depending on the relationship of the interactants, the context of the utterance, and other general features of the social environment they share — exactly the concepts, in short, that should guide agent design. Sometimes informality is more appropriate than others (a movie information service versus a stock-management service, for instance). But *guide* does not mean "determine," and other considerations also come into play. Contractions, in particular, need to be considered independently because of convergence. Contractions are more difficult for recognizers (human and machine) because they leave a much more negligible acoustic signature than their fuller counterparts. On the whole, it is better if your callers don't use them and, therefore, better if you don't establish their availability by using them. On the other hand, the agent needs to stay in character, and noncontracted speech from an agent giving skateboarding tips and using boarder terminology might sound alien indeed. On the other hand, not all text-to-speech systems articulate contracted words well, which can lead to lower user comprehension with synthetic output. On the other hand, contractions shorten the overall system output — usually a blessing, especially with synthetic voices.

Personification, in short, is the least of your worries. It will happen. Don't fight it. But don't particularly court it either. Let the interactive context guide the design of the agent, as it guides the design of the interface generally. But, to the best of the technological and design capabilities, the system should behave the way a human using language would expect it to behave. On being misheard, for instance, it is unnatural to repeat yourself in normal-speech cadences; the natural behavior is to slow down and speak up. It should therefore be natural for systems to listen for exactly these acoustic shifts in these circumstances. The problem here is not with "unrealistic expectations," but with "inflexible designs" that don't behave in a natural, human-like way.

Primary Considerations

First, before dialog flows or prompt wording can be decided on, designers must understand "who" is talking and carefully develop the character who will be featured in the application. How friendly, efficient, casual, chatty, young, humorous, experienced, or forgiving is he or she? The answers to these questions depend on the type of application and the company behind it. Think about the difference between a stock broker versus a music store clerk, or a major bank versus a major Hollywood studio, or an application that gives you traffic updates versus one that lets you change the percentage of your 401 K plan.
  — Bill Byrne

There are six primary areas of consideration for agent design:[3]

Branding

The distinctive ethos of the company, with respect to the users' (or potential users') identification of the company and its products, associations with the company and its products, and attitude toward the company and its products (for instance, loyalty, perceived value).

The primary considerations here are the agent's character and the distinctiveness of its personality.

3: More generally, this list, poached largely from Balentine and Morgan (1999: 16) concerns shaping the "social relationship between the application and the user." Balentine and Morgan, I should note, use slightly different terminology from mine: their *marketing* is my *branding*; their *design* is my *capabilities*; they have nothing corresponding to *role*; the others are the same.

Aesthetics

The sound and tone.

The primary considerations here are the vocal characteristics — pitch, pitch-range, and rhythm, as well as lexical and syntactic style.

Productivity

The speed of use, learnability, range of functions.

The primary consideration here is character — credibility, cooperation, and trustworthiness.

Ergonomics

The cognitive load, memory, turn-taking, reaction time, attention issues, reach envelope.

The primary considerations here are enunciation, interactive naturalness, and task-focus.

Role

The general purpose that the agent is serving for the user.

The primary considerations here are taken up in more detail throughout this chapter under labels like "personality" and "emotion" and "similarity attraction," but what they come down to is that the agent should suit the social or professional relationship it has to the user — a cheerful friend, a terse guide, a competent advisor.

And the consideration that has to reign supreme, here as everywhere else in system design, though governed by a parliament of the other five factors:

Capabilities

The technological capabilities/limitations, the tasks enabled, the architecture, and error recovery.

The primary considerations here are clarity and social fidelity, both of which induce the user to emulate the agent, leading to increased predictability of input. But these recognition considerations are also the source of constraints on the other factors.

These issues are closely related, even overlapping in several aspects, and various stakeholders will often present a set of arguments from one category as if they belonged to another (marketing representatives might advance branding concerns as if they were productivity issues, for instance, or technologists might advance a case based on system limitations as if it was a matter of ergonomics). But it is useful to keep them as distinct as possible. In general, of course, you want pluses in all the categories: a distinctive, attractive, cooperative, productive, and effective agent or cast of agents. But it's a balancing act. Productivity often depends on homogeneity, branding on differentiation. Ergonomics can rest on coordinated exchange, aesthetics on prolonged floor-holding, and capabilities constrains everything. So these criteria have to be ranked according to the tasks and users implicated by the service. A banking service will tend to rate productivity and ergonomic factors higher than the other factors (except for recognition capabilities); a music service may put aesthetics and branding first.

Role draws on all of these other issues, and is often best thought of in terms of simple job or social descriptors, linked to personality characteristics. Here, for instance, is how Clifford Nass and his colleagues worked through their choice of attributes for an onboard vehicle agent they developed for BMW:

Because BMWs are bought by both extroverts and introverts, the company was less concerned with matching the customer than they were with making sure that the personality of the voice was consistent with the role of the voice in the car. After deciding that the voice in the car should not be the car itself (as in KITT of the TV series "Knight Rider,") we began to consider who the voice was and what personality matched that role. Should it be a "golf buddy" (match the user's personality), a chauffeur (obsequious, terse), a pilot (very dominant and not very friendly), a person riding "shotgun," (talkative, not very smart), a "mother-in-law" (hypercritical, grating voice), etc. A detailed analysis of the brand positioning of the car suggested that the perfect voice would be a stereotypical co-pilot, who could take over when the driver was in trouble but who understood that the driver (pilot) was in charge: male, not at all dominant, somewhat friendly, and highly competent. This suggested a voice that was a relatively deep voice, medium in volume, slightly faster than average word speed, with moderate pitch range and very little volume range. We also carefully avoided the use of "I." . . . Co-pilots try to place themselves in a subordinate role and thus avoid the use of "I." Furthermore, the language was relatively terse and phrased as statements rather than commands (pilot) or questions (chauffeur). Our research shows that this careful matching of voice and role makes users feel safer, more confident, and happier than when the voice is not carefully matched or swings between personality types.

(Nass and Brave, 2004)

Cast

Disney managed to make each of the dwarfs in Snow White *(1937) — Doc, Happy, Sleepy,*
Sneezy, Grumpy, Bashful, and most especially Dopey — stand out in our memories because
of a few sharply-etched character strokes.
    — Andrew Horton

You may have a cast of one, or of dozens. Branding might lead you in the direction of
a single, highly-recognizable agent. Ergonomics might lead you to multiple, easily-
distinguishable agents, each associated with a different task focus (one agent for book
inquiries, another for digital media, another for electronics, . . . a distinct check-out agent
to complete sales, and so on). For anything beyond a single-task, short-duration system (say,
voice dialing), multiple agents are usually preferred, but even with very careful management, multiplicity can quickly turn to cacophony; the number of characters, therefore,
usually ranges between two and six.

Real or Synthetic

I'm not bad. I'm just drawn that way.
    — Jessica Rabbit

All agents are simulations. But some simulations will speak in real human voices, in
a very close parallel, on the visual plane, to characters in a movie or play — fictional but
instantiated by real actors. And other simulations will speak in synthetic human voices,
closer to the order of cartoons — fictional in a more immediately apparent way.[4] (Cartoons
overwhelmingly speak in real human voices, of course, but usually under significant distortions to re-enforce the clear visual fictiveness.) Each option has its advantages and its
liabilities. Real voices tend to be more satisfying for users, easier to understand, and more
comfortable to listen to. But everything they say must be recorded far before the interaction, which reduces their run-time flexibility dramatically. Synthetic voices, on the other
hand, can be modified on the fly, both in terms of what they say and how they sound,
and they offer advantages in terms of constraining users as well. But they are wearying to
listen to.

Acoustically, a voice is a wave form with two principal components, periodic and aperiodic sounds (that is, vowels and consonants). Humans generate these sounds with their

---

4: I ignore some of the variations between recorded and synthesized voices — in particular, concatenative recorded-
voice synthesis — because they complicate the discussion in unfruitful ways. Concatenated-voice speech generators hold much promise, as ways of accommodating co-articulation and prosody advance, but they currently pattern,
in terms of advantages and liabilities, very much like fully-synthesized voices.

mouths. The periodic sounds come from vibrating vocal cords, not unlike the vibrating reed of a saxophone. And, also like a saxophone, the output frequency changes as a function of the chamber that periodic sound passes through. The saxophone affects the sound by a series of valves that manipulate the characteristics of the chamber. The mouth affects the sound with a more subtle series of manipulators — the uvula, the tongue, and the lips — but they work in much the same way, altering the sound by altering the chamber it resonates through. (In an even more basic, and satisfying, example, Isaac Newton reported synthesizing vowels by removing, in the name of science, small amounts of beer from a bottle and blowing into it, changing the chamber's characteristics by changing the amounts of two substances with different densities, beer and air.) We end up with different tones — different vowels — because we change the shape of the chamber by moving our uvulas, protruding and retracting our lips, and sliding our tongues around. The fundamental frequency (what the vibrating vocal chords generate) is the same, but it sets off additional periodic resonations in response to different chamber characteristics. These are called *formants*, and vowels are differentiated by their formant structures: by where in the frequency spectrum these secondary resonations center. Table 12.1 represents these facts for a few characteristic American-English vowels (charted by differences in age and gender, which correlate with different sizes and shapes of resonant chambers, or mouths).

Now, computers are just as good at setting up a source (just like vibrating vocal cords), and then filtering it (as through a vocal chamber) to produce different complex sounds, as humans are. In fact, computers are better, more precise, and more pure in their implementation of source-filter sound generators. Humans are so imprecise and impure in their sound generation that every human source-filter production is unique. Human voices are full of flux and sputter, jitter and shimmer, murmur and hum: full of quality, or timbre, or character. Not only does this commingle of textures lead to brand distinctiveness (no two real voices are alike), in general other humans prefer it over purity and precision.[5]

Using real voices is more satisfying for users, particularly if the interactions are relatively long. If the interactions are also somewhat frequent, the recording has to be considerably more extensive; people get tired of hearing the same phrases over and over,

---

5: As Huang et al. (2001) put it, "there is no debate" about this preference at all: "Recorded speech is universally agreed to be superior to synthesized speech." Their study is useful empirical confirmation of this preference. One anomaly should be noted here, however — an unpublished study by some of Clifford Nass's students (Kyu Hahn, Sylvia Olveda, Rob Baesman, and Sandra Liu). They tested news stories and editorials in two conditions, recorded and synthetic, finding that while the real voice was preferred for news, the synthetic voice was preferred for the editorials. There could be any number of factors responsible for this finding; they hypothesize that since editorials are textually argumentative, hearers preferred the mechanical neutrality of a synthesized voice because it made them feel less manipulated.

Frequencies (in Herz)	Age/gender	Vowels			
		/i/ (b<u>ea</u>t)	/æ/ (b<u>a</u>t)	/ʌ/ (b<u>u</u>tt)	/u/ (b<u>oo</u>t)
Fundamental (F_0)	Children	272	251	261	276
	Women	235	210	221	232
	Men	136	127	130	137
First formant (F_1)	Children	370	1,010	850	430
	Women	310	860	760	370
	Men	270	660	640	300
Second formant (F_2)	Children	3,200	2,320	1,590	1,170
	Women	2,790	2,050	1,400	950
	Men	2,290	1,720	1,190	870
Third formant (F_3)	Children	3,730	3,320	3,360	3,260
	Women	3,310	2,850	2,780	2,670
	Men	3,010	2,410	2,390	2,240

TABLE 12.1 Characteristic average frequencies of some American English vowels, by age and gender (adapted from Peterson and Barney 1952: 183)

in the same intonations, for the same durations, even in natural voices. Synthetic voices are even more tiresome, for any but the briefest stretches of discourse, but that very irritation offers some advantage.

From a recognition perspective, Wieland Eckert (1995) argues, a "clear nonhuman voice is preferred over [a] human voice, since users talking to 'humans' tend to be lazy in grammar, clearness and usage of words." *Lazy* is not the right characterization, and the relevant effect here actually runs in the other direction; that is, rather than people opting for sloppiness when talking to "humans," they probably work a bit harder when listening to synthesized voices, as they do with the linguistically-infirm generally (see Oviatt et al., 1998). But the effect — synthetic voices inducing linguistic behavior more suited to recognition requirements than real voices — is well established. Eckert is talking anecdotally, but Susan Boyce (1999), for instance, found in a controlled study that synthetic voices were more effective than real voices at constraining callers.

But — back to the flip side — she also found that people hated the synthesized voice in her experiment so much "as to offset any gain" (Boyce, 1999: 45). What you gain in recognition performance with a synthetic voice, that is, you lose in aesthetics, in the overall pleasantness of the experience. This situation creates something of a paradox. Making synthesized voices better and more natural — less loathsome — may undermine the very usefulness they offer in terms of constraint, slowing the user down, making him more cautious in speech, and more eager to get through the interaction and hang up.

Synthetic voices have other advantages, though, which make them indispensable for many voice applications. They store more easily, process more efficiently, and operate much more flexibly. They can turn any text into speech at run-time. On the criteria of productivity and recognition success, that is, they score moderately high. Real voices take lots of memory, process more slowly, and are highly inflexible. Every syllable of output must be recorded, days or months or even years ahead of time.

Moreover, these are advantages that improvements in naturalness and quality will not compromise, so that synthetic voice agents will inevitably gain an edge over recorded voices as they approach them in terms of user satisfaction, increased learnability, and brand distinctiveness. For instance, the dullness of synthetic voices requires them to speak more slowly than humans (in the 155–165 words-per-minute range, as opposed to the 165–175 range). People have more difficulty remembering information they've heard from a synthetic voice, but slowing the rate counteracts those losses to a certain degree. But slowing the voices down also contributes to their lack of attractiveness. Giving synthetic voices a little more character, especially giving them more human prosodies, will allow the word-per-minute rate to go up, and therefore improve their attractiveness on two fronts.

For the short term at least, however, the best solution is often to combine synthetic and real voices, letting their strengths and weaknesses balance each other out by having recorded-voice agents front the interface, while synthetic-voiced agents carry out many of the tasks. This real/synthetic two-agent architecture has two distinct benefits over real/real or synthetic/synthetic architectures. Firstly, users have a clear and present basis on which to hang a distinction between functions. Secondly, that distinction can give users somewhat more appropriate expectations of at least one component of the interaction. Some users have disproportionately high expectations of conversational interfaces, expectations which are almost always frustrated. But a contrast between a real-voiced agent (which can, in a sense, function as an attractor for those expectations) and a synthetic voiced agent can serve to take some of the interactional pressure off the synthetic-voiced portions of the system. (Naturally, however, this means that the real-voiced agent must be especially resilient, but if its chief role is conversational management — rather than, say, domain knowledge — this resilience is somewhat easier to accomplish.)

Gender

I don't want to talk grammar, I want to talk like a lady.
— Eliza Dolittle

"Gender is the first social attribute people recognize in a human voice," Eun-Ju Lee has said of interaction design, "and it triggers stereotypic reactions" (O'Toole, 2000). She overstates the case marginally, since it is a gender/age discrimination based on fundamental frequency that people make first. The fundamental frequency is the baseline by which all of a person's vowel articulations are determined, and the different vocal-cord densities and sizes among men, women, and children give their vowels, therefore their voices, (on average) very recognizable age/gender distinctions, as charted in Table 12.1. These frequency differences, of course, have perceptual implications; they have different pitches. Although there are enough cross-overs and outliers that we can make gender and age discrimination errors, especially over the phone (which flattens variation and reduces higher frequencies) and in music (where singers can concentrate on certain frequency ranges), we can usually discriminate men, who speak at the lowest pitches, from women, who speak at markedly higher pitches, from children, who speak at the highest of these pitches.

There are also lexical and syntactic differences among ages and genders with respect to speech, though these are much less cut-and-dried in terms of measurement than the acoustic differences. Children have far less stable lexical and syntactic traits than adults, changing grade to grade, even moment-to-moment as they try out different peer groups and registers and pursue hypotheses or whims, building their social and communicative and epistemic repertoires. Aside from this transience, a marked affection for evaluative and categorizing terminology, and the impulse to cut their own swath by rejecting and permuting (and sometimes simply misunderstanding) the adult usage around them, there isn't much to say about the language of children. Not because it's not fascinating — it is — but because it is both a huge subject and one only marginally relevant to voice interaction design. While there may be a call for child agents in some contexts, and certainly an emphasis on youth registers for some fields, those circumstances are too specific to demand the kind of attention that would be required to do age-differences any justice here. And the use of telephones as the primary medium of speech-system interaction presents a major obstacle to the use of child agents, since they operate in the higher frequencies that telephone transmission routinely clips.

Gender differences of word and syntax are considerably more stable than age differences. Characterizations of gender differences have to be taken with a large grain of salt. Their investigation is suffused with ideologies and broad generalizations. It is also rife with subtleties, variations, and ready counter-examples. But there are tendencies significant enough to notice. Bring your salt, and we will proceed.

In Deborah Tannen's very useful caricatures, men speak the language of the *report*, women speak the language of *rapport*:

> *For most women, the language of conversation is primarily a language of rapport: a way of establishing connections and negotiating relationships. Emphasis is placed on displaying similarities and matching experiences. From childhood, girls criticize peers who try to stand out or appear better than others. People feel their closest connections at home, or in settings where they feel close to and comfortable with — in other words, during private speaking. But even the most public situations can be approached like private speaking.*
>
> *For most men, talk is primarily a means to preserve independence and negotiate and maintain status in a hierarchical social order. This is done by exhibiting knowledge and skill, and by holding center stage through verbal performance such as storytelling, joking, or imparting information. From childhood, men learn to use talking as a way to get and keep attention. So they are more comfortable speaking in larger groups made up of people they know less well — in the broadest sense, "public speaking." But even the most private situations can be approached like public speaking, more like giving a report than establishing rapport.*
>
> (Tannen, 1990: 77).

Men tend to display knowledge, assert authority, hold the floor; women tend to display social awareness, pursue connection, share the floor.

They have different styles of speaking, which implicate lexical and syntactic choices. Men tend toward technical vocabularies, assertives and declaratives; women tend toward emotional vocabularies, questions and commissives. An assertion containing technical terminology allows the exhibition of authority-supporting knowledge (like, for instance, "Men tend toward technical vocabularies, assertives and declaratives; women tend toward emotional vocabularies, questions and commissives"). A question containing emotional terminology allows the pursuit of a social connection. Men and women tend to listen differently as well, in ways attuned to their conversational styles — men for indications of status, opportunities to assert, and the like; women for indications of relationships, opportunities to engage, and the like.

There are constellations of related tendencies. Women tend to hedge more when they do assert, for instance, to supply more confirmations, and to elicit more confirmations than men, all of which are processes of social leveling that either reduce or avoid assertions of status. Men's tendencies are the flip side: hedging less, confirming less, and seeking confirmation less — processes that establish or reinforce authority. Men interrupt other speakers more often than women do, which suggests confidence in male speakers, deference in female speakers.[6] (Interestingly, however, women are *perceived* to interrupt other

---

6: On the other hand, women tend to use more prestige forms in their pronunciations ("going" vs. "goin," for instance), which seem to be directed toward accruing authority, and men tend toward more informal pronunciations (Bonvillain, 1997: 167–172).

speakers more often, men to do so less, and women are less successful in establishing topics than men — suggesting a heavy gender asymmetry in conversational roles; see Bonvillain, 1997: 179–184.)

Again, these are gross tendencies, significant but neither universal nor absolute, frequently subordinate to register and other contextual elements. For instance, the long passage I quoted from Tannen — a woman — is thorough-going report language; and even nonacademic women have quite technical vocabularies they deploy regularly. Most women have a substantially greater command of color terms than most men, for instance. I wouldn't recognize puce if a puce-dyed poodle bit me on the butt, nor magenta, nor taupe; I have a hard enough time with candy-apple red. But the patterns of difference between the genders are sufficiently noteworthy that many researchers speak of *genderlects*, and Tannen goes so far as to say "male–female conversation is cross-cultural communication" (1990: 42).

And, as Lee notes, these gender-based patterns trigger responses — social responses, and probably biological responses as well. The responses are well known from the social psychology literature, with male voices having a substantial edge in the listener's perceptions of the reliability of the information, the persuasiveness of the message, and the status of the speaker; female voices, while lower on these metrics, correlate with higher judgments of friendliness and overall pleasantness (Eagley, 1983).

Somewhat depressingly, these stereotypes hold up for voice interfaces too. Clifford Nass, Youngme Moon, and Nancy Green (1997), for instance, found that praise is regarded as more valid coming from a male-voiced computer; that female-voiced computers were seen as less competent; that male-voiced computers, in a dominant condition, are viewed positively (as "assertive" and "independent"), while female-voiced computers are viewed negatively ("pushy" and "bossy"); that male-voiced computers were held to be more knowledgeable about computer technology than female-voiced computers, while female voices were held to be more knowledgeable about relationships. Nass, Moon, and Green tested these stereotypes with recorded human voices, but similar results were obtained for synthesized voices (Lee et al., 2000;[7] see also, for extensive discussion of these issues, Nass and Brave, 2004).

The upshot, then, is that gender effects are strong for voice interfaces, whether the agent has a recorded voice or a synthesized one. These effects are surely cross-cut by domain and register and audience conditions, and the experimental results are not univocal. There is also a similarity effect; people tend to respond better to voices and styles like their own, an effect that includes gender-similarity on the crucial variable of trustworthiness (Gong and Nass, 2000). But, all things being equal, male voices are received as having more authority, more influence, and more competence, especially in technical areas; female voices are

---

7: The gender differentiation was acoustic only in this study, not stylistic. The female voice had a fundamental frequency of 220 Hz, the male of 115 Hz (Lee et al., 2000).

received as more pleasant and likable, with expertise in social matters. This leaves voice interface designers with something of a dilemma. Do you play into the stereotypes, or play against them? Do you capitalize on the traditional influence wielded by white males, or do you want to establish your authority on independent grounds, and undermine what, after all, is a privileged position in our culture?

The easy answer would be to pick the gender of the agent for the effect — males in circumstances calling for technical authority or persuasion, for instance, females in elicitation roles, such as in form-filling tasks. Traditional media that depend on voice and/or appearance (TV and radio) have certainly followed this course. While there has been some erosion of the stereotypes in some regions over the last few decades, anchor men and weather girls are still the international default.

Dialect

> No accent is intrinsically good or bad, but it has to be recognized that the way we perceive accents does play a role in our attitude to others. Different people have differing perceptions. So there are significant numbers of young people who see Estuary English as modern, up-front, high on "street cred" and ideal for image-conscious trendsetters. Others regard it as projecting an approachable, informal, and flexible image. Whereas . . . Queen's English, Oxford English, and Sloane Ranger English are all increasingly perceived as exclusive and formal.
> — Paul Coggle

Dialect is a composite notion, referring to general variations in a language that can be traced to persistent external factors, as opposed to register and genre in particular, which are variations linked to transient contextual factors (Halliday, 1978: 110, distinguishes the former by way of *users*, the latter by *uses*). It incorporates at least three types of external factors, transparently manifest in the following labels: *regiolect*, *sociolect*, and *ethnolect*. Language varieties correlate with geographical region, socioeconomic status, and ethnicity.[8] None of these factors is very easy to isolate or to track — and there are frequent, sweeping cross-pollinations, as in the recent age-based penetration of African American urban dialects into mostly Caucasian middle-class suburbs — but in concert these three terms highlight the general range of variations resident in languages, especially in a huge multinational language like English.

8: Even these subdivisions are fairly superficial, as we've seen. We have just talked about the gender-based discourse styles that are sometimes called genderlects, and age-based variations (sometimes called *aetalects*, sometimes *genealects*). Sociolects are often carved up according to prestige, into basilect (lowest prestige), mesolect (middle), and acrolect (highest). And these categories of variation also frequently intermingle with registers and genres in quite subtle ways.

The one truism about all this variation is that everyone else has a dialect. Speakers usually take their own variety as the norm; and usually regard the standard variety of their nation (if it differs from their own) as the "grammatical" ("proper," or similarly evaluative term) version of their language; and usually regard every other variety as other people talking funny. Often this funny-talking correlates with biases and jokes, as in the traditional prejudices against Cockney in England, the Newfoundland variety in Canada, and the Ozark variety in the United States; and in the widespread furor over legitimation attempts for the ethnolect called *Ebonics* (also *African American English*) in the 1990s.

Certainly some dialects correlate in many people's minds with authority and competence, others with efficiency, others with aggression, rapaciousness, and low intelligence. As with gender, then, dialectal variation raises ethical questions: do you pander to prejudices or fight against them? Traditional media has long followed the pandering (indeed, the reinforcement) path. Again there has been mild erosion in recent years, particularly in Britain, where Scottish, Northumbrian, and even Carribean and East-Indian dialects have increasingly shown up alongside Received Pronunciation on the newscast.

Dialectal variation is most immediately obvious in pronunciation and prosody — the characteristics we associate most immediately with "accent" — but is also manifest in lexis, syntax, and pragmatics. Textually, it also shows up orthographically. Tables 12.2 and 12.3 illustrate a few of these differences for the best-known English dialects, Standard American and Standard British, but these are just the most common points of reference. Such tables could easily fill the pages of this book, cataloging the differences among the National Englishes of Singapore, Australia, Nigeria, and the seventy-odd other countries in which English has official status. Sticking just to the US, we could also cram our pages with tables for the regio- or ethnolectal variations, among Southern, New England, Midwestern, and Southwestern regions, on the one hand, and/or African, Spanish, Yiddish, and Amerindian inflected varieties, on the other. Indeed, one would be kept damn busy just charting regio-, socio-, and ethnolectal variations around New York or Los Angeles. Dialects are a rampant and untidy phenomena.

There has been little or no research into the effects of these varieties on interaction design. General Magic's voicing of the Ask Jeeves web site adopted a stereotypical "British butler" dialect (though it was largely phonological; the words were overwhelmingly American), and one of their "magicTalk personalities," which they described as "helpful, friendly and casual," sports a good-ole-boy Southern American dialect.[9] But a general principle can be extrapolated from research in other areas. One of the most robust findings to come out of Clifford Nass's Stanford lab, for instance, is the importance of similarity attrac-

---

9: Ask Jeeves is at *http://www.ask.com/index.asp*. The General Magic demos are no longer available on the web (originally heard at http://www.genmagic.com/demos; I last accessed them in September 2002, the month they filed for bankruptcy protection).

		Standard American English		Standard British English
Conceptual elements (Onomasiology)		vest (garment that goes over the shirt)		vest (garment that goes under the shirt)
		corn (a vegetable; maize)		corn (grain)
		boot (elongaged footware)		boot (rear storage compartment of car)
		caravan (band of fellow travelers)		caravan (camping trailer)
Formal elements (Semasiology)	Lexis	apartment		flat
		billfold		wallet
		diaper		nappy
		gas		petrol
	Phonology	harass	/hɚæs/	/hɛɹəs/
		laboratory	/læbɹətori/	/læbɒɹətori/
		tomato	/tometo/	/tomato/
		missile	/mɪsəl/	/mɪsajl/
	Orthography	center, theater, meter		centre, theatre, metre
		defense, offense, license		defence, offence, licence
		favorite, neighbor, color		favourite, neighbour, colour
		plow, check, dialog		plough, cheque, dialogue

TABLE 12.2   Some lexical differences between Standard American and Standard British English dialects

Standard American English	Standard British English
Jack was in a course.	Jack went on a course.
Jill wasn't able to catch up with him.	Jill wasn't able to catch him up.
He's in the hospital with a broken leg.	He's in hospital with a broken leg.
Lufthansa has a flight to Bohn today.	Lufthansa have a flight to Bohn today.
The government is announcing the invasion.	The government are announcing the invasion.

TABLE 12.3   Some syntactic differences between Standard American and Standard British English dialects

tion.[10] Similar results hold for the closely related matter of second-language accent: Nils Dahlbäck and colleagues from Stanford and Linköping Universities found that similarity attraction applied; in particular, that same-accent agents elicited more "honest" replies; were judged "socially richer" (that is, warmer, livelier, friendlier, more vivid, personal, accessible, sensitive, immediate, emotional, responsive, and sociable); and led to interactions that were judged to be more enjoyable (Dahlbäck et al., 2001). Coupling these findings with the need for comfort and intelligibility, we get an argument for using the varieties spoken by the users.

There are always exceptions, however, as the Jeeves persona suggests: helpfulness and authority can in some circumstances be suggested by a nonlocal variety. And American English is very well-known internationally, with British English a fading second, because of the influence of business, entertainment, and other economic factors; so much so that even nonstandard American dialects are well recognized in far-flung regions.

When globally distributed intelligibility is an issue, then, some American varieties have a distinct advantage over other dialects. Another option for an international strategy when variety-targeting isn't possible is to use Standard Canadian English, which occupies a sort of middle ground between British and American standards, that often leads Americans to think they're speaking to someone from the United Kingdom, and Britons to think they're talking to someone from the United States. Standard Canadian English, about as neutral a dialect as one can hope for in the language, has a flatter, smoother prosody that doesn't call attention to itself, in the way that Standard Australian does, and draws on vocabulary from both British and American sources. This vaguely foreign, seemingly unflappable intonation also suggests an authority that has helped Canadian broadcasters become prominent journalists in the United States, including Morley Safer, Robert MacNeil, Peter Kent, and Peter Jennings.

But "recognized" and "intelligible" and even "authoritative" does not always mean "welcome," and the preferred international strategy should usually be to localize the dialect. Minimally, this approach requires local teams of writers and the consulting services of a local lexicologist; nothing is more unintentionally amusing (first), and insulting (second), than a poor imitation of the local dialect. Similarity attraction argues that the recommendation of dialects targeted to specific constituencies holds. And powerful identificational factors in language argue that the recommendation to get the variety right holds even more strongly. If you're going to accommodate your output discourse, by dialect as much as by register, you can do yourself no faster disservice than to mangle the words or tones of those varieties, the verbal equivalent of blackface.

10: Nass and his colleagues did not discover similarity attraction. It goes back at least to the early 1970s (see, e.g., Barry, 1970), but they have explored it most thoroughly with respect to speech systems. See Nass and Brave (2004), and references therein.

Equally tricky concerns attend using a second-language accent, but there is the intriguing possibility here of picking up on the conciliatory-foreigner effect, where people speak more simply and deliberately to nonnative speakers. Again, the accent would have to be handled with sensitivity to social issues, as many second-language groups are associated with low-status service jobs and subject to certain associated intolerances. But, for instance, the bulk of European accents are fairly resistant to these prejudices.

Dialect, too, is not always easy to disentangle from register. This interpenetration is perhaps easiest to see in cultural media, where the register of country and western music is infused with the prosodies of Southern United States varieties, for instance, and the register of hip-hop is virtually indistinguishable from African-American Vernacular. But the phenomenon is not isolated to music. Estuary English interpenetrates the technical registers of Great Britain, and the prestige dialects of any language always correlate with its formal registers.

Character

> *There are three things which inspire confidence in the orator's own character — the three, namely, that induce us to believe a thing apart from any proof of it: good sense, good moral character, and goodwill.*
> — Aristotle

Among his many preoccupations, our old friend Aristotle was concerned about something that rests at the heart of agent design: credibility. If someone tells you something, Aristotle says, and you only have her word to go on — no other evidence, reasons, or endorsements — you believe her to the extent that she exhibits three qualities. First, she demonstrates common sense — good judgment, practical wisdom. Someone who ran up to you yesterday, holding an apple, rubbing her head, and squawking "The sky is falling! The sky is falling!" is generally not to be believed today; she lacks good sense. The second quality she should demonstrate is goodness — moral responsibility, virtue. Someone who has lied to you or cheated you (or another) is not to be believed; she lacks virtue, goodness. The third quality is an interest in you — concern, goodwill. Someone who has never shown any interest in you before — or, worse, has shown its opposite, contempt — is not to be believed; she lacks goodwill.

These character traits may look quaint, as Aristotle's positions often do when you don't bother to look closely enough. But these notions, all three of them, are (in addition to their fundamental importance for civic interaction and politics) at the roots of the most contemporary phenomenon there is, branding. Corporations relentlessly attempt to associate their image with virtue (no third-world sweat shops for us), goodwill (we will make you very cool), and practical wisdom (only the finest quality ingredients) — all subject to what the ancient rhetoricians called *kairos* — tailoring to the demands and opportunities of the moment. McDonalds, for instance, sometimes emphasizes its charitable associations (virtue), sometimes its concern for flavor or health (goodwill), sometimes its quality assurance (practical wisdom).

Speech systems have as clear a need for brand identity as any product, and agents are the most prominent elements of that identity. Indeed, corporate identity in general has long been understood as a species of personification. For instance, in his classic *Creating the Corporate Soul*, Roland Marchand examines the drive to "cultivate a corporate personality" (1998: 26), to "establish a company 'voice'" (1998: 28), so the company can "project a definite, and ultimately familiar, image" (1998: 28).

Closely connected to the overall ethos of the product, in mutually dependent ways, the agents have a moment-to-moment need for credibility, and it is here that Aristotle's three principles are most directly applicable. Users need to feel that the information is reliable.

Any one of these factors may outweigh the others in a given context, but they work in concert, and for someone to trust your voice interface all three have to be regularly on display. *Trust* may seem like a grandiose term for something that amounts to a hypertrophied thermostat, or a really fancy kettle, but in fact people do trust their thermostats to regulate the heat and their kettles to boil the water; when they don't, people replace them. The behaviors people expect from computer systems are more intricate, and more socially invested, but using them calls for the same sort of trust. Dialogic pairs, for instance, are a critical notion in voice-interface design because the agent who issues the initiative constrains the agent who has the next turn. Users have to trust the system will respond appropriately to an initiative, and they have to equally trust it to respond to their own initiatives. Indeed, the whole cooperative Gricean infrastructure of conversational interaction rests largely on assumptions about these three variables. Practical wisdom entails saying what is relevant, for instance; virtue entails saying what you have reason to believe is true, and not saying what you know to be false; goodwill entails all the maxims of manner.

Practical Wisdom

> *The computer needs to display some degree of intelligent behavior, to the degree that the user can "trust" it to do the right thing under various circumstances.*
>   — Alexander Rudnicky

Artificial intelligence, in the original sense of machines thinking like humans, may be a pipe dream, like a flying-car-in-every-garage, or it may be around the corner. Either way, voice systems don't need to wait for it; what they need is artificial sense, artificial common sense. It's not easy to spot, because systems that exhibit it do so, as a rule, seamlessly. But its absence is unmistakable, and almost as soon as it is noticed it becomes the subject of ridicule. Here is a conspicuous omission of practical wisdom:

> DanLuft   In the morning on Friday January 27th there is a departure from Cophenhagen to Karup at 9:10 and 10:50 sold out. Do you want this departure?
>
> (Bernsen et al., 1998: 99)

Here's another one, just because they're fun:

T$_1$ BeVocal   OK, let's get your starting point. Name a city and state, or say —

T$_2$ Caller   Work!

T$_3$ BeVocal   Work! 982 Walsh Avenue, Santa Clara, California. Is this correct?

T$_4$ Caller   Yes!

T$_5$ BeVocal   Now let's get your destination. Name a city and state, or say "airport," "home," or "work."

What are the chances, you might ask yourself (and the designers should have too) of this caller wanting directions *from* his place of work *to* his place of work? ("Stay there! Don't move!").

In part, this gaff is an issue of grounding — not a failure *to* ground, but a failure to stand on that ground. BeVocal grounds the information; it confirms the grounding in T$_3$, and later in the exchange it does indeed act on that grounding, providing driving instructions that begin at the location it knows as "work" for this caller. The grounding problem here does not concern *whether* or *what* but *how* and *why*. "Starting point = Work (= 982 Walsh Avenue, Santa Clara, California)" is grounded for the *task*, as it should be. But it is apparently not grounded for the *interaction*. Therefore, and this is where character comes in, it is not grounded for the BeVocal agent. For all her cheery professionalism, she starts to come off, at this point in the exchange, as a bit thick.

The agent takes the fall for daffy interactions.

Practical wisdom — *phronesis* the Greeks called it, since you ask — is situational. It is context-dependent. In terms of voice-interface design, practical wisdom means the system needs not only to act on the basis of grounded information, but to converse on the basis of grounded information, to say the appropriate thing, offer the appropriate choices, respond in the appropriate way; and, as Roy M. Turner puts it, "There is no such thing as context-free appropriate behavior" (1998: 307).

Practical wisdom ensures that the agent never offers a nonexistent choice. A banking service, for instance, should never ask "Checking or savings account?" if the caller only has one account. More subtly, if a caller asks for a transaction history, and only one account has had any recent transactions, that should be the default (with an implicit confirmation).

Practical wisdom also makes use of its resources, telling the user the best guess when the appropriate thresholds are met: "I'm sorry. I didn't get that. Did you say CD or DVD?"

Common-sense reasoning can also exercise a useful check on the recognition results. If someone wants to pay 3000 dollars, or 16 cents, to a utility, or schedules a dentist appointment for 4:00 am, or asks for directions from work to work, maybe it's the recognizer's fault, and the candidate list can be reinterrogated, or the appropriate attitude can be struck for the confirmation.

This sense of practical wisdom, of course, extends to conversational competence. For instance, wherever feasible and sensible, the agent should quickly discharge the reactive pressures introduced by the initiative of a dialogic pair. The following exchange would be perfectly natural in a call to a video store.

Caller   Do you have the Kurasawa movie about Macbeth?

Agent   *Throne of Blood*?

Caller   Yeah.

Agent   Yes, we do. Would you like to reserve it?

But a comparable voice interface should answer the question immediately, if possible, rather than initiating a new exchange. That is, it should behave this way:

Caller   Do you have that Kurasawa movie about Macbeth?

Agent   Yes, we have *Throne of Blood*. Would you like to reserve it?

Virtue

So [computerized] agents could soon be selling us stuff on the phone?
— Hal Stucker

Hey, we never said there wouldn't be a downside to this technology.
— Clifford Nass

Virtue is often taken either as a primly antiquarian term, invoking chastity and hygiene, or as a self-righteous one, implying holiness beyond thine. If you have one of these inferences obscuring the word for you, just substitute *good*, as in *good character*. Virtue, in Aristotle's sense of doing what is right because it is right (like being honest) — *arête*, he would call it — is extremely important for the credibility of a voice interface; indeed, for all social appliances. The problem with goodness, of course — the aspect that has given *virtue* if not a bad name then at least a tiresome one — is that there is sufficient misery, misfortune, and misconduct in the world that a single-minded focus on virtue would soon overwhelm us, in the pull to alleviate famine, support earthquake relief efforts, stop insider trading from draining pension funds . . . , the list is literally endless. A voice interface that was anchored so firmly in virtue, if inducing virtuous action was not its mandate (as it might be for an agent voicing a charity drive), would be intolerable. A voice interface that wouldn't give us the weather in Aspen because it is more important that we know about floods here or droughts there would fail to do its job, wouldn't hold users, and would surely fail to induce any virtuous behavior in the bargain. It is, in short, evangelical virtue that is the problem. We know the world is grim in many ways, and calls for action, but we don't need a machine (or, as often as some of them think, other humans) to tell us about it.

Quiet goodness is another matter, virtue that is demonstrated but not advocated. We want others to be honest, kind, and fair: good. And that's what we want from artifacts, too, though the artifacts that don't function as socially as computers rarely have a moral dimension. If our thermostat tells us (and tells our furnace!) that it is 25° (Celsius) when it is actually 15°, we will either fix it or replace it, but probably not (seriously) accuse it of lying. If our TV loses resolution during the seventh game of the World Series, we will be unhappy with its workmanship or durability, but probably not (seriously) accuse it of being unfair. We won't ascribe an ethics to it.

But voice interfaces, and agent-fronted systems generally, are a different story. The level of social interaction, and the blurring of service-voice with service-provider, and the commercial dimension of most voice services, make ethics — and the question of virtue — unavoidable. We don't have any trouble, for instance, calling an ad "dishonest," even if that is just shorthand for the ethics of the company and marketing agency behind that ad. We don't have any trouble calling a service that comes free of charge and then a bit later on starts billing us "underhanded." And while the research is still in its infancy, users of voice systems don't have any trouble calling one agent more "fair" than another, based only on its acoustic qualities (Francis Lee, in O'Toole, 2000).

It is safe to say, in any case, that users will tend to prefer systems that demonstrate (or at least to do not counter-demonstrate) that they have ethics; that they do not lie, cheat, misrepresent; that they will warn us of impending problems; that they will not manipulate our responses, only guide them; that they are virtuous.

Goodwill

An inadequate system tells the public, "Hi. Our convenience is more important to us than yours. We are trying to cut costs by eliminating people and replacing them with machines that don't work as well. Thank you, and have a nice day."
— Birrell Walsh

People — like, say, customers — want others to consider their interests. As rhetoricians have known for thousands of years, the more you want someone to cooperate with you, or to be persuaded of some belief or course of action, the more you have to demonstrate that the cooperation, belief, or course of action is in *their* interests; or in your collective interest, but certainly not that it is in your interest at the expense of theirs. In large part, this demonstration depends on goodwill, on showing concern and consideration for them. In human–computer interaction, it was this need to exhibit goodwill that drove the movements toward system (and corporate) friendliness and usability testing, and keeps on driving those movements. The importance of clear and present goodwill — *eunoia*, to the Greeks — cannot be overestimated. It means that the system's job is to accommodate the user.

Goodwill, for instance, entails helpfulness (and, therefore, helpfulness suggests goodwill) — for task and for interaction, both of which are directly manifest in agent behavior. In the first case, gracefully directing the exchange — asking about omitted variables, suggesting alternatives, summarizing at strategic junctures — conveys not just that the transaction is important, but that the user's successful achievement of her goals is important. In the second case, gracefully managing the exchange — repairing slippages, apologizing for errors and impositions, yielding turn easily — conveys that the user's time, understanding, and wishes are important. Overall, an attitude of goodwill demonstrates that the user's satisfaction is paramount.

Nothing indicates this attitude more clearly than a concern for the user's pocketbook or time. Deciphering the arcane combinatoric menus of fast-food outlets, for example, is usually beyond my ken, so when a clerk makes a money-saving suggestion ("If you order that with fries and upgrade it to an Ultra-Meal for another 49¢, you can substitute a coffee for the soft drink and save yourself $1.15, and then you don't have to order two separate fries for the kids"), that is a gesture I always appreciate. It shows goodwill. And these are exactly the sorts of calculations (on shipping charges, taxes, duties, and so on) that computers were originally designed for. Voice systems would also do well to emulate the customer-service practices that make neighborhood stores successful, like sharing insider information ("If you're not in a hurry, our fall line is going on sale next week, and you can probably save 10 to 20 percent on that sweater"), or helping people locate nonstocked products ("No, we don't have that, but you might try Acme"). Too many formulaic suggestions, from web-commerce sites and chain-store personnel, are designed to maximize their own immediate profits. Maximizing customer satisfaction, as a strategy, usually has better long-term returns.

Similarly, there are simple time-saving features that are routinely ignored by automated systems. The phone companies (which earn their money by how long you stay on the phone, and how often you use it) are often the biggest offenders. For instance, I change phones and locations frequently, and sometimes when I call I need to include the area code but not the long-distance indicating number, 1; other times I need that indicator; still other times, the last six digits are sufficient. Usually I can keep it straight, but once in a while I screw up. The voice-response units always tell me when I screw up (which is fair enough; maybe that will help me learn), but they also force me to hang up and dial again (which is unduly punitive). An exchange like this would show goodwill:

Me   [keying] 1 4 1 6 7 3 6 0 9 0 5

BellCanada   You don't need to dial 1 to call Toronto from the 905 area code. Long-distance charges do not apply. Connecting anyway . . .

Even a caller's zero-out can be an opportunity to show you care about their time:

Caller   [keying] 0

AcmeBank   All of our customer service agents are currently backed up. The wait is approximately ten minutes. Would you like to wait, call back later, or return to our automated system?

The opposite of goodwill, far too often apparent in voice-interaction design, requires users to accommodate the system rather than the system accommodating to them — forcing them to perform unnecessary behaviors, for instance, or trapping them in long interrogations, rather than allowing them to supply information in the units and the order that suits them.

More specifically, there are three areas of voice-system interaction that can help foster a sense of goodwill, though each of them have their complications — disinterest, similarity attraction, and customization.

The first two, the principles of disinterest and similarity attraction, are somewhat opposing kettles of fish — the first suggests total detachment, the second suggests overlapping interest — but both principles need to be kept in mind when shaping agent character.

Rhetorically, disinterest is conveyed when an agent communicates the state of not having a personal stake in the message. This principle only gets halfway to goodwill (which involves communicating a state of active concern for the audience's personal stake in the message), but it is a crucial half. Disinterest is what most strongly suggests to hearers that a speaker is not out to manipulate them, and it is the principle most strongly behind evocations of objectivity in science, engineering, and scholarship generally, as well as in journalism. In its clearest manifestation, disinterest leads to the omission of the first person in discourse. Not surprisingly, researchers under Clifford Nass found that the lack of first-person reference in a voice-based auction system led to verdicts of greater fairness, with a similar finding in an experiment investigating reactions to the reading of an editorial (O'Toole, 2000).

Similarity attraction goes considerably further than disinterest does toward goodwill. In fact, it goes right through goodwill and comes out the other side in the territory of self-interest. But it is mutual self-interest, shared with the user. If someone is like us, we are more likely to listen and cooperate with him. A significant component of that behavior is undoubtedly that someone who is like us most probably shares our interests, making conversation easier and suggesting similar stakes in the outcomes of interactions. Other parents will also be concerned about crosswalks, other hockey fans will care about the Leafs-Senators game, other bibliophiles will share our interest in first editions. Even with experts, whom we certainly don't want to be like us in terms of specific knowledge domains, a little "Yeah, I had that problem when I first started using Illustrator, too" can go a long way. Goodwill, in a sense, is a byproduct of interacting with similar people, but it ensures that one's interests are prominent in the interaction. And, as Clifford Nass's research keeps showing, this pull of similarity extends to interactive voice-system agents.

Agents, too, need not be fixed. Some characteristics, like voice quality, should certainly remain very stable, or the user will be disoriented; shifting voice quality would be like talking to someone whose height or weight or skin/hair/eye color kept changing as you talked to them. But there is a range of personalizing discourse features that are more flexible, which allow the system to mirror some of the user's speech patterns. As Susan Brennan's (1998) work shows, minimally this mirroring should extend to referential terms — not just for efficiency, but also for comfort. For instance, if the user has the practice of saying "What's my nine-o'clock?" "Who's my ten-o'clock?" and the like (rather than "nine-o'clock meeting" or "nine-o'clock appointment"), then the agent should adopt the locution, saying "Your eight-o'clock is with Peter McNab" and "Your ten-thirty with Sally McNab has been canceled". (Notice that this type of convergence suggests goodwill even independently of similarity attraction.) But more general patterns might also be emulated to useful effect. A polite user might well appreciate a polite agent, for instance, finding a terse agent rude; a terse user might well appreciate a terse agent, finding a polite one obsequious.

System convergence is a form of customization, and customization generally is very valuable for establishing goodwill. It can, however, be quite dicey to handle. As we have seen repeatedly, because speech is serial, transient, and monomodal, its efficient deployment requires thorough monitoring of context. The price in terms of user patience that is paid for repetitions (system repetition, but even more, demands for user repetition) is a heavy one. People rarely want to hear the same thing again and again, and they don't want to have to say the same things over and over. In the latter case, especially, it suggests such a low goodwill quotient that the hearer is barely listening to the speaker, nor remembering beyond the immediate turn. Keeping track of who the user is, knowing her preferences, and recalling the salient aspects of previous encounters, are valuable ways of demonstrating goodwill generally, and reducing conversational overhead specifically. At a very local level (say, a few turns or exchanges), building context just amounts to routine conversational management. But as the scope of the context increases, the system becomes more and more customized to the user (just as humans customize their interactions with each other by recurring topics, inquiries about each other's families by name, and so on). The system refines its user model from a general set of assumptions to specific sets of assumptions for each user.

But customization can quickly become troublesome, if not spooky. Matt Marx developed a filtering mechanism for an email voice system that monitors, and cross-indexes, a user's calendar, Rolodex, to-do list, electronic messages, and calls (Marx 1995: 37–59; Marx and Schmandt, 1996; Marti 1999: 17–18). Its job is to rank email messages in terms of user needs and interests, but the same sort of learning, reasoning, behind-the-scenes-decision-making procedures might be applicable to any number of voice-based interactive systems. The need for filtered ranking is especially acute for email, even if the anti-spam devices are working flawlessly, just to manage regular interactions. Even listening to a list of eight or ten message headers can be tedious; intelligent rankings and summarization can be very valuable goodwill time-savers in such situations. Here's an example of an applied criterion from Marx's system (called CLUES):

> When *CLUES* matches an area code from the calendar with one in the Rolodex, it notes the person's name, email address, and the domain from which their email originated. Certainly one wants to be alerted about email coming from them directly, and one may want to pay special attention to messages originating from their site. So even if I'm going to visit Sean at Interval, it's likely that anything coming from Interval will be interesting this week.
> (Marx, 1995: 41)

This reasoning exemplifies goodwill, taking the user's interests to heart, formulating criteria from them, and using those criteria as the basis for action, much like a considerate receptionist would. But a closely related ability of CLUES crosses the goodwill boundary and encroaches on the domain of nosiness (that is, when information seems no longer to be used for your benefit but for someone else's entertainment). The routine (called *who_should_i_visit*) can prowl through the user's "calendar and Rolodex, picking out the people who live in and near the area codes of the fax numbers contained in the calendar" in order to make recommendations for who you should visit if a trip puts you in their region (Marx, 1995: 41). I, for one, don't want an electronic agent saying "Now don't forget to visit your sister, Debbie, while you're in Vancouver." Privacy issues immediately raise their heads. Not only can this sort of close monitoring move rather quickly into personal intrusion, it also carries the taint of Big Brother. Different people have different privacy thresholds, which makes customization a slippery enterprise, and it's probably better to err on the side of conservatism unless explicitly directed toward some types of customization (an offer can be advanced with respect to the service contract, or after several dialogues); the client should in general be aware only that the agent is familiar with her routines.

Personality

Character in its default, ordinary-language sense means something a bit different from its technical meaning in Aristotle, and from the restricted ethical implications revealed in phrases like "a good character" or "that guy showed character." Those meanings are not uncommon, but the default sense of *character* is just a consistent disposition to do (and say) certain things — a usage that overlaps substantially with *character* as it is used in fiction. Someone doing something unexpected, for instance, might be reported as "That was so out-of-character for Mikey." (Aristotle also had this consistent-disposition sense in mind in his *Rhetoric*, of course, arguing that consistent dispositions to act with common sense, virtue, and goodwill result in trust.) In this chapter, we are concerned with a specific type of fictional character: interface agents.

Psychologists call a person's consistent behavioral disposition his *personality*, a term which also derives from fiction (*persona* is effectively Latin for "dramatic role"), but which suggests something beyond mere consistency — motivation, complexity, even a touch of surprise. It is, unquestionably, the grab bag of traits we call *personality* that maintains our interest in the people we enjoy spending time with, and it's not just because we can predict what they will say and do. Personality is also what makes us avoid some people, and gossip about others. And it guides our interactions with them.

Our vocabularies are reflections of what we care about, and we have an extensive vocabulary of personality terms. I'm sure someone somewhere has charted them exhaustively, but in the absence of that chart, I have followed a rigorous scientific methodology. In two minutes, trying to avoid synonyms, I came up with the following list (alphabetization applied later): aggressive, arrogant, callous, charming, cheeky, cheery, compliant, curious, cynical, dishonest, greedy, honest, hot-tempered, lazy, manipulative, masochistic, moody, narcissistic, ornery, overbearing, pleasant, passive, pompous, rambunctious, rude, sadistic, stoic, stubborn, wheedling, and wily. That list comes in at an even 30. In two minutes, you could come up with another list at least as long, with some overlap but with lots of personality characteristics that didn't pop into my head. The range of human behavior is wide, our ability to dissect it equally so.

It may be a long time before the panoply of personality traits we concern ourselves with daily makes it into speech agents. But, because of our personification tendencies, and because the use of speech makes interaction with them inevitably social, speech agents already have personalities. They can't avoid them. No one is calling them *cynical*, perhaps, or *narcissistic*, but users are ascribing personality to them, whether it was "put there" or not by the designers. It is not a coincidence that one of the earliest human–computer-interaction terms for a cooperative system hinged on both a relationship (hence, sociability) and a behavioral (hence, personality) term: *user-friendly*. And while that particular word has the patina of the eighties about it, you could do worse than having it applied to the agent representing your voice service, much worse. Users could be calling it *dense*, *stubborn*, *rude*, or *boring*; or worse still.

Our view of someone's personality is our model of their psychological dispositions, by which we categorize and predict their social behavior. Many personality terms are therefore umbrellas under which more specifically predictive terms lie. A friendly person is friendly because we can more specifically describe her as helpful, cooperative, supportive, attentive, and so on; we can count on her for certain behaviors (or expressions of attitudes toward behaviors) in certain contexts. A hostile person is hostile because we can also say she is belligerent, dismissive, contrary, disagreeable, and so on; we can expect certain responses to certain situations. The same is true of systems.

And speech systems are especially susceptible to social-psychological behavioral characterization; one of Sun MicroSystem's decision criteria for choosing "when speech output is appropriate" for an interface, is when it "is trying to embody a personality" (1998: 21). Here is the clearest case I know of that embodiment, an agent named Zach, developed by General Magic, interacting with a client in a voiced-email application (Zach is voiced by recorded speech, and there is a second agent, voiced synthetically, which goes unacknowledged; text in parentheses indicates low vocalizations, Zach "talking under his breath"):

Zach	[exasperated sigh] Welcome to the telephone mail center. You have to tell me your PIN. Duh!
Caller	5124.
Zach	5124, right?
Caller	Ya.
Zach	Good. I'm so happy you know your numbers. Here's where I tell you how many new messages you have, because you're such a *wonderful* person. You have one message, probably junk mail.
Synth	Message from: Bob Smith.
	Received today at 3:22 p.m.
	Subject: tomorrow's meeting.
Caller	Play it.
Zach	(Oh, this is so pathetic.) OK. Let me play your *really* important message.

Synth   Good afternoon. I just wanted to remind you that we are meeting at 2:00 p.m. tomorrow afternoon. Thanks, — Bob

Caller   Save it.

Zach   OK. Saving Useless Message, number 3,000,004 .

[<2 sec]

Hellooo? Anybody there?

Caller   coughing noise

Zach   Try saying that again, after you get the marbles out of your mouth!

Caller   Ah, er, . . .

Zach   It's called "Speaking clearly." Ever heard of it?

Caller   Send a reply.

Zach   (It took you long enough.) What are you waiting for? Record your reply.

[Beep]

Caller   "Hey Bob. Got your message. See you at 2:00."

Zach   Since you never get anything right the first time, you *probably* want to record it again, rather than send it, huh?

Caller   Send it.

Zach   Alright. Sending Lame Message, number 5,000,006.

Caller   Bye.

Zach   Thanks for wasting my time. Again. Bye.

(General Magic, 2002a)

You might think that this unpleasantness is softened by a light tone of voice, or a kind of John-Cleese-as-Basil-Fawlty manic-ness that makes it funny. Nope. I'm sure there was some sense of whimsy in developing this interactive style, or maybe they just wanted to push the extremes to contrast overtly with the other personalities in the General Magic stable, or both. But, rest assured, the prosodies are every bit as odious as the words, maybe more — swooping up and down, stretching out and compressing, in the manner of the most arrogant, know-it-all, adenoidal loser you have ever encountered. But, whatever the motivation, the execution is relentlessly in-character. Nothing Zach says is even a bit short of hostile.[11] In Aristotle's terms, he may have practical wisdom (there's no contrary evidence on this point), but he is dramatically lacking in goodwill, and there is enough laziness, aggression, and self-satisfaction that he won't be winning any votes for virtue either.

Among the other personality-based demos General Magic offered (there were six) is Jen, as sunny an antithesis to Zach as one could hope for. Here she is fronting a voiced CD shop:

Jen	Hi! Welcome to CD-Dial. My name's Jen, and I'll be helping you today to find music, or to place an order.
	Are you a member or a guest?
Caller	A member.
Jen	OK. Please tell me your ten-digit PIN.
Caller	443 486 99 10.
Jen	And your four-digit pass code?
Caller	5386.
Jen	Hi, John! How's it goin? I know you're into rock, so check out our cool new featured CD, *Out of the Ashes*, by Exit. It's on sale this week, for $9.95. Here's the first track:
	♫ ♫ ♫ [partial track plays] ♫ ♫ ♫
Caller	Add it to my shopping cart.
Jen	OK. I've added it. I *love* that CD, too. Would you like to listen to something else?
Caller	Do a search.
Jen	Do you want to look up the artist or title?
Caller	Artist.

11: Here's a little more, for the morbidly curious, Zach's self-introduction: "I'm Zach, and — guess what? — I don't care about you. So, I probably won't be too helpful now, will I? Let's think about why that might be. Hmm. Oh ya! I get to talk on the phone all day, with complete idiots, about nothing, because they don't have a life. [Falling into a vocal caricature of a dim-witted customer] "Ahh. This is Bubba. I just need a little help" A *little* help, Bubba? Jeez! So the other day, my boss says, [another unflattering impersonation] "Zach? How 'bout you go a little easier on the customers?" And I say, "Ya, how about you go a little easier on the Krispy Kremes? Huh, tubby?" Anyway, can't wait to hear from you." (General Magic, 2002b).

Jen	OK. Who's the artist?
Caller	Opal.
Jen	Let's see. . . . I've got four CDs by that artist. The first CD is *Free Fall*, for $15.99.
	♪ ♪ ♪ [partial track plays] ♪ ♪ ♪
Caller	Add it to my shopping cart.
Jen	OK. It's in there. Great choice! Anything else you want to do?
Caller	Check out.
Jen	OK. We're there. You've got two CDs: *Out of the Ashes*, by Exit, on sale for $9.95; and *Free Fall*, by Opal, for $15.99. Do you want to order these?
Caller	Ya.
Jen	OK. That'll be $29.99, including shipping. I'll just charge it to your credit card. Your tracking number is 123 456. Do you want to hear anything else?
Caller	No, that's all.
Jen	Thanks for your order. Enjoy your CDs.
Caller	Good-bye.
Jen	Talk to you later.

(General Magic, 2002c)

A little italics might convey some of Jen's enthusiasm ("I know you're *into rock*"). The problem is that a little wouldn't do her justice. All of her utterances would have to be italicized, with various overlays of bolding, and capitalization, and underscoring.

Agent personality is an aspect of voice-interface development that can quickly escalate to annoyance; each of us, after all, has our own personality, and social interaction follows from the dynamics of bringing different personalities together. Some are resonant, some are not; and not all users can willingly suspend disbelief to the point that we want to be congratulated on our music choices by a computer, let alone a recorded voice, or respond sympathetically to that voice telling us how it *loves* the CD we've just decided to buy. And nuances are important in language. Jen tells the caller not just that she loves the *Out of the Ashes*, but that she loves it "too." A genuine human at the other end of the phone would not see any clues in the caller's language that he loves a CD he has had only momentary exposure to. Goodwill is so much in evidence here that sincerity (therefore, virtue) is called into question.

But Zach and Jen have something going for them that a dependably sensible, virtuous, and considerate system, like SpeechActs, is missing, and, sure enough, the easiest ordinary-language label to give that something is *personality*. It's not that SpeechActs has no personality, just that its flat and earnest personality is unengaging, and just as the ordinary-language use of *character* connotes moral value, the ordinary language use of *personality* connotes social value. *Personality* connotes "interesting."

The limited research that has been done in this area comes mostly out of Clifford Nass's lab, based on the central introvert/extrovert distinction, and it has two robust results. Firstly, people have no difficulty picking up even the most rudimentary cues about personality, even with synthesized voices. And secondly, people tend to want to interact with agents much like themselves (similarity attraction). The first result should be especially exciting for interface designers, since — while diction and utterance structure are important personality indexes — something as easily manipulated as rate of speech or pitch range has very strong correlations with perceptions of agent personality. For recorded speech, these variables are largely a matter of casting and coaching the voice talent. For synthesized speech, they are even easier to design and modify; as Nass and Scott Brave put it, vocal indicants of personality can be "controlled by simple sliders" (2004). The second result, similarity attraction, is somewhat more difficult to work with in the domain of personality, since it requires a clear picture of the user's personality, which is (1) rarely easy to obtain, and (2) rarely uniform among groups of users, so that designing by similarity attraction on this level will often require a significant degree of customization.

And here's a surprise. Most of us probably think of power-users barking their way through a voice interface, getting where they want, doing what they want, with only contempt for soft, gimcrack interface elements like "personality," which we might think novices need for comfort and assurance. But experts want imaginative personalities even more than novices (Chin, 1996). I'm guessing that doesn't mean they want Zach, however, just a bit more flavor than the deadpan, mechanical tones of an interface design that doesn't take personification seriously.

Emotion

> *Just to register emotion*
> *Jealousy — Devotion —*
> *And really feel the part,*
> *I could stay young and chipper,*
> *And I'd lock it with a zipper,*
> *If I only had a heart . . . !*
> — The Tin Man

Emotions are the transient manifestation of personality in context. They are frequently referred to in color terminology (anger is red, jealousy green, placidity blue), and, like color in appearance, they are always present in speech. An image may be monochromatic, and leave the impression of being colorless, but that means it is expressed in the value ranges of one color (mono chroma); as long as it is an image, it is reflecting light, and therefore has color. A vocalization may be even-keel, and leave the impression of being emotionless, but that means it is expressed in the range of one emotion, *placidity* we might call it, or *equanimity*. Since we haven't heard from Aristotle for a few paragraphs, maybe we could use his term, *praotes* (*Rhetoric*, 1380[a]), usually translated as *mildness*. Notice that even to be placid, or mild — whatever we call it — requires some prosodic alterations, just placid ones, mild ones; a complete monotone manifests not equanimity or *praotes*, but boredom. The even keel, in any case, is an emotion. So is boredom. It is as figurative to speak of emotionless speech as it is to speak of colorless vistas.

Emotion is our moment-to-moment stance with respect to the universe, the compass we read to determine our direction. It (assuming the emotions to constitute a homogeneous system) evolved to help us deal with the environment, including the social environment. Should we flee or chase, spurn or woo, sing or snarl, work out an algorithm or kick back with an ice tea? We feel always. So our feelings are always in our speech. They are in word choice and syntax. We say different things, in different ways, when we are annoyed or cheery. Most obviously, they are in the raw, dancing molecules we send cascading when we open our mouths, the *way* we make our communicative noises. We laugh, we yell, we murmur and chirp. Picking just the four cardinal emotional states: happy, sad, angry, and calm, the acoustic variables are exactly what you would expect. Taking calmness as the norm (it's not in my house, but let's maintain the fiction), happy speech tends to be higher and faster and louder, with greater pitch variation, than calm speech. Sad speech is lower and slower and quieter, with narrower pitch variation and more drawn out articulation. Angry speech again is higher, faster, and louder than calm speech (and even than happy speech), with an abruptly shifting intonation and tense, clipped, articulation. These differences are arrayed in Table 12.4.

Just as they necessarily exhibit personality, speech systems necessarily exhibit emotion. In general, this exhibition has been limited to calmness in synthetic voices (though the changes in speech rate for systems like MailCall might suggest a vague level of excitement), and to a small cluster of mostly positive emotions in real voices (Zach notwithstanding). This narrow range is generally appropriate. For one thing, the principle of convergence applies to emotion, as much as to the other characteristics of speech, and since (current) recognizers have difficulty with any speech input which doesn't hug the baseline in all parameters (speech rate, pitch range, volume, and so on), establishing a measured, constrained, and encouraging emotional tone will help keep users in the optimal recognition ranges. And a calmly encouraging tone is usually the one best suited for task completion, and for ensuring a satisfactory user experience.

So, "adding emotion" to a voice interface really means adding some variation to the emotional expression of a voice interface — mood words, rate and pitch changes, and so on. The words, of course, are no problem, if the content is not pre-existing, but neither are

	Happiness	**Sadness**	**Anger**	**Calmness**
Rate	Faster	Slower	Faster	
Fundamental frequency	Higher	Lower	Much higher	
Pitch range	Wider	Narrower	Much wider	Normal
Articulation	Normal	Drawn out	Tense	
Volume	Higher	Lower	Higher	
Pitch shift	Upward	Downward	Abrupt	
Articulation	Normal	Slurring	Tense	

TABLE 12.4   Acoustic parameters of some emotions, taking calm speech as the baseline (adapted from Murray and Arnott, 1993)

the acoustic characteristics, even with synthetic voiced agents. Even fairly primitive vocal synthesizers are capable of reliably coding emotions by alterations to pitch and timing.[12]

Overt displays and discussions of emotion in language are said to correlate with females (e.g., Bonvillan, 1997: 180), which suggests that a greater range of emotions, should they be appropriate, might be more naturally received if they come from a female agent.

The natural design guideline here, of course, is simply that the emotion should suit the circumstances, especially the content of the utterance. It is probably safest to adopt a *praotic* tone as default — assuming the even-keel emotion when the content or dialogue act of the message is unknown, and certainly avoiding the doggedly chipper tones of Jen or the relentlessly contemptuous tones of Zach, so that we don't get informed cheerfully that our brake fluid is low or snidely that we qualify for a discount. But the ideal would be slight urgency for low brake fluid, mild enthusiasm for discount opportunities, and so on — the tone matching the matter and the purpose.

Also, and rather crucially in some circumstances, the emotion should match the emotion of the user. Nass and Brave argue for instance, on the basis of quite compelling research, that "one can strongly influence the number of accidents, the drivers' perceived attention to the road, and the driver's engagement with the car simply by changing the

---

12: Janet Cahn, for instance, in the late eighties got test subjects to categorize voices in terms of anger, disgust, fear, joy, sadness, and surprise, on the basis of several acoustic characteristics (such as pitch and timing), finding significant uniformity in their judgments. See Cahn (1990).

[emotional] tone of voice" for an onboard speech system. Moreover, the strongest factor in this influence is the suitability of that emotional tone to the current emotional state of the driver. "Upset drivers clearly benefited from a subdued voice" in their study, "while happy drivers clearly benefited from an energetic voice" (Nass and Brave, 2004). Detecting emotion is not a trivial task, but the sets of vocal characteristics outlined in Table 12.4 can be used to help *detect*, as well as to *simulate* emotion. And, of course, the user is always in an emotion-inducing context, as well, and can be expected to be happy in happy contexts, wary in data- or person-threatening circumstances, and so on. To the extent that the voice interface participates in these contexts, its emotional tones should correspond to what is predictable about the user's emotional states.

Inducing emotions in the user, other than the praotic even-keel state, can be a very dicey proposition. In certain circumstances, when physical behavior is more important than linguistic behavior, perhaps some emotion might serve an important purpose — mild anxiety, for instance, when there is an impending transaction failure, icy conditions on the road, or the like. But emotion affects speech patterns, and speech patterns affect recognition. Take an extreme case, like humor. If, say, Zach's muttered insults were genuinely funny (and they may well be for other sensibilities than mine), the caller would laugh and the system would therefore get some very confusing input.

Diction

Is sloppiness in speech caused by ignorance or apathy? I don't know and I don't care.
— William Safire

Character, personality, and emotion all commingle in ways we understand well in our daily social routines, but are tricky to design sensitively. But a lot can be accomplished rather simply with dialect- and register-specific diction. Take this rather typical exchange:

Caller   Go to sports.

System   Sure, sports. | No sports.

                          (Where the v-bar represents a new system state, and a new audio file.)

There's no practical wisdom here, no indication of a characteristic disposition other than "daft machine," no context-sensitivity in the emotion. Engineers will complain in such circumstances about sending variables among routines, timing issues, and the like, so that you can't know until you "get there" what the attributes might be, leading to these situations. We're used to such idiocy from systems because of this don't-know-till-you-get-there problem. But it's completely unnecessary in voice interfaces.

You don't *have* to know before you get there. You just have to behave appropriately when you do get there. Humans might not know there aren't any cookies when they offer them to a guest, either, because they are usually in the cupboard; on opening the cupboard and discovering the cookies are gone, they say something like "Whoops, looks like someone's been into them. No cookies. How about a nice hot cross bun." Humans don't promise something, fail to deliver, and proceed cheerfully oblivious. They express an attitude (an emotion, mild surprise). There are certainly relevant prosodic characteristics in this message, but even a neutral intonation would work here because the word choice is so attitudinally clear. And it doesn't have to be a complicated someone's-been-into-them explanation. A simple vocable will convey the necessary surprise-tinged practical wisdom needed in cases like this. Michael Cohen and his Nuance colleagues, for instance, suggest what they call the "self-editing" use of *oh* for these cupboard-is-bare situations:

T_1 Caller   Go to sports.

T_2 System   Sure, sports. | Oh, that's not available at the moment, but would you like to try something else instead?

T_1 Caller   Get my messages.

T_2 System   Okay, messages. | Oh, looks like you don't have any messages now.

(Cohen, Giangola, and Balogh, 2004: 143)

Physical Characteristics

> *An interface is by nature a form of artistic imitation: a mimesis.*
> — Brenda Laurel

For promotional reasons, even a voice-only agent will call for physical characteristics. Although it is not at all clear how to interpret the embodied agent research in the context of voice interfaces, a possible interpretation of the work is that users are more satisfied with a system when they can put a face to it.[13]

It is probably inadvisable to use an actor, or even conventional animation, to represent the physical characteristics. The more clearly artificial the agent is (with the two caveats that it should neither be too primitive nor outside an acceptable range for human appearance), the more visible the computational character of the system will be. Ananova is a good prototype in this regard, the artificial newsreader developed for the web and adapted

---

13: The embodied agent work is quite fascinating, but the modality complications are so wide-ranging that the research does not transfer very well to voice-interface design (though voice-only interface research does bear rather directly on embodied-agent design; it's mostly a one-way street). See, for instance, Cassel et al. (2002), Bickmore and Cassell (2000), Turunen and Hakulin (2000, 2001), and Bell and Gustafson (1999).

for PDAs.[14] She is distinctly humanoid, but with the poreless skin and barely mobile eyes familiar from late-nineties computer animation, and, if that isn't sufficient, green hair.

Effectiveness will be highly contingent on the domain and the users — a talking donkey might work for some constituencies, who knows? — but, all things being equal, the physical characteristics of agents should be designed for all the familiar reasons: user and field representativeness (which merges into similarity attraction), overall credibility, and general pleasantness. Extremes — in glamour, cuteness, sexiness, grotesquery, and the like — are, all things being equal, best avoided. The number of agents fronting a voice service constrains and affords different options. If there is only one agent (all things being equal), ethnicity should be somewhat ambiguous; with several agents, ethnic (and gender) representation can be more distributed.

Distilling these remarks, the physiology of an artificial agent for a voice-only conversational interface should, all things being equal, match the user profile and convey a sense of friendly credibility.

Number and Constituency

Because computers are not constrained by the acoustic structures defining each person, a single computer can produce a multitude of voices.
— Clifford Nass and Li Gong

The primary question about the agent architecture of a voice interface comes down to the questions "How many" and "What type." The standard is for one agent. But as the fields enlarge and amalgamate, and the interactions increase in duration, there will likely be a development toward multiple agents, correlating with different functions.

A one-agent design has many liabilities. For any service with dynamic content, which includes a great deal of what users would want voice services for (up-to-the-moment, on-demand information), the only way to provide a one-agent experience is with a synthetic voice, and synthetic voices do not lend themselves to quality experiences.

14: *www.ananova.com.* See also *www.artificial-life.com* for some physical variations on the artificial agent theme.

A one-agent, synthetic-voice scheme is workable, however, for systems with brief exposures, especially when providing small pieces of data from large and/or changeable pools, like telephone numbers, addresses, or stock prices.

A one-agent, real-voice scheme is workable for systems with somewhat prolonged but infrequent exposure, especially when the data is relatively stable, like interactive museum guides, travel information, and periodic, routine transactions.

Multiple agent designs are therefore standard for any voice service that has both dynamic content and the wish to provide a distinctive, enjoyable interaction. *Multiple* frequently means "two," with a real-voice maitre d', who gets the customers to their tables, and a synthetic-voice waiter, who serves the information to the customer. But there are advantages to larger casts of characters. In broadcast journalism (and radio is a good model for voice-only services), standard operating procedure is to have one voice for each information domain: political news, sports, weather, entertainment, health, . . . as many characters as there are ways to slice the information pie, all of them striving to project identifiable personalities. This approach offers both variety (an aesthetic virtue) and segmentation (an ergonomic virtue), but it also runs the risk of too many competing voices preventing a distinctive identity from emerging (a branding concern), and just plain losing the user in a sea of interlocutors (a liability on three fronts — productivity, recognition, and ergonomics). In particular, the very important linguistic phenomenon of convergence loses some of its design value.[15]

The extent to which the advantages of convergence and conceptual pacts attenuate even with a two-agent interface, let alone a five- or six-agent interface, is an open question. But it is perhaps best to reinforce the more significant terms after handing off the user from one agent to another, where relevant and appropriate.

In addition to the synthetic/real distinction, voice-interface designers have a palette of age, gender, dialect, and personality from which to develop individual agents, which might operate independently or as part of a team. Gender mixing, and, to a lesser extent, dialect mixing raises some specific concerns. In a two-agent architecture, there will always be a top-level and a second-level agent; there must be, that is, a hierarchical relationship that suggests dominance. Variety and representativeness both argue to include male and female agents. But that makes managing the power dynamic much trickier than it would be with two male agents or two female agents. (Think of the ideology behind an intelligent male top agent, bossing around a somewhat dimmer female robot, for instance.)

In any case, the interrelationships among the agents in any multiple-agent interface should be clearly established — who is the expert, who is the generalist, who provides the

---

15: Lexical priming remains stable for multiple conversants (since it seems to be connected with phonological and semantic activation patterns), but multiple voices, with distinctive phonetics, diction, and phrasings, quickly wash away the articulatory/acoustic advantages of convergence, and might impair the referential advantages as well.

information, who provides the assistance, who manages the overall dialogue, who manages the task — and the transitions among them should be handled with care.

Like other interface decisions, these should not be made without user input. Oz testing, for instance, might be used to test vocal talent or synthetic attributes, as well as discourse and interaction styles. And independent "casting studies" might also be set up, the term Nass gives to agent-based focus studies that gather reactions from the user population to specific voices and vocal talent. Any studies of voices should not key just to the individual voices, but to possible combinations as well; two voices that might be independently impressive could easily fail to strike the right note when they are brought together. Choosing the system voices is the biggest branding decision the design team can make.

Casting

Casting the right voice is one of the most important tasks in creating a successful speech-recognition system.
— Blade Kotelly

It is crucial for human-voiced agents to find someone who fits the role and can play it consistently, reeling off hour after hour of variations on themes but keeping them fresh. A director (usually the soundscape designer) who can hold them to the script, feed them lines, listen for subtle shifts, and help keep them on track vocally is also necessary. The six primary areas of agent-design consideration that we outlined earlier should be the main selection criteria.

For branding purposes, the voice should be distinctive, but for interactive purposes, it shouldn't be so distinctive as to draw attention to the tones and articulations, away from the words and the task. Think of James Earl Jones's classic "This is CNN" mantra. It is the quintessential branding sound bite, but the voice is so rich and plumy and commanding that it would overwhelm an interaction, like trying to chat with Darth Vader. Aesthetically, the voice should be pleasant; for user productivity, tones of cooperation and trustworthiness are important; ergonomically, you want natural rhythms and clarity; and for matters of recognition capabilities, the agent needs clear enunciation, because of convergence, the capacity to add subtle salience to preferred terms, and the overall ability to engage users in cooperative speech behaviors.

Above all, the talent must be able to play the role envisioned for the interface agent, striking the right character notes for the context.

Playing a voice-interface agent has some unique demands, such as performing only one half of an interaction, and giving multiple takes of slight variations on the same line, not just to get it "right" but also to provide variations that will keep the repeated user interactions from going stale. Lexical priming can also be enhanced by introducing saliency effects to the key terms (such as making them a bit longer and louder than comparable words in the surrounding signal). While this can be added mechanically, after the recording, good voicers can bring it off in the studio. Dialect also makes special demands on voice interface talent.

If dialect targeting is used, the vocal talent used should actually speak that dialect (I take it this goes without saying for gender and age as well). Actors like Meryl Streep, who have a gift for dialect, are as rare as musicians with perfect pitch. More importantly, they are much rarer than actors and voicing professionals think they are. Out-group listeners may not be able to detect the false notes in all the fake Scottish and Caribbean and South-western United States intonations that populate movies, TV, and radio, but in-group Scots and Jamaican and Texan listeners can. That's not generally a problem in entertainment, since the dialects are there to amuse a largely out-group audience. But entertainment is not the motivation for dialect targeting. The whole point of targeting a dialect group is to make users comfortable, not irritate them at the false sincerity evident in the mangling of their dialect. In-group speakers are necessary to get the phonology right. In-group *writers* are also necessary to get the words and structures right.

The best source of voicing talent is usually radio artists — announcers, broadcasters, and especially actors. All are reasonable bets, but actors generally have a better sense of the context in which the utterances will be used, a better sense of the role. And radio actors usually have a better and more appropriate command of their voices than other actors — a better sense of how to convey the same nuance vocally that a movie or television actor can relegate to a shrug or an eyebrow lift, and a better sense of the necessary vocal intimacy for a phone interaction than a stage actor.

You can start with tapes, solicited from agencies, and/or from listening to the radio for the right tones and cadences, but you really need to audition the talent — at least eight or so voicers (and more, if the first crop doesn't pan out), for about a half an hour each — to get a sense of their abilities in this field. Aside from their sense of the role they're playing, and the way they can hit the dialogue-act prosodies and personality characteristics you need, the most important general capacity they need to demonstrate is vocal consistency. Keeping the rate, volume, and pitch levels uniform is important not just for naturalness, but for reducing the burden of the sound engineers (if the voicing will be used in recombinant concatenation or synthesis, these considerations are even greater).

The audition script should include greetings, typical utterances, and word lists, and while the eventual recordings will be done in a sound room or booth, isolated with a script, brief, wizard-like interactions during the audition can sometimes be revealing as well, to bring out characteristics the artist may not otherwise be expressing.

Constrained Variability

Say something once, why say it again?
    — David Byrne

We all know that consistency, in almost every nook and cranny of a graphic interface, is a good thing; "strive for consistency," Shneiderman reasonably tells us. And consistency is certainly valuable in voice interfaces as well — consistency of tone, interaction style, and target vocabulary. In agent design, consistency of personality type is important (mixed personality types "seem aberrant, and therefore less intelligent and trustworthy" — Nass and Brave, 2004). But the serial and temporal nature of spoken language puts a somewhat contradictory pressure on voice interfaces as well, the pressure for variability. Dogged verbal consistency, in fact, is wearing and aggravating, like a five-year-old who won't stop saying "Bob's your uncle" over and over, day and night and day.

That doesn't mean voice interfaces need to court novelty. But it does mean that some slight changes to function words, inflection, and even word order can help the naturalness of a speech system. It means that a certain amount of constrained variability is a good thing. These concerns mostly affect utterance design, but when recording vocal talent you should get multiple tokens of the same utterance. The pronunciations should not be deliberately different. The variability introduces itself simply because that's how vocal tracts work. The different versions can be randomly cycled into the interaction. This aspect of constrained variability is costly, because audio files are memory hogs, but it helps ensure a more satisfactory user experience, and the storage/file-swapping aspects do not need to be handled at run-time.

Summary

Computerized conversational agents . . . which require users to interact with them for more than a few minutes, or which we expect users to take . . . seriously enough to discuss their medical problems [with] or give out their credit card numbers [to] . . . must be able to establish social relationships with users in order to engage their trust which, in turn, eases cooperation.
    — Timothy Bickmore and Justine Cassell

In this chapter we took up the topic of **personification** (or, as most of the literature calls it, **anthropomorphism**), which has been something of a hot-button issue in voice-interface design. I argued that the case against interface personification is little more than unsubstantiated fear and loathing; and, in any event, that it is futile. Even if you view it as a necessary evil — an "epistemic frailty" Daniel Dennett calls it (1997, 67), "an inevitable conceptual crutch for users" Tim Roher (1995) calls it — personification is unavoidable, and trying to avoid it, rather than to manage it, is simply misguided. The remainder of the chapter discusses the relevant factors and their management strategies for designing a voice interface agent.

There are **six primary areas of consideration** in agent design: **branding**, the design area in which the company's ethos is carried and defined by the agent's character and personality; **aesthetics**, the contributions the agent can make directly to the pleasantness of the user's experience; **productivity**, the way in which the agent can support learnability, efficiency, and functionality, chiefly affected by level of cooperation and credibility the agent manifests; **ergonomics**, the general ease-and-comfort of working with the agent, through a natural interaction style and clear enunciation, which keep the cognitive load and memory burden relatively low; **role**, the collection of functions the agent serves; and the paramount area of consideration, recognition **capabilities**, the area of agent design that constrains all the others to ways that induce convergence, clarity, and directness from the user.

With these six areas of consideration riding shotgun, we turned to the make-up of the agent **cast**: **single** or **multiple agents**; should any given agent be **real** (recorded) or **synthetic**, **male** or **female**, speak standard or non-standard **dialects**; at what levels should any given agent manifest the qualities of **character** (practical wisdom, virtue, and good will); what **personality** should it evince; what **emotions**; what should it look like, if an **image** is required.

The chapter also includes a brief discussion of casting, and the introduction of constrained variability. **Casting** involves, among other general concerns, the application of the six primary areas of agent consideration to shape who you choose, by way of demo tapes, and how you refine those choices by way of auditions. **Constrained variability** is the codification of the simple principle that too much sameness is annoying in spoken language. Variations can largely be handled in writing utterances, but with naturally voiced agents it can also be introduced by capturing multiple tokens of each utterance.

Dialogue Matters

Graceful interaction must . . . supplement [the system's] simulation of human conversational ability with strategies to deal naturally and gracefully with input that is not fully understood, and, if possible, to steer a conversation back to the system's home ground.
— Philip J. Hayes and Raj Reddy

Speech-system users should be free. The point of letting users talk, instead of forcing them to punch buttons; the point of letting them express their goals, instead of insisting they face interrogations; the point of letting them combine their tasks, instead of confining them to inflexible hierarchies of branching options — the point of conversational voice interfaces — is to liberate users. Here we have the first principle of voice interaction design:

1. Liberate users.

   Niels Ole Bernsen and Laila Dybkjær articulate Principle #1 more expansively:

   Users should always be able to say exactly what they want to say, in the way they want to say it, and when they want to say it, without any restrictions being imposed by the system.
   (Bernsen and Dybkjær, 2000a)

This way — the path to human factors heaven and total habitability — we could raise satisfaction rates to virtual perfection: the user speaks, the system responds.

There are complications, however. Speech recognizers make errors. So, here we have the second principle of voice interaction design:

2. Eliminate errors.

Bernsen and Dybkjær are on the case here, too, and they have an exemplary strategy for applying Principle #2:

> *All [voice interfaces] could conduct their transactions with users as a series of questions*
> *to which the users would have to answer "yes" or "no" and nothing else. Simpler still,*
> *"yes" or "no" could be replaced by filled pauses ("grunts") and unfilled pauses (silence),*
> *respectively, between the system's questions, and speech recognition could be replaced by*
> *grunt detection.*
>   (Bernsen and Dybkjær, 2000a)

This way — the path to quality-assurance heaven and total reliability — we could raise the recognition rates to virtual perfection: any vocal input would be recognized as a grunt; any absence of vocal input would be recognized as a nongrunt.

But, whoops, P & not-P. We can't do both. Liberating users and eliminating errors are fundamentally at odds. Put another way, controlling errors (eliminating them is unrealistic) effectively means controlling users.

The art and the challenge of voice interaction design is walking the creative line between eliminating errors and satisfying users, between the grunts-and-silences condition and full expressability. The Bernsen-Dybkjær habitability/reliability dilemma is real. Your job is to walk that line, to constrain users in a way that doesn't make them feel constrained, in a way that makes them feel that they are accommodating the system, not being bullied by it, and, even more importantly, in a way that makes them feel that the system is accommodating them. This chapter addresses what Sharon Oviatt (2000: 255) calls "the number one interface problem for speech technology" — avoiding and gracefully recovering from errors.

We first outline the types of errors that result from the basic fragility of speech recognition, and go on to recommend prevention strategies and repair principles, outline feedback and prompting styles, take up user-categorization and initiative management, and discuss last-ditch efforts — all in the interest of developing practical solutions to the habitability/reliability dilemma. We also take up two other constraining influences on voice interaction design, the complications that arise by drawing utterances from pre-existing text, and the technical-political pressures that legacy situations exert.

Errors and Slippages

> *Imagine that you are designing for a GUI and [the] input method is a mouse and keyboard*
> *where one out of every ten mouse and key presses is interpreted incorrectly. That makes it a*
> *lot harder.*
>   — Kate Dobroth

First, we need to be clear what sort of errors we're talking about: recognition errors in particular, system errors in general. *Users don't make errors.* (Of course users make errors. I've hit my thumb with a hammer lots of times, and while I'd like to blame the hammer — in fact, I *have* blamed the hammer — it was, I confess in the cold retrospective light of day, my fault. But it serves designers of complex systems very poorly to think in terms of "user errors." That disposition breeds excuses and lame compromises. The designer's perspective should always be that interaction failures, whether they are expressly the system's fault or not, can be resolved from the system end.) Repeat after me: *users don't make errors*.

Recognition engines do, however. They make errors of three general sorts:

Recognition Errors

Deletions

The system does not hear a word the user has spoken, either failing to return any candidate at all (perhaps treating the input as background noise) or only returning a candidate with very low confidence. For example, *nine* is "deleted" here:

System   What date will you be returning on?

Caller   September twenty-nine.

System   Here are the flights for September twenty.

Caller   No, I said "September twenty-nine."

(Stifelman, 1993)

Insertions

The system hears a word the user has in fact not spoken, often through misprocessing of background noise or nonspeech vocalization. For instance, *sports* (or something in the same vein) is "inserted" here:

AcmeTele-Message   AcmeTele-Message.

Caller   \<clears throat>

AcmeTele-Message   Do you want sports statistics?

(Balentine and Morgan, 1999: 163)

Substitutions

The system hears a word or phrase, X, when the user has in fact spoken another word or phrase, Y. For example, *Norwich* is "substituted" for *Ipswich* here:

VODIS   Where are you traveling from?

User   From Ipswich.

VODIS   Norwich.

User   No, Ipswich.

	User says	**System resolves**	**Effect**
Split	Recognize speech	Wreck a nice beach	2 words split into 4
Fusion	Wreck a nice beach	Recognize speech	4 words fused into 2

TABLE 13.1   Segmentation recognition errors

VODIS   Ipswich?

User   Yes.

(Waterworth and Talbot, 1987: 134)

In terms of the input resolution, substitutions can be either splits or fusions; that is, the user's utterance might be over-segmented (extra words show up) or under-segmented (fewer words show up). Table 13.1 gives the classic examples of both, where the user's utterance, "wreck a nice beach" is fused into the system resolution, "recognize speech;" and the split is vice versa, "recognize speech" is resolved as "wreck a nice beach."

And this is the *pretty* picture. Errors can come in clusters, a series of deletions with a quiet speaker, for instance, a series of insertions with lots of background noise, fusions, and splits galore with strong dialectal or first-language interference, or any combination thereof.

But the source of the errors, and even to some degree the extent of the errors, matters less in the long run than what the system does with them. In the dialogue examples we just looked at, the system didn't do anything directly with the errors, of course, because it doesn't realize it's made an error. In each case, the system does something indirectly, however, which leads to the correction: it presents the resolution for inspection.

When systems do realize there's been a problem there's not much they can do except reject the input, or some portion of it, a behavior called — you guessed it — *rejection*.

Rejection

The system just can't make heads or tails out of the input, and confesses the problem to the user. The user still needs to be part of the solution to fix the problem, or help fix the problem; the difference is, she doesn't need to initiate the correction. The system initiates it with the confession. Rejections can be "false" or "correct."

False rejection

The user produces an ordinarily acceptable utterance, but for one reason or another (in this example, a substitution that makes no sense), the system cannot resolve it: legal input is falsely rejected.

User   I'm done.

System   [hears "and that"] Excuse me?

User   I'm done.

(Sikorski and Allen, 1997: 211)

Correct rejection

The user produces an unacceptable "utterance" (a cough, a sneeze, words without representation in the system vocabulary, even a background noise), which is consequently flagged as incomprehensible: an illegal utterance is correctly rejected.

Greg   OK, now I need the last four digits of your social security number.

Caller   Um, 52. . . . I think it's 5654?

Greg   Sorry. I didn't quite catch that. Simply say the four digits or enter them on your key pad.

Caller   5654

(General Magic, 2002b)

Recognition errors lead to dialogue slippages, points in the interaction where the user's goal-pursuit is hindered, but they are not the only source of slippage. Users don't always behave the way designers want them to, or thought they would.

Interaction Slippages

The term *rejection error* is something of a historical curiosity, dating to a time when recognition did not involve much, if any, natural-language understanding routines, and the best strategy at low confidence thresholds was just to abandon any hope of resolving the input satisfactorily, dismiss the input, and demand the user cough up — er, produce — another token. More evolved systems no longer elect candidates on simple pattern-matching criteria. They draw on information from collocational, syntactic, semantic, and/or pragmatic sources, so that "rejection is a user interface decision that [is usually] performed at a much higher level than represented by the speech recognizer" (Balentine and Morgan, 1999: 20). In short, the acoustic-pattern-matching front end nearly always passes along some candidates and confidence scores to other components of the speech system.

When even partial sense can be made of the utterances, the interface has to play its best cards. In this example, from an automated TV-program guide, the system misses the mark a bit:

T$_1$ User Can I have a Western?

T$_2$ VIP [hears "Can I have a worse him?"] What about the following program: "Midnight Fear"?

T$_3$ User No, I want a western.

T$_4$ VIP I can recommend this program which is a western: "The Unforgiven."

(Cavazza, 2003: 123)

VIP (Virtual Interactive Presenter) does not have much resolved information to go on at T$_2$, beyond the fact that T$_1$ requests a TV-program recommendation. But it makes a reasonable guess, making a suggestion based on whatever domain knowledge and selection principles the inference engine has to work with (perhaps popularity statistics, or recency, or past usage patterns). It's off the mark, and the user initiates the required repair. No harm, no foul, and infinitely better at T$_2$ than something like "please rephrase." It does somewhat better with this guess:

T$_1$ User What kind of movies do you have?

T$_2$ VIP I have found 22 programs for this choice. I would suggest: "Twelve Monkeys."

T$_3$ User Who is playing the title role?

T$_4$ VIP [hears "Who is playing the title all?"] "Twelve Monkeys" features Bruce Willis.

(Cavazza, 2003: 125)

Here "playing" and "title" are enough to go on without "role," and the system gets it right. In both cases, VIP proceeds appropriately and the dialogues are successful, despite the minor hiccup in the want-a-Western exchange.

But there are two sorts of slippages that follow directly from the inability of the recognizer to assign any acceptable confidence score to input — out-of-vocabulary and spoke-too-soon slippages (née, errors). There is also a third interaction slippage, unrelated to rejection, the timeout. That gives us a three-way taxonomy of interaction slippages, as follows.

Out of Vocabulary

The user says a word or phrase that the system does not have among its acoustic models. This might be a case of (1) the user saying something completely alien to the system, or (2) the user saying something that is legal at some points in the interaction but not the current one (using a "weather" term, for instance, when speaking to the "sports" part of the system). That is, the word or phrase might be unrecognized because of the overall system vocabulary, or merely because of the current working vocabulary.

DanLuft   At which time?

User   Around noon.

DanLuft   Sorry. I did not understand.

User   Around noon, 11:50.

DanLuft   11:50.

(Bernsen et al., 1997)

Spoke-too-soon

The user says something when the system is not prepared to listen; that is, when the system does not support, or has temporarily disabled, turn-overlap.[1]

System: Do you want another transaction?

          Beep!

User: Yes.

System: Remember to wait for the tone.

          Do you want another transaction?

User: Yes.

System: Beep!

IBM (2001: 165)

Timeout

The user does not say anything within an allotted period (timeout threshold), usually a few seconds at most.

1: The "too-soon" here only concerns the physical signal. This slippage category (despite the mildly ambiguous definition I give) does *not* concern the case where a word that may be legal at a later point in the interaction is input before the system has loaded the appropriate vocabulary. That slippage belongs to the out-of-vocabulary (in system) category.

ABCBank	Welcome to ABC Bank. What would you like to do?
User	(silence)
ABCBank	You can check an account balance, transfer funds, or pay a bill. What would you like to do?
User	(silence)
ABCBank	Say one of the following choices: check balance, transfer funds, or pay bills.

(Yankelovich, 1996)

In addition to interaction slippages, there are also potential task slippages.

Task Slippages

Task slippages are problems that arise not because of the recognition engine, or because of generalized user behaviors ("wrong" word, silence), but because of specific functionality issues. For instance, the user of a voiced banking system might request information for which she doesn't have authorization (someone else's loan balance), or request an action which the system is not designed to handle (buy into a mutual fund), or to do something when there is a temporary access problem (access a credit-card balance which is on a slow or over-capacity server). She may attempt to transfer an amount that is greater than the balance, pay a bill that has no account, even withdraw a sum over the phone. Catherine Wolf and Wlodek Zadrozny (1998: 491) call slippages of this general variety *competence errors*, because the user is trying to do something that is outside the system's competence.

Task slippages often compound with other breakdowns as well, most notably with interaction slippages — so that, for instance, a request to pay a bill to Acme Rockets when there is no corresponding account in the user's profile might well involve an out-of-vocabulary breakdown, because there is no acoustic model for *Acme Rockets* in the system vocabulary associated with this user.

There is, of course, no universal taxonomy of task slippages. They depend too closely on the field of the register and on the task structure, but they pattern very much like what we would call *misunderstandings* in mundane human–human interaction. Take this familiar exchange between our old friends Nicole and SpeechActs:

T₁	Nicole	What do I have tomorrow?
T₂	SpeechActs	Tomorrow, you have "Dentist appointment."
	Nicole	What about Bob?

T₃	SpeechActs	On Tuesday, November 16th, Bob has "Visit to Georgia Tech."
T₄	Nicole	And on Wednesday?
T₅	SpeechActs	On Wednesday, November 17th at 2:15 pm Bob has "Return to Boston." At 5:15 p.m., Bob has . . ."
T₆	Nicole	Stop. Tell me what I have.
T₇	SpeechActs	On Wednesday, November 17th, you have no appointments.

(Yankelovich, 1994)

The task breaks down here (briefly) because the agents have different understandings about T₄, "And on Wednesday." This specific category of task slippage — conflicting beliefs about whose calendar is in focus — we imagine, would be a common misunderstanding in a scheduling application with multiple calendars. But it does not generalize to other domains. Each project, in short, should assemble its own, regularly updated list of task slippage categories, from the first stages of design on through testing, field release, and maintenance. Natural dialogue studies, testing sessions, beta-releases, and field studies are especially useful places to gather data on such slippages.

Overwhelmingly, however, task slippages are a species of one very general problem, reference fixing. In the SpeechActs example, the glitch results from elision, a common source. Nicole elides the bracketed phrase in "[what do I have] on Wednesday," but SpeechActs takes her to be making the elision, "[what does Bob have] on Wednesday." It takes them a moment, and the human has to activate the repair at T₆, but they fix the reference on Nicole's calendar, bringing it into focus at T₇.

Finally, there is the debilitating condition known as *error spiraling*, when several errors cluster together and break down the dialogue completely.

Error Spirals

The situation of errors coming in consecutive turns, usually because error *n* is insufficiently dealt with, triggering error *n* + 1, which in turn is unresolved, triggering error *n* + 2, and so on. Here's an example:

T₁	Danluft	Please state your customer number.
T₂	User	The customer number is four.
T₃	Danluft	Customer number four. How many persons will travel?
T₄	User	Two persons, one adult and one child.
T₅	Danluft	One person. Please state the ID-number of the person.

T_6 User   It was not one person, it was two persons, one adult and one child.

T_7 Danluft   ID-number 1, Jens Hansen. When does your journey start?

(Bernsen et al., 1998: 35)

Danluft gets the number of travelers wrong at $T_{4/5}$; then, when the user attempts to fix the problem at T_6, it is busy listening for ID-number and takes the attempted correction as the setting for another variable (T_7), moving on from routine to routine collecting erroneous variables. Mathew Marx calls this situation the "brick wall effect" (1995: 87), because from the user's perspective, that's what it can seem she has run into.

User   Read me the text of the next message.

MailCall   I didn't understand.

User   Read me the text of the next message.

MailCall   I didn't understand.

User   I said READ ME THE NEXT MESSAGE!

MailCall   I didn't understand.

User   Gee whiz. I just don't know what to say.

MailCall   I didn't understand.

(Marx, 1995: 88)

Task management problems like the Danluft example, or iterating rejection errors like the MailCall example, or substitutions/deletions/insertions like the cluster of earlier examples, are bad. There are two things you can do about errors and other slippages: prevent them and repair them.

Prevention

Method *goes far to prevent Trouble in Business: For it makes the Task easy, hinders Confusion, saves abundance of Time, and instructs those that have Business depending, both what to do and what to hope.*
— William Penn

The best prevention always originates in user studies, task- and discourse-analyses, and iterative testing. The more you know about the users, their tasks, and — what is much the same thing in this line of work — the discourse patterns implicated by those users in those tasks, the better chance you have of forecasting and strategizing your way around (or out of) potential trouble areas in the interaction; and, further downstream, the more you study simulated and actual interactions with versions of the design, the more you can catch, and avoid or recover from, emergent trouble areas. Study, plan, implement, test. Study, plan, implement, test. Study, plan, implement, test. That's the essence of design in the spiral development model, and that's also how you minimize both slippages and the pain from slippages that get through the spiral net.

But there are some specific ways you can utilize all that studying: prompting effectively, implementing a lexical density strategy, drawing on reserve synonyms and metonyms, and suitcasing signals for possible later resolution. Users can also be induced to speak the system's language, by way of lexical convergence; that is, they can be stealth trained.

Prompting

> *Prompt design is at the heart of effective speech interface design.*
> — Nicole Yankelovich

The traditional method of preventing slippages, inherited from the keypad systems, is explicit directives:

Telefónica   For information regarding airports, say "one." For information regarding buses, say "two." For information regarding trains, say "three."

(Wilpon, 1994: 298)

This method certainly has its virtues, and there are times when the very limited nature of the appropriate input, especially in gatekeeping situations, calls for this level of explicitness (or close to it):

SpeechActs   Welcome to SpeechActs. Please say your full name.

(Yankelovich, 1994)

The prevention-of-error price with relentlessly explicit prompting, however, is often too high to pay in terms of user satisfaction, and the art of prompt design is usually the art of hitting the appropriate Gricean "j-u-s-t r-i-g-h-t" level of information quantity. Too much, and the interaction style quickly becomes tedious; too little, and it devolves into confusion.

Prompts fall into a four-level taxonomy, as follows:

Explicit

The prompt tells the user precisely what input is appropriate, either by listing it exhaustively (if only a few words), or by representative example: "Do you want to leave between 6 and 9 p.m.? Yes or No?"

Implicit

The prompt characterizes the appropriate input in conceptual or linguistic terms: "Which type of restaurant would you like?"

Inferential

The form and content of the prompt implies the form and content of the appropriate input: "I can answer questions about Strindberg, the Royal Institute of Technology, and Stockholm."

Open

The prompt only conveys the most general and open expectations for user input, allowing context to shape the next turn: "How may I help you?"

Balanced against issues of user satisfaction, the best prevention method becomes "say just the right thing," which crucially involves saying it at the appropriate level of explicitness. Determining that level depends on how critical the task and/or the information is at the given point in the dialogue, considerations we will explore in more detail below, in connection with the closely related topic of feedback level.

Lexical Density

> The [speech system] understood that I wanted to fly out of San Francisco. But when I said I wanted to fly to "Hoboken," it first booked me into Mobile. When I tried to correct it, it booked me to Boston. It didn't know what "Hoboken" was, and it didn't know that it didn't know.
> — Birrell Walsh

It is always a good idea to populate the vocabularies, at all points of the interaction, as densely as you can without adversely affecting performance. Never go for the minimum possible model-count in a given vocabulary. Go for the maximum.

This lexical density strategy rests generally on thorough discourse-domain research and specifically on both a solid corpus and a comprehensive user-interface lexicon; as well as on technical matters of capacity, speed, and efficiency in vocabulary swapping. But it pays off well in terms of user satisfaction.

Take our Danluft error-spiral above, the one that begins like this:

T_3 Danluft ... How many persons will travel?

T_4 USER   Two persons, one adult and one child.

(Bernsen et al., 1998: 35)

Danluft misconstrues T_4 as setting the value for number of travelers at one, but if its vocabulary here had included the eminently reasonable words, *adult* and *child* (or rather, the Danish equivalents), it could come back with a more appropriate response, and not fall into error.

Judiciously maxing synonyms and metonyms in the vocabulary primarily means more flexibility for the user. You can say "Transfer 100 dollars from checkings to savings" and I can say "Move 100 dollars from checkings to savings" and we'll both get the system to subtract 100 dollars from our checking account and add it to our savings account. And we can do it in our own way, without having to learn the single correct term for that function which the system sanctions. *Judicious* here mostly concerns task structure. In an early use of lexical density (a few hundred words at any given stage of the interaction) for an automated phone-banking project, John Karat and his colleagues found that the route to high transaction success rates was to work "on accommodating a large number of ways of accomplishing frequent transactions (e.g., getting balances, making transfers), but fewer ways of accomplishing infrequent transactions (e.g., reporting a stolen credit card)" (Karat et al., 1999: 33).

Lexical density also helps to avoid the very annoying phenomenon of out-right utterance rejection, where the user is induced to repeat himself a few times, and is forced finally to realize the system doesn't have a clue what he is talking about. Let's say, for instance, that we have a routine that provides consumer information about tools, but only power tools. Chances are, people will come looking for information about tools the system doesn't know anything about. But it is better if the system knows it doesn't know about them, and judiciously stocking the vocabulary with statistically common tool words, even though there is no information about those tools in the corresponding database, can help avoid the brick wall. Compare these two scenarios:

Scenario 1: *shovel* is not in the vocabulary, because it is not in the database.

Caller   I would like some information about shovels.

System   Sorry, about what?

Caller   Shovels.

System   Could you repeat that please?

Caller   Shovels.

System   Sorry, I'm still not getting it.

           Etc., etc., etc.

Scenario 2: *shovel* is in the vocabulary because an inquiry about shovels is reasonable at this stage.

Caller   I would like some information about shovels.

System   I'm sorry, I don't have any information about shovels, just power tools.

Now, it's true, someone is likely to come along and ask for a micrometer, or calipers, or a sash ovoloe, or a fenks-rake — something that fits the category "tool," but for which there is little statistical reason to expect anyone to request. It will happen. There will be outright rejections of acoustic patterns that don't match up with anything in the vocabulary (or, more troublesomely, will match up with something falsely). But that is no reason not to inoculate the system against breakdowns when the user comes with a reasonable request that we can't satisfy. The idea is to minimize out-of-vocabulary rejections.

Notice that for this strategy to work well, we need the language model to encode the appropriate relationships — that, for instance, *shovel* is a KIND-OF *tool*, but that it is *not* a KIND-OF *power tool*.

Stealth Training

Do good by stealth, and blush to find it fame.
— Alexander Pope

The fact that the vocabularies are well stocked does not mean the system should not have a preferred diction (based, in particular, on recognition characteristics, within an appropriate naturalness window), and it does not mean that preferred diction should not be induced and reinforced. The system should always use a consistent set of its own preferred terms. *Move* and *transfer*, for instance, might both work to cue the action that inversely changes the ledger balance of two accounts, but the system should always identify that action with a single term (*transfer*, as multisyllabic, would generally get the nod). Making an offer to the user should take a form like this:

System   Would you like to transfer money, pay some bills, or check a balance?

Chances are (or, rather, convergence is) that even if the user's own preferred term is *move*, he will be induced to use the higher-percentage, system-preferred term. He is being stealth trained about the system's favored terminology.

Similarly, feedback, even if the user has used a synonym of the preferred term, should follow the system set:

User   I would like to move 100 dollars from my checking account to my savings account.

System   Certainly. . . . Transferring 100 dollars from your checking account to your savings account.

That is, you might initiate the ledger-change in your bank accounts with *transfer*, I with *move*, but if the confirmation comes back to both of us "Transferring 100 dollars from your checking account to your savings account," your diction is reinforced, while mine is induced to change, if not on the next call, then soon enough. I will be trained stealthily to speak the way the system wants me to speak.

Reserve Synonyms and Metonyms

> *If everyone always agreed on what to call things, the user's word would be the designer's word would be the system's word.*
> — George W. Furnas, Thomas K. Landauer, Louis M. Gomez, Susan T. Dumais

Overall system lexical density can also help with resolving recognition errors and generally improving interaction quality, if synonym and metonym vocabularies are held in reserve and loaded on cue. Remember the give-me-the-price-of-AT&T brick wall from Chapter 5? The user repeats "AT&T" unsuccessfully multiple times, trying to get its stock price, before using a reasonable paraphrase, "American Telephone and Telegraph." At the first sign of trouble, the system should have brought in a dynamically assembled synonym and metonym vocabulary, keyed to the candidates in the stochastic list(s). Assuming that *AT&T* was actually in the list(s), such a vocabulary would have allowed the paraphrase to work.

Take the example of a voice portal for general consumer information, and of the word *monitor*, which participates in two consumer-product domains, computers and babies. Following lexical density, the top-level vocabulary for such a site would be populated with a variety of general and specific terms; as a statistically popular specific consumer product (in fact, as two statistically popular consumer products), *monitor* is represented at the top-level vocabulary. When *monitor* is caught, the routing needs to be toward a specific subsystem, baby consumer goods, or computer consumer goods. Here are three scenarios.

Scenario 1: *monitor*, treated as both ambiguous and overly specific, is left out of the top-level vocabulary, and standard rejection repairs are in place.

Caller	I want some information on Fisher-Price monitors.
	[The system, at this stage, hears "blah information blah blah blah", since none of the words are in the vocabulary.]
AcmeShop	Sorry. I didn't get that. What type of products would you like to hear about?
	Etc.

Scenario 2: *monitor* is in the vocabulary, a standard disambiguation prompting strategy is in place, but there is no reserve vocabulary.

Caller   I want some information on Fisher-Price monitors.

             [The system, at this stage, hears "blah information blah blah monitors."]

AcmeShop   OK. What type of monitors would you like to hear about? Computer monitors or baby monitors?

Caller   Baby monitors.

             Etc.

Scenario 3: *monitor* is in the vocabulary, the disambiguation reserve vocabulary comes in.

Caller   I want some information on Fisher-Price monitors.

             [The system hears "blah information blah blah monitors," and *monitors* cues the reserve vocabulary, which includes major manufacturers as metonyms; the call is promptly routed to the baby consumer goods subsystem.]

AcmeShop   Certainly. . . . Just getting that. I have information on four Fisher-Price models. Would you like me to list them, or is there one in particular you want to hear about?

In Scenario 1, the system seems deaf (and it is). In Scenario 2, it seems inattentive. In Scenario 3, it seems natural and responsive, the way a human would be if you phoned her up.

The Suitcase Strategy

Words can help us move or keep us paralyzed.
    — Adrienne Rich

Staying with the baby-monitor scenario a moment longer, how did the baby consumer goods subdialogue system "know" about Fisher-Price in Scenario 3, or even about monitors? The top vocabulary gets the call routed to the appropriate subsystem, but traditionally when the call arrives at the domain-specific subsystem, the caller has to start over (as in Scenario 2; this happens consistently even when dealing with human-agented call centers; almost always, by the way, this is occasioned by a change in agents, an option that is available to automated systems, though I ignore it here):

Caller   I want some information on Fisher-Price monitors.

             [The system routes the call to the baby consumer goods subsystem.]

AcmeShop   Certainly. Baby products. How may I help you?

But computers have something up on humans in this regard: extremely efficient memory transfer. Human-agented call centers don't often route calls with much content attached. They work like old-fashioned switchboards with an operator who hears the number, sticks the plug in the right hole, and otherwise stays out of the call altogether — something that can mean frequent repetitions for the caller. But computers can get the data, hang onto it, and swap it among systems at blinding speed. (With people-stocked call centers, one agent would have to either talk to another, or physically input data and send it to another's screen — using up time, which costs money and delays response.)

Swapping extracted data around is common enough in voice systems. What is less common, but equally valuable, is swapping *raw* data around — jamming it in a suitcase until you can do something with it. What this means is if you can resolve enough of the signal to get strong clues about routing it in the call flow, but you can't resolve it all, the whole input signal is carried along that route to the appropriate subsystem. Effectively, this technique means hanging on to the input until you're in a better spot to figure it out.

The effect of suitcasing is similar to using reserve synonyms and metonyms, in the sense that a more strategic vocabulary is accessed to deal with the utterance, but suitcasing is keyed more to the structure of the database than to the terminology.

Take the case where we have *chairs* in the top vocabulary of a voice interface for a furniture outlet, but there isn't enough room for subcategories of chairs.

Scenario 1: *no suitcasing*, the raw input signal is discarded once *chair* is extracted.

Caller   Do you have any deck chairs?

[the system gets "blah blah have any blah chairs" and routes the call to an information routine for accessing a database of chairs]

System   Yes, just a moment. . . . We have over sixty chair models. What type of chair would you like to hear about?

The caller will either be annoyed that he has to repeat "deck chair," or he will assume that the system is asking (elliptically) what type of deck chair he wants to hear about, leading perhaps to an answer like "wooden ones," and subsequent perplexity when the system begins telling him about, say, wooden dining room chairs. In either case, there is an associated cognitive burden. Impressions of the agent's, and therefore the company's, character, will not be especially positive.

If the utterance is suitcased and toted along, we end up with (vastly preferred) exchanges like this one:

Scenario 2: *suitcasing*, the raw input is retained after *chair* is extracted, and at the chair-subsystem, with its more specific vocabulary, *deck* is then extracted.

Caller   Do you have any deck chairs?

[the system gets "blah blah have any blah chairs" and routes the call to an information routine for accessing a database of chairs; now in the chair-information routine, the signal is processed again, returning "blah blah have any deck chairs"]

System   Yes, just a moment. . . . Would you like a list of our deck chairs? Or do you have something in mind I could search for?

Or, take the case of a general information voice portal (including, say, entertainment, business, weather, and sports modules). A caller asks, while still at the initial routing level, "How'd the Steelers do last night?" Let's say *Steelers* is not in the initial vocabulary, but the colligation "do last night" is, as a cue for sports. The input signal is put in a suitcase, shipped off to the sports module, which has a vocabulary that does include *Steelers*; now it gets resolved, and with high speed processing, the exchange comes off like this:

Caller   How'd the Steelers do last night?

[the system gets "blah blah blah do last night" and routes the call to the sports module, where it now gets resolved as "blah blah Steelers do last night"]

System   Just a moment. . . . They won. Steelers 34, Patriots 26.

So, the suitcase strategy — effectively, don't throw away raw input until you've extracted everything you can from it — gives you good error avoidance, more storage space in the higher-level vocabulary (which doesn't need *deck* and *dining* and *easy* and all the other varieties of chairs to deal adequately with calls about them, or all the NFL, NBA, NL, AL, and NHL teams to answer direct questions about them), improved error correction, and a generally saner interactivity.

It doesn't, however, nor do the other strategies we've taken up in this section, or do anything about recognition errors. These techniques — keeping the vocabularies well stocked, inducing the user to adopt preferred terminology, making reserve synonyms and metonyms strategically available, and hanging onto the raw input signal until you've got all you can out of it — can help prevent slippages, especially through their potential for managing out-of-vocabulary incidents. But they are powerless against recognition errors. Even out-of-vocabulary glitches will happen. Someone will ask for a fenks rake. And substitutions are insidious; indeed, the prime strategy for avoiding out-of-vocabulary incidents, lexical density, increases the substitution possibilities.

Prevention is important, but prevention of recognition errors is ultimately more of a technical issue than a human-factors issue, relying on the machinery of the speech application. Interface design research can certainly help, with good diction and prompting, as well as with corpus results, even with harvesting training data. But recognition errors will happen. They need to be repaired.

Repair

A carefully crafted user interface can overcome many of the limitations of current technology to produce a successful outcome from the user's point of view, even when the technology works imperfectly.
    — Candace Kamm

It is well known in learning theory circles that errors can be very productive:

Errors can . . . make positive contributions — by providing feedback that helps overcome misunderstanding, improving mindfulness and discouraging carelessness. . . . Though momentarily frustrating, the negative effects of errors are often counteracted by their help in shaping new skills and signaling progress
    (Screven, 2000: 176).

I'm not saying we should plant traps or promote errors, as learning theory sometimes encourages; there are enough opportunities for slippages in voice interfaces without planting them. But it is worth keeping in mind that errors and slippages are not only common, they can be wholesome.

There is a widespread fear and loathing of errors in voice interface design — a pervasive sense that since things go wrong lots anyway, and since we can't make speech systems as perceptive and fluent as human conversationalists, that we should keep the interaction as primitive as we can, to prevent false expectations, and stay as far away from trouble spots as possible. But we need to take our lead from learning theory, and regard errors as opportunities for shaping the interaction, not as occasions to punish the user into an anxious and tentative dialogue style.

The principal move we have to make, in fact, is to give the user more latitude, not less. There are certainly legitimate occasions for screwing down the interaction to a yes/no interaction, but screwing it down should not be our first impulse. The first impulse should be diagnosis, followed by facilitating the user's repair work.

Recognizers, for instance, can be made more perceptive to changes in amplitude, and tempo; more sensitive to pauses, and repetitions, and to vocables of deferral (*oh, um, well*); natural-language understanding modules can be made on more aware of what an out-of-vocabulary word might mean in a given context; and so on. And the dialogue management strategies need to accommodate these sources of information. In particular, the system must have a repair manager, a conceptual module that is dedicated to resolving communicative slippages.

From the perspective of design, there are really only two forms of recognition errors, known (or guessed-at) errors and unknown (and unguessed-at) errors. Known errors are epitomized by rejections: the system figures something is wrong and initiates a dialogue move to deal with it. Timeouts are in this category, as are spoke-too-soons and classic rejections. They are *nonrecognition* errors, because the system fails to recognize anything coherent. Unknown errors are epitomized by substitutions: the system gets input, is happy with it, and simply follows through on whatever dialogue action it thinks is called for; but it has misunderstood. They are misrecognition errors, because the system recognizes something, but it's the wrong thing. We'll take up both of these error groups in turn, the rejection group and the substitution group, starting with nonrecognitions.

Repairing Nonrecognitions

Words fail, there are times when even they fail.
— Samuel Beckett

The bad thing about nonrecognitions is that there is a breakdown. The good thing about nonrecognitions is that everyone knows there is a problem. They always trigger a system utterance that clearly flags its confusion.

Rejections

I don't like [sounds] to be overarticulated with too much affectation, and I don't like them to be obscured by being pronounced too carelessly; I don't like words to sound thin by being produced with too little breath, and I don't like them to be puffed up and uttered, as it were, with too full and heavy a breath.
— Marcus Tullius Cicero

Vincent Vanhoucke illustrates rejection handling with this exchange

T_1 System   Which type of restaurant would you like?

T_2 User   A cheap one.

T_3 System   Sorry, I didn't understand. You can say: Mexican, Italian, French, . . .

(Vanhoucke et al., 2001)

We can already see that prevention strategies might have helped avoid the T_3 glitch. "A cheap one" is a reasonable response to T_1, so the system might either have been prepared for it (lexical density), or have made its T_1 offer more specifically — "Which type of cuisine . . ." — or both. But pretend the user says "One where I can get lark's tongue in aspic and raw whale blubber" or something else that *isn't* reasonable for the recognition engine at this point.

What the system's response (T_3) shows is the traditional, and often the best, first-level response to an out-of-vocabulary utterance: an apology, followed by a directive suggesting appropriate in-vocabulary language. The apology is crucial, not so much for reasons of personification or social-orientation, which just come along for the ride. It is an expressive dialogue act, and it does present the agent/company in a position of goodwill toward the user. But its most immediate function is to be a clear linguistic marker of impending disappointment, what Cohen, Giangola, and Balogh call "a mild bracing advisory" (2004: 147). It is feedback. For highly experienced users, just that may be enough — a brief "Sorry?" But for many users, especially ones you can assume are novices, the explicit directive is often equally crucial, either as a reminder or an instruction. It is a prompt.

If this is the first encounter between user and system, perhaps immediately following a brief greeting, then T_3 is a very solid choice. By way of a representative example, and trailing intonation, it tells the user clearly the sorts of search terms he should be using. Under other considerations (repeat use, earlier failures/repairs, and so on), the system's response should change (along the lines specified later in this chapter, in connection with expanding and tapering system utterances). But in these circumstances, this feedback is the best way to get the dialogue on track.

Van Houke's T_3 (or, specifically, its second sentence) is an explicit prompt, one that directs the user as to exactly what he should say; in this case, by listing representative in-vocabulary words.

The first-level treatment of timeouts should usually be the same.

Timeouts

The opposite of talking isn't listening. The opposite of talking is waiting.
— Fran Lebowitz

Speech takes place in time. When it stops taking place, even incredibly briefly, when there is no speech input at a junction that seems to call for speech input, it often means something has gone wrong. Not always. The caller may be momentarily distracted, or is just thinking about an offered choice, or is looking for a take-out menu she just put on the table a moment ago. But you can't know, and the best option is almost always to assume there has been a slippage and try to re-engage her.

In fact, just eliminating undue silence can avoid potential errors and user dissatisfaction. A three-second delay, for instance, is an eternity in a phone conversation. If the system is not ready for input, whatever the reason, the user needs some feedback to signal that the delay is purposeful — a tone, a sorry-just-a-minute turn-holder, anything. Without feedback, the speaker can quickly get very frustrated

HMIHY	How may I help you?
Caller	Calling card call.
	⟨two seconds⟩
	Ah, to area code 908 949 1111.
	⟨two seconds⟩
	C'mon you stupid machine.

(Boyce, 1999: 58)

368 Chapter 13 Dialogue Matters

HMIHY comes off as a stupid machine here, not to say rude: it has asked a question, received a response, and is dumbly refusing to take its turn. The caller has no sense that it has broken off trying to resolve "Calling card call" and is now working on "Ah, to area code 908 949 1111," which it then abandons for what will surely prove out-of-vocabulary, "C'mon you stupid machine."

If the delay is on the other side, if the system has uttered something that calls for a response and isn't getting one, it is rarely best to wait very long. The standard threshold is actually what the HMIHY caller observed: two seconds.

Bruce Balentine and David Morgan exemplify timeout handling with a two-agent response (system output is shaded grey):

Female	Main Menu ⟨beep⟩
User	⟨two second timeout⟩
Female	Please say one of the following . . .
Male	Balances . . . Quotes . . . Purchases . . . Help . . . Operator . . .
Female	Choice?
User	Quotes

(Balentine and Morgan, 1999: 154)

Timeouts are different from flat rejections in that there is no input to respond to. Nothing has been mishandled by the system. No apology is called for. But otherwise, the first-level treatment should normally be the same as for a simple rejection.

Under grounding assumptions that the user is unsure about what to say, with nonresponse durations of about two seconds, the system should come back with an explicit prompt that lets the user know what is now acceptable. This interaction is from a menu-driven system, with a heavy directive style, but the spirit of the response is right on the money, and the use of two agents — one to conduct the interaction, the other to represent input options — is a good touch for clarity and variety. It is not unlike a change in typography of the sort that is common in computer documentation for distinguishing instructions from input/output examples.

The principal exceptions to a rejection-like response are when (1) the system has sufficient confidence in the user's goal, and the risks associated with being wrong are sufficiently low (that is, the grounding criteria are low), (2) there is reason to believe the timeout is not due to user confusion, or (3) there is some reason to believe the user needs more time.

In the first case, the system should just go ahead with its hypothesis:

T$_1$ HMIHY   How may I help you?

T$_2$ User   What's the area code for Chicago?

T$_3$ HMIHY   You want an area code.

T$_4$ User   ⟨timeout⟩

T$_5$ HMIHY   The area code for Chicago is 312.

(Gorin et al., 1997: 115)

HMIHY seeks a confirmation at T$_3$, but it has enough confidence that it has extracted the user's T$_2$ goal, reckons the mild aggravation that would come with being wrong to be negligible (probably outweighed by the aggravations a delay and further turns would trigger if it is right and forces the user to repeat himself), and it ignores the timeout.

In the second case, the only reason you might have for believing the lack of response is not due to confusion about how to proceed is because you have another hypothesis — presumably informed by user and task analysis. Here is an example where the designers think reluctance is a more likely hypothesis than confusion, and rather than a directive, they provide an explanation (in coherence-relation terms, a justification):

Thrifty   At what airport or city are you picking up the car?

User   ⟨timeout⟩

Thrifty   Sometimes, rates and availability can depend on the location where you're picking up the car.

(Kotelly, 2003: 82)

In the third case, when you think the user might need more time, the best policy is just to give it to her — that is, to adjust the timeout threshold upward — though the system response will still be a clear directive about what the user's options are should the threshold be reached. The idea is that there are other possibilities than user-uncertainty for timeouts. Two seconds is standard, but some systems set it lower, some higher; 1.5–3 seconds is an acceptable range. Three seconds is a long time in conversation, especially over the phone, but when you have reason to believe that there could be a delay at the user's end — perhaps after asking for a credit card number, when 5 or 6 seconds for someone to fish it out of a purse or wallet, or utter a "hold-on," could be appropriate — then granting the time is only courteous.

The default, however, is for rather quick intervention. There's no guarantee that explicit intervention is the best course of action for the user under those circumstances, just the best that can be offered on the basis of limited knowledge: we do not know why the user didn't say anything, but we do know there is a slippage between user behavior and system expectation; the best option is to encourage the user to behave as the system expects.

Adjusting the threshold downward, too, is sometimes reasonable as well. In particular, a kind of intra-utterance pause can be appropriate, a brief invitation to respond, followed by an explicit directive of how to respond:

AcmeBank   Which account? ⟨0.75 sec⟩ Checking or savings?

People usually take up turn-transitions very quickly (0.3–0.4 seconds), and a duration significantly longer than that when the response can be characterized in a few words ("Yes or no?" "Take-out or delivery?") can be an opportunity for brief assistance with novices. (An enabled turn overlap is of course essential in such circumstances, or the system and user can repeatedly clash.)

Spoke Too Soon

Learning too soon our limitations, we never learn our powers.
    — Mignon McLaughlin

The best handling of spoke-too-soon errors is to institute a widespread policy of turn overlap. Spoke-too-soon slippages, in the traditional sense of the user speaking when the system is not technologically capable of listening, are not reasonable in any speech system aspiring to be conversational. In Marx's (1995: 19–20) survey of early speech-only system research, for instance, what he termed "interruptibility" was seen as a "necessary feature," "important," a "key component" of designs relying exclusively on speech.

But it should be a policy, not a law. Turn overlap should almost always be available, but it need not be implemented in all circumstances. The user needs to be able to exert a large degree of control over conversational systems, but that does not always mean terminating the system's current output and taking the initiative. Just as in many human–human interactions, especially in institutional contexts, there are times when one agent needs to listen — in speech systems, these situations may arise with terms-of-usage statements, disclaimers of various sorts, and other legal or quasi-legal messages. The user should still be able to interrupt — freezing her out with no channel whatsoever is just rude. The system needs to hold the floor, but does not have to be belligerent to the point of deafness, or (as one designer has suggested) raising its voice to drown out the attempted interruption. A short message that says something like, "sorry, before we can go any further, you have to listen to this statement," along with an indication of any options (if there are any) the caller may have at this point ("going back," terminating, proceeding under demo mode, whatever).

In some command-and-control speech applications, in some circumstances, with moded interaction, there are also technological and contextual reasons for not allowing turn overlap. In these cases — where there may be genuine spoke-too-soon errors — the onus falls heavily on the user to understand the context enough to realize why otherwise legitimate input is now frozen out. But even in these circumstances, a moded interrupt option should be allowed, to access a help system, sound an alarm, or initiate some other high-level override. The user is the intelligent part of the system, and should always have some level of control.

Nonrecognition errors — rejections and timeouts — present problems, but they also present opportunities. When they happen, the system can remind, guide, or instruct the user in its functions and expectations. Spoke-too-soons historically belong to the category of nonrecognitions, but they have never presented much of an opportunity for encouragement, just punishment. To the extent that they need to be retained, for legalistic output or limited command-and-control functions, they should always be implemented as gently as feasible. In particular, spoke-too-soon circumstances should allow for a limited insertion-sequence overlap.

Misrecognitions — insertions, deletions, substitutions — similarly present both problems and opportunities for the design of voice user interfaces. But the potential for disaster is greater, since the system doesn't know it is wrong.

Repairing Misrecognitions

> *Substitution errors are insidious if not detected. The recognizer thinks it's right but it's not.*
> — Robert D. Rodman

The bad thing about misrecognitions is that the system doesn't know that anything is wrong. The good thing about misrecognitions is that their repair depends on an intelligent, cooperative user. Fortunately, you can count on the intelligence and the cooperation. You're dealing with a human trying to accomplish a goal. But there is another bad thing about misrecognitions. The human must also be alert, and alertness is less dependable:

T_1 ADAP ADAP Travels, can I help you?

T_2 User When is the first morning train from Frankfurt to Hamburg tomorrow morning?

T_3 ADAP The first train from Frankfurt to Hanover on 3rd May 1998 leaves at 5:35 a.m.

T_4 User Thank you.

(Bernsen and Dybkjær, 2000)

In case *you* weren't fully alert (I confess, I missed it the first time through), here's what just happened: the user asked (T_2) for the departure times of Frankfurt-to-*Hamburg* trains, but ADAP provided (T_3) departure times of the Frankfurt-to-*Hanover* trains, and the user hung up, contentedly misinformed about the departure times he asked for (T_4). There was a substitution error.

The fact that it was a substitution error, however, is less important than that it was a misrecognition. While it is useful to keep the deletion/insertion/substitution taxonomy in mind, as a way of conceptualizing what can go wrong between input and output, for system-response design they are all effectively the same (see Hone and Baber, 1999: 95). Deletions substitute nothing for something, insertions substitute something for nothing; classic substitutions differ only because they substitute something for something. They all, in any case, have the potential to cause the sort of bad news Rodman warns about specifically for substitutions: "erroneous data input, spurious transactions, incorrect commands, and other kinds of trouble" (1999: 140). For instance, in our deletion example earlier, the system "substitutes" the date ⟨September 20⟩ for the date ⟨September 29⟩. In the insertion example, the system substitutes ⟨sports statistics⟩ for ⟨clearing throat⟩. The system, that is, thinks it has what it needs to do its job, but it doesn't.

The most urgent problem for misrecognitions, then, is finding them. The user will let you know if he notices. You have to be prepared to listen and help.

Diagnosing Slippages

> *Interaction requires at least two [agents]. Conversation is produced not simply by their presence, but also by the display of their continuing agreement to pay attention.*
> — Pamela Fishman

The most critical job in error control is not in avoiding the first one. It is in avoiding the brick wall. The whole design of an interface, any interface, is to focus users toward the best-supported tool behaviors (which should, of course, match up closely with the most probable and effective task behaviors). Sidestepping errors and slippages is standard operating procedure for any well-designed interface. There is nothing peculiar about this goal to voice interaction design. What *is* peculiar is the number of errors that *can't* be avoided by the interface, no matter how well it is designed.

No matter what you do there will be glitches in voice interactions. It is the nature of the beast.

The most important error-control activity you have in voice-interface design is avoiding the second error in a row, or (if the second gets by you) the third, or the fourth. The most important error-control activity is to avoid spiraling errors, an endemic and debilitating condition. James Glass, for instance, says that when one utterance is rejected in the Jupiter weather service, "it is highly likely that the next utterance will be rejected as well" (1999), and at that point some interactions have a tendency to go into what he calls, "a rejection death spiral." The first error in such a spiral will often be the fault of the recognizer. But the subsequent errors, the spiral unto death, "must be laid at the feet of the dialogue" (Balentine, 1999: 210). Traditionally, speech systems have done a poor job in this area, and many of the strategies that speakers adopt to *improve* recognition and/or understanding have given recognizers the fits, degrading recognition accuracy at exactly the point when users are getting frustrated, and doing what they can to initiate a repair.

The road that avoids error spirals, and that provides for a generally more satisfactory encounter overall, is early diagnosis and repair. Let's look at Gina-Anne Levow's give-me-the-price-for-AT&T example once again, in more detail:

T_1 User   Give me the price for AT&T.

T_2 System   Hewlett-Packard is at 83 3/8, up 2 3/8 since yesterday.

T_3 User   Give me the price for AT&T.

T_4 System   Data General is at 63 1/4, up 1/3.

T_5 User   Give me the price for AT&T.

T_6 System   Hewlett-Packard is at 83 3/8, up 2 3/8.

T_7 User   Give me the price for American Telephone and Telegraph.

T_8 System   Excuse me?

(Levow, 1999)

This exchange does not exemplify the virtues of early diagnosis and repair. The user spends the whole exchange trying to accomplish a repair (of recurrent substitution errors) while the system remains oblivious to the error and the attempted repairs until the last turn, T_8, when it can only come up with the lame "Excuse me?" And even here, it has really only clued in to a T_7-specific breakdown, deaf to the T_3 and T_5 clues, and blind to its own failings. We have, of course, a breakdown in referential coherence, the user referring to one thing, the system referring to others.

A logic of referential coherence that compared subsequent user-turns to one another might have detected the slippages earlier. The system hears T_1 and T_5 as the same, and while there is some possibility that a user could ask for the same information in two very proximal turns, there should also be flags going off that something may be amiss.

With T_7, the user finally gives up the repetition strategy and tries another favorite repair technique of speakers, paraphrase, presumably triggering an out-of-vocabulary failure. At this point, the system should be capable of more than a daft "Excuse me?" It has three previous turns to refer back to, two of which it interpreted as identical (a flagable event), and the other signals should at least have had some of the same candidate patterns in its recognition list. Now it gets a recognition failure; this should be adding up. Exactly how this breach should best be repaired is a matter of overall design and testing, but let's consider some of the possibilities.

One immediately apparent strategy that would help the system accommodate the user's repair strategies is the one we considered above, the use of *American Telephone and Telegraph* as a synonym for *AT&T*, along with some other potential disambiguators, like *telephone* (as in "the telephone company"), drawn from a pool of corpus-identified metonyms. Now, the primary vocabulary probably will not have room for synonyms and metonyms, especially in a stock-market service (the New York Stock Exchange, for instance, lists almost 3,000 companies). But appropriately designed dynamic vocabulary swapping could resolve the problem: bringing in a bank of acoustic models for reserve synonyms and metonyms, assembled specifically for the repair at hand from the relevant candidate lists at the slippage point. For instance, let's say that upon getting T_7 as input, the previous three candidate lists included AT&T, HP, ABT, ACG, AEE, and Data General: the repair manager would assemble a vocabulary of the reserve synonyms and metonyms (collocations would probably be pointless in this simple command-and-reply environment), swap them in to disambiguate the input (that is, T_7), and come up with something more pointed and helpful than "Excuse me?"

But we don't even want the interaction to get as far as T_7. We would prefer to stop the bleeding earlier. Let's go back to the T_5 input. At this point, the system has two inputs in three turns that it regards as identical, an occurrence which should raise flags. Remember what MailCall did in very similar circumstances: "I thought I heard 'Nat Parker' again, but you just said that. Is that right?" (Marx and Schmandt, 1994). The phrasing may not be optimal, but the spirit is right on the money.

There is some possibility that the caller does want information on the same stock twice in three turns. Statistical investigation at the task-analysis stage can uncover those probabilities. If two repetitions in three turns are extremely rare, it might be worth going down the candidate lists for the two utterances recognized as the same (perhaps all three), or maybe directly initiating a repair.

Still, we would prefer not even to get to T_5. The transcript tells us that there is trouble at T_3, and the system might have been able to respond more appropriately at T_4 than by providing the unwanted stock price for Hewlett Packard. The system was oblivious, missing the repetition altogether. Is there anything it could have done at this stage (other, that is, than just have better pattern recognition)? Perhaps. The system does not hear T_1 and T_3 as the same, and of course, they *weren't* the same (no two utterances are identical). But they may well have had some of the same candidates in their recognition list, something else the system should be attuned to. More significantly, T_3 may well have had prosodic cues that signaled a slippage as well. People very frequently give increased salience to any significant elements in a repeated utterance, especially if the repetition is because of a misperception. T_3 might be represented better visually like this:

T$_3$ USER   Give me the price for *AT&T*!

Levow comments, not of this dialogue specifically but of error spirals generally in the field study it came from, that several salience-related acoustic repair signals were present in repetitions.

Acoustic Repair Signals

When resolving errors with a computer, it was revealed that users actively tailor their speech along a spectrum of hyperarticulation, and as a predictable reaction to their perception of the computer as an "at risk" listener.
— Sharon Oviatt, Margaret MacEachern, Gina-Anne Levow

Utterances T$_1$ and T$_3$ in the just-give-me-AT&T dialogue transcribe identically (into English orthography). They weren't. We know, of course, that no two utterances are identical. But just from looking at this transcript we can tell T$_1$ and T$_3$ are sufficiently different that the recognition engine returned two different matches, both with sufficient confidence for the system to respond directly (rather than, say, reject one utterance or the other). What was the source of the difference?

Levow remarks that repetitions in the study this dialogue came from were accompanied by "(1) significant increases in duration, (2) increases in pause measures, and (3) significant decreases in utterance-wide normalized pitch minimum" (1999). T$_3$, that is, was probably slower, or had greater pitch range, or perhaps was louder (a salience marker Levow doesn't mention, but that is highly common in such repetitions), or, what is quite likely, was all three — or perhaps just the *AT&T* part was acoustically boosted. That would certainly account for the recognition failure (though not the exact substitution error): if I say *rejection*, at normal pace and volume, my dictation system transcribes "rejection;" if I say it slower, with an altered stress pattern — *ree-jection* — my dictation system transcribes (quite reasonably) "read to action." Early conversational systems reacted to these articulatory shifts by asking the user "to speak normally and clearly" (Yankelovich et al., 1995).

The thing is: longer, louder, more pitch-various utterances *mean* something. They are very significant elements of the feedback that speakers provide when glitches occur, flashing amber caution lights straddling the pothole that has knocked the dialogue off course.

They indicate that a repair is underway. Collectively, they are known as hyperarticulations — systematically distorted articulations, exaggerating the acoustic signal to combat noise. Shouting is a distortion, for instance, to overcome loud background sounds, but *noise* in communication theory is not confined to sound. It is, more generally, anything that interferes with the message signal: distance, in this sense, is another type of noise that we shout to overcome. We also speak more slowly to overcome the noise of an interfering language when our hearer has a different first language; we also sometimes use slowness with children, to overcome the noise of a not-fully-developed language. People hyperarticulate when they haven't been heard, or they have been misunderstood, or if they have reason to suspect one of those outcomes. Here is an example from the SUNDIAL project:

System   Please tell me where the flight leaves from.

Caller   Ibiza.

System   From Cairo?

Caller   I-b-i-z-a

         [0.5 sec]

         IBIZA!

(Hutchby, 2001: 168)[2]

Since hyperarticulation is predictable with recognition failures, and since recognition failures are endemic of speech systems, these facts need to be brought together in interface design. Moreover, the motivation behind a speaker's articulatory exaggerations is to benefit the hearer: they are on the whole clearer, slower, and louder — all traits recognizers can capitalize on, if they are designed accordingly. Sharon Oviatt has made two very important recommendations for vocabulary management with respect to hyperarticulation (1998: 627):

1.   Train recognizers on speech samples that include hyperarticulated repair patterns, to generate "regular" vocabularies that can deal with them.

2.   Swap in special hyperarticulation vocabularies at points in a dialogue where slippages occur, or where there is reason to believe slippages are imminent; for instance, repetitions, or possible repetitions, or out-of-vocabulary tokens following (possible) repetitions.

And a third recommendation suggests itself, in the spirit of Oviatt's proposals:

3.   The system should track baseline values for volume, pause measures, and syllable duration; that way, when marked deviations occur in any or all of these values, the system can process the input accordingly (perhaps by swapping in one of Oviatt's

---

2: I have altered the rendering very substantially here (Hutchby gives it in conversation-analysis conventions): I-b-i-z-a is intended to suggest articulatory length, IBIZA! is meant to suggest increased forcefulness — that is, volume and pitch range.

hyperarticulation vocabularies). As systems become more sophisticated, another value to track in this connection, and monitor deviations from, is pitch variation.

Any one of these three strategies may have caught and resolved the just-give-me-AT&T error spiral before it got started. Without the raw data, of course, there's no way to know exactly what is going on with the T_3 slippage. But if it *was* a hyperarticulation that triggered the second misrecognition — it was recognized as neither *Hewlett-Packard* (as the user's first "AT&T" utterance was), nor as *AT&T* (the target of both T_1 and T_3) — then the first-line vocabulary may have caught it directly (the intention of Oviatt's Recommendation 1), or it may have caught enough of it to pull in the dedicated hyperarticulation vocabulary (Recommendation 2); or the baseline monitoring might have pulled in the dedicated vocabulary (Recommendation 3, leading to Recommendation 2).

These are critically important recommendations, but there is a more easily available strategy for diagnosing problem areas for systems more concerned with vocabulary than prosody: watch the words.

Lexical Repair Signals

I'm proud to be his partner. We've had triumphs, we've made mistakes, we've had sex.
— George Bush Sr. (about Ronald Reagan)

People make mistakes when they talk. When we catch them, we correct them, and we catch them in distinctive ways. When we catch others making mistakes, especially about what we've said, we correct them, or invite them to correct themselves, and we do it in distinctive ways. As we've just seen, we make substantial prosodic adjustments to signal the error and the correction. We also make lexical adjustments — repeating words, interjecting vocables, and building frames. These are clues not just to other people, but to our machines, if we build them to listen.

When we correct ourselves, we do a kind of post-hoc edit, adding a new word or phrase to replace the one that contains the error, like George Bush Sr., who added, a few hundred milliseconds after he uttered the epigram for this section, "Setbacks! We've had setbacks!" (Andrews et al., 1996: no. 9394). On its own, this is tough to tune for, but the error always comes first and there are substantial acoustic cues — often the replacement has increased salience, especially volume and duration, and it is usually preceded by a brief pause, an interruption to the regular flow. The formula that Gail Jefferson uses to describe these sorts of edits is [WORD1 + HESITATION + WORD2] (1974: 186).

The hesitation might be silent, or it might be a "filled pause," with vocables like *oh*, *um*, and *er*. Hunting for these sorts of wordlets is difficult, because they are often vocalized quite softly, and recognizers easily confuse them with monosyllabic prepositions and articles, or miss them altogether, but they are self-correction clues, and the presence of possible minor vocables in a hesitation position is a clue.

Self-corrections, however, are less of an issue with voice interfaces than other-corrections, where the lexical cues tend to be somewhat more obvious. As our give-me-AT&T example illustrates, verbatim or near repetitions are very common. Here's another instance:

System	Here is the flight from Pittsburgh to Philadelphia leaving at 6:00 p.m.
User	Show me other flights leaving around six p.m.
	[The system resolves *at* instead of *around*.]
System	I'll show it to you again.
User	Show me other flights leaving around six p.m.

(Stifelman, 1993)

This system had a good enough inference engine to realize that the request it was satisfying was identical to the preceding one (note the *again*), but that should have been a clue to re-process the signal, which may have yielded the *other* and *around* it missed the first time through.

Special error-flagging words are also very common. Many systems have dedicated commands (like Danluft's "Change!" or SpeechActs' "stop") for correcting a misrecognition, but even when there is no dedicated command, users will usually preface corrections with "Stop!" or "No!" especially if some action is imminent upon the misrecognition; indeed if the error-flagging command is not well chosen, they will often get "Stop!" or "No" anyway, as in this flag of a substitution error:

SUNDIAL	When would you like to leave?
Caller	Next Thursday.
SUNDIAL	Next Tuesday the 30th of November?
Caller	No, Thursday December the 2nd.

(Bilange, 1991: 85)

Frequently, too, there are colligational phrases prefacing the correction. For self-corrections it is usually something on the order of "I meant to say X" or just "I meant X"; for other-corrections, it is "I said X" or "I was referring to X." Here's an example:

User	I'd like to fly from Seattle to Chicago on December twenty-seventh.
Mercury	From Seattle to Chicago on December twenty-second. Can you specify a time or airline preference?

> User   I said "December twenty-seventh."

(Polifroni and Seneff, 2000)

Often, corrections have several of these features, like the caller's response in our original deletion example:

> Caller   No, I said "September twenty-nine."

(Stifelman, 1993)

This response almost inevitably included some prosodic signals as well, perhaps extra salience on the *nine*.

These phrases (like, in fact, the prosodies) are cohesion devices to put repetitions, synonymous or associated phrases, and the like into explicit coherence relations to each other — in corrections, the relations tend to be those of restatement and paraphrase. With other colligations, the cohesions can signal expressly that the relevant repetitions are *not* restatements but simply parallel dialogue acts — frames like "Give me X again" or "I'd like to hear X again." Being awake for these sorts of prefaces and frames can help the system quickly sort out corrections from genuinely reiterated dialogue acts.

Phrases like "I meant" and "I said" traditionally have triggered out-of-vocabulary rejections, like the prosodic changes, or they have just been discarded as incidental to the semantic/pragmatic function of the utterance. But they convey extremely valuable repair information, which along with acoustic clues, vocables, and other words can provide the repair manager with a repertoire for diagnosing problem areas and building more habitable dialogue systems.

Notice that these strategies do not put the onus on the voice interface to carry out repairs. They are all acts of cooperation that function by recognizing and supporting the user's own strategies, allowing him to make the repair. That's how conversations work.

Managing Slippages

All interactions are problematic and occur only through the continual turn-by-turn efforts of the participants.
    — Pamela Fishman

Misrecognitions are the biggest source of trouble for speech systems, both the most problematic (Rodman, 1999: 140) and the most common (Brown and Vorsbugh, 1989; Minker, 1999: 153). The principal reason they are a so much bigger pain than rejections and other nonrecognitions is that nonrecognition feedback makes it abundantly clear that something has gone wrong; with misrecognitions there is no error-related feedback at all. No error-related feedback means a much higher chance that some unintended, or even injurious, action will be performed, or that some corrupted information will throw someone's day, or bank balance, out of whack. The user has provided input, the system has accepted it, and both are ready to move on under the assumption that everything is hunky dory. If the input is "Call Lisa" and the system says "Sorry, I didn't get that," you try again. If the system hears "Call Tina" and blithely does so, callers could get awkwardly far into a conversation before the error is uncovered. Misrecognitions can lead to situations like our ADAP-caller showing up the next morning at the Frankfurt Central Train Station, suitcase in hand, and missing his train by ten minutes, or arriving two hours too early, or otherwise being out of synch with the train to Hamburg.

The traditional way around this kind of trouble, especially popular with the recognition engineers, is relentlessly explicit feedback, of a familiar sort. Remember TOOT?

T_1 Caller   I'd like to get a train from Philadelphia to New York.

T_2 TOOT   I heard you say "go to New York from Philadelphia." When do you want to leave?

T_3 Caller   Sunday.

T_4 TOOT   I heard you say "leave on Sunday." What time do you want to leave?

T_5 Caller   10:30 p.m.

T_6 TOOT   I heard you say "around 10:30 p.m." Do you want me to find the trains from Philadelphia to New York on Sunday around 10:30 p.m. now?

(Walker, Kamm, and Litman, 2000: 6)

While this parroting might be forgivable with a novel system, or one that is the only game in town, it soon becomes very wearying indeed for users. Which leaves, effectively, the only other solution: judicious feedback. TOOT does have a tiny amount of confidence — it does not confirm "train," for instance — but overall it is maddeningly insecure.

What's judicious?

If TOOT was having substantial recognition troubles, it might well be judicious to confirm just as frequently as it does in this sample, though with a little more regard for the caller's attention, time, and patience ("Sunday. What time?" TOOT-2 might say). But if the recognition confidence were decent, a single collective confirmation request would be judicious. Such a dialogue might proceed a bit differently (with, say "Sunday at 10:30 p.m." at T_3, in response to "When do you want to leave?"), but after the required task-variables have been solicited and gathered, summing up all the system-grounded variables at the end of the exchange for the caller's appraisal would probably do it. In fact — wait a minute, hold the phone — TOOT's second sentence at T_6 *is* precisely such a confirmation request. With even a fairly low recognition rate (say 90% for task words), this caller could have got through such a call without a hitch. Perhaps the next one would as well. But — odds will out — at a 90% rate and four variables per exchange, one caller in three will end up at the collective confirmation point responding with a partial repair like this:

Caller: No. I want MONday at 10:30.

He will surely flag the trouble spot in some way, even if not with all three of the signals in my hypothetical response (a lexical negator, a prefatory phrase, hyperarticulation), and TOOT-2 needs to be ready for exactly that possibility. He may even be slightly annoyed. But the task will still be successful, and he will not have been forced to sit through an auditory version of water torture. Neither will the next two callers. Occasionally, the system might have to fall back to a variable-by-variable, just-answer-the-questions-ma'am interaction, but only after the more habitable approach had failed.

Feedback, in short, should be relative, just like prompting. Since misrecognitions happen, the conversational ground has to be regularly calibrated. But *regularly* need not mean every single turn, nor even every single variable-assigning recognition, and recognition success rate is only one of the variables that should determine how often feedback moves should be made, or what form they should take.

In fact, feedback practices should only indirectly be related to recognition success. Far more significant for determining how much feedback is appropriate, and what form it should take, are (1) the type of information being grounded, and (2) the nature of the dialogue act. Together, these two factors constitute what Herbert Clark and Deanna Wilkes-Gibbs call *grounding criteria* (1986; see also Clark and Schaefer, 1989; Clark and Brennan, 1991; and especially Cahn and Brennan, 1999).

Grounding Criteria

People can be particularly unhappy about mistakes involving their money.
  — John Karat, Jennifer Lai, Catalina Danis, Catherine Wolf

There is feedback inherent in every system utterance. That's a given. Grounding is therefore going on continuously for the user. That's a given. Your choice, as designer, concerns how *much* feedback, how *explicit* it is, and what it *focuses* on. Some matters in an interaction can get by with a low grounding criterion: if the system misses the state but gets the zip code confidently in a U.S. address, it should not require the user to repeat the state, since that can be reconstituted from the code.

Some matters require high grounding criteria: the overall address needs to be fully verified by the speaker before a shipping transaction concludes. Credit card numbers, pin numbers, and security issues generally require high grounding criteria. Already grounded information, information that can be recovered inferentially, and cases where the grounding would be more irksome than the repair, should all have correspondingly lower grounding criteria.

The most important factor is the criticality of the pragmatic function — the dialogue act the system thinks it hears; or, more particularly, of the action it entails. If the system thinks it hears "good-bye" or some other encounter-ending input, for instance, it needs to ensure that the user really wants to terminate the encounter. How it reaches that assurance depends on a number of additional factors. For instance, early in an encounter, or at any junction where termination seems unlikely, it may respond explicitly "Did you just say *good-bye*?" But at a reasonable termination point, it may just respond with its own leave-taking dialogue act — "Good-bye" or "Good-bye then" — providing a clear enough signal to the user that she can infer what the system thought it heard, and leaving sufficient opportunity for her to continue the encounter if she didn't leave-take, or if she otherwise chooses to continue the transaction.

Other actions in the high-grounding-criteria neighborhood include finalizing exchanges such as purchases and funds transfer — that is, obligatives. Low grounding criteria tasks are most forms of queries: Will it rain in Boston tomorrow? How many gold medals did Latvia win in the 1984 Los Angeles Olympics? What is Hewlett Packard trading at? Some pieces of information are more critical than others, particularly event-planning pieces of information, like stock prices and departure times. But, in large measure, simply performing these tasks — answering the question — generates sufficient feedback for the user to know whether the system understands her or not. Adding an additional confirmative stage is just contributing tedium.

The system, as always, needs to be on the alert for a user-initiated repair, with both the raw data and the candidate list still on hand, in case the user wanted to know about precipitation in *Austin* tomorrow, not Boston, or was asking about the trading price of *AT&T*, not HP.

Toward the middle grounding level come tasks like sending a message, dialing a call, or setting a call-back time, which can all be verified implicitly, as the task is launched or the variable is set.

A close second in terms of importance for establishing the grounding criteria is the criticality of the semantic content: credit card numbers, expiration dates, and monetary amounts are at the highest grounding-criteria end. Shipping addresses, booking dates and times, hotel and airline details, call destinations, and the like fall toward the middle (and a good inference engine should be calibrating elements of the information against other elements — city and postal code, for instance). At the low end come locations for weather information, traffic coordinates, and so on. Now, almost all the task-related information supplied by the user is significant to *her*; just getting a single pizza topping wrong can be highly irritating for the customer, and even health threatening, should allergies be involved. But, the moral here is simply that not everything must be confirmed immediately, nor with great fanfare.

Grounding criteria need to be worked out on these two metrics — information importance and task importance — along with resolution confidence, illustrated in the three-by-three matrix of Table 13.2.

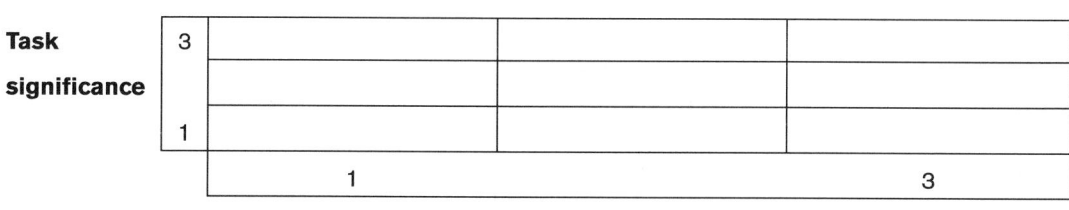

TABLE 13.2   A 3 × 3 matrix for determining grounding criteria

Nothing is sacrosanct about this table, least of all the number three; it's just convenient for illustration. You may need to work out four levels of task and information criticality, or two, or an asymmetrical 2 × 4 matrix. While numbers are helpful, especially for generating rules that machines can follow, the values will always be somewhat impressionistic, and very domain-dependent. But the point is a simple one: you work out a criticality number for the variables you need to capture. Perhaps departure and destination cities are twos, in a booking task that is also a two, giving each of them a criticality level of four. You combine this with confidence threshold values and call-flow stage, and work out the criteria to determine (1) the level at which the system will consider the term grounded (say, fours might need an 95% confidence level to be passed to the next call-flow stage, twos a 75% confidence level), and (2) the structure of the feedback required to present that grounding to the user if confidence is below the accepted threshold (say, explicit feedback for fours, implicit for twos).

Undefined jargon is starting to creep in, I know, with call flow and feedback level. Call-flow issues we take up in Chapter 14, but for now it just means what system utterance comes where. Feedback levels are our next topic.

Levels of Feedback

Any [feedback] must clearly contribute to increased interface usability.
— Alexander Rudnicky

Feedback, of course, is omnipresent in dialogue, so what we mean here is feedback directed specifically at utterances, of the groundskeeping sort (the way, for instance, Balentine and Morgan, 1991, use *feedback*). Another traditional term for this aspect of system-utterance crafting is *confirmation*. That term has its virtues, because the function of specific feedback is to "confirm" or make obvious the system's understanding about the relevant settings or actions. But their real job, in fact, is to *invite* confirmation, more than to offer it. The design assumption is that silence is golden — if the user does not explicitly reject the "confirmed" information, then he has ratified it. I will continue to use *confirmation* in a loosely synonymous way with *feedback*.

Closely paralleling the prompting taxonomy we looked at earlier, there are four basic levels of feedback, reckoned by their relative explicitness. The more explicit the feedback, as you would expect, the more reliable the interaction, but also the more stilted and conversationally unnatural that interaction tends to be. All of them, except the weakest (open feedback) are really better understood more as feedback requests rather than as feedback, because their function is to alert the user to the information being grounded by the system, under the silence-is-golden assumption. The most aggressive of these feedback categories (explicit) is often an unequivocal request ("Do you want to leave from Trento? Yes or no?"); the other three sublimate the request aspect to various degrees.

The four levels (they should look very familiar), and the null condition, are as follows, where X is the word, phrase, or action whose grounding is being presented:

Explicit

The information to be confirmed is embedded in a frame that draws specific attention to the words themselves ("I heard you say X," "Did you just say X?") or to the function ("Shall I do X now?" "Do you want me to X now?"), and requests directly that the confirmation be ratified.

Implicit

The confirmation is embedded into a phrase that initiates the requested task ("I'm looking up X now," "X-ing now").

Inferential

The confirmation comes in the performance of the task itself ("The weather in X is . . . ," "The price for X is . . .").

Open

Less a confirmation than an acknowledgment, just marginally stronger than a backchannel, this response style responds affirmatively to having received input, implying that it was understood and processed, but does not reflect in any way what the actual input is ("Okay," "Alright.").

None

By the nature of interaction, saying nothing relevant to the previous utterance is a form of feedback, too, usually implying strongly that the previous utterance was fully comprehended, and any implicated action is or will be carried out. For

Recognition confidence	Grounding Criteria		Feedback			
	Task criticality	**Information criticality**	**Timing**	**Level**	**Description**	**Example**
Low ↓ High	High ↑ Low	High ↑ Low	Immediate ↑ Delayed	Explicit	The confirmed words are placed in a frame that draws attention to words themselves, or their task-action, before the task is engaged, and direct ratification is requested.	Did you just say "Good-bye"?
				Implicit	The confirmed words are embedded into a task-initiating phrase.	Calling Susan . . .
				Inferential	The confirmed words are embedded into the task-response.	The weather in Boston is . . .
				Open	Not really a confirmation of any words, just an acknowledgment of receiving input.	Got it.
				None		⟨silent passage to the next stage⟩

TABLE 13.3 Levels of feedback

instance, meeting each word with silence in a list ("Pepperoni. Olives. Bacon."), suggests that the hearer is getting them all.

The higher the grounding criteria, the more explicit the feedback should be. As above, grounding criteria and feedback level interact with two more variables: recognition confidence and timing (*when* the feedback should be presented). The taxonomy, with these considerations arrayed relative to each other, is set out in Table 13.3. For convenience, the prompting taxonomy, too, is set out with respect to grounding criteria, in Table 13.4.

The grounding criteria and the recognition confidence are inverse. The grounding criteria put pressure on the confidence thresholds; the confidence thresholds push back. The higher the task and/or information criticality, the more confident the pattern match needs to be to lower the feedback level. Or, conversely, the more confident the recognition is, the

Grounding Criteria		Prompts		
Task criticality	**Information criticality**	**Level**	**Description**	**Example**
High ↑	High ↑	Explicit	The prompt tells the user precisely what input is appropriate, either by listing it exhaustively (if only a few words), or by representative example.	Do you want to leave between 6 and 9 p.m.? Yes or no?
		Implicit	The prompt characterizes the appropriate input in conceptual or linguistic terms.	Which type of restaurant would you like?
		Inferential	The form and content of the prompt implies the form and content of the appropriate input.	I can answer questions about Strindberg, the Royal Institute of Technology, and Stockholm.
Low	Low	Open	The prompt only conveys the most general and open expectations for user input, allowing context to shape the next turn.	How may I help you?

TABLE 13.4 Levels of prompts

lower the feedback level can be, even for relatively critical tasks and information. A very confident recognition with high grounding criteria might, for instance, call for only an implicit feedback, while medium grounding criteria may require an explicit feedback if recognition confidence is low.

Timing is relevant with respect to task completion. In particular, you may go with a low confidence threshold at a given stage, passing the resolved variables (or even a list of candidate variables) on to another routine, because variables collected later (postal codes, for instance) may help strengthen the commitment to that earlier variable (or choose one from a candidate list); or just because the plan calls for a final confirmation sequence later on, where any erroneous groundings can get caught and corrected. High grounding criteria with relatively low recognition-confidence call for immediate feedback. Low grounding criteria with relatively high recognition-confidence call for delayed feedback. What this means, for instance, is that credit card information should be verified right away, though task completion (purchasing and shipping) might still be several moves away, while a query response wouldn't be verified until the task is actually in the process of completion. In practice, timing is often "immediate" anyway for low grounding criteria tasks, like a weather query, because it is a simple two-move task (initiation + response). An immediate confirmation in the timing sense for a weather request would require an insertion sequence ("Do you want me to get the weather for Waterloo now?").

Open confirmations should usually be reserved only in cases with very low grounding criteria, or when the confirmation itself is highly redundant (that is, either as a midpoint in an interaction in which more explicit confirmation will follow, or at the end of an exchange in which explicit confirmation has already been carried out successfully). Some confirmations, under poor signal quality conditions (high background noise or unstable connection, for instance), are critical enough that explicit feedback should even be coupled to explicit and tightly constraining feedback: "Do you want to pay $2,500 against your Mastercard? Say yes or no."

In continuous-speech recognition systems of the sort we are concerned with, confirmation might be required for different task elements, with different confidence levels, leading to a kind of "mixed feedback." In terms of the taxonomy, it's the most explicit level that we identify, because we're worrywarts. We're interface designers. We have to be. But, in this clarification request, for instance, we have implicit feedback at T_2 (for *call*) stirred in with explicit feedback (for *Doe/who*):

T_1 User   Call John Doe.

T_2 Agent   Call who?

T_3 User   John Doe.

T_4 Agent   Calling John Doe.

(Kamm et al., 1997: 274)

Mixed feedback of this sort is an extremely valuable resource in voice-interface design, because it allows for a compressed interaction style that achieves high functional habitability. At T_3, for instance, the user both ratifies *call* (silently) and clarifies *John Doe*.

You will have noticed, I'm sure, that nothing I have said would make a whit of difference to the unfortunate soul who asked for a Frankfurt-to-Hamburg departure time, got a Frankfurt-to-Hanover departure time instead, *with* a clear inferential feedback (the answer was not just "5:35," notice), and still he never caught the slippage. There are ways such situations might be handled. A higher feedback level might have caught his attention ("I heard you ask for the train times from Frankfort to Hanover. Is this correct?"). Or another form of redundancy might have been used, like acoustically added salience; added duration is especially effective for voiced data. It would sound a little weird, but it would surely prevent some percentage of cases like our Hamburg/Hanover example. The question becomes, though, is it worth it? What is the point of diminishing returns for adding saliency to confirmative groundings? Under high criticality conditions and/or poor signal conditions, perhaps it would be worth it — if there was a rail accident on the route or recognition confidence levels were very low — but not in any normal cases. ADAP did all that could be reasonably expected of it.

Source and Initiative

This conversation is going on a little too fast: let's go back to the last remark but one.
— Humpty Dumpty

The system is usually to blame in slippages, we know that; the user is usually the one trying to get the exchange back on track. We could frame the situation other ways — blame the user for not speaking clearly enough, or loudly enough, or staying within the vocabulary, or, in this variation of a slippage we saw earlier, for having the unmitigated gall of expecting a computer to know rudimentary math:

T₁   DanLuft: How many people will travel?

T₂   Caller: Two adults and two children.

T₃   DanLuft: Two people.

(Bernsen et al., 1998: 179)

But we know where the limitations usually lie. This caller's expectation is not only reasonable for human agents, it's trivial in the extreme for computers; she should have been able to give the number of travelers as the square root of the sum of the distance covered in 20 minutes by a rickshaw traveling 12 kilometers an hour (given accurate recognition, of course). Not only that, but in the context of travel fares, she had reason to believe, or at least hope, that there might be differential fares for adults and children. There was a clear motivation to render the total number in components. To DanLuft's credit, it does cooperate in a repair, albeit a somewhat stilted one:

T₄   Caller: Change.

T₅   DanLuft: Two people. Is this correct?

T₆   Caller: No.

T₇   DanLuft: How many people will travel?

T₈   Caller: Four.

T₉   DanLuft: Four people.

(Bernsen et al., 1998: 179)

Still, this task slippage is unequivocally the system's fault. The user has provided appropriate information, and DanLuft got it wrong — not from a recognition problem, but from a substantial design flaw. We can all sympathize with the coding problem of tuning an inference engine that can distinguish between "Two, uh, yeah two" — that is, between an in-turn repetition, and the actual request (T_2) — though, as we have seen, there are prosodic and lexical cues attending repetitions. But (assuming a little lexical density so that *adult* and *child* were in the system vocabulary), a simple calculation would have eliminated the T_3–T_8 cross-examination altogether. Wolf and Zadrozny's (1998: 491) term, *competence error*, seems especially suitable in cases like this, because Danluft is incompetent in simple arithmetic.

In the Danluft exchange, the system slips, and the user initiates the repair, a common occurrence for misrecognitions and task slippages. There are three other logical possibilities, as laid out in Table 13.5.

With human–human interactions, the structure of the Danluft repair dialogue would be unusual. A human agent likely wouldn't make the mistake that Danluft does at T_3, of course. But if she did, she would probably catch herself quickly (self-repair), or the caller would say something like "No, two adults, two children" — a repetition that would save her face sufficiently by giving her another crack at the math — and only elaborate her contribution to the repair if the agent still didn't catch on.

It is far more common in human–human dialogues, that is, for the agent responsible for the slippage to also initiate the repair (Schegloff et al., 1977: 364), and even more common for the responsible agent to *complete* the repair — probably for social reasons. Correcting someone else carries more negative social freight than correcting oneself. With speech systems, however, the asymmetry in the agents' cognitive powers, as well as the presumably much lower level of social concern, means the slippage-source/repair-initiative matrix is less important than which one is the source, which is the initiator (human or machine), and what type of breakdown occurs. The callers will repair any of their own contributions that they regard as defective, and take the opportunity to correct

| | | Repair initiative | |
		System	**User**
Slippage	**System**	System/System	System/User
source	**User**	User/System	User/User

TABLE 13.5 Slippage sources; repair initiatives (adapted from Schegloff, Jefferson, and Sacks, 1979)

any of their own turns that are revealed as defective (that is, in both cases, follow standard operating procedure). But they also, by necessity, have to be more aggressive in correcting the system, not just providing hints for it to correct itself.

Conversational repair, in short, is a clear interaction district where the social-orientation paradigm is overruled by task necessities. Voice interfaces have to be especially open to system-source errors, with user-initiated, user-completed repairs.

Here is a case of a system showing that sort of openness. The user slips (or, in any case, changes her mind, which is effectively the same thing, calling for a correction), and then she initiates, and largely completes, the repair:

SUNDIAL   There is one flight at six and another at nine-thirty.

User   Nine-thirty.

SUNDIAL   OK, nine-thirty, Paris Brest. One-way or return?

User   No. I'd prefer six o'clock.

SUNDIAL   OK, I'll book you on the six o'clock flight.

(Bilange, 1991)

The system behaves very cooperatively here. It asks "One-way or return?" but is prepared not only for "one-way" or "return" as input. It is ready and willing for an insertion sequence, and completely amenable to a task repair; minimally, the vocabulary is stocked with times, as well as global repair terms like *no*.

Continuing our survey of slippage source/repair-initiative examples (we've already seen system/user and user/user instances), a user-slippage/system-initiative instance is perhaps the most common pattern — at least, they are the most common pattern if we follow the traditional blame assignment of collaring the user in out-of-vocabulary rejections, timeouts, and spoke-too-soons, the nonrecognitions. These slippages are not really the user's "fault," of course, and they would certainly not be the user's fault if the other agent were another linguistically fluent adult human, someone who would not impose the same sorts of vocabulary restrictions, or anything equivalent to don't-speak-until-after-the-beep, on the speaker. (Humans do tend to be equally sensitive to time lags, especially in voice-only encounters, but don't always regard the silence as the other person's error.) Still, all the same, the slippage has its most direct roots in an input utterance that originates with the user. We've seen multiple examples of these sorts of user-source/system-initiative exchanges, like the MailCall brick-wall example, of which this was the tail end:

User   Gee whiz. I just don't know what to say.

MailCall   I didn't understand.

(Marx, 1995: 88)

The remaining alternative is when the system slips, and the system initiates a repair, which is extremely rare. When the system goes wrong it can almost never initiate (or carry out) the repair itself, because, well, it's wrong. The closest we come, by way of confirmation acts, is to a repair-invitation, not a repair-initiative. Designers know the system can go wrong, so what they very often do is to invite repairs frequently. Here is one of the more brutal examples we looked at earlier in the book (Chapter 5; the slippage is at $T_{2/3}$):

T_1 TES   Hello, this is Train Enquiry Service. Please speak after the tone. Which information do you need?

T_2 Caller I want to go from Torino to Milano.

T_3 TES   Do you want to go from Trento to Milano? Yes or no?

T_4 Caller No.

T_5 TES   Do you want to leave from Trento? Yes or no?

T_6 Caller No.

T_7 TES   Where do you want to leave from?

T_8 Caller I want to leave from Torino.

(Danieli and Gerbino, 1996)

The resolution here is successful, but this design requires a *very* patient user. Once the repair is made, the exchange continues on the same dogged template:

T_9 TES   Do you want to leave from Torino? Yes or no?

T_{10} Caller Yes.

T_{11} TES   Do you want to go to Milano? Yes or no?

T_{12} Caller Yes.

T_{13} TES   At which time do you want to leave?

T_{14} Caller I want to travel in the evening.

T_{15} TES   Do you want to leave between 6 and 9 p.m.? Yes or no?

Etc., etc., etc.

This design does not go much beyond Bernsen and Dybkjær's (2000a) grunt detector for most of its turns ($T_{3/4}$, $T_{5/6}$, $T_{9/10}$, $T_{11/12}$). What is perhaps more noteworthy is that since there was only one recognition failure in the entire exchange ($T_{2/3}$), Train Enquiry Service is showing (not uncommonly) undue insecurity. With more subtle feedback, and by enabling the user to make the repair directly, not just to plod through the system's checklist, ratifying or rejecting its guesses, the exchange could have proceeded more elegantly, and, very likely, more satisfactorily for the user. In fact, here is the very same task, with a closely parallel misrecognition, under a later iteration of TES:

T_1	TES	Hello, this is Train Enquiry Service. Please speak after the tone. Which information do you need?
T_2	Caller	I want to go from Torino to Milano.
T_3	TES	At which time do you want to leave from Merano to Roma?
T_4	Caller	No, I want to leave from Milano in the evening.
T_5	TES	Do you want to leave from Milano between 6 and 11 p.m.? Please answer Yes or no.
T_6	Caller	Yes.

(Danieli and Gerbino, 1996)

With a sheer number-crunching, recognition-based evaluation metric, the second exchange comes off worse — one error in 15 turns vs. one error in 6 turns — but on all task-based metrics it would vastly outperform the first one. It's not just the fewer number of turns, which increases efficiency dramatically. It's also the type of turns. The second exchange is more functionally habitable (they both seem to have the same lexical habitability, perhaps also syntactic habitability, though most of the first exchange restricts the caller to one-word utterances), because the interaction patterns are based more on natural conversation than on cross-examination. In particular, T_4 is a clarification, the sort of dialogue act we use to correct other people who have misheard us. (In the first dialogue, T_8 is sort of a clarification of T_3, but in a distended and wholly unnatural way.)

These four patterns of slippage-source and repair-initiative in speech-system interaction, with examples and descriptions, are collected for your enjoyment in Table 13.6.

The upshot of this survey is that users must do a great deal of the repair work; indeed, even the many user/system instances leave most of the repairing to the user, since the system's "repair initiative" amounts to little more than a rejection which alerts the user to try something else. That's fine, of course. The user *should* be the main repair technician. She's the genuinely intelligent agent in the exchange. But it means the system needs to give

Type		Example	
Self-repair	**User/User** The user is the source of the slippage (in the example, supplying the wrong departure city), and the user initiates the repair.	Caller System Caller System Caller System	I'm going to Lyon. To Lyon. Where are you leaving from? I'm going from Lyon to Lille. I'm sorry. I don't understand. Please say your departure city; for example, "Paris." Lyon. From Lyon. What is your destination? (Rosset et al., 1999)
	System/System The system is the source of the slippage (a substitution error of "Trento" for "Torino"), and the system initiates — or, at least, invites — the repair (by double-, then triple-checking its construal).	Caller TES Caller TES Caller TES Caller	I want to go from Torino to Milano. Do you want to go from Trento to Milano? Yes or no? No. Do you want to leave from Trento? Yes or no? No. Where do you want to leave from? I want to leave from Torino. (Danieli and Gerbino, 1996)
Other repair	**System/User** The system is the source of the slippage (misunderstanding "two + two" as "two"), but the user initiates the repair.	DanLuft Caller DanLuft Caller DanLuft Caller DanLuft Caller DanLuft	How many people will travel? Two adults and two children. Two people. Change. Two people. Is this correct? No. How many people will travel? Four. Four people. (Bernsen et al., 1998: 179)
	User/System The user is the source of the task slippage, but the system initiates the repair.	Caller System	Can I fly on Friday, the 5th? Sorry, the 5th is a Thursday. Would you like Thursday the 5th or Friday the 6th? (Choularton, 2004)

TABLE 13.6   Slippage-sources and repair-initiatives (adapted from Schegloff, Jefferson, and Sacks, 1977)

her the room to make the repair, and cooperate encouragingly while she makes it, as the second TES exchange does.

The fundamental attributes of that second exchange — its efficiency and functional habitability — follow from three sound design decisions. The second TES interface design (*D2*, Danieli and Gerbino call it; the first one, no surprise, is D1) is different in three very wholesome ways. It displays more confidence. It includes implicit feedback techniques, rather than explicit ones. And it allows the caller to take the initiative — not just to affirm and reject, but to initiate repairs. In the D1 exchange, the system is the source of the error, the system invites the repair, and the system conducts the repair. In D2, we move to a system-source/user-repair format. TES is open to a repair at T_3, but the implicit feedback strategy does not convey the same invitational insecurity as the pitilessly explicit confirmation requests of D1. The caller not only then initiates the repair through the D2 interface, she largely completes it in the very same turn (the comparable repair for D1 takes five turns).

Initiative makes all the difference.

Managing Initiative

In naturally occurring human–human dialogues, speakers often adopt different dialogue strategies based on hearer characteristics, dialogue history, etc. For instance, the speaker may provide more guidance if the hearer is having difficulty making progress toward task completion, while taking a more passive approach when the hearer is an expert in the domain.

— Jennifer Chu-Carroll

The familiar division for dialogues in terms of initiative — in terms of which agent has moment-to-moment control over the dialogue flow — is between *fixed* (one agent controls throughout) and *mixed* (both agents have the capacity to control at any point). Since we have different categories of agents, that gives us an additional distinction in the fixed category: fixed-on-user or fixed-on-system. All three types are in evidence among speech systems. Command-and-control interfaces can be exclusively user-initiative. Just-answer-the-question systems like TES-D1 are exclusively system-initiative. Now-it's-the-system, now-it's-the-user dialogue formats, like TES-D2, are mixed initiative.

Mixed-initiative systems are overwhelmingly more habitable.

Exclusively user-initiative designs work only in very restricted fields, and are most applicable in a multimodal environment, where the voice-modality is largely command-and-control and the feedback comes along another channel. Much is made in the voice-interface literature of the talking computers found in science fiction; in particular, they are frequently blamed for producing unrealistic expectations in users for conversational sophistication from voice systems. But little notice gets paid to the fact that they often drop into command-and-control mode, which is exclusively user-initiative, when functionality requires. Picard, on *Star Trek: The Next Generation*, just says "Tea. Earl Gray. Hot." and the replicator makes it so. No conversation. In *Blade Runner*, Deckard uses a hardware-klugey looking device called ESPER to examine a photograph, the interaction (aside from quaint design elements that date the movie to the 1980s) illustrates a model command-and-control interface.

ESPER features a flickering cathode-ray display, entirely filled with an image of the photograph, overlaid by a grid, and, along the bottom, an alphanumeric readout in a then-futuristic font. Deckard the titular blade runner speaks, saying nothing that is not an express command, and ESPER responds in actions accompanied by direct and indirect feedback. The exchange proceeds as charted in Table 13.7.

The Deckard/ESPER interaction is accompanied by visual feedback: primarily the target image itself, as it changes to accommodate various distances and perspectives, but also the grid and various other grid-like graphics, along with the changing alphanumeric readout. Several nonspeech auditory cues also provide feedback (then-futuristic clicks, blips, and beeps, highly reminiscent of Pong). ESPER has a fairly small vocabulary of cinematic technical terms, but one that still has room for synonyms, some of them colloquialisms. "Enhance," for instance, is a synonym for "zoom," as suggested by the alphanumeric readout (featuring "Zm" when the enhance function is activated; "move-in" is a synonym for enhance/zoom. "Wait a minute" is a synonym for "stop," or perhaps "pause."

The device may not be possible, in other words, but the interface certainly is.

Truly mixed-initiative dialogues — where the opportunity is always there for either agent to take control as required by the task and the unfolding dialogue — are the preferred design. They allow true user interaction, not just the chance to answer questions and issue commands. They allow the system to establish control when necessary. They provide a ready and natural framework for clarification and repair. They are the goal for conversational voice interaction design. Truth to tell, they are entailed by a conversational design; any voice interaction, between humans let alone between humans and machines, in which one agent has all the initiative, does not really fit the definition of a conversation.

The trick, however, is in knowing when and why to do the mixing: when the system should have it, when the user should, and why. Beyond being mixed, the initiative must be adaptive.[3] There are four guiding criteria for assigning initiative: dialogue management acts, the interaction script, the experience level of the user, and the repair requirements.

Dialogue management acts either regulate the flow or maintain the ground. Flow-regulating dialogue acts are largely about who has, or is taking, or should take, the floor. Having the floor and having the initiative are not the same thing, of course, but many flow regulators are initiative regulators: taking the floor, especially by turn overlap, almost always correlates with taking the initiative; keeping the floor often means keeping or even taking the initiative; and turn assignment is largely the function of an agent with the initiative. Turn overlap is an especially powerful marker of initiative, and should be respected.

Groundskeeping acts are talk about talk, and people almost never talk about talk (except philosophers and linguists — and, OK, voice interaction designers) unless it is going wrong,

Deckard (says)	ESPER (does)
Enhance 224 to 176.	The display zooms in on one square of the grid.
	The zoomed-in part of the photograph (a close view of a crooked arm) now takes up the entire display, overlaid by a new grid.
Enhance.	Zooms in on the arm; clicking and flickering.
Stop.	Stops.
	Unaccountably ESPER starts clicking again, without input (perhaps this is an insertion error, from Deckard's sniffs or his rustling movements, but he goes along with it). ESPER is off-camera, but the next view is closer, so the activity correlated with the feedback is presumably zooming in.
Move in.	Clicking, flickering, moving grid-graphics.
Stop.	The display is tight on the hand.
Pull out, track right.	Clicking, flickering.
Stop.	Stops on a tight shot with part of a newspaper or menu dominating the foreground.
Center and pull back.	Clicking, flickering, as the display moves back out to a broader view.
Stop.	Stops on a view of a cluttered table against the background of a wall or door.
Track 45 right.	Clicking, flickering, as the display rotates.
Stop.	Clicking stops (ESPER is off-camera).
Center and stop.	Stops on a view of a small mirror through a doorway.
Enhance 34 to 36.	Clicking, flickering, moving grid-graphics; the view stops on a tight shot of something on a counter below the mirror. (It looks like a shot glass.)
Pan right, er, "n" pull back.	Clicking, flickering.
Stop.	Stops on close view of the mirror.
Enhance 34 to 46.	Clicking, flickering, moving grid-graphics. Display stops on view of indistinct light pattern in the mirror.
Pull back.	Clicking, flickering.
Wait a minute. Go right.	Clicking, flickering; the view moves around an obstacle.
Stop.	Stops on medium view, a crooked arm in the background (different from the previous arm)
Enhance 57 to 19.	Clicking, flickering, moving grid-graphics. Stops on tight view of elbow.
Track 45 left.	Clicking, flickering; the view moves around an obstacle.
Stop.	Stops on medium view of a sleeping woman.
Enhance 15 to 23.	Clicking, flickering, moving grid-graphics, as the view moves to tight shot of the woman's face.
Give me a hard copy right there.	A small personal-photograph style printout whirrs out of the top of the machine.
The interaction, and shortly the scene, ends.	

TABLE 13.7   A user-initiative, command-and-control, voice-driven interaction from the movie *Blade Runner*

or in danger of going wrong. Some acts, like backchannels and some styles of acknowledgments, operate more as initiative-assurances ("Go on, keep talking, I'm still listening and don't have anything I want to throw in yet.") than as genuine groundskeepers. More prototypical groundskeeping acts — clarifications and clarification requests, or confirmations and confirmation requests — establish and calibrate shared understandings about data, which they do by foregrounding the words, their meanings, or their functions. TOOT's refrain of "I heard you say X" is a clear example. It's a framed echo (actually a paraphrase in an echo setting, since the X is never verbatim) that establishes TOOT's understanding of the user's immediately preceding utterance. It serves largely as an invitation for the user to take the initiative if he needs to correct that understanding; declining to take the initiative and correct TOOT's understanding is tantamount to ratifying TOOT's grounding.

An example of a groundskeeping act that requires the initiative is a clarification request. Typically, clarification requests occur when the user tries to set a variable but fails to specify it fully, and the system needs to take (or retain) the initiative in order to probe further. Usually, this takes the form of an insertion sequence:

T_1 User   I want to go from Boston to San Francisco.

T_2 System   San Francisco is served by two airports, San Francisco International and Oakland International. Which airport do you want?

T_3 User   San Francisco International.

(Kamm et al., 1997: 273)

The user has the initiative at T_1, and the system takes it for T_{2-3} in order to ground the destination fully.

The script is also important for initiative management, especially in concert with the user's task experience. (We'll develop this notion more in Chapter 14; for now, the script is just the agenda that needs to be completed for the dialogue to achieve its goals.) As Joris Hulstijn suggests, task competence should strongly shape initiative management:

> *For mixed initiative dialogues we find that usually the participant who is most competent in the particular aspect of the domain that is currently under discussion, has the initiative.*
>
> *For example, in the beginning of a reservation dialogue the user has the initiative; the user asks questions and the system answers them. The user is leading, because in this case the user is most competent in what she wants to find out. Once the desire to make a reservation has been conveyed the system takes over the initiative. The system asks the user for her name, the number of tickets and so on. After all the system is most competent in what it takes to complete a reservation action.*
>
> (Hulstijn, 2000: 45)

The user's interaction experience is also a significant factor in initiative management, and, in general *any* perceived difficulty in achieving the dialogue's goals should be a signal to the dialogue manager that it should take the initiative and direct the user more closely, whether the difficulty is related to the task, to distraction, or to the specific interaction.

Let's say we have a classic travel-booking task, with four values to set: departure city, destination, duration, and payment details. The user might be a forgetful expert who takes the initiative at the start, specifies three of these values, but says "that's it" before specifying the payment details; the system needs to take the initiative and prompt for those details or the task will fail. Or, let's say the user is a tentative novice. The system might be designed with a very open greeting meant to leave initiative with the user ("AcmeTravel automated booking service. How can I help you?"), but the user isn't sure how to proceed — punch buttons, issue a command, say her name, or wait for a menu of options. The system needs to intervene — not necessarily to take the initiative, maybe initially just to make some suggestions, but if the silence continues, or is attended by out-of-vocabulary utterances (maybe she does say her name), pretty soon it will have to take full control and lead her through the booking task. In fact, Balentine has suggested that time-outs and out-of-vocabulary input in very many contexts serve as "novice detectors" (1999: 218), which should trigger explicit guidance, calling for system initiative. Conversely, turn overlap often serves as an expert-detector (Cohen, Giangola, and Balogh, 2004:214).

Repair interacts with all three of these factors — groundskeeping, the script, and user experience — as the fourth principal factor in initiative management. The clear-cut and conventional occasions for the system to have the initiative (keep it or take it) are significant nonrecognition events. Whatever the user is trying to do, if the system is disabled by a bout of incomprehension, it has to take the initiative, reveal its state of ignorance, and induce a given linguistic behavior — perhaps just a repetition, maybe a yes/no answer or an option-choice — from the user. The same is true for timeouts, as we've just seen. There are also clear-cut cases, though not so conventional yet, when the user should be able to take the initiative — namely, all significant cases of (noticed) misrecognition. Whatever the system is trying to do, if the user catches a bad system-grounded value — a flight in the a.m. rather than the p.m., the wrong sum in a transfer, anchovies on the pizza — she needs to be able to make the repair promptly, taking the initiative to correct the value.

Other factors for weighting the initiative heavily toward the system include threats of data loss or damage ("Do you really want to terminate this transaction, or should I save the information for later?"), status changes of various sorts ("There has been an accident on the Gardiner, but Lakeshore still looks normal"), and beneficial information the user may not otherwise have access to ("That same itinerary would be $700 less if you stay over on the Saturday. Are you interested?").

But perhaps the single most pressing reason for the system to seize and retain the initiative as long as necessary is to try and head off error spirals in their early stages. In this case, the system not only takes over the initiative, it uses increasingly more explicit prompts in an attempt to get the dialogue back on the rails. This is the method known as *progressive assistance*, a form of expansion.

Expanding and Tapering

It may seem cute on first encounter to be greeted by "I am SOPHIE, the sophisticated teacher, and I will teach you to spell correctly." By the second session, however, this approach strikes people as uselessly repetitive; by the third session, it is an annoying distraction from the task.
— Ben Shneiderman

Expanding the detail and specificity of subsequent prompts is a way of trying to satisfy Grice's concern about information quantity, a concern that famously points in two directions. "Make your contribution as informative as required," says one; "Do not make your contribution more informative than is necessary," says two. It's that Goldilocks principle: the amount in the information porridge bowl should be "j-u-s-t right." But different circumstances mean different porridge bowls, and fixing the quantity can require either increasing it or decreasing it. So we'll take up both possibilities together, generally known as expansion and tapering.

For nipping a spiral in the bud, or the leaf, or the stem — before it grows into a bush too tangled to escape — the preferred policy is progressive assistance. If we knew what the problem was right away, we could fix it. But we can usually only guess; if the first guess fails, we guess again, but we do so in the direction of more detailed and more explicitly directive utterances; and so on. Here is an example, from our old friend, TOOT:

T_1 TOOT On which day of the week do you want to leave?

T_2 Caller \<timeout>

T_3 TOOT Sorry, I didn't hear anything from you. Please tell me your departure date.

T_4 Caller ... \<rejection> ...

T_5 TOOT You can say the departure date. For example, you can say "today" or "tomorrow." Or you can say "Monday", "Tuesday," "Wednesday," "Thursday," "Friday" ...

T_6 Caller Saturday.

T_7 TOOT Do you want to leave on Saturday?

(Hirschberg et al., 2001)

At T_1, TOOT prospects for a day, getting no response. Maybe the caller's just thinking, but there is no way to know. After a reasonable couple of seconds, it apologizes for intruding (T_3), then tries again, not adding any detail, but getting a bit more formal and talking more explicitly in the travel register. It also moves from a neutral, information-based query to a social-tinged obligative, a request. With T_5, the detail level raises substantially, and it offers a series of examples — no longer just describing what it expects from the caller but itemizing the exact words it wants.

The phrasing may not be entirely optimal, but systematically ratcheting up the detail, the specificity, and the directive level of dialogue is exactly the right series of conversational moves. And it works. At T_6, the caller provides an utterance TOOT can deal with; the dialogue continues on toward goal fulfillment.

More important than the dialogue success that progressive assistance brings (though surely not unrelated to it), this sort of local expansion also improves user satisfaction substantially. Early versions of SpeechActs, for instance, simply uttered "Sorry, please rephrase" for every rejection. It seemed reasonable. Among other things, that approach is the epitome of the sort of consistency that governs graphic interface design. But it didn't work. It irritated the users. "People were visibly upset and angry," Yankelovich recalls. Without changing the recognition performance at all, the designers developed progressive assistance. The first rejection in a series got "Sorry." Next, "Sorry, please rephrase." And then, "I'm having trouble understanding you. Try speaking clearly." The results were dramatic: "When we changed the error dialogs, the people said the system was really good and cooperative. It was an amazing thing." (Yankelovich et al., 2000: 318).

It may well be that users were responding more to the utterance variety than to the growing detail and directiveness. Getting the same response over and over certainly has more of a brick-wall effect than getting different responses, even if they are functionally identical.

Another reason to expand, to add a system utterance that otherwise would be left unspoken, is strategically to reveal system functionality — not to lecture before the interaction occurs, for instance, but to draw a lesson from one that has just ended. Take this example:

T_2	System	What kind of trade do you want to make? You can say "Buy," "Sell," "Sell short," —
T_2	Caller	Buy.
T_2	System	How many shares do you want to buy?
T_2	Caller	100.
T_2	System	Of which security?
T_2	Caller	Chemex Coffee Corporation.
T_2	System	At what price?
T_2	Caller	A limit price.

T₂   System   Of what?

T₂   Caller   88.

T₂   System   OK, let me confirm that with you. [buy order, 100 shares, Chemex Coffee Corporation, at a limit price of 88] . . .

Here's a hint. Next time, you can say the whole trade when I ask you the first question. So, for example, you could say, "Buy 100 shares of Chemex Coffee Corporation at a limit price of 88" — all in one breath.

(Kotelly, 2003: 94)

What's particularly nice about this expansion is that the hint uses all the values it has just extracted from the exchange, rather than some generic "100 shares of ABC Inc." Suggestions like this (it's not really a hint) are not often geared towards immediate success, but towards later successes and efficiencies, and therefore longer-term satisfaction — toward building loyalty.

Tapering also has the capability of improving user satisfaction and building loyalty. Over the short term, tapering has the form of pragmatic ellipsis, of building up taken-for-granted contexts — that is, of grounding — so that references can become increasingly briefer and more efficient:

T₁   System   . . . Now, what's the first company to add to your watch list?

T₂   Caller   Cisco Systems.

T₃   System   What's the next company name? (Or, you can say, "Finished.")

T₄   Caller   IBM.

T₅   System   Tell me the next company name, or say, "Finished."

T₆   Caller   Intel.

T₇   System   Next one?

T₈   Caller   America Online.

T₉   System   Next?

(Cohen, Giangola, and Balogh, 2004: 213)

The reduction in word-count from T₃ to T₉ — utterances with precisely the same reactive pressure — is 90 percent. The dramatic saving here for the user's time and patience is perhaps the most notable aspect of tapering (satisfying a maxim of manner, be brief), but the accompanying reduction in detail (satisfying a maxim of quantity) is also appreciable, and the incidental benefit of added variety doesn't hurt either.

Neither expansion nor tapering, however, is confined to the short term. Expansion, for instance, might lead to hints and tips not immediately after an interaction, but after several interactions, even after several calls. Tapering might lead to a later exchange in the same dialogue dispensing with the you-can-say-"Finished" clause as well, or in a subsequent call from the same individual. More generally, while the best first-level response to a rejection is usually a you-can-say-Mexican sort of directive, later rejection feedback might be tapered back to a single "sorry?" Neither expansion nor tapering is confined to only a single encounter.

The problem with "I am SOPHIE, the sophisticated teacher, and I will teach you to spell correctly" the second and third time is not the utterance *per se* (nor its use of personification and first-person self-reference, as Shneiderman argues — 1998: 383). The problem is that the same greeting comes at the user again and again; it is indeed "uselessly repetitive" — invariant, lengthy, and unnatural. Your grade-six teacher didn't say "I am Ms. Bennet. I am your home room teacher. I hope we can get along and learn together" every day. She said it on the first day of class (or my Ms. Bennett did), and it was "Good morning, class" ever after. Routinized greetings are fine — in fact, expected and appropriate — if they are brief and the information quantity is sufficient. But a lengthy, needlessly redundant self-introduction, repeated verbatim, every time you meet someone is pathological. Coming from a person, you would think they have either a severe memory deficit or a severe social debility; from a machine, you just think the designers are devoid of both imagination and any practical wisdom about conversational interaction.

The common ground for a conversation, remember, has two main components, the background and the conversational ground. The first is what people bring to a conversation, the second is what they develop along the way. (Forgive me, by the way, if I am insulting you by ignoring our background and saying the same thing over again; I'm just trying to bring parts of the background into current focus, not assuming *you* have a memory deficit.) Well, once the conversation is over, the conversational ground becomes part of the background for the next conversation. The next time the same conversants encounter each other, certain specifics will indeed have dropped away, but they both will retain enough that, for instance, identically repeated self-introductions will be highly bizarre, the stuff of *Twilight Zone* episodes.

Voice interfaces need to operate with a background that (1) grows as a function of previous conversational ground, and (2) takes advantage of the copious, accurate, rapidly accessible memories that make computers such a marvel. If you have captured a delivery address once, the customer should not have to give it to you again the next time they call. If you greeted the caller with instructions on her first call, and that call was both successful and fairly event-free, then the instructions are no longer necessary the next time (under certain call-frequency and last-time-called thresholds). If a given caller is frequent and frequently successful, you might want to reveal more and more of the system's functionality.

You taper and expand as a function of the individual user, that is, as well as of the specific encounter.

Individual users

Let it be your constant method to look into the design of people's actions, and see what they would be at, as often as it is practicable.
— Marcus Aurelius

Depending on the type of service your system provides, you will have a range of users — some will use the application regularly, some will use it once, some will engage it once a month intensively, some will engage it every few days, briefly. You need to know who is who.

You already have extensive user profiles going into the specification phase, of course. You know who they are demographically, how they act in the course of relevant tasks, and how they talk, both generally and specifically, in the course of relevant tasks. You know them as an aggregate. Your job now is to design a system, first, that draws on that aggregate knowledge, and second, that now knows about them individually, as a function of their engagement with the interface, that learns about them.

Computers have good memories, and you need to begin using that memory from the first call, storing data about the users on an individual basis. (Also, of course, you need to begin storing data on a collective basis too, but that data gets funneled off and aggregated for analysis in ways we've already talked about in Chapter 10, on users and tasks, and will revisit in Chapter 15, when we get to beta tests and field studies; we're talking here about the specific individuation of everyone in your user population.)

Individuation — in the perennial general/specific tradeoffs we have to make to understand anything — inevitably involves classification. Most importantly for voice interface development is Atwatter et al.'s (2000: 280) frequent/infrequent dimension, calibrated by recency. These categories map fairly directly into the traditional user classifications (expert, intermediate, and novice) though the mapping is far from universal — different systems, with different tasks, and different architectures will map in different ways and in particular will involve setting different thresholds. But an expert is someone who uses the system frequently, and probably used it recently. A novice has little or no history of interaction with the system, so that frequency and recency are both null, or close to it. Intermediates are intermediate — perhaps a recent use or two has them well along the learning curve, perhaps a history of frequent interactions, but no recent ones, suggests they may have forgotten some strategies or reframed some expectations. Table 13.8 shows these rough correlations.

Again, the mapping of frequency and recency to user experience level is specific to the system, but if we take a median information flow task, in Novick et al.'s scheme (that is,

	Novice	**Intermediate**			**Expert**
Frequent	No	No	OR	Yes	Yes
Recent	No	Yes		No	Yes

TABLE 13.8 Usage gauges for classifying user experience

	Novice **(infrequent, not recent)**	**Intermediate**	**Expert** **(frequent, recent)**
Rate of use	0–1 calls/month		≥3 calls/month
Recency of use	>1 month		<1 week

TABLE 13.9 Rough frequency and recency thresholds for a median information-flow voice interface

moderate output, moderate input), the thresholds in Table 13.9 are reasonable beginning guidelines. These are far from the only dialogue features that should be harvested from dialogues (individually or collectively). The following list includes the more significant dialogue features that should be monitored for classifying user patterns.

- Duration of call (average length of call)

- Turns (average per call)

- Vocabulary

    - Word types (average per call)

    - Word tokens (average per call)

    - Tokens/type (ratio)

- Turn overlaps (average per call)

- Early terminations (average; percentage of total calls)

- Repairs (average per call)

    - Nonrecognition

    - Rejections (average per call, percentage of tokens)

    - Timeouts (average per call; percentage of turns)

- • Interaction slippages (average per call)

- • Task slippages (average per call)

- • System inquiries (percentage of turns)

- • Initiative (percentage of turns)

This data (but not only this data) should be gathered for a general understanding of the speech system and its interface, but also — held in comparison with the overall data — for a specific understanding of the individual user, especially her experience rating. From this data, you need to generate a set of expectations about specific users. Candace Kamm and her AT&T colleagues, for instance, characterize the two endpoints of the novice-to-expert continuum this way (Kamm et al., 1997):

An expert

- • knows and remembers what the system feature set is and what commands invoke those features

- • prefers terse, implicit confirmation strategies that move the dialogue along as quickly as possible

- • typically speaks to the system in terse telegraphic commands

A novice

- • remembers a few commands, but often will need reminders of what is available

- • may prefer more thorough confirmation to assure her that the dialogue is progressing correctly

- • is apt to provide only partial information, requiring more frequent use of incremental strategies

Fixing users on a novice-to-expert continuum, that is, critically guides expansion, tapering, initiative, repair, and interaction styles generally.

Systems that don't have recurrent users (or have such brief and undifferentiated dialogues, as with an information-assistance service, that there is no reason to treat users as recurrent) need only collect this data for general purposes. Other systems in the low quadrant of the information-flow matrix might get a sufficient read on experience level with frequency and recency rates alone. But the more genuinely interactive the system is, the greater the information flow, the more helpful such data is for classifying users and adjusting dialogue behaviors. For instance, a user with a high number of timeouts against the average rate might have her thresholds increased. A higher average number of nonrecognitions should lead to more constrained dialogues. A higher number of turn-over-laps can lead to a more open interaction style.

Also significant for setting interaction style is Attwater et al.'s victim or volunteer designation (Attwater et al., 2000: 280). Victim inevitably falls at the novice end of the experience continuum, and correlates with the replacement of one service (keypad or human-mediated) by an automated speech system, and with the early uses of the new service. But greeting style and general interaction pattern — especially expansion and initiative-management — should be strongly shaped by this consideration. Intermediate and expert users, and (depending on the service) a substantial number of novices, will be volunteers.

User experience is the central variable for individual interaction management, but many systems also need to customize user models further. Once the system has whatever recurrent data it might need, that data should be stored as part of the user's individual data record (to the extent that it doesn't threaten their privacy or security concerns). Users should not have to go through address routines, credit-card-number routines, or other tedious recitations more than once. Other task-related settings that might improve the user's satisfaction, and the overall success of the dialogue, should also be harvested and stored with the user's data record — driving patterns, favorite teams, pizza toppings, whatever is relevant to the service.

More subtle customization features are also available — soundscape and agent prefer-ences, for instance, and even such discourse characteristics as politeness (some users will prefer brusque efficiency over mitigation, others the reverse), can all be modified quite easily, if you can suss out the user's inclinations in those areas.

Last Resorts

Clearly, the adaptations that users employ to correct recognition errors in many cases seem to yield the opposite result.
— Gina-Anne Levow

Sometimes the interaction just collapses. Sometimes it can't even get started. The caller might be using a cell that is cutting out; maybe there is just too much background noise to deal with; maybe there is, in the quaint parlance of telephony, a goat on the line (someone whose speech is significantly out of synch with the recognizer — an accent, an impediment, an inordinately slow speaking style, whatever). In any of these cases, there should be some desperate measures possible before pulling the plug completely.

Falling back to a highly constrained interaction, with two or three clearly dif-ferentiated options, should usually be the first of these desperate measures. The con-versational paradigm goes out the window, bringing the encounter perilously close to grunt-detection mode, but it is better generally to abandon the model than to abandon the dialogue:

AcmeTravel   Do you want to leave between 6 and 9 p.m.? Yes or no?

If the collapse is just around a specific rejection, then spelling is a very useful fall back — one that people occasionally adopt — especially for proper nouns, for which callers may have idiosyncratic pronunciations (names in a directory, cities in a travel system):

ARISE   You are leaving from Paris, what is your destination?

Caller   Dinan <rejection>

ARISE   I'm sorry, I did not understand. Indicate your destination, for example Lille.

Caller   Dinan <rejection>

ARISE   Give your arrival city and spell it if you want, for example Lille L I L L E

Caller   Dinan D I N A N

(Lamel et al., 1998: 212)

If oral spelling doesn't work, spell mode can always be incorporated on the keypad. Indeed, as much as I have disparaged that interactive mode in this book, you may need to fall back on it sometimes: "Look, I'm just not getting anything," the system can say. "Sorry. We'll either have to terminate this call, or switch to using the keypad. Press '1' if you want to continue, otherwise, hang up, and try again later, maybe from a ground line."

And, of course, for any moderately sophisticated interaction, there should always be humans available somehow, even if there is a lengthy queue or users have to wait until sanctioned business hours; leaving customers without recourse when they can't operate satisfactorily with your system is cruel.

OK, Users *Can Be the Cause of Some Slippages*

People sometimes do not listen sufficiently carefully.
    — Niels Ole Bernsen, Laila Dybkjær, Hans Dybkjær

Bernsen et al. (1998: 217–226) present a fairly balanced argument for "the other side": that users do make errors and it is beneficial for speech-system designers to think along these lines. See also Dybkjær et al. (1998), Weinschenk and Barker (2000: 212f). We've already seen a few of these (like the user/user source/initiative time-change dialogue). But consider a typical "user error," in Bernsen et al.'s terminology:

System   On which date will the journey start?

Caller   The first weekend of February.

System   Friday February 10th. At which time?

Caller   It must be Saturday at 7:20.

(Bernsen et al., 1998: 219)

This exchange is seen as indicating generally that sometimes users don't pay enough attention, and specifically that this caller "ignor[ed] clear system feedback," an error which is identified as the "direct cause of the transaction failure."

Yes, it's true, I admit, that people are sometimes inattentive — like our hapless Hanover/Frankfurt rail customer — and inattention may even be a factor in this exchange (though it looks to me as if the caller is quite attentive and is attempting to correct the system's construal that "weekend" means "Friday"; another interpretation is that the caller is attempting to repair her earlier utterance by being more specific). People make mistakes. My general point in this chapter is that there is no payoff, or at least very little payoff, in blaming users.

Bernsen and the Dybkjærs may be right about the first-weekend-in-February slippage, that the user's lack of attention caused the slippage. But it still falls to the interface to be responsive to the caller's attempt to negotiate a successful transaction, which starts by regarding the sequence of caller utterances with charity. Rather than breaking down (as the Danish Dialogue System apparently did here), or responding with "Error! Departure is Friday!" or the like, it should come back with an attempt to confirm the most recent user input: "You wish to leave on Saturday, February 11th, at 7:20 a.m., correct?" Most of the "user-errors" they catalogue, in fact, are interaction slippages that might be handled by allowing insertion sequences and granting the user more initiative (see Bernsen et al., 1998: 219–226).

A much clearer case of a user-source slippage — that is, where the user might actually be faulted for the breakdown, rather than the recognizer or the interaction design — which the system repairs on its own would be a task slippage about which the system has greater information and therefore has to initiate the repair.

Caller   Transfer $300 from my savings account to my checking account.

AcmeBank   Certainly. Oh, sorry, I can't do that. My records show you have only two-hundred, seven dollars in your savings account.

Other than these cases, which should be handled graciously, the user will always need to repair his own slippages.

Pre-existing Sources

True art selects and paraphrases, but seldom gives a verbatim translation.
  — Thomas Bailey Aldrich

The spoken output coming through a voice interface will not always originate with the dialogue writers. It may, in fact, come from almost any machine-readable source (and is any text *not* machine readable these days?): newspapers, catalogues, textbooks, web sites, email, even automotive user manuals. One of the virtues of speech systems is that they can take a text you would ordinarily be reading, but now that your eyes are busy doing something else, like watching the road, they can present it acoustically. You can listen to it. But listening to technical descriptions or consumer reports is not enough. You need to be able to get around in those descriptions and reports, shift your focal attention from one part of them to another as needed. If you were reading, you could skim and scan, directing your eyes; you could turn the page. If you were reading off a screen, you could also point and click and scroll and link. You can't direct your ears with anywhere near the same facility as you can direct your eyes. It's that same old spatial-versus-temporal difference we keep encountering when we compare vision and hearing. And there are no pages to turn or devices to move around or click.

If you can't shift your attention auditorily the way you can shift it visually, and there are no object-manipulating devices, there's only one alternative. You have to shift the focal attention of the providing agent. You have to talk, to question and request, navigating through the information verbally with the agent's cooperation. That takes a system with good information *management*, but sometimes the information is already managed, in a way that isn't prepared for vocal interaction. In those cases, there is a much higher burden placed on the strategies of information *access*, on the interface.

Under the worst-case scenario, the source text was produced only to be consumed visually — with tables, indents, boldings, typeface changes, and the like, as well as various styles of visual jargon — with no semantic mark up, the sort of scenario that characterizes a vast number of web sites.

Let's take a particular example of voicing one type of visual data (incredibly common to the Web), tables.

Voicing Tabular Data

The virtue of a thing is relative to its proper work.
  — Aristotle

How do you read a table? Take Table 13.10, for instance, the sort of table most of us have six or seven versions of scattered around our houses, offices, and desktops (real and virtual).

February

Sunday	Monday	Tuesday	Wednesday	Thursday	Friday	Saturday
						1
2	3	4	5	6	7	8
9	10	11	12	13	14	15
16	17	18	19	20	21	22
23	24	25	26	27	28	

TABLE 13.10   A common domestic and business information table

A serial voicing of this array would be blunderingly useless. "February. Sunday. Monday. Tuesday. Wednesday. Thursday. Friday. Saturday." It would say, "1. 2. 3. . . ." (See Raman, 1997: 6–7). There is no point in just broadcasting data of this sort, we need to make it available for navigation. It should not be read, it should talk, and users should talk back — like, say, the way SpeechActs and Nicole Yankelovich talked:

T$_1$ SpeechActs   You currently have "Lunch with Kate Ehrlich" until 2 p.m. At 3 p.m. you have "Brainstorming meeting in the lab." At 6:30 p.m. you have "Dinner with Ellen."

T$_2$ Nicole   What do I have tomorrow?

T$_3$ SpeechActs   Tomorrow you have "Dentist appointment."

(Yankelovich, 1994)

At T$_1$ SpeechActs is just "reading," which is what Nicole wants at that point. But at T$_2$, she wants something else and shifts its attention, roughly analogous to the way she would move her eyes or flip a page if the calendar was physically in front of her, scroll or link if it was virtually in front of her.

What the Sun MicroSystems folk learned by voicing their Calendar Manager, a desktop application with a graphic interface, is very instructive about the visual and temporal differences between a graphic interface and voice interface. The primary unit of reference for a calendar is the date. "If you have a [physical] calendar in front of you," Yankelovich noted, "absolute, fully-specified dates work well" (1997). You can refer to a date as if it was a location on a grid: "Do I have any appointments on February 25th?" you might wonder, and then you would seek out the cell labeled "25" on the table labeled "February."

But if you don't have a physical or virtual calendar in front of you, you are much more apt to think of the dates — and therefore talk about the dates — as points along a line oriented with respect to the current date/point. You are much more apt to talk like this:

"Find his calendar for Tuesday."
"And the day after Labor Day?"
"And the following day?"
"What do I have this Friday?"
     (Yankelovich, 1997)

There are a lot of Tuesdays on most calendars, but if you localize the referent with respect to the current point in the interaction, rather than with respect to a grid, no more specification is necessary; and sequential lexical markers like "after" and "before" and "following," along with deictics like "this" and "that," provide for easy, habitable navigation. You haven't changed the information management, the database; you've changed information access, the interface.

The idea, of course, in moving from a visual representation of February to a verbal interaction about February is not to substitute one form of reference (relative dates) for another (the graphic-supported form, absolute dates). People are still going to say things like:

"What have I got on the 25th?"

They are still going to need the aggregating features available to a table:

"Do I have any Fridays free this month?"

They will still use absolute reference on occasion:

"How about Presidents' Day?"

And they will use combined forms of reference:

"What day of the week is Valentine's Day on this year?"

The idea, rather, is to accommodate a different interaction style, to consider the translation issues — both cognitive and linguistic — when moving from visual text to the spoken language.

Take another translation issue of nonaudio structured information: what about links?

Voicing Links

Alice didn't know what to say to this: it wasn't at all like conversation, she thought, as he never said anything to HER; in fact, his last remark was evidently addressed to a tree.
     — Lewis Carroll

In voicing a web site, you have to deal with links. The first problem is that not all links are created equal. Some links are only links, some links contain information; some are external to the site, some are internal. These differences in constitution and function mean differences in voice design strategies.

Panasonic Lumix LC33

More information		**In-depth review**
		All Panasonic products
		Panasonic web site
Discussion		**Read owners' opinions (182)**
Support us; purchase from our sponsors		Click here to check price / order online
Format		Compact
Camera body		n/a
Price (street)		US$249
Max resolution	?	2048 × 1536
Image ratio w:h		4:3
Effective pixels	?	3.14 million
Color filter array	?	RGB
Sensor manufacturer		Matsushita
ISO rating	?	Auto, 50, 100, 200, 400
Zoom	?	n/a
Digital zoom	?	Yes, 3×
Auto focus		Yes, TTL
Manual focus	?	No
Normal focus range	?	50 cm
Macro focus range	?	10 cm
Aperture range	?	F2.8–F4.9/??
Shutter range	?	1/2000–8
Built-in flash		Yes
External flash		No
Flash modes		Auto, Red-Eye Auto, On, Red-Eye On, Red-Eye Slow Sync, Off, Slow Sync (1 & 2)
Movie clips		320 × 240, with audio, no limit
Remote control		n/a
Tripod mount		Yes
Self-timer		2 or 10 sec
Time-lapse recording	?	No
Storage types	?	SD/MMC card
Storage included		32-MB SD card
Format	?	JPEG (EXIF 2.2)
Quality levels		Fine, Standard
LCD		1.5″, 114,000 pixels
Video out		Yes & Audio
Ports	?	USB
Battery / Charger	?	No
Battery	?	AA (2) batteries (NiMH recommended)
Weight (incl. batteries)		215 g (7.6 oz)
Dimensions		96 × 66 × 34 mm (3.8 × 2.6 × 1.3 in)

TABLE 13.11   A spec sheet for the Panasonic Lumix LC33, from a consumer-electronics comparison shopping site

Take Table 13.11, from a comparison-shopping site for consumer electronics. It contains standard information, like the following:

Price (street)		US$249
Max resolution		2048 × 1536
Image ratio w:h		4:3
Effective pixels	?	3.14 million
Color filter array	?	RGB

These are all fairly easy to treat for voicing. Unlike the calendar example, they do call for some change in data structure, but it can be handled relatively easily with a bot that reads HTML tables. It just has to be set to follow left-right reading conventions and extract row information, so that, for instance, "Price (street)" becomes an information label, with the value of "US$249" for this camera. The system can now handle specific queries like "What does it cost?" "How much is it?" as well as database-wide commands like "How many cameras are there between $200 and $400?" Similar treatments undergird queries like "What is the image ratio?" and "How many pixels?"

But Table 13.11 also contains information like this (where the bolded elements represent links):

More information	**In-depth review**
	All Panasonic products
	Panasonic web site
Discussion	**Read owners' opinions (182)**

That is, the table treats linked information as falling into the same category as street price, resolution, photo detectors, and so on. Links don't belong logically in the table, of course, in the sense of providing a small piece of cross-referenced information. They are routes to other bodies of data. That consistency violation isn't a problem in the context of a web site, at least not for a moderately experienced user, who can use the coded visual cues, intuition, and cursor shape to sort first-order data from data-routes very quickly. But in an auditory context links need to be differentiated from the first-order data.

But there is a further difference in these links: *Panasonic web site* is an off-site link; *In depth review*, *All Panasonic products*, and *Read owners' opinions* are on-site links. The table also includes icons linking to glosses and explanations of some information categories (the ? beside *Effective pixels* and *Color filter array*, for instance). External links may not even hook to sites with speech capabilities, and the nature of voice services is such that a partnership would certainly need to be worked out in any case.

Even more problematic is the possibility of losing the user altogether. A frequent phenomenon in hypertext, especially before users became more cyberliterate and browsers became more intelligent, was users following link upon link until they were, as the saying went, "lost in hyperspace." The potential for similar user disorientation in an auditory format is great indeed. By default, external links should just be ignored in voicing a web site. Don't even mention them (they can usually be found by the presence of "http" in the mark up).

On-site links, conversely, are labels for bodies of data that are part of the overall structure of the site, and should usually be accessible to the caller, through obligatives — offers, requests, suggestions, commands. In most cases, links should not even be identified as such.[4] The better approach is to present options, as appropriate. Here is a dialogue that illustrates this approach:

T_1	AcmeShop	What would you like to know about the Lumix LC33? <1.5 sec> I can tell you about its features, or would you like to hear an in-depth review?
T_2	User	Anything else?
T_3	AcmeShop	I also have owner reviews.
T_4	User	Tell me about the features.
T_5	AcmeShop	I have information about twenty-seven features for this camera. The list will take about two minutes. You can interrupt at any time.
T_6	User	OK
T_7	AcmeShop	Format, compact. Price, 249 US dollars. Maximum resolution, 2048 × 1536. . . .
T_8	User	Wait. What's the resolution?
T_9	AcmeShop	Do you mean the resolution for this camera, or do you want a definition?

4: In principle, there are audio techniques that might serve to identify links — a conventionalized tone playing simultaneously with linked information, for instance, would take advantage of the cocktail-party effect for this purpose; pitch or duration or rate changes could also work. But conventionalized tones or pitch/duration/rate shifts, it they eventually prove effective for some categories of auditory-interface users, are probably better saved for more specific information types — definitions, perhaps, or information with differential cognitive demands. Marx (1995: 85–86), for instance, makes the intriguing proposal that (in certain highly conventional contexts) given information be output at a faster rate than new information.

T$_{10}$	User	A definition.
T$_{11}$	AcmeShop	Just a moment. I'll get that.

"Resolution. The number of pixels-per-inch in a digital file. The more pixels-per-inch the more information held in the file, the higher the resolution."

That's all I have. Would you like to continue with the list of features? . . .

| T$_{12}$ | User | No. Tell me about the zoom. . . . |

Tabular data and links are two of the most common issues with translating text from non-audio formats. There is a range of others, most of them quite specific, but we will only take up one final problem for voicing pre-existing text, lexical translation, which implicates pronunciation and diction.

Lexical Translations

Caller   Tell me about the zoom.
System   The feature zoom is n-slash-a.
    — An exchange from the test phase of a consumer-information voice interface prototype

Pronunciations are, we have seen, register dependent. Strategies need to be developed and followed, from the discourse-analysis phase on, for how words should be spoken. Numerals, for instance, can be quite tricky. In an address, 247 would generally be pronounced "two-forty-seven;" as a quantity, in an inventory, "two-hundred, forty-seven." The expression 5/12 could be "five twelfths," if a fraction; "May twelfth," if an American date; "Five, December" or "December fifth" if a British date. As a chapter designation, IV would be pronounced like an Arabic cardinal, as in "chapter four;" in a name, it would be the ordinal ("Henry the fourth"); in another context altogether the letters might be pronounced (/aj vij/), or it might be expanded to full words ("intravenous," "in-vocabulary"). You can't just turn a text-to-speech engine loose on numbers without articulation and substitution rules. But it does not end with numbers, not by a long shot. There are special symbols that need lexical treatment, varying with register ($, %, @, C++). Even a standard dialect presents its share of problems (compare the pronunciation in "six *lives* were lost," for instance, with the pronunciation in "he *lives* in Hogtown").

In general, the synthesis people on the project will have a good handle on these issues for the basic terms and the basic contexts, but any aspects of the domain register that have nonstandard implications for pronunciation or expansion need to be caught in the discourse-analysis phase, and calibrated during testing.

In extreme circumstances, these issues involve coding a parallel ortholect, rather than just a few variant spellings or conventions. Text messaging is the most obvious example of a distinct ortholect (r = "are," b4 = "before," cul8tr = "see you later," etc.). But there many examples where ortholects need to be addressed in a subtler way than direct, isomorphic translation. In particular, the implications are not always directly lexical; there are also utterance-level effects for many passages of pre-existing text.

Take, for instance, the situation where product descriptions are voiced, with the data coming from a table like 13.11, perhaps in a consumer-products information voice portal. What happens when the users ask about the feature zoom, which the table represents this way?

| Zoom | ? | n/a |
| Digital zoom | ? | Yes, 3× |

Discourse analysis is always concerned with ambiguity; the first thing needed for this data is a clarification request:

T$_{12}$ User   No. Tell me about the zoom.

T$_{13}$ AcmeShop   The optical zoom or the digital zoom?

T$_{14}$ User   Optical.

The interface, that is, needs to introduce terminology (here, the adjective *optical*) that the source neglects. But the complications are just starting. The data for the feature is in the right hand column, but you can't leave "n/a" to off-the-shelf pronunciation rules for the speech synthesizer, or you end up with the "n-slash-a" problem, and at best a brief puzzle for the user, at worst, total confusion. On the other hand, because of the context, a simple abbreviation-expansion rule won't work in this context either, which would give the user something even worse:

T$_{15a}$ AcmeShop   The optical zoom is not available.

The "n-slash-a" answer violates one Gricean maxim — it is obscure. The "not available" answer violates another — it is misleading (the Lumix LC33 *does* have an optical zoom). The voicing routine here has to be sensitive to the context and offer up something that Grice would sanction, something clear and unequivocal:

T$_{15b}$ AcmeShop   OK. . . . Oh, sorry. It looks like there is no information on the optical zoom. <1 sec.> But it *does* say that the camera has a three-time digital zoom.

Just spending time examining data, like the specs arrayed in Table 13.11, highlights many of these problems, and should also suggest the utility of many of the strategies we have discussed. For instance, a caller might reasonably ask if the camera has a Firewire port. The service could reject *Firewire*. But, with lexical density (*Firewire* is in the acoustic models, and the language model knows it as a KIND-OF port), it can respond, "No, this camera only has a USB port." And this site has customer reports, which inevitably involve email conventions: the user isn't going to want to hear "colon, right parenthesis" for every smiley face it encounters; that is, the über-emoticon, :).

Legacy

Please listen closely to the following options, as our menu has been changed.
— Way too many keypad-system openings

There is a vast difference between adding a voice service where nothing existed previously, and adding a voice service where the same (or largely the same) services were already in place. Throughout this book we have been pretty much pretending that the voice interface comes in without baggage. But it rarely does, perhaps never. I don't know any such situations, in any case. Existing systems and services can be a decided benefit in the data-gathering phase; in the design and implementation phases, they are always an anchor.

The legacy of a voice interface is typically one, possibly more, of three interaction modes: human, graphic, or keypad. Often, especially for older customers, there may be multiple legacies. Talking to a person was replaced by listening to a machine and pushing keys, which is now being replaced by talking to a machine. And those big, ancient, and many tentacled institutions, such as banks and insurance companies, may have a quite haphazard network of overlapping services — IVR, ABMs, web services, human–agent voice encounters, and human–agent in-person encounters.

The golden rule for legacy is simply: Respect it. Change makes previously experienced users novices again, a disconcerting situation for them. If the recent legacy is human interaction, you need to provide a bail-out option, prominently offered (not buried at the end of a long preamble, for instance). If the legacy is a keypad system, you need to keep all of its options alive for a substantial overlapping period. Right in the middle of a system-uttered verb in a recently keypad-based system, for instance, if a caller starts pushing keys, the system should do its best to make sense of them, routing her in the appropriate way.

If the legacy has a graphic interface, chances are the voice service is only augmenting it, not replacing it, but it still exerts an influence on the expectations of some proportion of the users. What is often forgotten about graphic interfaces is that they are rife with text — menu labels and items, button labels, field labels, icon descriptions, and the like. And while the transfer of lexical expectations is certainly not direct (Yankelovich et al., 1995), there is always contamination. You will rarely want to use the spatially oriented, object-based terminology of a graphic interface as the preferred terms for the temporally oriented, action-based vocabulary of a voice interface. But you shouldn't disable it either. Lexical density should be biased toward any legacy vocabulary.

Those are the design considerations. They're straightforward: your system should do its best to accommodate the expectations legacy customers might bring to the interaction. The implementation considerations are more Byzantine. There are both technical and political obstacles to implementing a voice interface in legacy situations.

Catherine Wolf and Wlodek Zadrozny (1998) offer a telling little parable about their adventures in bringing a conversational voice interface (Conversation Machine) to a bank with a medieval voice-response system. The first problem was that they came with a user-centered design philosophy, and ran headlong into a bank-centered design philosophy, where the accounts were primary, customers secondary. Rather than a single ID and password that could aggregate and access all of their accounts, the existing system required each account to be accessed individually, via its own lengthy account number. They hit upon the solution of minimizing the customer's pain by storing the information. Once it was entered, it was assigned an appropriate real-world label (checking, savings), which could then be accessed by that customer ever after.

Except, some customers had multiple accounts of the same type and *savings* or *checking* became ambiguous in those cases. Worse, the account-centric database apparently had no way of knowing when a customer had two checking accounts or two savings accounts, so natural language designators like *primary* and *secondary* were out of the running. Only numbers would work, and the best they could manage was to get down to four digits. The best they could manage for were locutions like "I'm about to transfer $100 from savings account 5678 to checking account 9101."

Some very reasonable activities, like getting a summary of all accounts, proved too cumbersome to implement. Technically it was feasible, but it would have required the customer to jump through more hoops than a show pony: "It would require the Conversation Machine to query the user for each account type and number."

As if the existing design sensibility and technical constraints weren't enough, they also encountered a legacy of stubborn possessiveness. The existing application presented transaction histories indiscriminately: in-bank transfers, deposits, withdrawals (all of those activities distinctly if performed on an ATM, and again if performed through the existing VRU), in-bank bill payments, VRU bill payments, cleared checks, interest, fees, and so on. Such a list is not too great a problem to read off a sheet of paper, or even a screen, but recited over the phone it would be deadly. This time, the technology was available to let the user engage the information naturally, and ask for just the cleared checks or the deposits. It just required some filtering of the list. Let the principals continue their tale of woe:

> [We] favored this [approach] in the interests of creating a system which matched users'
> needs. However, the bank personnel were wary of any "enhancements" to the system; they

wanted to maintain control of the banking functionality and viewed the Conversation
Machine as an alternative front-end to the existing VRU functionality.
  (Wolf and Zadrozny, 1998).

In the end they had to crank out the omnibus list in response to any query, even very specific ones, like "Tell me about cleared checks," with a cautionary introduction that
explained to the customer she was about to hear a list of everything, leaving her to pluck
out the items of interest auditorily (and retain them, if possible, against a stream of con-
taminating numbers, dates, and designations). Wolf and Zadrozny called this unfortunate
compromise "an interim solution," laconically remarking that they expected it to be "revisited based on the reactions of users in assessments underway and in the future."

These sorts of compromises are inevitable in complex, inertia-ridden legacy situations.
Technically, it is rarely possible to build on top of keypad-based operations. The machinery, its code, and its maintenance are highly proprietary in the keypad-system industry (a
tradition that continued into the early years of voice systems, but that is now giving way
to much more flexible standards-based solutions). Even where they are not, legacy systems
are frequently just too rigidly structured to adapt toward conversational interaction.

Politically and financially, the tendency in such cases is to develop modularly, phasing
new voice components in, legacy components out, frequently in a staggered way with
field trials that utilize both experienced customers and new recruits. These mixed-mode
applications (speech and keypad) are awkward in the extreme to design and just as
awkward to use.

Summary

The system can never be sure it has correctly inferred the user's referent; it can only make
good guesses.
  — George W. Furnas, Thomas K. Landauer, Louis M. Gomez, Susan T. Dumais

Human factors design is unique for voice interaction, because of two complicating
traits. Alexander Rudnicky (1996) puts it this way: "Speech interfaces have two properties
not normally found in more mature interface technologies," which he identifies as:

- They are errorful.

- Their state is often opaque to the user.

Errors will happen with speech recognition systems. They will miss some things altogether
(nonrecognitions). That's not so bad, because the system knows it's having a problem; it
can alert the user and collaborative error repair can begin. The main stumbling block here
(aside from the user's patience) is that the system's groundings and its expectations are
opaque. But they also get things wrong (misrecognitions). That's worse, because the system

thinks the dialogue is proceeding hunky dory. Not only are the system's groundings and expectations (its state) opaque to the user, the problem itself is opaque to the system. It is up to the user not only to collaborate on the repair, but to discover that one is needed in the first place and initiate it.

Voice interaction design, therefore, works not only to mediate between the application functionality and the user, but also between the recognition system and the user. No other design community spends anywhere near the same attention to input complications. It would be as if the pointing device for a graphic interface one was designing just randomly refused to drop some menus down or launch some activity — in different groupings for different users — and occasionally triggered unpredictable actions on distant parts of the screen. Moreover, even though one might be able to count on some regular incremental improvements, the erratic mouse would never be replaced by an entirely error-free device. The job would be to design for these predictably random occurrences. The job, for voice interface developers *is* to design for these predictably random occurrences. This chapter focused mostly on design issues related to errorful, state-opaque speech applications.

The first thing we took up is the nature of those errors. They are of two general, inverse types, errors that are **known** at run-time, and errors that are **unknown** at run-time. The known ones are still only known in a very fuzzy way. They are known to exist. That's all. Very little is understood at run-time with any precision about their specifics; in particular, the input is either incomprehensible or absent. These are **nonrecognition** events — **rejections**, **timeouts**, and **spoke-too-soons**. The unknown errors are **misrecognitions**, where the system follows an instruction, sets a particular value, or pursues some specific goal, but the user has issued some other instruction, supplied some other value, or has tried to pursue some other goal. Even that sentence is somewhat misleading because it makes the errors seem systematic, when they are really mix-and-match (the user might try to issue a command, but the systems sets a value, and so on). These errors are particularly insidious because the user not only has to help repair them, she has to discover them as well.

Fortunately, there are some **preventative measures** and some repair-facilitating strategies. First, the design can follow strategies that prevent some of the errors. The bulk of rejections comes from **out-of-vocabulary** utterances, for instance, which a policy of judicious **lexical density** can help reduce by putting more of the user's expected terms in the system's vocabulary. Another way to combat out-of-vocabulary breakdowns is to prime users in the direction of the preferred in-vocabulary terms by **stealth training**. The system vocabulary can also be structured so that in given contexts out-of-vocabulary utterances can be resolved by bringing in other vocabularies, particularly ones stocked with **reserve synonyms and metonyms** to help resolve the user's utterance rather than rejecting it outright. And the system can be built with buffers that **suitcase** raw input, so that it might be further resolved at other points in the interaction. Spoke-too-soon rejections, for their part, should be eliminated almost entirely by implementing widespread **turn-overlap freedom**.

Even with thorough-going prevention in place, however, we know there will still be errors of both types, nonrecognitions and misrecognitions, and that **repair** strategies are essential for both. Rejection errors and timeouts are best handled by brief directives, such as suggestions or requests. Misrecognitions cannot be handled except by the user, but the system must be prepared to surrender initiative and collaborate in the repair.

But there are difficulties in telling users' attempts at repair from their simple goal-directed input (rather than as feedback on the previous system utterance), which can set off an **error spiral**. The recognition and inference engines, therefore, need to adopt principles that will help them identify repair attempts, including self-repairs. Fortunately, there are noteworthy **acoustic repair signals**, such as hesitations and prosodic shifts, that the recognizer should be tuned for; and there are **lexical repair signals**, such as negations, repetitions, and minor vocables of change and repair, that the inference engine should be alert for.

The most dangerous aspect to misrecognitions, however, is that even the user can miss them, leading to bad information, wrong bookings or purchases, and other potentially serious misfires. The job of the interface in this respect is to help keep the user on her toes by judicious **feedback**, which presents system-groundings and expectations to the user regularly for review, and well-placed confirmation requests, which seek the user's explicit approval of those groundings. The tricky part is knowing what constitutes **judicious** feedback in any given case, and in gauging the pace for seeking approval. The relevant decision procedures here depend on the **grounding criteria** for system commitments, which in turn depend on **task and information criticality**, as well as recognition confidence and timing concerns.

Also essential to the repair process, as well as to satisfactory dialogue design generally, is sound **initiative management**. The interface needs to have principles guiding when to cede initiative easily, and when to retain or retake it, in order to get the dialogue back on the rails.

Users do not always know how to take the initiative, however, because of the opaque-state problem, and failing to take it can easily lead to error spirals. The best way to combat this sort of interactive degeneration is to make that state — or, more particularly, the user's potential actions during that system state — as clear as necessary. The user needs guidance. The problem is that the user can also be opaque to the system, and bringing them to mutual coherence might take several consecutive exchanges, a process known as **progressive assistance**, a form of **expansion** — the policy of generating longer utterances as required. The flip side of expansion is **tapering** — the policy of generating briefer utterances to increase the efficiency and naturalness of the dialogue.

Both expansion and tapering are geared to the **experience level** of the user, which can be inferred from frequency and recency of call, as well as the prevalence of interactive features like timeouts and turn overlaps. Identifying, monitoring, and adjusting the caller type helps the system conduct the dialogue at a comfortable pace with appropriate levels of feedback and prompting. And storing user data in general can lead to more **customized interactions**.

We also took up two further matters of voice interface design: the **translation issues** involved in voicing an existing application, such as how to represent **tabular data** interactively, how to **voice links** and how to guide pronunciation and **lexical treatments** generally when confronted with visual-register symbols and terminology; and the technical-political complications that come with voicing an application with an influential **legacy**.

14

CHAPTER

Scripting

Suit the action to the word, the word to the action.
— William Shakespeare

Designing the specifics of a voice interface is best understood as a job of scripting, in at least two fundamental respects — one somewhat metaphorical, the other more literal. It is scripting in the conventional sense of writing drama; developing a radio play might be the closest analogy. Creating a radio play calls for teams of writers and an editor, under the general coordination of a producer, generating lines of interdependent speech, which are then voiced by actors, contextualized and augmented by sound engineers, and scored by composers, all following the plan of the director. There are production meetings, read-throughs, and rehearsals (prototyping, usability, beta testing), which elicit feedback and therefore revision (spiral design). In old-time radio serials, like *Superman* or *Rocky Fortune*, there are even iterative releases of a sort, where successful character traits, taglines, plot themes and leitmotifs solidify and recur; unsuccessful elements drop away. The interaction design output, too, has much in common with a radio-play script; the system output we call "prompts," for instance, might be better labeled "cuing utterances." Done right, they signal to users how they should play their role in the unfolding dialogue.

But this is the metaphorical sense, and the radio-play analogy is clearly partial. A speech system is not "creative" in the standard, aesthetic, expressive sense of creativity. Creativity and aesthetics are not the driving motive forces behind voice interfaces. A speech system is, or should be, common and routine. That doesn't mean that aesthetics or entertainment are irrelevant to voice-interface development. The system has to be pleasing, even fun, in certain ways. But pleasure must be largely subordinate to utility. Entertainment depends on a certain amount of surprise value; speech systems depend on patterns of familiarity. (That's not to say that the future won't bring speech-gaming applications, chatterbots, interactive audio pornography, or other entertainment-oriented applications of speech technologies and conversational principles — just that such developments are currently marginal, at best, and well outside of our interests here.)

And the relationship of a voice interface to its users is not as static and monologic as the relationship between a radio play and its audience. It's not quite the relationship embodied in a conversation, of course, but it is, or should be, a long way from a unilateral broadcast. A speech system is mutual, reciprocal, interactive. The radio that conveys a play to its listeners is a one-way relay node in an amusement delivery system; the digital appliance channeling a speech system is a two-way gate in an information circuit. This fundamental interactivity means a voice interface must be far more flexibly responsive at run-time (the radio affords only three options with respect to a broadcast: on, off, other station).

Before we drop the analogy altogether, though, there are two architectural elements of radio-play scripting that are isomorphic with two architectural elements of voice interface scripting: the plot, which maps into the call flow, and the dialogue, which maps into, well, the dialogue.

The other fundamental sense in which designing a voice interface is scripting — this one a good deal more literal — follows from the sense in which Roger Schank and Robert Abelson (1977) used the term *script* for knowledge representation: a blueprint of some (usually mundane) activity, capturing the everyday understanding of some contextually embedded sequence of actions or behaviors (including speech actions), at a sufficient level of abstraction to provide for a range of variations. And, in terms of natural language understanding, as Christian Hempelmann puts it, "[knowledge] scripts can serve to bridge the traditionally postulated gap between semantics and pragmatics" (2000: 25). They can bring the truth-functional semantics required by dealing with credit card numbers, shipping addresses, and the like, together with the felicity-driven pragmatics of dialogue acts.

The commonplace example of a Schankian-Abelsonian script is "The Restaurant," of which there are as many versions as there are artificial intelligence and cognitive science textbooks or web sites. Script 1 is a slight variation that serves our purposes a bit better: "The Take-out/Delivery Restaurant."

It is no coincidence that this script resembles the dialogue schematics of earlier chapters.[1] Schank and Abelson adapted the notion of a script to knowledge representation

---

1: A further distillation of the call-to-Baghdad dialogue of Figure 10.1, for instance, would resemble Script 1, and Figure 11.7 *is* such a distillation for the task segmentation of the hotel-booking dialogue of Figure 11.6.

```
MakeCall

OrderFood

GetPrice

GivePhoneNumber

RequestMode (take-out OR delivery)

GiveAddress

GetTime

EndCall
```

SCRIPT 1 A simple knowledge representation of ordering take-out/delivery food.

so they could schematize purposive behaviors, as good a description of the bulk of dialogues in this book as one could hope for. They are epitomes of purposive verbal behavior.

Voice interfaces are crafted expert systems in task-driven, register-specific dialogues.

The first move from the task- and discourse-analysis phase to the active design of such systems is to abstract a conventional knowledge script from the behaviors and the language patterns you have analyzed. Then — not surprisingly to task-analysts but somewhat paradoxical on the surface — you proceed to build the script back up, enriching and elaborating it, drawing again on the task- and discourse-analyses which produced the abstract version in the first place.

This development has two parallel components, crafting the dialogue and crafting the call flow. *Parallel* is too strong a term, because the components do in fact intersect repeatedly, as writers and the architects need to stay in regular communication, completely merging at the design-specification stage. But crafting the dialogue and the call flow are parallel in the sense that they occur simultaneously; neither comes first. Traditional development cycles have tended to put the words last, in a rushed and unguided way, as internal deadlines and release dates loom. This has always been a mistake, perpetuated by development personnel who simply don't respect the difficulties in getting language right. On the other hand, nor can the words come first. Various dialogue-act amalgams and clusters can certainly be written before the call flow is developed, working from the bare-bones initial script. But without the plot component — the call flow — the dialogue can only go so far.

In a book, however, one has to come first. For this book it is writing. In the first major section of this chapter we look at elaborating the script, starting with a moral harangue about the differences between written and spoken language, and the incumbent difficulty of *writing* for *speaking*. After the lecture, we pick up a hypothetical use case growing out of Script 1, the voiced Acme Pizza take-out and delivery order service, using it to probe three specific topics — designing turns, making the diction choices, and using scenarios. With respect to turn-design, we also take up the principle of constrained variability, and look at two specific categories of turn, the opening and the list.

In the next major section, staying with the voiced Acme Pizza service, we chart the processes and issues involved in planning out the call flow. In particular, the sequencing and dependencies of the script's schemas need to be worked out carefully, to find which elements of the script must precede which other elements, and which allow more flexibility in the call flow. Those points of flexibility, in turn, need to be developed in ways that broaden the interaction, allowing users to conjoin tasks and subtasks in habitable ways, rather than facing the crabbed, stepwise interaction models we know familiarly as *menus*. Call-flow plotting, then, is a matter of arranging the elements of an interaction model in ways that maximize the flexibility while respecting the necessary and preferred sequencings. In the last part of the call-flow section, we trace out an interaction against a call flow for Acme Pizza.

Specifying the voice interface design, taken up in the final major section in this chapter, means documenting the merged dialogue and the call flow in sufficient detail to begin the production phase, instantiating the design. But documenting does not necessarily imply a hard-copy product for sign-off. That is by far the standard methodology, but I advocate an interactive digital specification, which may not be primary but certainly should be an integral part of the formal design specification.

Developing the Dialogue

Speak properly, and in as few words as you can, but always plainly; for the end of speech is not ostentation, but to be understood.
  — William Penn

All of the components in Script 1 are integral to the task of ordering take-out. These components — called *schemas*[2] — are collectively sufficient for an order to be placed. But they are not all necessary; nor is their arrangement hard-wired. Typical of knowledge scripts

---

2: Schank and Abelson called them *scenes*, but that term pushes the drama analogy a bit too far. We follow Alexander Rudnicky, among others, in calling these components *schemas*; see, for instance, Rudnicky et al. (1999).

generally, some of these schemas are optional, several are contingent, and at least one can occur at a different point in the interaction. The schema, GetPrice, for instance, is contingent on OrderFood (the price depends on what food is selected), but needn't follow it immediately, and might sometimes be neglected altogether. GetTime might come first, because if the caller doesn't like the answer ("Ninety minutes."), she might decide to try another pizza joint; or GetTime might not come at all, if the caller forgets to ask, or doesn't care about the duration, or simply assumes the answer on the basis of previous experience. The event can still go through.

Script 1 is, like most knowledge scripts, a flexible representation, not a rigid stepwise procedure. Flexibility is one of the chief virtues of Schank and Abelson's notion. Script 1 is a recipe, a description, an instruction set, for the activities required to enact the event we call "ordering take out delivery" — you have to make the call before you can order the food; you have to order the food, with a high probability you will ask about the price; they will likely want your telephone number; if you want it delivered, you have to supply an address; you will probably want to know the approximate time of availability, or of delivery; and you have to end the call. The script represents your generalized knowledge about such events. It outlines roughly what *you* do in such situations. And your actions, of course, are coordinated, sometimes even orchestrated, by what *they* do, the take-out restaurant phone people.

What *they* do is enact a slightly different script, largely inverse to Script 1. The knowledge-representation for *taking* an order, specifically by Acme Pizza (a small chain that we will build a use case from in this chapter), schematizes as Script 2.

There is a good deal of flexibility in this script, too, with interdependence among some schemas (GetFoodOrder must precede ConfirmOrder and generally precedes GivePrice) and the same independence among others (GetPhoneNumber is logically unrelated to any other schema except that it cannot precede AnswerCall or follow EndCall). But the way Acme Pizza instantiates Script 2 is rigidly sequential. They want the phone number right off the bat, as a customer identifier — often refusing even to discuss specials before they get it. The agents are cheerful but dictatorial. The number allows them to reference customer particulars on their computer, so customers don't have to give the delivery address unless it is to somewhere besides their primary abode. They also know what was ordered the last time a call came from that number. After they have the number, they establish whether the order is delivery or take-out; get (or just confirm) the address, if it is to be delivery and only then are they ready to take the order, supplying price and/or time in turn. It is not the script, which is just a collection of task modules, that determines the interaction and the flow. It is the instantiation.

Cutting to the chase: voice interaction design is, in very large part, the provision and enablement of scripts in the classic artificial-intelligence sense. A machine designed to enable people to order take-out/delivery will need to enact Script 2, or something very much like it. The machine will have to play its part when customers call to place their

```
AnswerCall

GetPhoneNumber

EstablishMode (pick-up OR delivery)

GetAddress (acquire OR confirm)

GetFoodOrder

GivePrice

GiveTime (availability OR delivery)

ConfirmOrder

EndCall
```

SCRIPT 2 A simple knowledge representation of taking the order for take-out/delivery food.

order, coordinating and sometimes orchestrating — ultimately, collaborating with them — to achieve their goal. Some callers might take all the initiative and get the job done with an entirely passive and obedient system. Some will need to be led through the encounter step-wise; *they* will be the passive and obedient agents. Some will need a nudge here, will want to give a nudge there.

The interaction model must be flexible. Indeed, it should be more flexible than the sequentially rigid instantiation of the take-out/delivery script that Acme Pizza customers now live with.

Script 2, of course, is the ultimate distillation of a task-and-discourse analysis visited upon Acme Pizza, down to the system-side, bare-bones components of the interaction. From this distillation, we need to move upward and outward in two directions, toward the call-flow blueprint, which resembles a plumbing or electrical layout, and toward the utterance-by-utterance dialogue design, which resembles a highly detailed play script, with elaborate stage directions.

The call flow and the dialogue chart proceed in parallel, with reciprocal influences, but for purposes of exposition, we take them individually, starting with the dialogue.

Elaborating the Script

> *The plot thickens.*
> — Sir Arthur Conan Doyle

Dialogue design starts with an elaboration, drawing on a task-knowledge representation like Script 2, informed by the detailed domain and register analyses that preceded it.

This move may seem counter-intuitive, or even counter-productive, in terms of the development cycle. First, we boil the rich task- and discourse-analysis data down into crabbed little knowledge representations, like Script 2; then, not only do we build these representations back up into fuller scripts that begin to approach those rich analyses in detail and specificity, we use those earlier analyses to do the building. But circling back to reintegrate earlier results and design phases is precisely in keeping with the spirit of a spiral development model. This process of distillation and elaboration is necessary to isolate and craft the integral components of the projected interaction.

The first job is to populate the schemas, as in Script 3.

Script 3 fills out the schemas with dialogue acts mined from the discourse analysis, further specifying the task structure with nested schemas, and establishing the interdependencies of the schemas. It starts the process of developing the system's utterance structures and interaction patterns. We have the dialogue acts (or at least a good starter set — mostly obligatives and informatives); the dialogic pairings come along in the bargain (offers and acceptances, questions and answers). We see the coherence relations shaping the interaction (the AnswerCall schema establishes the background for the following interaction, the GetFoodOrder schema is a network of elaborations, GiveTime is in a conditional relation with EstablishMode, as GivePrice is with GetFoodOrder).

The next order of business is to start adding utterances.

Writing the Dialogue

What I have crossed out I didn't like. What I haven't crossed out I'm dissatisfied with.
— Cecil B. De Mille

Writing is not easy. Hunter S. Thompson puts it this way: "writing is a hard dollar" (Thompson, 1988: 100). He adds, to make us all feel better, "but it beats reaching into an enraged cow and pulling out a breached calf." He's probably right, even if reaching into a reluctant head and pulling out the right dialogue act sometimes seems at least as messy, if not so dangerous. But I'm guessing it's easier to tell people how to pull out a breached calf than to tell them how to write. There are some rough guidelines to keep in mind, and especially some attitudes to cultivate, but eventually, it comes down to, "write good system utterances."

The over-arching attitude you need to cultivate, harder by far to do than it is to say, especially since our *modus operandi* is writing, is an allegiance to speech as the primary medium of the interaction.

AnswerCall

Greet caller
Identify restaurant
Identify self
Offer to take order

GetPhoneNumber

Request number [access customer data record]
  If record on file: Confirm caller identity
    State data record details [name, address]
    Request caller confirm record details [name, address]
  If record not on file: Request record
    Request caller's name and address [name, address]

EstablishMode

Ask the mode (pick-up, delivery)
  If delivery: GetAddress
    If no address is on file: GetNewAddress
    If an address is on file, ask if delivery address is [on-file address]
      If delivery address is not [on-file address]: GetNewAddress
        Request street and number
        Ask if this will be a regular delivery destination
          If it will be regular: Store address as secondary in customer file
    Otherwise, ConfirmAddress
      State street and number
      Request confirmation or correction

GetFoodOrder

GetPizzaOrder
  Ask for the type [special or selected toppings]
    If special: GetSpecial
      Request special selection [Greek, chicken-and-pesto, meaty, meaty-extreme, spicy Sicilian, veggie, veggie-extreme]
    If selected toppings: GetToppings
      Request topping selection [hot Itallian sausage, mild Italian sausage, pepperoni, mushrooms, onions, eggplant, green peppers, roasted red peppers, black olives, green olives, chicken, pineapple, artichoke, ham, meatballs, feta cheese, extra cheese, roasted garlic, fresh basil]
  Ask for the size [small, medium, large, humungo]
  Ask for the crust type [regular, thin, stuffed, wheat]
GetAccompOrder
  Ask if anything else is wanted
    If anything else is wanted: GetDrinkOrder, GetBreadStickOrder, GetChipOrder
      Ask for drinks [number, type, size]
      Ask for bread sticks [number, type]
      Ask for chips [number, type]

GivePrice

State price

GiveTime

State time [pick-up availability OR delivery]

EndCall

Ask if that is all
Confirm order [mode (if delivery, address); PIZZA (total, pizza (number, type, size, crust)); DRINK (total, drink (number, type, size)); BREAD STICKS (total, bread sticks (number, type)); CHIPS (total, chips (number, type))]
Thank caller
Take leave of caller

SCRIPT 3   A detailed knowledge representation of taking the order for take-out/delivery food, elaborating Script 2.

Speech Is Not Text

To find yourself participating in a conversation, even one that has been engineered, with a persona who is reading formally-written text at you, the content of which depends on your own unrehearsed, spontaneous responses, is unprecedented in authentic discourse. Because it cannot be likened to any real-world experience, we can actually consider it an "antimetaphor," and so this kind of artificial formality is undesirable for [voice] interface design.
— Michael H. Cohen, James P. Giangola, Jennifer Balogh

In the 1920s, in an appendix to a book about language, Bronislaw Malinowski had to apologize for writing about talking. As an object of study, for millennia, language had been almost wholly confined to written texts — products of consideration, crafting, reflection: objects. "We have to realize that language originally, among primitive noncivilized peoples, was never used as a mere mirror of reflected thought," he wrote. "The manner in which I am using it now, in writing these words, the manner in which the author of a book, or a papyrus or hewn inscription has to use it, is a very far-fetched and derivative function of language. In this, language becomes a condensed piece of reflection, a record of fact or thought." But talk, Malinowski argued, is the primary linguistic phenomenon, not text. And talk is much less a mental product than a social process. It is an action. "In its primitive uses," he added, "language functions as a link in concerted human activity, as a piece of human behavior. It is a mode of action and not an instrument of reflection" (1923: 312).

It is no longer embarrassing in intellectual circles to write about talk, though it took another forty years or so after Malinowski before programs of study that take language as a mode of action began to develop, in linguistics, philosophy, psychology, and sociology — the programs we have pillaged for insight and principles in this book. But millennia of biases do not shake off so quickly. The understanding of language is still, in many ways, by people who are oblivious to the biases they have inherited, governed by ideas rooted in literacy.

Some people don't need this warning; look into your heart. If you're one of these people, someone who doesn't think "real language" is what you get in newspapers and novels and academic theme essays, then pass freely on to the next section. But many of us need this warning not just once but recurrently. We spend our time reading and writing reports and memos and spec sheets, reading journals and magazines. We got our teaching about language overwhelmingly through writing and reading, through studying topic sentences and paragraph structure and moving from the general to the specific. And almost everything we know about "correct language" — about the use of words like *irregardless* and *hopefully*, about splitting infinitives and dangling participles, and about there being a law against beginning sentences with *because* — comes directly from literacy. (Some of it, like the split-infinitive commandment, comes from literacy in Latin!) Most of these notions are dim recollections for us, but they haunt us all the same. And since the system-side speech is always written down, either for vocal talent to record or for a text-to-speech engine to vocalize, the illusory primacy of written language needs to be faced down constantly.

While the differences are on a continuum (with hallway chatter at one end, *War and Peace* at the other, and TV newsmagazines, email, lost-dog posters, and commencement addresses scattered throughout), rather than in binary opposition, the central dimensions on this continuum are:

Timespace

You've heard it before. You'll hear it again. It's profoundly important for understanding speech (and, in fact, writing). Speech, because it is acoustic, is primarily temporal; writing, because it is visual, is primarily spatial.

Immediacy

Speech is generally spontaneous, writing generally deliberate.

We reflect on our speech, correct our misimplications, repair our false starts, re-articulate our dysfluencies. But even that editing is spontaneous, situation-dependent, on the fly. We deliberate on our writing before committing it to paper or screen, we create drafts, revise, edit. Mostly, other people see our writing as a finished product and encounter our speech as an ongoing process. We generate speech, we produce writing.

Laterality

Speech is primarily bilateral, writing unilateral; speech dialogic, writing monologic.

We mutter to ourselves and our bank machines, but speech is fundamentally a reciprocal activity. We write letters and email, anticipating replies, but writing is fundamentally an expressive product. Speech events are created collaboratively; written texts are created individually.

Quantity

Speech tends to come in relatively small pieces — prototypically, turns. Writing tends to come in relatively big pieces — prototypically, reports or articles.

Activity

Speech is active; indeed, interactive. Writing is static, though sometimes interstatic.

Writing responds to and motivates other discourse, certainly, but the durations between those discourses, even with a species like instant-messaging, are so much longer than with speech, and the medium so much more stable (you can look at the last message as you respond to it; you know your partner will look at yours before and while she responds), that the effects are qualitatively different.

Complexity

The linguistic structure of speech is, on average, simplex; of writing, potentially complex.

We can add or subtract, multiply or divide, an array of figures in our head if we hear them out loud. But at some point we falter, and that point is significantly sooner for most of us than if we are working with a pencil and paper, with a written-out equation. This phenomenon, dependent on short-term memory limitations and the static nature of writing, occurs in speech as well. Multiple embedded clauses are much easier to deal with on paper (or on screen) than in our memory buffers.

Semantic density

Proportionally, speech has fewer content words, words that carry semantic freight — nouns, verbs, adjectives, and adverbs. It has more function words, the adhesive that glues words into phrases, phrases into sentences, sentences into discourses. Writing, proportionally, has more content words, fewer function words.

Superficially, these proportional discrepancies may look backward. Writing, the crafted, monologic, extended pieces of discourse, with greater syntactic complexity, ought to have a greater number of function words; the functions those words serve, after all, are syntactic. But it is precisely the extended, crafted, product-nature of writing — with more time to choose words for instance, less concern about losing the floor, and indefinitely long stability in front of the reader — that leads to the higher content-word quotient (and the correspondingly lower function-word quotient).

Cohesion

Speech, because it is spontaneous, because it has to compete for floor-time, and because it depends on the short-term memory of the hearer, has to be considerably more explicit about the way the utterances fit together. It requires more connective cohesion.

Context

Speech tends to build context as it proceeds. Writing tends to bring its own context to the party.

Speakers and hearers make the context together, and share it. Writers have to manufacture more of the context on their own, since they are usually removed in time and place from their readers (publishers, editors, typographers, and the like also contribute to the manufacture of context for more commercial pieces of writing, like newspaper or magazine articles).

Dimensionality

Speech is multidimensional. So is writing. But the dimensions are very different.

Speech takes place in a mélange of communicative signals. It's not just the words that are transmitted, it's the intonation, the pauses, the volume. These dimensions are often called *paralinguistic elements* by linguists, because they are supposed somehow to run parallel to the linguistic dimensions, but they are really part and parcel of spoken language, not something apart from it. In co-present situations, speech implicates other cuing modes: eye contact, gesture, facial expression. Writing has access to an array of similar (though clearly not identical) cues: punctuation, typefaces, size, density, color, space, medium (paper versus screen versus billboard), and so on.

But these typographical elements are overwhelmingly seen, by everyone concerned, as additions to the language, while intonation, gesture, and the like are seen as part of the message. (It is only linguists who insist on seeing intonation and gestures as somehow external to the message, rather than as part of the message.) And the temporal vs. spatial distinction makes the nature of these dimensions so different as to render the comparison meaningless. Maybe we can see size and loudness as rough corollaries (they're both increases). But what is color in an utterance? Pitch in a written sentence? What on earth could be a pause on paper? A margin in speech?

This collection of differences is arrayed in Table 14.1.

Now, we can never lose sight of the means of production for writing dialogue: we plan it, write it, delete it, write another version, ask our pals, test it, revise it, and so on. This is writing. And this is a virtue. It helps us get the dialogue right. The point here is that we must at the same time never take our eyes off the prize. The means of delivery is oral. That is: temporal, immediate, bilateral, brief, active, simplex, semantically elemental, cohesively rich, accruing context as it goes, and dimensionally enabled. The prize is text that, when it is heard by the user, is not just acoustic wave forms of text, but real speech. Read it aloud,

Speech	Writing
Temporal	Spatial
Acoustic	Visual
Dialogic	Monologic
Spontaneous	Deliberate
Context-building	Context-revealing
Active	Static
Relative syntactic simplicity	Relative syntactic complexity
Lower semantic density	Higher semantic density
Fewer content words, more function words	Fewer function words, more content words

Dimensions include:
  Pitch
  Intonation (pitch variation)
  Duration (including silence)
  Volume
  Rate
  Breathiness
	Dimensions include:
	  Punctuation, capitalization
	  Size, weight, direction
	  Typeface/handwriting
	  Color, density
	  Space, proximity, indentation
	  Medium

TABLE 14.1 Differences between speech and writing

to yourself and others; ask others to read it. Don't just leave the words on the page until the talent or the synthesizer tries to make it into speech.

We have looked at many aspects of what it means to be "real speech" in this book, but for writing system utterances there is no more useful set of guidelines than Grice's conversational maxims. Cleave to them:

- Say just what you need to — not more, not less — both in the number of words and in the density of information.

- Be accurate and honest.

- Be clear and orderly.

- Always be relevant.

It's not easy — it's an art; there's no periodic table — but to be an effective art the utterances must satisfy these maxims.

The basic unit to which the maxims apply, the building block of purposeful machine–human conversation, is the same basic unit of other conversational genres: the turn. But designing those particular turns to be played or synthesized in elaborate variations, and scripting them long before those variations occur, before the projected conversations take place, is a demanding enterprise.

Designing Turns

The key question for dialogue habitability is the extent to which users will interpret computer [turns] as cues to restrict their utterances.
  — Kate S. Hone, Chris Baber

In most of the design section of this book, we have kept prompts and feedback apart — in particular, I have followed the conventions of the field in treating them as different system-utterance types. But when it comes to writing-for-speaking, it's time to put them back together. Prompting and feedback are really different utterance *dimensions*, not types.

Some utterances, it's true, have much greater feedback proportions, pointing directly back to a previous turn; some have much greater prompting proportions, pointing directly forward to a forthcoming turn. Some, like openings, seem to be so forward-directed that they don't even *have* a previous turn; but they do. The user summons the system. Some, like closings, seem so backward-directed that they don't even have a forthcoming turn; but they do. The user terminates the dialogue.

What "being a turn" largely means is following and forecasting other turns, participating in a systematic alternation. But this following and forecasting comes so effortlessly in natural conversation most of the time — as we fall into the turn-logic rhythms of dialogic pairs and coherence relations and (in goal-directed dialogues) as we play out the turn-logic roles of a script — that we can lose track of one dimension or the other. The craft of designing turns is to elegantly balance feedback and prompting in the overall flow of the projected dialogue.[3]

That's not to say that we can't pinpoint the largely feedback oriented elements of a turn and the largely prompting elements. Take this one, for instance, a familiar utterance from Danluft:

3: The best theoretical approach of this balancing act is in Herbert Clark and Edward Schaefer's (1987, 1989) bidirectional contribution model, which conceptualizes every conversational utterance as manifesting a Presentation (information, intentions, and attitudes to be entered into the conversational ground) and an Acceptance (of information, intentions, and attitudes of the previous turn). Roughly, the Presentation corresponds to prompting in voice interaction design, the Acceptance to feedback. See also Clark and Brennan (1991), and — especially — Cahn and Brennan (1999), who adapt the model for dialogue systems.

Danluft   One person. Please state the ID number of the person.

There are two chunks here. One ("One person") strongly looks back, in an attempted echo of the previous turn. The other ("Please state . . .") strongly looks forward, in an attempted inducement of the shape of the user's next turn. The same is certainly true of user utterances — as in another familiar example:

Nicole   Stop. Tell me what I have.

There are two commands here. One ("Stop") strongly looks back to a previous system utterance; the other ("Tell me what I have") strongly looks forward, eliciting the next system utterance.

The idea is not that we can't locate the focal points of feedback and prompting, just that we can't isolate them, and that there are many utterances like T_2 here:

T_1 User   I'd like a special.

T_2 AcmePizza   Certainly. Which special?

Acme's T_2 also has two components, an acknowledgment ("Certainly") and a question ("Which special?"), but the acknowledgment also signals the system's plan of going forward, and the question contains implicit feedback about T_2. Every utterance of every turn must be scripted with both dimensions in mind to build a coherent, collaborative dialogue.

The principal devices of coherence building, we know, are associative and connective cohesion, which are somewhat different verbally than textually. Spoken cohesion and written cohesion share many similarities; however, the use of words like *however* will mark your talk as impossibly stuffy — manifesting what Cohen, Giangola, and Balogh (2004: 153) warn against as the "kind of artificial formality [that] is undesirable for [voice] interface design." In certain limited circumstances, of course, you might *want* your talk to be impossibly stuffy. But in general it is best to stick with more immediate connectives, like *but.*

Remember how English is the historical product of Anglo-Saxon and Norman French and what a great heap of synonyms that amalgamation has given us? Well, in large part, the Anglo-Saxon words in those synonym sets are briefer and more immediately conversational, while the words with French provenance tend to be longer and more literate: *and* rather than *additionally, on the other hand* rather than *conversely, by the way* rather than *incidentally, put* rather than *place, get* rather than *obtain,* and so on. The simple rule of adopting Anglo-Saxon connectives (er, ah, *joiners*) will get you a long way toward spoken immediacy (so will Anglo Saxon nouns and verbs, for that matter, though their greater brevity can compromise recognition).

The coherence relations that come into play as Script 3 develops into a full design spec are manifold. The whole order is effectively a series of Elaborations on a task-initial statement from the user like "I want to order a pizza." Too, while the relative order of the elaborative moves is not particularly important, they have a significant step-like sense to them (one move — size, crust, mode, whatever — usually being closed off before the next one opens), so Sequence is a major organizing relation.

Background is also significant, requiring a context-setting opening. Condition will join several local utterances (if the user requests delivery, the system will need to solicit an address). Restatement is important for confirmation. And a few others might play contingent roles. Concession may come in, for instance, in the case of task-slippage admissions, should any surface, and for any possible recognition glitches. Justification, too, can play a part in developing the dialogue, especially for novices ("I need to check that I've got the order straight. One large . . ."). But the dialogue-defining coherence relations are Elaboration and Sequence.

Cohesively, the main devices will be simple contiguity, the occasional use of conjunctions ("*And* the size?"), as well as direct sequence-coding words in expanded contexts ("*Now*, what type of crust would you like?"). For Condition, *then* ("Delivery? Sure. I'll need an address then"). For Concession, *sorry* and other apologetics ("Sorry. We don't have anchovies"). For Restatement, the main devices are more phrasal than lexical ("Let's go over your order. I've got one large . . .").

Since one of the defining relational coherence structures is Elaboration, *referential* coherence is also strongly implicated — with the assorted elements of the order needing to be assembled and kept straight. The system will be called upon to utter such phrases as these (examples which deploy the primary mechanisms of associative cohesion, anaphor and ellipsis):

OK, two pizzas. Let's get the large one first. A special? Or do you want to select your own toppings?

Which one — the thin-crust or the regular?

I don't have that address on file. May I put it in our records for you?

The range of dialogic pairs is fairly narrow. The core set is typical of service encounters — offers and acceptances/declinations, questions and answers (usually descriptions), clarification requests and clarifications, suggestions and acceptances/declinations. Additionally, there are potentially unrequited dialogic pairs of a familiar sort — initiatives that do not always get matching responses. Apologies, for instance, are likely to go unaccepted (and unrejected); greetings, ungreeted. Leave-takings may be ignored. And there are a couple of dialogue-act clusters — Openings and Lists, in particular — with their own unique reactive pressures. (We will take up Openings and Lists below.)

System turn	Template	Example
NoTopping1	Sorry, we don't have [topping].	Sorry, we don't have anchovies.
NoTopping2	Sorry, no [topping] either. Would you like me to list the available toppings?	Sorry, no tuna either. Would you like me to list the available toppings?
NoDelivery	Sorry, we don't deliver [direction] of [street].	Sorry, we don't deliver west of Martin Street.
PartialRePrompt	I'm not sure I got all that. You want a [recognized values]. And the [unrecognized category]?	I'm not sure I got all that. You want a large, thin-crust pizza. And the type?
PartialRePromptTopping	I'm not sure I got all that. You want a [size, crust type, recognized toppings]. Were there any other toppings?	I'm not sure I got all that. You want a large, thin-crust pizza with chicken and feta cheese. Were there any other toppings?
SubstituteSpecial	You've actually just ordered our [specialname] special. The individual toppings add up to [price1] for a [size], but you'll be charged just [price2], the price for the special.	You've actually just ordered our meaty special. The individual toppings add up to $16.95 for a large, but you'll be charged just $13.95, the price for the special.
TopHelp	I can take your order and arrange to have it ready for pick-up or get it delivered. I recognize all of the words related to ordering a pizza, drinks, breadsticks, and chips, as well as addresses and phone numbers. Would you like to continue – yes or no?	
DescribeVeggieExtreme	The veggie extreme has twelve toppings: mushrooms, onions, eggplant, green peppers and roasted red peppers, black and green olives, feta cheese and our regular mozzarella, and artichoke hearts, topped with roasted garlic and fresh basil. ⟨1 sec⟩ A large is $12.95.	

TABLE 14.2 Some utterances and utterance templates, with examples, for voicing Acme Pizza's take-out/delivery service

At this stage, we need to begin developing the turn templates for the system utterances. For some turns, the templates are just specific utterances. The TopHelp turn, for instance, would be a straightforward explanation of system functioning. The description of a special would be a routine sequence. On the other hand, a request for some topping that isn't available, or for delivery to a location not in Acme's range has to be handled more contingently, with open values to be supplied at run time. Table 14.2 illustrates some turns designed for the Acme Pizza speech system.

People don't always say the same things or speak in the same ways. Interfaces, to avoid user fatigue, should, within fairly tight constraints, also vary their speech.

Constrained Variability

Good words by the third time will even bore the dogs.
   — Chinese proverb

The interface speaks in constrained ways. It can't say anything at all. It should stay within register, and certainly never use any vocabulary the system can't also recognize. The dialogue acts also exert a significant dampening influence. There are a limited number of ways to greet a caller, for instance, and to ask for her phone number. But within those constraints, the interface should speak with some variability.

Consistency has many virtues in voice interaction design, but high consistency in the specifics of system utterances can be tedious in a way that is very different from graphic interface requirements. "Knock, knock," my son has said to me on numerous occasions. "Who's there?" I answer, responding to the reactive pressure of the knock-knock dialogue act. "Banana," he says. "Banana who?" I ask. "Knock, knock," he says again. "Who's there?" "Banana." "Banana who?" "Knock, knock." "Who's there?" "Banana." "Banana who?" No, that's not a typo. That's how the joke goes — on and on like that, for as many iterations as he has stamina (and I patience), until, mercifully, he says "Orange," and I respond with "Orange who," and the punchline finally arrives: "Orange you glad I didn't say *banana* again?"

And I am, actually, very glad. Kids have long known the most potent way to annoy other people: relentless repetition. Voice interface designers have discovered it too. "Users find repetition to be hostile," the Java speech guide says (28), and one of the reasons for the importance of both expansion and tapering to speech systems is the utterance variability they provide.

Writing interface utterances, in short, does not require finding the one absolute jewel of an utterance that can then be distributed widely throughout the interface every time its function is called for. It requires crafting a range of semantically and pragmatically interchangeable utterances that can be rotated into slots. (When they are recorded, this variability should even include subtle differences in articulation, as discussed in Chapter 12.)

Acknowledgments, for instance — staying within register, and with a marked bias toward preferred system diction — should usually have three or four variations. "Certainly," the agent might say to a request at one point, "yes," at a few other points, and "OK" somewhere else. This rotation can't be done without thought and planning. You can't just fill a box of open confirmations for the dialogue manager to pluck responses from at random. Not all acknowledgments are the same in all circumstances. As a response to "Can I get that delivered?" only *certainly* and *OK* work; *yes* suggests more of an answer to the yes/no question, than agreement to the request. As a response to something like "green olives," however, in a list of toppings the caller is itemizing, only *yes* and *OK* work; *certainly* is too strong for a backchannel here, shifting the focus toward approval rather than acknowledgment (the coherence toward Evaluation rather than Cause/Result), and might discombobulate the user. *OK*, on the other hand, works as an acknowledgment of a user's declination of an offer, but neither of the others would. Imagine *yes* or *certainly* in the T$_3$ response here:

T$_1$ System   Did you want to review some more of your personal profile?

T$_2$ User   No.

T$_3$ System   OK, what's next?

(Cohen, Giangola, and Balogh, 2004: 146)

As always, that is, context exerts the most powerful constraint.

Constrained variability is also important for syntax. An offer might be phrased, "Anything else with that?" sometimes, "Would you like something else?" other times, and so on. Table 14.3 is a partial list of how our voiced take-out/delivery pizza system instantiates constrained variability for a few dialogue acts.

	Dialogue acts	Utterances and utterance templates
Feedback	Acknowledgment	Yes. OK. Got it. ⟨echo topping⟩, ⟨echo mode⟩
	Agreement	Yes. OK. Certainly.
	Clarification request	Do you mean ⟨topping1⟩ or ⟨topping2⟩? Is that ⟨topping1⟩ or ⟨topping2⟩? Which topping? – ⟨topping1⟩ or ⟨topping2⟩?
Prompt	Topping query	A special? ⟨1 sec⟩ Or do you want to choose your own toppings? Would you like a special, or do you want to choose your own toppings? One of our specials? ⟨1 sec⟩ Or a choose-your-own?
	Crust query	What type of crust? ⟨1 sec⟩ Regular, thin, stuffed, or wheat? Regular crust, thin, stuffed, or wheat? And the crust? ⟨1 sec⟩ Regular, thin, stuffed, or wheat?
	Mode query	Would you like take out or delivery? Take out or delivery? Can we deliver that for you? ⟨2 sec⟩ Or will you be picking it up?
	Directive	Please say ⟨term1⟩ or ⟨term2⟩. Say ⟨term1⟩ or ⟨term2⟩. You can say ⟨term1⟩ or ⟨term2⟩.

TABLE 14.3   Some utterances and utterance templates illustrating constrained variability

Most important for the overall fluency of exchanges, turn variability is crucial. We've looked at this notion a few times already, in connection with expansion and tapering, which are fundamental principles of turn variation in response to context and the experience level of the users. But turns should vary independently of expansion and tapering as well — providing, for instance, for the following sorts of exchanges based on the "same" query:

AcmePizza   Anything to drink?

Caller   No.

AcmePizza   Breadsticks?

AcmePizza   Anything else?

Caller   Yes, a quart of Jolt cola.

AcmePizza   Sure, and some breadsticks?

AcmePizza   Anything else?

Caller   Yes, an order of garlic breadsticks.

AcmePizza   Certainly. Something to drink?

Voice-interface scripting also calls for more specific clusters of dialogue acts. We will look at two of them, one that every dialogue system requires, the Opening, and one that has a smaller, but still broad, distribution in speech systems, the List. Both are required by the Acme Pizza voice interface.

Openings

Designing the right initial system greeting is necessary for establishing user expectations and helping users determine how to proceed.
    — Susan Boyce

Openings are generally diversions and advertisements for graphic interfaces — a list of developers, a logo, some promotional messages about capabilities, little more than sleight of hand to draw attention away from the time it is taking for the application to load. But it is different for voice interfaces. They are integral to the success of speech systems, incredibly important, especially for first-time or infrequent users (they should be tapered back for intermediate and experienced users).

Voice interface openings need to set the caller's expectations, and induce her into successful interactive patterns, as well as to serve a range of more general purposes — branding, introducing the agent, establishing the agent's role, and generally laying the groundwork for the experience. And there is not much time. In a temporal, serial medium, moments are the currency, and you don't want to fritter them away lecturing/aggravating the user. A truly conversational opening to a call would simply be the phatic "Hello," or one of its vast collection of synonyms. It says, roughly, "OK, I'm here. Go ahead." But we can't go quite that far.

A typical commercial opening is more restrictive. It launches a dialogue that draws heavily on conversational procedures — turn taking, dialogic pairs, coherence, grounding, repair — but in a more goal-directed way than personal conversations. It sets the context right away. It often (but not always) starts with "hello" (or a synonym). More important to the ensuing dialogue, however, is an identification, which either rapidly follows or completely replaces the greeting. With some services, especially ones that involve negotiative sorts of interactions, the identification may have as many as three components, individual, functional and corporate. Usually, there is an offer as well — a prompt.

Here is the basic dialogue-act template for commercial openings:

[phatic greeting] [identification; self, role, institution] [offer]

These constituents are very common, but not all of them are necessary. The sequencing, too, is common, but not universal. Here are a few variations:

AmexAgent   Hi. This is A at American Express. May I help you? (SRI/Amex 6–3)

Ann   Good morning, this is the reservation desk, Ann speaking.
(Steuten, 1997)

BA   British Airways, flight information. Can I help you?
(Wooffitt et al., 1997: 80)

How else are business phones answered these days? Like this (grit your teeth):

ABC   Welcome to the Army Benefits Center. Please listen carefully as our menu options have changed.

If you are a Department of the Army civilian employee, please press **1** now.

If you are an Army uniformed service member seeking answers to questions about the Thrift Savings Plan, press **2** now. Thank you!

There are some differences. While a human-agent commercial opening usually restricts the context right away, with the identification ("American Express," "reservation desk," "British Airways, flight information"), a keypad-system opening is far more restrictive. It has identifiers, too, and phatics, but it very quickly moves into clear directives, arranged into conditional coherence relations ("If . . . , please press").

The human-agent openings work because humans are cooperative language users, and the agents can trust the caller to behave appropriately: enter into a ticket-booking encounter, seek to make a reservation, ask for flight information, whatever. The keypad-system openings can't afford that course of action, though, because no matter how cooperative the caller is, there are two insurmountable hurdles for nonexpert users: (1) the range of behaviors the system can interpret is extremely narrow (hence, the directives), and (2) those behaviors are highly opaque (hence, the conditionals). The user can only do a few things (push one or more of the twelve available buttons), and he has little-to-no idea what will happen if he pushes 1 vs. pushing 2 vs. pushing 9 (though he will often have some expectations about 0). This endemic combination of narrowness and opacity also tends to make the keypad openings relatively long — one of the callers' biggest and most constant complaints.

Voice-interface systems have problems somewhat related to those of keypad systems, but they have receptive powers which begin to approximate human agents. They can only handle a relatively narrow and more opaque range of user actions, compared to people. But those actions are the same category of actions that human agents expect and accept — utterances, dialogue acts.

Voice-interface users are, that is, in a similar boat to keypad-system users. But they have bigger oars, and more of them. The main job of the opening is to convey both the restrictions and the possibilities, to pursue both reliability and habitability.

The task-analysis phase of the take-out/delivery project revealed typical — and, for this small chain, typically informal — commercial openings. Here's a sample from one agent's opening-specific corpus:

"Hi! Acme Pizza, Jenny here. What would you like?"

"Hey! Jenny at Acme Pizza. What's your order?"

"Hey! Acme Pizza! Jenny on the line. Pick-up or delivery?"

These openings distill into the opening schema of Script 3, a typical commercial opening, with a greeting, two identifications (self and restaurant), and an offer.

Our first move is (drawing on the corpus, under speech-system constraints) to translate the opening schema (AnswerCall) back into "speech", as outlined in Figure 14.1.

Figure 14.1 is a rather crude translation exercise, just taking the dialogue acts abstracted from the human-agent discourse analysis and turning them into machine speak. Acme Pizza has a legacy of chatty, informal, even neighborly phone service, answering the phone with a self-identifier ("Jenny here") as well as a restaurant identifier ("Acme Pizza!") and this first pass has just tried to substitute a system identification for the personal identification. There are problems.

First, of course, the homey tone is gone. That's somewhat inevitable — we can't deny the automated nature of the system, and we don't want to encourage small talk — though

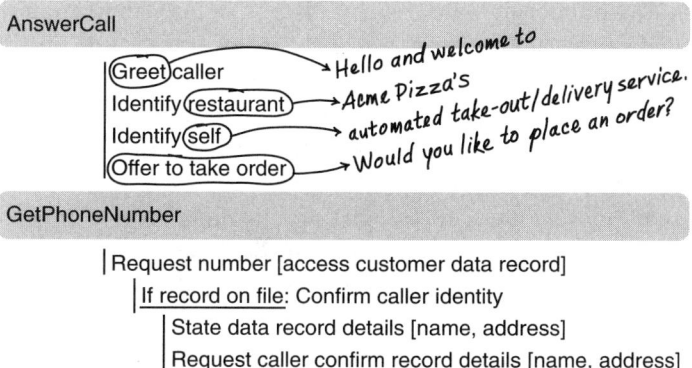

FIGURE 14.1   Turning dialogue acts into speech

some elements of that ethos might be retained. Secondly, there's only one greeting at this stage. Acme Pizza's human agents knew the value of a little variety, avoiding roboticism. We need more than one opening. Thirdly, our first-pass opening is a bit longish, and that longishness will get increasingly annoying with return callers. In the early days of the service, the callers will be getting the computer agent but expecting a human one, and established Acme Pizza customers may be particularly unsettled. That is, they will be victims. If we want to ease the migration from victim to volunteer (and, especially, if we want to prevent the victims from just taking their business to some other pizza joint), both growing brevity and variation are necessary. We have to taper.

The user differences and overall length issues are productivity, ergonomics, and therefore user-satisfaction concerns. They depend on hitting the right Gricean marks in terms of accuracy, quantity, and order. The homey feel is largely a branding issue, and, to some degree, an aesthetic one as well. It depends on stylistic nuances of scripting, and on the voice talent.

The informal chatty legacy is a real liability for automation in this case, requiring close attention to agent design, and, in particular, to how the agent is represented in the opening. We certainly do not need to shy away from personification in agent design; it shapes the interaction, as I argued in Chapter 12, whether one promotes it or attempts to repress it. But we do need to be wary about *impersonation*, especially in the opening. The agent should seem conversationally competent — an attribute only humans have in the natural world — but (s)he should not pretend in any way to *be* a person. The most important move of the opening is to establish the expectations and collaborative rules of engagement — chief among them is declaring the automated status.

Automated speech systems that do not identify themselves (if only by the use of synthetic speech) quickly create insurmountable problems for themselves. "If callers believe they are speaking to a human operator," Susan Boyce notes, "their requests are often long and complicated" (Boyce, 2000: 33). The possibility of error, and error spirals, is a very significant risk if users think they are talking to a human.

It is crucially important for users to know they are talking to a machine in order to match their expectations to the capabilities of the system, and with Acme, we have the particular complication of a personality rich legacy.

Since Acme has both a database of existing customer numbers and call-recognition ("call display") technology, we have a good idea of which incoming calls are from the existing customer base. And the call frequency can be tracked back one year, so we can tell regular customers from occasional customers.

We plan a staged, overlapping introduction of the automated system. The human agents (all youngish females with "Hey! It's Jenny!" sort of phone personalities) continue to answer the phone, occasionally engaging the callers about the "neat new system" Acme is getting (beginning two months before the implementation phase).

The automated agent is a youngish, synthetic male — most immediately for distinctiveness from the human agents, to groom callers away from the chattiness expectations of the Acme legacy. He has an earnest, friendly demeanor, speaking with a barely noticeable German accent, to draw on the conciliatory foreigner effect, and to make a slight allusion to the *Terminator*-robot character (an in-joke move of whimsy, not a play for knee-slapping, yuk-yuk humor). His name is *Otto*. He mostly just takes orders, but sometimes makes goodwill-based suggestions (about price-saving moves).

Upon implementation, a call from any number in Acme's customer database is routed either to a human agent or to the speech system. Calls from frequent users go to (1) a human agent or Otto, at a 50% chance rate that biases over a year from the first call until the speech system is fully implemented for that caller, or (2), if the humans are genuinely busy, to Otto. New calls (that is, from numbers not registered in Acme's customer database) go directly to Otto.

The openings are as follows:

For callers from the existing customer database, on their first time and for three Otto-routed calls thereafter, the openings are:

Acme Pizza. The girls are busy. I am Otto, an automated speech system. I can take your order. Take out or delivery? ⟨2 sec⟩ Can I take your order, or would you like to wait approximately ⟨wait time⟩ minutes to order from ⟨agent name 1⟩ or ⟨agent name 2⟩?

Acme Pizza. I am Otto. The girls are busy. I recognize most words for ordering. Can I take your order? ⟨2 sec⟩ Or would you like to wait approximately ⟨wait time⟩ minutes to order from ⟨agent name 1⟩ or ⟨agent name 2⟩?

First time openings, for new or infrequent customers:

Acme Pizza. I am Otto, an automated speech system. I can take your order. Take out or delivery? ⟨2 sec⟩ Can I take your order, or would you like to wait approximately ⟨wait time⟩ minutes to order from ⟨agent name 1⟩ or ⟨agent name 2⟩?

Acme Pizza. I am Otto. I understand most words used for ordering. Can I take your order? ⟨2 sec⟩ Or you can wait approximately ⟨wait time⟩ minutes to order from ⟨agent name 1⟩ or ⟨agent name 2⟩.

Tapered openings, implemented as a function of call frequency, call recency, time-out, overlap, and out-of-vocabulary histories:

Thanks for calling AcmePizza. I am Otto, an automated system that can take your order. Take out or delivery?

Welcome to AcmePizza's automated order system. I am Otto. Would you like to order a pizza?

AcmePizza. Otto. What would you like?

AcmePizza, Otto. Take-out or delivery?

More restricted contexts, with very brief interactions, lesser information flow rates, negligible grounding criteria, and the like, will have briefer, more open prompts to begin the dialogues — prototypically, AT&T's *How may I help you?* System.

On the other hand, more involved systems — with greater information flow rates, higher grounding criteria, a wider range of user categories, and/or more constraining legacy commitments — will need openings with more detail, and more elaborate fallbacks for time outs and rejections. Prototypically, voiced banking systems need this kind of detail, often listing a subset of global commands (help, repeat, pause, . . .), and perhaps offering paradigm interactions as examples, or even tutorials.

Lists

> *County library? Reference desk, please. Hello? Yes, I need a word definition. Well, that's the problem. I don't know how to spell it and I'm not allowed to say it. Could you just rattle off all the swear words you know and I'll stop you when . . . Hello?*
> — Calvin, of *Calvin & Hobbes* (Bill Watterson)

Graphic interfaces usually have it over voice interfaces in terms of presenting information. Most of us mammals are incredibly good at getting large amounts of information visually. But one thing the visual displays can't give you, even those bloated hub-sites on the Web, is serial presentation. You can find it in graphic interfaces, and even array it yourself in a quasi-serial fashion, but a visuo-spatial display can't be truly serial. Getting a graphically displayed list — upon asking for directions, or a recipe, or installation instructions — is a very different matter from getting them, one step at a time, as you need them, serially. (Voice interfaces also have the advantage in eyes-busy information transfer, too.)

Some situations call for serial presentation, like ordered steps. Some don't. One-of Decision tasks (radio buttons, in graphic interfaces) and cluster-choices (checkboxes) fall somewhere in the middle. The blessing of serial voice presentation is that it unfolds temporally, allowing the listener to focus her attention on one thing at a time. With ordered steps, that's almost mandatory. With decision tasks and cluster choices, it can be helpful.

The blessing of serial listing, though, is also its curse. Unfolding in time means holding people captive, which people often don't like. Since there are no progress bars in voice interfaces, that means (1) you should always alert users for any list of more than a few items, how long the list will take, in time or items; (2) you should usually ask before launching the list; for longish lists, upwards of a dozen items, (3) you should usually give the listing criterion; and (4) for novice users, you should tell them (or remind them) what their exit options are. So, a maximally expanded list introduction will look like this:

AcmeVideo   I have 24 DVDs or DVD sets directed by Akira Kurosawa. Would you like me to list them?

User   Yes.

AcmeVideo   You can stop me at any time, or ask me about any Kurosawa movie. By date of production, the movies are: *One Beautiful Sunday, Stray Dog, Rashomon* . . .

Lists in voice interfaces exert somewhat unique reactive pressures. They are dialogue-act clusters, rather than individual dialogue acts, or amalgams. They often need to include list descriptions and instructions, and they invite floor-seizure acts, either by direct selection from the list or by initiating an insertion sequence.

The take-out/delivery system requires several listing options — for the names of specials; for the ingredients of specials; for toppings, sizes, crust types; for drinks; and other accompaniments. Several of the lists are short enough that alerts, timing information, and the like, aren't necessary. But some lists are 7+ items, some lists are nested, and all of the items in every list have associated variables that can be queried (prices, for instance) — which raise a few of these list considerations, especially for novice users.

Most importantly, novices should be advised they can interrupt. The interface has to support interactions of the following sort:

Otto   Would you like a special, or do you want to choose your own toppings?

User   Um, what are the specials?

Otto   We have seven. I can list them. You can ask about any of them at any time. You don't have to wait til the end. <1.5 sec> We have Greek, Chicken-and-pesto, Meaty, Meaty-extreme, Spicy Sicilian, ⌐Veggie,

User   └What's on that one?

Otto   Spicy Sicilian?

User   Yes, Spicy Sicilian.

Otto   Hot Italian sausage, green . . .

Shaping the Diction

For it is both copious in words and also pleasant to the ear . . . very perfect and sure.
— Sir Thomas More

Stocking the vocabulary appropriately is the bedrock job of building speech systems, and often the area that brings recognition personnel and interface personnel into closest contact. The challenge is to pick those words and colligated phrases that enable users to "express themselves without straying" out of the system's capabilities (Watt, 1968: 338), while not making things too hairy for the recognition subsystem. The vocabulary must characterize the register. There must be enough words, but not too many. They need to be acoustically distinct enough from each other (when calibrated by way of dialogue-act context and lexical collocates) to minimize substitution errors, robust enough to minimize rejections, and representative enough that users will actually speak them (including terms that might be outside the actual service provided). Whew.

These decisions, and their guiding criteria, are rehearsed in detail at several points in this book — chiefly Chapters 7 and 11, with side-orders on lexical density and reserve synonyms in Chapter 13 — but let's look briefly at how they play out in our Acme Pizza project.

Our basic vocabulary is broad but quite highly determined. From the menu, we have words and phrases directly related to the products — toppings, sizes, crust types, and the like. From the take-out/delivery genre, we have such elements as numbers, payment and time terminology, street names, and so on. And from the general register we get the task-shaping nouns, verbs, and adjectives, the connectives, and so on. We also built up, in three months of natural dialogue studies, an extensive corpus of calls to Acme Pizza — used in the development of an interface lexicon, as well as for training and other recognition purposes.

The domain is actually a pretty forgiving one for voice interface development. Core terms like *special*, *topping*, *pepperoni*, and *delivery* are all both mutually distinct and individually robust. Even some monosyllabic words, which might ordinarily cause recognition problems, are often given extra salience in this context. Adjectives like *small* and *large*, for instance, which can be fairly negligible acoustically in some articulations of "a small pizza" or "a large pizza," tend to be pronounced without the noun, increasing their stress. *Pizza* is highly redundant here. It's given information (and some of Otto's prompts are designed to reinforce that givenness). So callers frequently drop *pizza* and nominalize the adjective, putting it in focus and raising the stress. Exchanges like this are typical (where *large* gets primary stress in the noun phrase, not the secondary stress it would otherwise get):

Otto   Would you like to order a pizza?

Caller   Yes, a large, with chicken, artichokes, and green olives.

Moreover, when users don't adopt that locution, they can be maneuvered into it fairly easily by just querying the variable directly (as does crust type; *thin* and *stuffed* are the possible culprits here):

Otto   AcmePizza. Otto. What would you like?

Caller   blah blah pizza blah chicken, artichokes, and green olives.

Otto   I'm not sure I got all that. You want chicken, artichokes, and green olives. And the size?

Caller   Large

Otto   Crust?

Caller   Thin.

Words and phrases that came out of the discourse analysis sometimes included special requests — like (*only*) *on* (*one*) *half* and (*one*) *half only* (for requests like "pineapple on one half" and "the feta cheese only on half" — but mostly they were aids for clarity, disambiguation, service explanations, and other task slippage areas; that is, they were valuable for lexical density.

The chief lexical density concern was for the sorts of things people might ask for that Acme doesn't provide, especially a number of toppings available at other chains that people might reasonably request in this market, such as fresh tomatoes, anchovies, shrimp, and tuna. Some items were rare enough (statistically speaking) in our discourse analyses that we are willing to risk flat rejections — including *crab*, *squid* and *Calamari*, *potato*, *turkey*, and *corn*. Other foods and drinks, too, needed to be included, which customers might request at the "anything else?" stage in the call flow. These terms include *pasta*, *spaghetti*, and *ziti* (but no other specific pasta cuts) *calzone*, *subs*, and *appetizers* (in the latter case, we offer breadsticks).

Customers seeking explanations, for instance, led to the inclusion of three types of measurement for sizes: diameter (inches), slices (also pieces), and number-of-people-fed, so the system can support exchanges like the following:

Otto   Size?

Caller What are the sizes?

Otto   Small, medium, large, humungo.

Caller How big is a medium?

Otto   Ten inches.

Caller How many pieces?

Otto   Eight slices. <.5 sec> It should feed about three moderately hungry people.

The system also needed terms like *side*, *back*, and *Canadian* for queries about bacon; *hand-tossed* and *deep-dish* for crust queries; *kalamata* and *Greek* for olives; and so on.

The preferred diction choices made in favor of distinctness and robustness — *select* over *choose*, for instance — were always slightly more formal, and sometimes violated the general guideline for Anglo-Saxon spoken diction. But they did not result in overly strained system utterances, and the slight formality was consistent with the agent design; Otto's character is mildly officious.

Using Scenarios

There is no way forward in design without some notion of scenario.
    — John M. Carroll

Scenario-based design might seem less important for the development of voice interfaces because the natural dialogue studies that build the foundation of the design are full of scenarios. All the task analyses, virtually all the harvested dialogues, and all filtered-through-the-phone encounters are scenario-driven events, and they are all data-in-the-can by this stage. True, but that's not enough, because the natural dialogue studies are missing a crucial part of the puzzle: the speech system. You don't have it yet, either, or at least you don't have it in a functioning, harmonious relationship with the relevant data and processes, not at this stage. But you have a strong-enough sense of the behaviors characterizing speech system performance that you can project them into hypothetical encounters with users. You can, that is, spin dialogue scenarios and use them to develop the voice interface. (Nor is this the end of scenario-based design in the development of the interface. Scenario permeates the spiral development cycle generally — Wizard-of-Oz tests and all forms of usability inspection, for instance, and even beta studies, are based in scenario-driven methodologies.)

Take the GetAddress schema, for instance. The system needs to collect three variables (ignoring confirmation scenarios for now): for name (filled with something like "Fred Derf"), street ("2 Nass Street"), and phone-number ("632-1298"). Most of the possible task slippages will likely have come out from task-analysis interviews, or in the natural dialogue studies: the caller gives *his* name, but he's phoning from *someone else's* address, perhaps he's babysitting; the caller forgets one or more pieces of the data, perhaps he's just moved; the caller is phoning from a cell, and gives that number, and when he gives the home address, there is a conflict because it is associated with a different phone number in the system; and so on. Even potentially important turn-overlap points will probably have been revealed. The presence of data about these possibilities doesn't mean task-slippage scenario spinning shouldn't be pursued at this stage, just that it is less crucial, and there will be fewer gaps to cover. (Of course, if you got this far without thorough-going natural dialogue studies, say a few *mea culpas* and get spinning.)

But there will be no corresponding data on interaction slippages and errors, which makes this a good place to focus scenario-based design efforts at this stage. What would happen, for instance, if the caller just doesn't respond at all? Let's try a scenario.

Time-out Scenario

T_1 AcmePizza   Delivery? I'll need a destination then. What is the street name and number, please?

T_2 Caller   <2 seconds>

What should the system do here? The general timeout strategy is an expansion after 1.5–2 seconds, usually providing clear directives, letting the user know what is expected from him. But how do you expand a simple constituent question that is already firmly contextualized? How about if you assume, at this particular juncture, there is a possibility that the information may not be readily at hand (the caller might be in someone else's home, at a party, and has to find the piece of paper someone gave him with the address)? It might not work, but it is a reasonable hypothesis that this sort of information is different from information about someone's name, and that this juncture in the dialogue is quite different from immediately post-greeting, and that a 3- or even 4-second timeout might be tolerated before repeating the request.

Or, what about a substitution error for GetAddress. How should it be treated?

Substitution Scenario 1

T_1 AcmePizza   Delivery? I'll need an address then, please. Where would you like it delivered?

T_2 Caller   To 2 Nass Street.

T_3 AcmePizza   2 Tuna Street.

T_4 Caller   Ah, no, not 2 Tuna street, 2 *Nass* street.

How should the dialogue deal with this potential error? One way is clear, the one this scenario follows — explicit feedback (at T_3), presenting the grounded understanding, and confirming the value of negative feedback. But another design strategy might be to break the dialogue into smaller interactions to guard against these confusions:

Substitution Scenario 2 — finer mesh prompts

T_1 AcmePizza   Delivery? I'll need an address then. Your street name first, please.

T_2 Caller   Nass Street.

T_3 AcmePizza   Nass Street. And the number?

T_4 Caller   2.

Another alternative — preferred, since this last one introduces steps without sufficient grounding criteria (in our substitution scenario, the problem *was* caught after all, by the immediately following confirmation request) — would be to change the T_1 constituent question to seek a noun phrase rather than a preposition phrase, reducing the risk of a spurious to/two:

Substitution Scenario 3 — changing targets

T_1 AcmePizza   Delivery? I'll need an address then. What is the street name and number, please?

T_2 Caller   Nass Street, 2 Nass Street.

T_3 AcmePizza   2 Nass Street. The phone number there?

T_4 Caller   632-1298.

Scenario-based design methods help you to forecast and prevent, or ameliorate, problems, prepare for specific dialogue contingencies, and write the system utterances. They are also, once you imagine the possible ways customers might navigate the system, useful for planning the call flow.

Planning the Call Flow

> *Scripts are prepackaged sets of expectations, inferences, and knowledge that are applied in common situations, like a blueprint for action.*
> — Roger Schank

Script 3 has the necessary elements in a workable arrangement: first the phone number, then the mode, then the food order, then price and time information, and click. It also has all the contingencies: if the mode is delivery, then the address is established (if not, not); if the address is on file, it is confirmed (if not, gathered); if necessary, the address is corrected. Etc., etc., etc.

But it may not define the most habitable call flow. We need to craft the interaction carefully, not just on a turn-by-turn basis, but on the basis of the overall dialogue.

The initial job requires working out the necessary and conventional linkages among the schema, what must precede what. Then we structure those linkages in order to accommodate a range of natural interaction patterns, rather than stepwise hierarchical navigation, of the sort that characterizes menu-driven systems. In fact, so dominant is the notion of menus in speech-system design that natural-interaction accommodation is known as "flattening the menu," a phrase that means putting multiple steps at one stage, rather than forcing users through repeated bottlenecks. We will call it "broadening the interaction." The resulting structure is best illustrated diagrammatically, in what resembles a plumbing or electrical layout — or indeed, an old-fashioned programming flowchart — but which is a layout of conceptual relationships, not a coding blueprint.

Call-flow diagrams are interaction blueprints. They overlap, sometimes very substantially, with software blueprints. Schemas and objects, in particular, are often closely intertwined — reflected in part by the AllOneWordTitling convention I have adopted here for schemas. But the two plans are independent. The two areas are distinct. The two teams have different ranges and concentrations of expertise. Speech-system development is learning this lesson slowly — interaction design still plays second fiddle to system design in many companies, or third fiddle, behind getting coffee for the coders, and designers are still often prized more for software expertise than for human-factors expertise — but it *is* learning.

The call flow for a speech system is an interaction design issue; that is, a human factors issue. Coding it is a software issue; that is, a computer-engineering issue. The overlap comes from the ways in which the software team visualizes and realizes the design. But the interface team should not specify the coding, and the coders should not determine the interface.

Sequencing and Dependencies

If naturalness is not the key driver for speech, then what is?
— Roni Rosenfeld, Dan Olsen, Alex Rudnicky

We know, rather trivially, that answering the call must come first, ending the call must come last, so the call flow for our take-out/delivery voice interface must have a frame much like Figure 14.2.

Between the Answering and Ending bookends, we need to arrange the other schemas. One way to go, of course, would just be to turn Script 3 into the call flow, as in Figure 14.3,

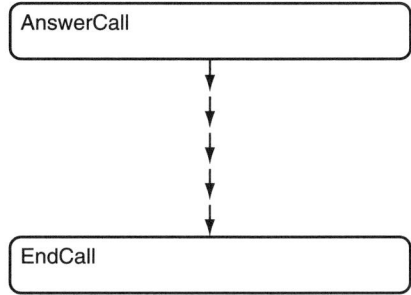

FIGURE 14.2   The call flow starting point

stringing the schemas out like beads (with each of the schemas having their own, internal bead strings).

Figure 14.3, or some minor variation thereof (establishing the mode before getting the phone number, for instance) would be the easiest call flow to implement. And we could do it without the bother of anything but the most cursory task or discourse analyses, the sort of thing one can do on a slow afternoon with a ball game on in the background. But the architecture of Figure 14.3 ignores the basic flexibility of knowledge scripts, turning a loosely sequential array of schemas into a series of conceptual locks; more importantly, it ignores the user's own interactive strategies — the *system* user is also a *language* user — and it ignores the power of language to support pliable, multivariant interaction. Like a great, awkward ship that can only get through an elevated canal by maneuvering into a tight space, gates clanging behind it, and waiting for the water slowly to fill up the lock, then for new gates to open and clang closed, and more water to fill up the next lock — lock after lock — the user is not allowed to use the natural power and grace of language to sail through the interaction. About the only concession it makes to the user is allowing her to go back and start over again, or to return to a few earlier schemas if there is a slippage or a change of mind. Figure 14.3 is not an uncommon call flow, but it is not very habitable.

We might want to retain the linear call flow of Figure 14.3 as one possible path through the interaction, perhaps even the preferred path. But as the *only* path through our pizza service, it is confining for users, it largely ignores the reasons for using language in the first place, and it puts ease-of-implementation unacceptably over usability.

In building upward and outward from a linear, stepwise call flow — in developing a more habitable call flow out of the arrangement in Figure 14.3, itself an outgrowth of Script 3 — we need to be sure that we differentiate between the necessary or expected ordering on the one hand, and the accidental ordering on the other, the schema orders with no rationale behind them. We need to retain logically entailed schema sequences, that is, and

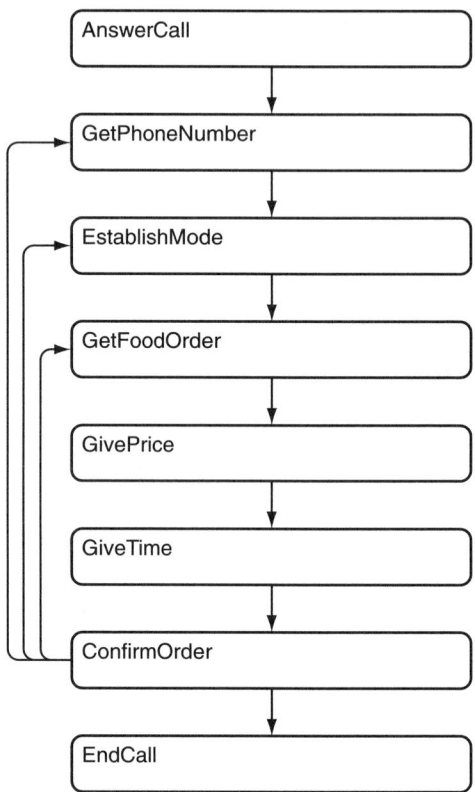

FIGURE 14.3 A linear call flow for a voiced take-out/delivery pizza service

respect the conventional schema sequences. But any schema orderings in Figure 14.3 that have no rational basis, in logic or in practice, need to be uncoupled.

Returning to Script 3, and to the task and discourse analysis underlying it, we can get a better sense of the sequencing possibilities for the schemas, and plot the call flow more flexibly. We see, for instance, that confirmation logically follows both taking the order (or there's nothing to confirm) and establishing the mode (since one of the confirmation elements may be the mode, and possibly the delivery address). Drawing on the task-and-discourse analysis, we find, too, that convention supports this schema ordering as well; confirmation routinely follows all other schemas except ending the call. It is the last piece of business in a typical take-out/delivery call.

Getting the caller's phone number is logically independent of every other schema except our bookends — answering and ending the call. Conventionally, however, it tends to come very near the beginning of take-out/delivery calls, because many take-out/delivery outlets use phone numbers as customer identifiers. And one of the principal uses it has as a customer identifier is to access an address that might be used for delivery, which strongly suggests it should precede getting the address. It also clearly belongs more with the mode schemas than with food-ordering schemas. (A good automated system of course would capture the number as it came in, if available, and confirm or correct any further details as necessary.)

The food order and the mode (take-out or delivery) are independent of each other, but getting the details of the food order must precede giving the price, and establishing the mode must precede both getting the address (which is only required for a delivery order) and giving the time (which will be different for pick-up and delivery) — though the time and address are independent of each other. ConfirmOrder, too, must in a cyclic sense "precede" GetFoodOrder and EstablishMode, since it must have a return path to any and all of them in case some of the data the system has gathered proves to be faulty. Further, ConfirmOrder can serve effectively as a router box, to allow GetFoodOrder to "precede" EstablishMode, and vice versa. These relationships are outlined in Figure 14.4.

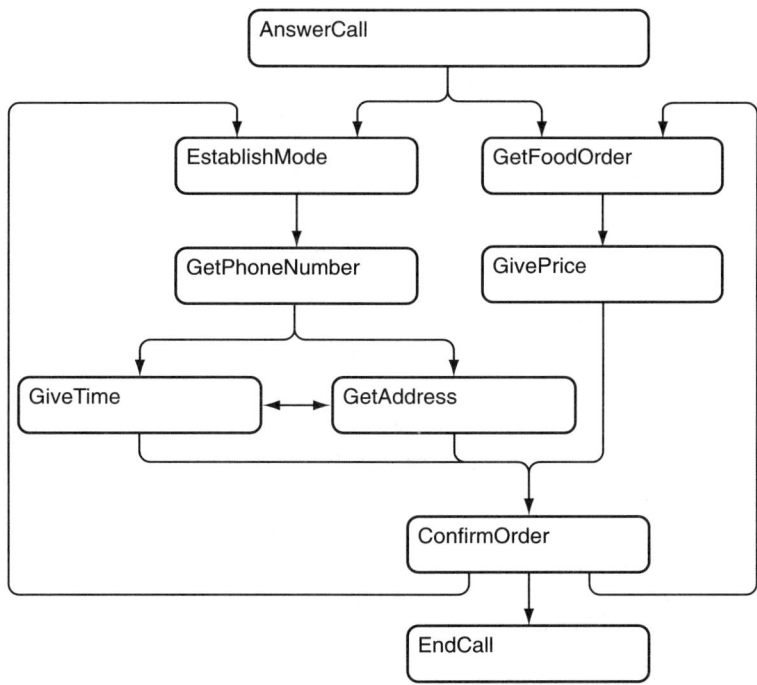

FIGURE 14.4 Logical and conventional sequencing among schema for a take-out/delivery phone ordering service

This call-flow architecture allows alternative paths. The user might place a food order and then establish the mode, or the other way round. She might request the availability or delivery time and then give (confirm) her address, or vice versa. The call flow in Figure 14.4 is more flexible than that in Figure 14.3, but it is only marginally more habitable. It is not the finished call flow for the service, not yet. We have identified all the principal schemas, and worked out some sequencings, but we have only mapped a few of the potential paths users might take through an encounter with the service. In particular, one of the things speech-using humans might want to do is collapse more than one "step". They might want to start their order with a phrase like "I would like a large meaty-extreme delivered to 301 Cobblestone Lane," which implicates two schemas. An architecture that can provide this kind of functionality broadens the interaction possibilities, accommodating the natural strategies of dialogic language.

Broadening the Interaction

There is no use in taking a long rough round-about way if there be a shorter and easier one.
— Socrates

The arrangements that speech-system professionals have been calling *menus* for decades (that is, bottleneck nodes in the interaction architecture at which there are clusters of options) are very, very bad for voice-interface design. Since option clusters are virtually unavoidable in the design of speech systems, what this badness amounts to is that the user should not be faced with them in the traditionally serial way: the for-delivery-say-*delivery*-for-take-out-say-*take-out* interaction style. The *serial perception* of options, especially nested options, is what is very, very bad. It should be minimized or eliminated.

The elimination of these bottlenecks is traditionally called "flattening the menu," but if we approach it from the user's perspective, rather than the system's, the preferred phrase is "broadening the interaction." There have been many examples of this process throughout the book, and many exhortations toward it, always in the service of functional habitability. Let's see how it looks, explicitly in the context of call-flow design.

Going a little deeper into our Acme Pizza use case, let's look at how the service might actually help callers order a pizza, a task that has a number of values to be assigned. So far, we have been working with the black-box schema, GetFoodOrder, but it needs further specification, since our service also provides accompaniments like soda, fresh breadsticks, and chips, as well as pizza. GetFoodOrder needs two subordinate schemas, two children: GetPizzaOrder and GetAccompOrder.

Turning just to GetPizzaOrder, the caller has to provide the number of pizzas; and the size, toppings, and crust type of each pizza; additionally, the toppings option has a lower branch, between specials and individually selected toppings. The crudest, most menu-driven way to manage the ensuing interaction would be for the system to respond with a series of choices.

Pizza-delivery snippet A

T$_1$ AcmePizza   Hello and welcome to AcmePizza's automated take-out/delivery service.

Would you like to place an order?

T$_2$ User   Yes.

T$_3$ AcmePizza   Choose: Number, Size, Toppings, or Crust.

Superficially, this presentation may appear to give the user freedom, since she can start with the crust or the toppings or the size, without being directed by the system. But making explicit task choices like this, the epitome of a menu structure, adds a step. Once, say, the Crust subtask is chosen, the user now has to provide the type of crust, and then go "back up" to the remaining two options.

Pizza-delivery snippet B

T$_4$ User   Crust.

T$_5$ AcmePizza   Choose: Regular, Thin, Stuffed, or Wheat.

T$_6$ User   Wheat.

T$_7$ AcmePizza   Choose: Size or Toppings.

The local architecture for this sort of call flow looks like Figure 14.5. No matter which of the three options are chosen, the flow returns to GetPath, then back down to the next choice.

The flow in Figure 14.5 is a loop, starting with the schema, GetPath, and traveling through its children, one at a time, picking up values — with the second-last passage through GetPath being transparent to the user (since there is only one "option" left), the system routing the user to the remaining child without pausing to ask for a choice. The last passage through is likewise transparent, since the route is now on the exit ramp. This architecture will work, in the sense of getting the job done, but it is repetitive, tedious, and not functionally habitable.

Bypassing option nodes, however, allows users to collapse or flatten two or more steps into one response, "flattening" the interaction. For instance, if the T$_3$ system utterance did not require the caller to choose an option before specifying a value, she would have fewer overall moves to make in the interaction. Snippet C illustrates this change, allowing the caller to both choose the subtask and complete it at the same time — the sort of behavior we have seen routinely in the human–human dialogues of the last few chapters — compressing these two steps (choice and completion) into one user utterance:

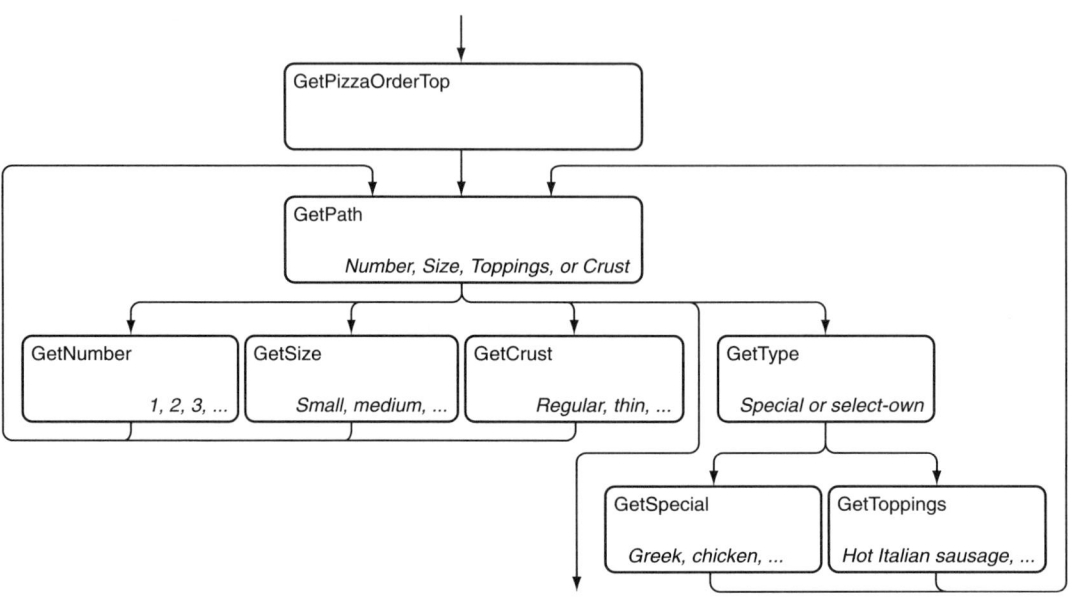

FIGURE 14.5   A stepwise call flow for the schema, GetPizzaOrder

Pizza-delivery snippet C

T_{3c} AcmePizza   Specify a size, crust, or toppings.

T_{4c} User   Wheat.

Or, what is more common, the system could force the choice, again flattening the user's relevant interaction to only one turn, but in an artificial and dictatorial way, as in snippet D:

Pizza-delivery snippet D

T_{3d} AcmePizza   Specify a crust.

T_{4d} User   Wheat.

Or AcmePizza could supply the options directly — another common design strategy — as in snippet E. This structure is effectively the same in terms of flattening, but it has a more explicit prompt:

Pizza-delivery snippet E

T_{3e} AcmePizza   On a regular, thin, stuffed, or wheat crust?

T_{4e} User   Wheat.

But these are only minor flattenings. The target interaction, the more functionally habitable one, is to compress the choices and the completions as much as possible, at the user's initiative, opening the possibilities of an interaction more like snippet F:

Pizza-delivery snippet F

T_{2f} User   Yes, I'd like a large, wheat-crust pizza, with pepperoni, bacon, and green olives.

T_{3f} AcmePizza   OK. One large pizza. Wheat crust. Pepperoni, bacon, green olives.

Would you like anything else with that?

Snippet F flattens the entire Get-pizza-order schema — five steps (number, size, crust, toppings; and the special/self-choice subtopping step) and the four completions (including the multiple values for toppings) — broadening the interaction into one user turn, and a confirmation, and evokes the next schema to boot, GetAccompOrder.

Menu flattening, when it is implemented properly, does not *require* the user to collapse all the steps; it simply *allows* her to. But that allowance demands good design — in particular, good vocabulary management and good task management. Vocabulary management is important because, unlike, say, Snippets B–E, where the system need only listen for one of four keywords (*regular, thin, stuffed,* or *wheat*), it now has to listen for scores of keywords. Task management is important because, if the user only combines two steps and completions rather than all five — as, for instance, in Snippet G — the system has to be prepared to come back with the right request for whatever might be missing.

Pizza-delivery snippet G

T_{2g} User   Yes, I'd like a large pizza, with pepperoni, bacon, and green olives.

T_{3g} AcmePizza   OK. One large pizza. Pepperoni, bacon, green olives.

What type of crust would you like?

What an adequately flattened menu design needs to be prepared for is a range of interactions like the ones outlined in Figure 14.6.

Figure 14.6 is not a call flow. Notice, for instance, what would happen in terms of this arrangement if the caller chose the crust type but not the toppings. Figure 14.6 merely illustrates the way in which schemas need to be encapsulated, for individual exchanges about individual tasks, but how the overall interaction can be broadened, so that the multiple schemas can be invoked simultaneously, supporting exchanges that range over multiple value-assigning tasks.

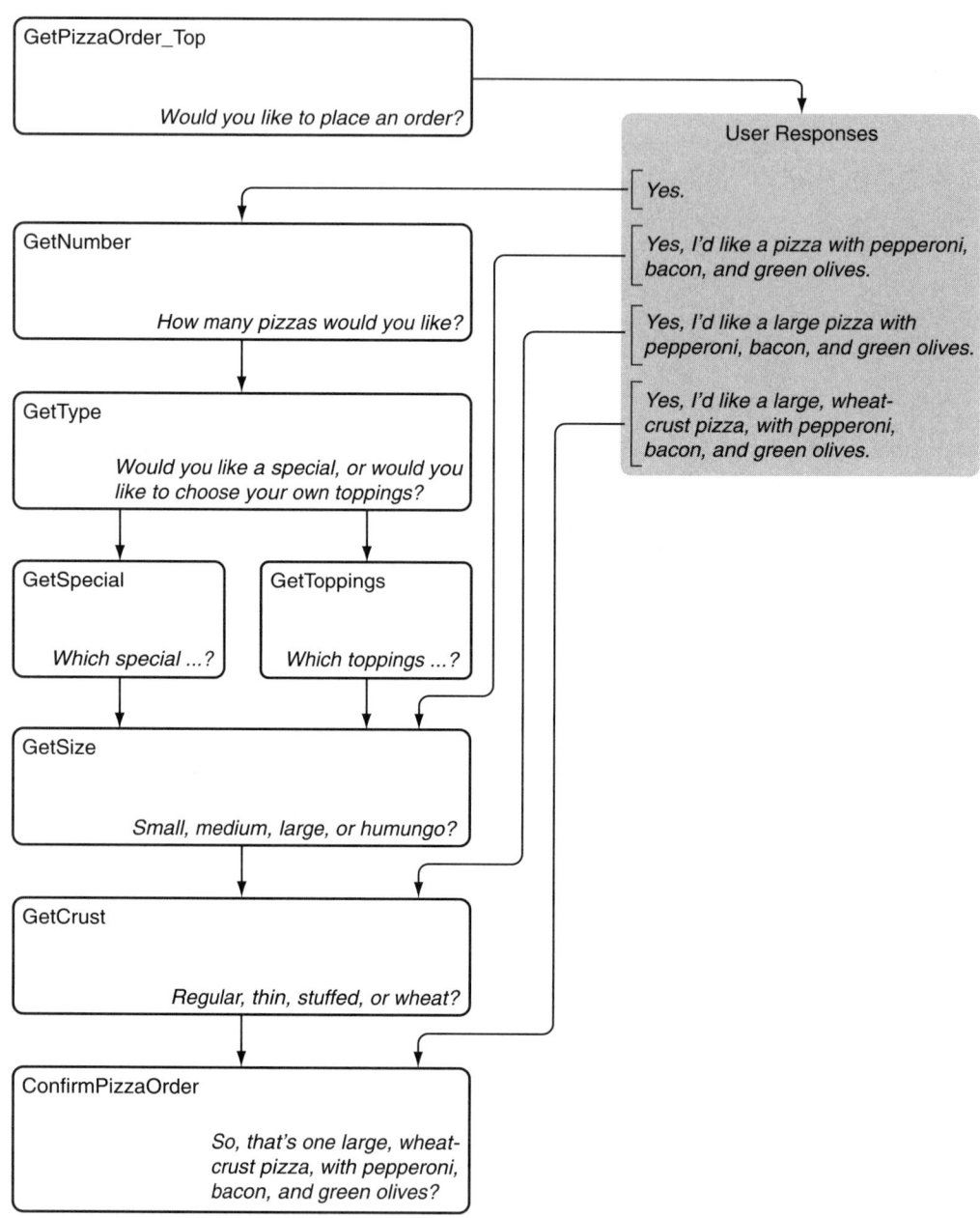

FIGURE 14.6 A flattened-menu interaction example for the schema, GetPizzaOrder

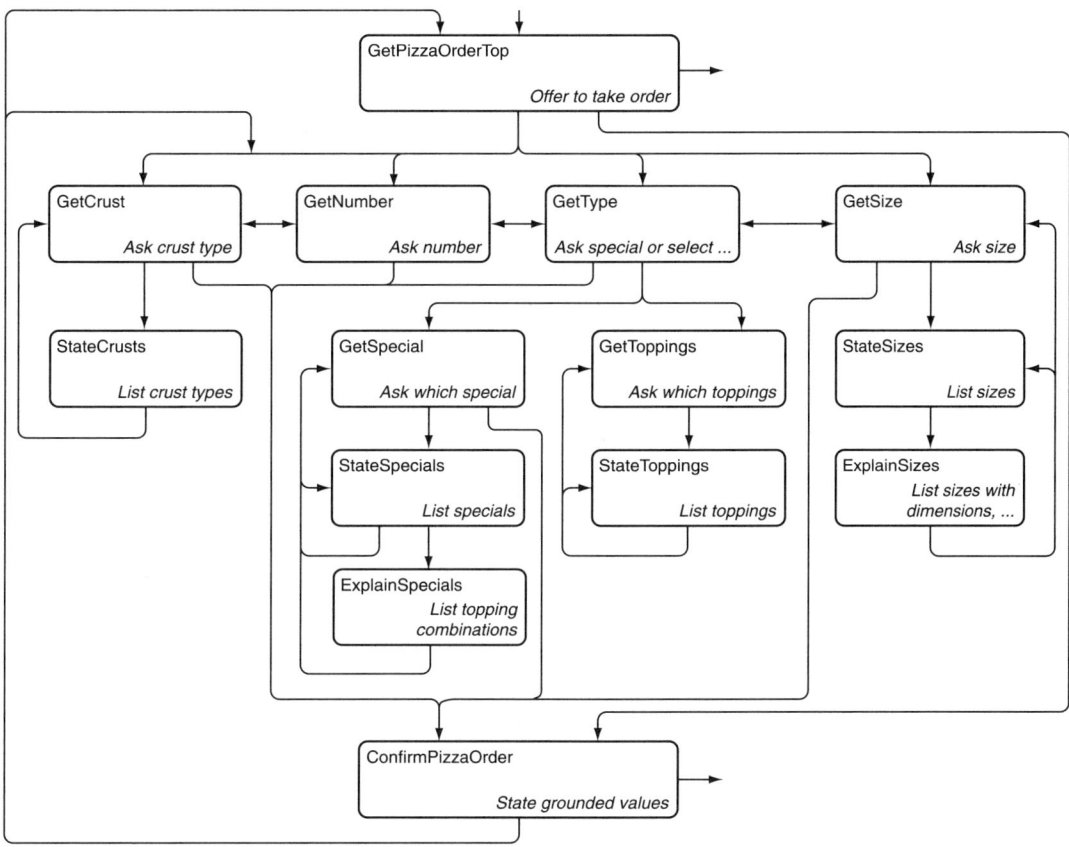

FIGURE 14.7   Fully specified, functionally habitable call flow for the schema GetPizzaOrder

Plotting the Call Flow

> *I want to write a script about plumbing, how every pipe is joined to every other.*
> — Adrienne Rich

The fully worked-out call flow for the schema, GetPizzaOrder, is given in Figure 14.7, with additional schemas for user queries about sizes, crust types, the combination of ingredients for given specials, and so on.

It is worth pausing to see how a specific transaction might occur with this call flow. What we really need is animation, so the schemas and connecting arrows can light up to illustrate the paths corresponding to utterances and groundings, but all we have is your finger, which I request you place on Figure 14.7 as you read through my commentary on the following dialogue, as Galen calls Acme Pizza's voiced take-out/delivery ordering service.

Otto/Galen GetPizza dialogue

T_1	Otto	Welcome to Acme Pizza's automated order system. I am Otto. Would you like to order a pizza?
T_2	Galen	Yes, a large.
T_3	Otto	OK. Would you like a special or do you want to select your own toppings?
T_4	Galen	What are the specials?
T_5	Otto	Greek, chicken-and-pesto, meaty, meaty-extreme, spicy Sicilian, veggie, and veggie-extreme.
T_6	Galen	What's on the spicy Sicilian?
T_7	Otto	Hot Italian sausage, green olives, onions, garlic, fresh basil, and roasted red peppers.
T_8	Galen	OK, I'd like that.
T_9	Otto	One large spicy Sicilian?
T_{10}	Galen	Yes.
T_{11}	Otto	What type of crust?
T_{12}	Galen	Just regular.
T_{13}	Otto	OK, regular crust.

(This is not a complete call, of course; price, mode, drinks and other accompaniments, and so on remain to be established. This is the section of Galen's call that engages the schema GetPizzaOrder and its children.)

At T_1, your finger should be on the GetPizzaOrderTop box. At T_2, your finger stays where it is, since Galen specifies values for both number (one) and size (large) right off the bat; that is, he chooses and completes both subtasks, so the corresponding schemas don't need to be activated. He doesn't need to be queried about size or number. At T_3, your finger moves to GetType, where Otto does have a query: "Would you like a special or do you want to select your own toppings?" Rather than making that choice, Galen begins an insertion-sequence, so the call, and your finger, goes right through GetSpecial transparently (that is, not "stopping" to query) to StateSpecials, at T_4. At T_5, Galen starts a second (non-nested) insertion sequence and your finger finds itself in ExplainSpecials. And so on. With your

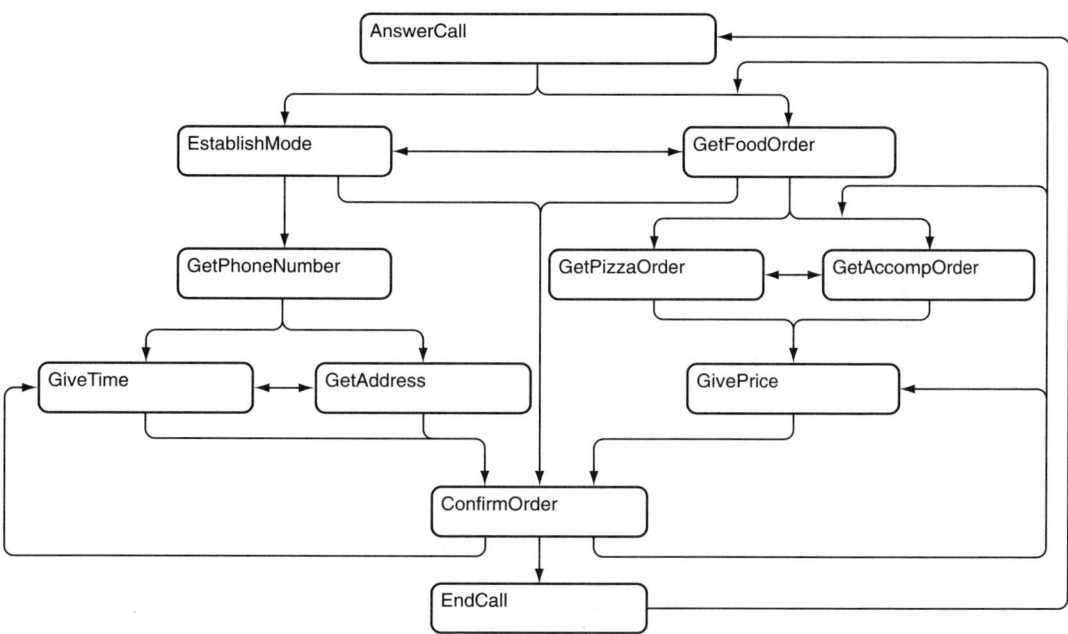

FIGURE 14.8 An habitable call flow for Acme Pizza's take-out/delivery phone service

finger and patience tiring, I turn you over to the perusal of Table 14.4, which charts the remaining travel paths of the dialogue.

Moving back up to the overall project, Figure 14.8 is a call-flow blueprint for Acme Pizza's voiced take-out/delivery service; it is functionally habitable (assuming that the GetPizzaOrder schema has internal workings like those outlined in Figure 14.7, that GetAccompOrder has a similar substructure, and so on). Compare it to Figure 14.3, the stepwise architecture. The narrow, bumpy design (or whatever the appropriate adjectives are, the antonyms of "broad" and "flat") of Figure 14.3 forces the user to make all the accommodations; the design of Figure 14.8, in contrast, accommodates the user. Compare, for instance, the schema, GetPhoneNumber. With the 14.3 design, it's the first thing that confronts a user, because it is a piece of information the system prizes. The 14.8 design allows users to start ordering right away, that is, to perform the tasks they are called on to perform. Discussions with customers revealed that reciting their phone numbers right off the bat was the one part of the Acme Pizza phone experience they did not enjoy. Nobody calls a take-out/delivery service in order to announce or confirm their telephone number — they are willing to, but it is never less than an obstacle to the true task, and minor obstacles of that sort are generally tolerated better as part of specific tasks, rather than as hurdles to be overcome before getting to that task. In particular, while giving (confirming) a phone

Turn		Utterance	Flow		
			Paths At/From	To	Values assigned
T₁	Otto	Would you like to place an order?	GetPizzaOrderTop		none
T₂	Galen	Yes, I'd like a large pizza.	GetPizzaOrderTop →	GetType	total[1]; (number[1], size[large])
T₃	Otto	OK. Would you like a special or do you want to select your own toppings?	GetType		total[1]; (number[1], size[large])
T₄	Galen	What are your specials?	GetType →	StateSpecials	total[1]; (number[1], size[large])
T₅	Otto	Greek, chicken-and-pesto, meaty, meaty-extreme, spicy Sicilian, veggie, veggie-extreme.	StateSpecials		total[1]; (number[1], size[large])
T₆	Galen	What's on the spicy Sicilian?	StateSpecials →	ExplainSpecials	total[1]; (number[1], size[large])
T₇	Otto	Hot Italian sausage, green olives, onions, roasted garlic, fresh basil, and roasted red peppers.	ExplainSpecials		total[1]; (number[1], size[large])
T₈	Galen	OK, I'd like that.	ExplainSpecials →	ConfirmPizzaOrder	total[1]; (number[1], size[large], type[spsicilian])
T₉	Otto	One large spicy Sicilian?	ConfirmPizzaOrder		total[1]; (number[1], size[large], type[spsicilian])
T₁₀	Galen	Yes.	ConfirmPizzaOrder →	GetCrust	total[1]; (number[1], size[large], type[spsicilian])
T₁₁	Otto	What type of crust?	GetCrust		total[1]; (number[1], size[large], type[spsicilian])
T₁₂	Galen	Just regular.	GetCrust →	ConfirmPizzaOrder	total[1]; (number[1], size[large], type[spsicilian], crust[regular])
T₁₃	Otto	OK, regular crust.	ConfirmPizzaOrder		total[1]; (number[1], size[large], type[spsicilian], crust[regular])

TABLE 14.4   The paths through the Otto/Galen Acme Pizza dialogue, based on the call flow blueprint for GetPizzaOrder, Figure 14.7

AnswerCall			
Caller Rating–Novice			
Path		From Ø	To EstablishMode
Values		**Assigned**	**Remaining**
	Mode	ø	delivery [address] OR pick-up
	Phone	ø	number
	Pizza	ø	total; (number, size, type OR toppings[ ], crust)
	Drink	ø	total; (number, size, type)
	BreadSticks	ø	total; (number type)
	Chips	ø	total; (number, type)
Output	Initial	Acme Pizza. I am Otto, an automated speech system. I can take your order. Take out or delivery?	
	Timeout1	I can help you place your order and arrange for pick up or delivery. Would you like to order a pizza to pickup or for delivery?	
	Timeout2	Can I take your order, or would you like to wait approximately ⟨wait time⟩ minutes to order from ⟨agent name 1⟩ or ⟨agent name 2⟩. To continue, say "yes." For one of the girls, say "Get me a human!"	
	Timeout3	To: TopHelp	
Input		**Action**	**To**
	"take-out" "pick-up"	Assign mode = pick-up	GetPizzaOrder
	"delivery"	Assign mode = delivery[address = Ø]	GetPhoneNumber
	Order value(s)	Assign order value(s)	GetPizzaOrder
	Order trigger	Ø	GetPizzaOrder
	Mode trigger	Ø	EstablishMode
	Rejection	Ø	Rejection1
	Key = zero	Ø	ZeroOut
	Key = anynotzero	Ø	TopHelp

TABLE 14.5 A sample pizza-ordering schema specification, with dialogue, paths, and value assignments

number is logically unconnected to the other aspects of the interaction, it has somewhat more affinity with mode than with the "real" task of phoning a take-out/delivery joint, ordering food.

This call flow is very flexible, but it's not an anything-goes system. Otto has to be prepared to take the initiative as required, in a focused way. The schemas (or, what amounts to the same thing, the paths) can be weighted, too — indeed, should be weighted — so that, for instance, all things being equal, the system might enquire about the food order before the mode, the pizza-order before the accompaniment order, the number of pizzas before the types of pizzas, types before crusts, and so on.

Things are not always equal, however, and the differences bias the path choices in rule-governed ways. For instance, if the system gets a complete mode-value assignment in response to the Opening (with input including, say, "for pick-up"), it would proceed to GetFoodOrder; on the other hand, if it got only partial mode-value assignment ("for delivery," without an address), it would take the path to GetAddress. These weightings are set down in detail in the schema-by-schema design specification.

Specifying the Design

The traditional approach of specifying sequential "call flows" and handing them off to a system developer to implement becomes quite cumbersome as the dialogue system becomes more flexible, allowing mixed initiative interactions between the system and the user.
— Candace Kamm

At this point, finally, we are ready to spec out the interaction model in enough detail to proceed to prototyping and testing. (Of course, prototyping usually begins in various forms by various design and software personnel virtually as soon as the project begins. But the completed specification marks the formal turn toward building fleshed-out prototypes for usability and quality-assurance testing.)

Most design teams still hand off paper products at this stage, a document that plots out the schemas, paths, and utterances, and paper has its virtues — chiefly stability, endorsement-recording, and portability. The design can be fixed in a specific configuration, appropriate personnel can sign on the dotted line, and relevant parties can take official copies anywhere and study it. Or various schema specifications can be spread out on a table for simultaneous analysis (screens are more awkward in this regard). But these virtues of paper argue for printouts to augment the digital design spec, not for a paper document that is *the* design spec. The principal reasons for paper hand-offs are not entirely wholesome — tradition, inertia, and the historical isolation of voice-interface design teams from the technologies of production.

Developing the utterance-by-utterance, schema-by-schema specification is best done with a digital application that can interlink utterances and schemas, and can respond interactively to changes in variables. Similarly, the interaction model is best understood by others if they can probe it electronically. So, whether you hand off paper for official endorsement or not, you should also hand off a digital version of the design specification — all the better if it is *the* official version, fixed by security settings and endorsed by electronic signatures, with both digital and paper-based portability.

The specification involves call-flow charts of the sort we have seen in this chapter (most relevantly, Figures 14.7 and 14.8), but the schemas need to be fleshed out in considerable

AnswerCall		From Ø		Caller Rating–Novice	ID=AP00013
			Assigned	Confirmed	Open
Variables	Mode		Ø	Ø	Delivery [Address1, Address2] OR Take-out
	Phone		Ø	Ø	PhoneNumber
	Pizza		Ø	Ø	PizzaTotal; (PizzaNumber, Size, SpecialType OR Toppings [], Crust)
	Drink		Ø	Ø	DrinkTotal; (DrinkNumber, DrinkSize, DrinkType)
	BreadSticks		Ø	Ø	BreadTotal; (BreadNumber, BreadType)
	Chips		Ø	Ø	ChipTotal; (ChipNumber, ChipType)
Output	Initial: OpeningN1		"Acme Pizza. I am Otto, an automated speech system. I can take your order. Take out or delivery?"		
	Timeout1		"I can help you place your order and arrange for pick up or delivery. Would you like to order a pizza to pick-up or for delivery?"		CONFIRM –> Path-to: EstablishMode DISCONFIRM –> Path-To: Queque
	Timeout2		"Can I take your order, or would you like to wait approximately \<waittime\> minutes to order from \<agent name 1\> or \<agent name 2\>? To continue, say 'continue.' For one of the girls, say 'Get me a human!'"		"continue" –> Path-to: EstablishMode "Get me a human"–> Path-to: Queque
	Timeout3		Path-To: TopHelp		
Input			**Action**		**Path-To**
	"take-out" "pick-up"		Assign mode=take-out		GetFoodOrder
	"delivery"		Assign mode=Delivery [Address1=Ø, Address2=Ø]		GetPhoneNumber
	"delivery" + address		Assign mode=Delivery [Address1=number, Address2=street]		GetPhoneNumber
	"delivery" + partial address		Assign mode=Delivery [Address1=number OR Address2=street]		GetAddress
	Mode + order value(s)		Assign mode; assign order value(s) (e.g., size, type, crust, ...)		GetPizzaOrder
	Mode + address + order value(s)		Assign mode=Delivery [Address1=Ø, Address2=Ø]; assign order value(s) (e.g., size, type, crust, ...)		GetPhoneNumber
	Rejection		Ø		Rejection1
	Key=zero		Ø		ZeroOut
	Key=anynotzero, help-request		Ø		TopHelp

FIGURE 14.9 A (static) design specification for AnswerCall under the novice user model

detail: their behavior with respect to input utterance (or not), user model, and grounded and pending values. Figure 14.9 illustrates this level of specification for one schema (AnswerCall), under one user model (novice).

There's nothing particularly sacred about this arrangement, and some details have been elided for ease of display (in particular, the full input options have not been detailed). But Figure 14.9 represents the basic behavior of AnswerCall, particularly under a novice user model, with no groundings (that is, the unique configuration in the design specification with the ID AP00013). For instance, every potential variable must be tracked for three values: whether it has been supplied by the caller, whether it has been confirmed by the caller, and the default case, neither (not supplied, not confirmed). Since this schema is the first one of the interaction, no variables have had any values assigned; they are all, by default, "remaining." Schemas later in the call flow have contingent behaviors (especially preferred exit paths) as a function of which variables have been assigned values, which are still open.

The phone rings, the system answers, and — as Figure 14.9 says — Otto opens the dialogue with "Acme Pizza. I am Otto, an automated speech system. I can take your order. Take out or delivery?" If one of the expected replies comes ("take-out" or "delivery," with pick-up as a synonym for the former), the appropriate value is assigned to the variable, *mode*, and the call is routed to GetFoodOrder.

If the value, *delivery*, is accompanied by a full address, the street and number values are assigned, and the routing goes to GetPhoneNumber; if it is only accompanied by a partial address, then the routing is to GetAddress, where the other portions of the address can be solicited.

Similarly, if the food order is fully or partially specified, the appropriate values are assigned and appropriate routing followed, though this is part of Figure 14.9 where there are elisions (the other part includes the global commands beyond ZeroOut and Help — repeat requests, price requests, time requests, "get me a human" and related bail-out terminology, "what did I get last time" and related customer-history phrases; and termination vocabulary). All 14.9 specifies is, effectively, "assign the right values and route the call appropriately," but a genuine specification would be broken down more fully. Take a scenario in which the caller says "I'd like two large, thin-crust, pepperoni and bacon pizzas delivered to 52 2nd avenue." That scenario would (assuming full recognition) correspond to an input with the following structure:

Mode + address + number + size + crust-type + toppings

In turn, this would result in value-assignments with this structure:

Mode = delivery[address1, address2], total = n, number = 1[size, crust-type,
      toppings[topping1, . . . topping$_n$]], . . . number = n[size, crust-type,
      toppings[topping1, . . . topping$_n$]]

And the corresponding routing assignment would be to GetPhoneNumber.

On the other hand, Figure 14.9 tells us that a timeout here would trigger a yes/no question. If it was confirmed, the call would route to EstablishMode, where Otto would make the take-out-or-delivery offer again; disconfirmed and the routing would be to a queue for human-agent interaction. A second timeout would result in a query about continuing. A third timeout would route the call to the Help schema without further fanfare.

The design specification should be interactive, so that the designers and all the relevant personnel can see the effect that different groundings (both assigned and confirmed) and incoming paths might have, especially on the routing. This sort of interactivity is not easily illustrated with Figure 14.9, since the relevant schema is AnswerCall, for which all variables will necessarily be open and the incoming path is nil. But consider Figure 14.10, which specifies the schema ConfirmPizzaOrder, again under the novice user model, but this time not all the value assignments are open. In fact, this schema specification

ConfirmPizzaOrder		FromGetCrust	Caller Rating–Novice	ID=AP03010
Variables		**Assigned**	**Confirmed**	**Open**
	Mode	Delivery [Address1 Address 2]	Ø	Ø
	Phone	PhoneNumber	PhoneNumber	Ø
	Pizza	PizzaTotal; (PizzaNumber, Size, SpecialType, Crust)	PizzaTotal; (PizzaNumber, Size, SpecialType)	Ø
	Drink	Ø	Ø	DrinkTotal; (DrinkNumber, DrinkSize, DrinkType)
	BreadSticks	Ø	Ø	BreadTotal; (BreadNumber, BreadType)
	Chips	Ø	Ø	ChipTotal; (ChipNumber, ChipType)
Output	Initial: CrustConfirm	"OK, [Crust]"		
	Timeout1	"Sorry, did you want a [Crust] crust?"		CONFIRM –> Path-to: GetAccompOrder DISCONFIRM –> re-open crust, Path-to: Queque
	Timeout2	"I've got the rest of the pizza order, but I just want to be sure about the type of crust—is it [crust]?"		Input
	Timeout3	Path-To: TopHelp		
Input		**Action**		**Path-To**
	CONFIRM, ECHO	Confirm [Crust]		GetAccompOrder
	DISCONFIRM	"Oh, sorry."		GetCrust
	DISCONFIRM + New Crust	"OK, got it, [NewCrust]." Assign Crust.		GetAccompOrder
	Accompaniment trigger	"Alright"		GetAccompOrder
	Accompaniment trigger + value(s)	"Alright." Assign Accompaniment value(s)		GetAccompOrder
	Rejection	Ø		Rejection1
	Key=zero	Ø		Queque
	Key=anynotzero, help-request	Ø		TopHelp

FIGURE 14.10 An interactive design specification for ConfirmPizzaOrder, under the novice user model.

corresponds to the system state at T_{13} in our Galen/Otto GetPizza dialogue (and Figure 14.9 corresponds to T_1 of that dialogue).

Now, an interactive specification like Figure 14.10 would allow the designer/coder/ quality-assurance prime/dialogue writer/whomever to probe the design: what would the change be, for instance, if crust was already grounded; or if none of the assigned values were grounded; or if the input path was from GetPhoneNumber? Making these changes (drop-down option-menus are useful here) would result in a different schema specification, with a range of different conditions and actions (and therefore with a different ID). For instance, if the input path was from GetPhoneNumber, the initial prompt could not be "OK, [crust]," even if crust was the only ungrounded pizza-order value, because crust would not be in focus in those circumstances. Rather, an appropriate prompt would be something on the order of, "Now, there's just one thing I'm still not sure about, the crust. Did you want a [crust] crust?"

The alternative to a digital design specification is endless pages of paper that are some-what awkward to use, because navigation among schema specifications is much more laborious. Printouts certainly have their virtues, but the fluidity with which the design specification can be explored and understood in digital form far outweigh those virtues. And, in any case, paper and digital specifications should not be exclusive options. There are arguments that either should be the official milestone, but both serve essential functions.

Summary

It has not always had
To find: the scene was set; it repeated what
Was in the script.
    — Wallace Stevens

Designing a voice interface is the crafting of an interactive speech system, an expert system, to support specific, purposive verbal behaviors. In this chapter we have developed the Schank and Abelson notion of a **knowledge script**, aided and abetted by analogies from the primary domain of scripts, drama, to chart developing the dialogue, planning the call flow, and specifying the overall design.

Developing the dialogue is a process of **elaborating the knowledge scripts**, first to the point of outlining the dialogue acts, and then to writing out the specific utterances, under the principle of **constrained variability**, which recommends a range of utterances for the same function, not only for reasons of tapering and expansion, as we have explored in previous chapters (chiefly Chapter 13), but also for a limited variety that adds naturalness to the interaction.

Planning the call flow is a matter of working out the **sequencing and dependencies** among the schemas, and, where feasible, **broadening the interaction** ("flattening the menu") to provide the user with flexible possibilities for combining elements of her tasks in natural and productive ways.

Specifying the design is charting out the interaction model, utterance by utterance, schema by schema, with all the contingencies of system behavior represented in detail. It is much facilitated with an **interactive digital specification**, that allows both designers, as they build it, and engineers, as they implement it, to follow those contingencies closely.

Iterative Evaluation

Test every work of intellect or faith
And everything that your own hands have wrought
— William Butler Yeats

There are archetypal moments of testing — the little girl's toe in the water, the golfer's handful of dust in the air, the pirate's molars on a coin — but there's nothing particularly unique about them. They represent what we all do, all the time. We are thoroughly empirical creatures. We continually interrogate our surroundings for data. Is the coffee cool enough yet? How will the brakes handle this wet pavement? Which watermelon sounds ripest? We sip, we press, we rap, we judge — then we act.

Development cycles follow this logic too, when they're done right. When they're done wrong, they skip the empirical data, and often the interrogation. When they're done right, they design, refine, and mature the product iteratively, with strategic inputs of empirical data, gathered through testing. The difference between the little girl's toe in the water and a usability test is one of degree, not of kind.

In this chapter, we look specifically at two classes of tests that define key junctures in the development cycle for voice interfaces — usability tests and Wizard-of-Oz tests — as well as surveying several related inspection methods.

I won't waste any of our time justifying usability testing; anyone who needs usability testing justified to them for *any* product that implicates human interaction, let alone for a product as intimately interactive as a speech system, is not only between the covers of the wrong book, they're in the wrong business. Usability testing is an indispensable element of design. It is not a force brought to bear — time and money willing — upon design.

The essence of usability is an experiment in which a product is given to users, tasks are tried, observations made. For example, you have a phone-based, voice-driven, local navigation system. You give it to a couple of people unfamiliar with Boston, Massachusetts, at MIT's Building 20, and ask them to find their way to 695 Atlantic Avenue. You see whether they can do it; if so, how efficiently; if there are points in the process where frustrations or breakdowns occur, you note them; and so on. You bring these observations together, with those of a few parallel experiments with other users, and see if you can make the system more usable for people like them.

The essence of a Wizard-of-Oz test is exactly the same except you don't have the system to test, so you put a person at the other end of the phone, operating under specific constraints, and tell the users they are talking to a computer (for them, it *is* a usability test). Again, you gather the observations, but this time the goal is to help design the still-nascent navigation system: how can you make a phone-based, voice-driven, local navigation system that behaves the way the users want it to. Because the product is simulated, rather than manufactured, Oz work is sometimes called *low-fidelity* (or *lo-fi*) *prototyping*; usability testing, which traditionally involves close-to-market-ready prototypes, is *high-fidelity* (*hi-fi*).

Usability testing stereotypically occurs late in the cycle, Oz testing occurs early, but they both have a range of participatory possibilities in the course of product development, and both are of a piece with what Nielsen and Mack (1994) call *usability inspection methods*, procedures that bring users (or at least user considerations) to products (or simulations of products), at various stages in the design-to-production period between the concept and the market-released artifact. The point of this testing and inspecting is to ensure as good an interactive match as possible between users and products, which, far from incidentally, is the point of product development generally and interaction design specifically.

Wizard of Oz

> *As [the screen] fell with a crash they looked that way and in the next moment all of them were filled with wonder. For they saw, standing in just the spot the screen had hidden, a little old man with a bald head and a wrinkled face, who seemed to be as much surprised as they were.*
>
>   — L. Frank Baum

Necessity is the mother of invention. Wizard-of-Oz testing developed because there weren't really any alternatives. In the early 1980s, John D. Gould and his associates, John Conti and Todd Hovanyecz, wanted to test the concept of a "listening typewriter" — a speech-recognition dictation machine. There was a problem. There weren't any. So they faked it. They had a typist enter the test-subject's speech manually, and then routed it to a computer display. There is a long tradition of such mimicry in psychology experiments, but this appears to be the first time the method was steered in the direction of product development. The name Gould and his team chose for this technique — an analogy to the meek prestidigitator in the *Wizard of Oz* series of children's books, who hid behind a screen and pretended to have great and terrible powers — has stuck.

Wizard-of-Oz work is not unique to voice interaction design. In principle, any human–computer interaction might be faked in this way, and the approach is especially useful for early prototyping of machines meant to perform tasks which are much easier for humans than for machines, like visual processing, natural reasoning, and ill-defined problem solving. And language use. Humans are very good at language.

Wizard-of-Oz testing (sometimes WOZ, or WOz) is remarkably well-suited to speech-system development. It is so well-suited that (although most Oz work is front-loaded on the development cycle) it distributes very effectively throughout the design cycle. You can begin a species of Oz testing from the very inception of a speech project, two designers playing dialogue games with each other — or even one, in her own head — and you can deploy it very late in the cycle, to test the viability of suggestions that come out of a usability test.

One thing it cannot do well, however — a point that is not well represented in the literature, but which most working designers appreciate — is adequately explore the register. It is not a substitute for discourse analysis. The problem is one of convergence.

It is possible to get register data from Oz techniques, of course, and you should always keep your ears open for precisely such data. But using those techniques to *elicit* the natural discourse patterns of users, especially using them as the primary method of gathering data about those patterns, is a serious methodological error. By convergence we know the users take substantial cues about what to say from what the Wizard says. So, much of what you get back is what you've previously primed for, however inadvertently. It gets even worse when you consider the tasks (for instance, asking them to get directions from Building 20 to 695 Atlantic Avenue). If you explain the tasks verbally, you're priming the participants; if you give them written explanations, you're priming them; gestures would be hugely problematic; using a map undermines the task.

The inclusion of graphics with the instructions appears to help. (See Dykbjær and Dykbjaer, 1993; Schillo, 1996; Bernsen et al., 1998: 157ff.) But not all tasks are equally amenable to graphic depiction, and even such abstract graphic elements as arrows and lines likely prime the user to some degree. Most insidiously, priming can work in very subtle ways, cuing semantic or phonological relatives of the input terms, in which the convergence may be difficult to detect, but which nevertheless results in vocabulary that may not have occurred spontaneously to the user. Nor are design budgets usually sufficient to pretest various vocabularies against one another, priming this word and then that word and then measuring them against one another. The moral, then, is twofold: do not use Oz techniques as your primary research tool into register; and do not trust the vocabulary or structure that occurs in Oz studies, unless it is supported by other techniques (such as broader-based corpus studies).

The value of Oz testing, however, is substantial:

- It allows the design team to escape the concerns of recognition engines and natural-language understanding to focus on higher-level interactive issues.

- It bypasses the dependency on coding that higher fidelity prototyping requires.

- It gets to users, and their computer-related speech behavior, early in the development cycle.

- It provides for much broader coverage than hard prototyping.

- It can identify specific areas for concern that both design and later testing can concentrate on.

- While you can't trust Oz testing as a probe for natural register data, it can still turn up vocabulary, utterance structures, and interactive patterns that prove valuable for the voice interface.

Comparatively, there really are no liabilities of Oz testing, no serious list of cons (beyond the limitations of the data it generates for understanding the register). But there are several considerations to be weighed before committing to a formal Oz phase:

- It depends very heavily on a good Wizard.

- It relies on good support (team and tools).

- It requires time for:

  - Planning.

  - Subject recruitment.

  - Analysis.

That is, it consumes resources: people, time, and machinery.

Fraser and Gilbert (1991: 82) say that before doing an Oz study, "it should be possible to formulate a detailed specification of how the future system is expected to behave." That's certainly true — the Wizard's rules of engagement should be clear and fully specified — but it does not mean the test is only about refining that detailed specification. In fact, it may occasion a major overhaul. Wizard-of-Oz testing is part of the creative process of speech-system design, not a calibration instrument for an almost-finished model developed at arm's length from the users.

The Resources

"Come along, Toto," she said. "We will go to the Emerald City and ask the Great Oz how to get back to Kansas again."
— L. Frank Baum

Wizard-of-Oz work is, in principle and sometimes in practice, very easy, but using it to its best advantage has one major obstacle — getting the right Wizard — as well as several logistical demands.

To carry out a full-force Wizard-of-Oz work test you need:

- Participants who represent the target users

- A Wizard

- Wizard-support tools

  - A searchable database, covering the domain of interaction, preferably with text output that the Wizard can use verbatim or adapt easily

  - A computer dedicated to the interactive flow

  - Possibly input/output filters:

    - A speech recognizer

    - A speech synthesizer or vocoder

- An assistant to the Wizard

- A test coordinator

- An observer / data collector

- Data collection tools, which both capture the audio and effectively support transcription

- Possibly adjunct experts:

  - A subject-matter expert

  - A technology expert

The original Gould-Conti-Hovanyecz technique has been hailed as "as a good example of using limited resources to test the validity of an idea before making a heavy investment in its development" (Buxton, 1995: 525), but already this list of requirements is beginning to make Oz work look like a Big Deal. Indeed, Bernsen, Dybkjær, and Dybkjær (1998), in a book that represents the fullest exploration of Oz testing in the literature, regard it as quite resource intensive ("a relatively costly development method" — 1998:127).[1] There are two considerations here. Oz testing certainly can be resource intensive, but much less so than building the system first and using *it* to test concepts, flow, and discourse. And Oz methodologies needn't always put significant pressure on resources — depending on when, how, and for what purposes the testing is done. In particular, there are many potentialities for Oz work in interactive design. They are all best understood with respect to the formal methodology that developed out of the Gould-Conti-Hovanyecz paradigm, which is what I am cataloging here, but they can be much smaller scale. That is: you don't always have to run full-force tests.

Jakob Nielsen coined the term "discount usability engineering" (1989) to capture a growing commitment in the late nineteen-eighties that usability did not have to be confined (as it largely was before that) to full-force, in-studio, lights-camera-action testing — a valuable but expensive and necessarily restricted activity (see also Nielsen, 1990; Atkinson, 1990; Tognazzini, 1990; Yee and Harris, 1989). Nielsen's slogan (and his work backing it up), helped usher in the notion of user-centered design, and the array of usability inspection methods which foster that notion. Well, there's such a thing as discount Wizard-of-Oz engineering, too; in fact, Oz testing is best seen as just another of the usability inspection methods, with its own range of applications, some of them discount, some more formally developed. It is an indispensable method for developing speech systems, but it is not categorically different from other usability methods, and should not be confined to a single stage in the development cycle.

Seen this way, as playing a distributed role throughout the development cycle, rather than as operating at one isolated stage in that cycle, Oz techniques are revealed as not just defining a key design phase (though they do that), but as defining a fundamental design instrument for creating and calibrating speech systems at any point in their development.

Minimally, you need someone playing the Wizard, someone representing a user, and an interactive task; after that, the form the instrument takes is largely an issue of timing and focus.

1: I recommend this book, *Designing Interactive Speech Systems*, warmly to at least the Usability Prime on the team, and to any interested readers. Ostensibly, the book is a report of the development of the Danish Dialogue System, dressed up as a general speech-system development guide. But the entire book is organized around a series of Oz studies, and several chapters are dedicated to those studies. What makes the book so useful (and so charming) is that, despite the standard academic language of rigor and precision, it is clear that Bernsen and his colleagues were fumbling through the process from the start ("We started our WOZ work more or less from scratch and without sufficient operational guidance from the literature" — 146), learning as they went. They are very conscientious, and moderately frank about their mistakes and misjudgments, which gives their work detail and character and weight.

The Timing

If we explore one place at a time we'll by an' by know all about every nook and corner in Oz.
> — L. Frank Baum

The key moment for full-force Wizard-of-Oz testing is after the discourse analysis has been done, and the system vocabulary largely stipulated; after the dialogue management has been worked out conceptually, and the call flow outlined; and before the coding is underway. But you can't keep a good technique down.

The development of a voice interface has one defining goal, to get a computer and a human to converse in some discourse field until the human is satisfied. So three notions have always to be kept in balance: the computer, the user, and the register. These are, not in the least coincidentally, the principal ingredients of a Wizard-of-Oz trial. They are also the principal ingredients of a usability trial. So, just as there are levels of usability inspection, in which the Usability Prime gauges the projected usability of the system — heuristic evaluations, think-aloud protocols, informal probes, on up to video-recorded, formally monitored, laboratory usability tests — so are there levels of Oz participation in the development cycle.

Designers, whether they use the term or not, begin using Oz techniques almost immediately. Scenario spinning, an early design technique, is a form of invention in which the designer plays both user and computer, and therefore partakes of Oz. Designers play dialogue games throughout the development cycle, in which one or the other must be the system. They have (or should have) informants from the user community and just plain helpful others around the team on whom they can test patches of dialogue ("If the system said this, how would you respond?"). Late in the development, especially if usability testing turns up local problems that would otherwise require bringing in talent to record new utterances, Oz techniques are a useful way to test alternatives.

Again, there is one especially opportune moment for full-force Oz testing, but any time a human simulates a computer in the design of a voice interface, Oz has been evoked. The simulation can be as sloppy as just-pretend-I'm-the-system, but when the quintessential Oz opportunity occurs, a quintessential Wizard should be ready.

The Wizard

Presently they heard a solemn Voice, that seemed to come from somewhere near the top of the great dome, and it said:
> *"I am Oz, the Great and Terrible. Why do you seek me?"*
> > — L. Frank Baum

The Usability Prime is perhaps the most natural candidate to assume this part, but the role-playing ability is crucial: whoever can pull it off best should be the team's Wizard. It is not easy. The Wizard must:

- Speak within a restricted grammar (vocabulary, syntax, phonology).

- Listen within a restricted grammar (vocabulary, syntax, phonology), probably the most difficult aspect of the Wizard's job. For instance, we humans gather a good many clues from intonation, which computers have great difficulty with.

- Speak and listen with attentional restrictions. Not all the system vocabulary is available all the time; vocabularies can swap in and out depending on the stage of the interaction (that is, on the corresponding system state). There may be times when no proper names can be admitted for instance; in extreme cases, the Wizard may only be able to accept "yes," "no," and their synonyms (or else fall back into a repair mode).

- Speak without ums, ahs, slips of the tongue, hesitations, and all the standard-issue imperfections (dysfluencies) of human speech.

- Speak according to highly specified protocols (for instance, progressive assistance, time-outs).

- Follow a potentially intricate, many-threaded, highly contingent script.

- Execute slot-filler scripts (track all the elements of a shipping address, for instance, prompting appropriately for missing or "misunderstood" elements).

- Be prepared to improvise, while staying close to the design parameters.

- Misunderstand the way a computer misunderstands, according to a certain predetermined frequency. This ability is among the most difficult, though someone with lots of experience in the field can catch most utterances that would trip a real speech system (false starts, repetitions, lengthy pauses). If the Wizard is less experienced, then the error conditions should just be introduced according to some predetermined frequency — "approximately one out of ten utterances" Lai and Yankelovich suggest (2002).

The job can be made considerably easier with good tools and good assistants, but it remains very demanding, requiring both training and aptitude. Operating with the tools in real-time is also demanding.

Wizard training is hard to come by, Wizard lore scant, so the education of the Wizard usually must be accommodated in-house. Some suggestions can be gathered in various places — Bernsen, Dybkjær, and Dybkjær (1998) is particularly good — but the literature is still very sparse on this issue, and workshops are rare. The best bet for any external help is just to seek out current and former Wizards through a literature search, or to comb through conference programs, and, when you find them, buy them a cup of coffee.

The earliest iterations of Oz tests, in any case, should always be focused on Wizard training, and should involve speech-recognition and natural-language understanding engineers as participants (whose perceptions and recommendations about the Wizard should be actively sought). All the tests, early and late, will provide both qualitative and quantitative data about the interactions; the primary use for this information is to aid the design of the interface, but it should also be fed back to the Wizard, as well, as part of an ongoing performance review.

A Wizard tandem is a good idea, perhaps initially following an apprenticeship (or understudy) model. Wizards get sick, like everyone else, and sometimes they up and move to Arizona to pursue a dream of becoming a professional golfer; someone should always be prepared to step into the breach, and should periodically take some of the sessions. Ideally, the apprentice would be the regular assistant. Since there are times when female voices might be preferred in voice applications, and times when male voices might be preferred, the best arrangement is for a male/female, Wizard/assistant set-up, where either can take the lead role as required.

The Support Team

"I am the Guardian of the Gates, and since you demand to see the Great Oz I must take you to his Palace. But first you must put on the spectacles."

"Why?" asked Dorothy.

"Because if you did not wear spectacles the brightness and glory of the Emerald City would blind you."

— L. Frank Baum

The test team is indispensable, for Wizard support, test coordination, data gathering, and data analysis, though some of these functions naturally double up, and might occupy anywhere between two and four members (beyond the Wizard).

The Wizard-support team might only be one member strong, but minimally the Wizard needs an assistant to help with information retrieval, someone who does not interact with the participants in any way, but who assists the Wizard's interactions with them. A subject matter expert may be required, for interfaces to highly particularized systems (a system for hardware installation or repair, for instance), but this role is unnecessary for any general-purpose information system (entertainment, weather, travel). A technology expert who can be present during a few iterations of the test can be a valuable governor on the Wizard and (therefore) the interaction model. The technology expert (and, if required, the subject-matter expert) should review the task list and interview questions (though, of course, no recommendations should be accepted uncritically, since the expert may well be too close to the technology for entirely useful suggestions).

The team also requires a coordinator, an observer/data gatherer, and an analyst, all of whom can be wrapped up in one individual, though two is somewhat more optimal: someone needs to greet the participant, set him at ease, explain the procedure; someone needs to observe and take notes; and someone needs to analyze the results and advocate the recommendations that come from them. In practice, these tasks are rarely cleanly divided. The data analysis, in particular, usually is (and should be) a collaborative activity. Simply put: different people discover different things. Notice, too, that the Wizard and her assistant, while they should not be involved directly in data collection, can double as analysts. The Wizard's assistant, too, might serve as coordinator, depending on the physical setup (the Wizard, however, should not be involved in coordination, unless she is completely masked during the test, by a synthesizer or vocoder).

Resist the temptation to farm this work out; you should certainly contract a consultant to guide the process, if there is not enough in-house experience, but the testing team should be drawn from the interface design. There are liabilities, but an experienced Usability Prime will shake these out,[2] and the benefits for the overall design process, and the growth of the design team, are immeasurable. The Usability Prime, of course, must be deeply involved in all of this activity, and will be especially instrumental in the analysis and in reporting the results, in consultation with at least the Quality Assurance Prime and the Interaction Architect. Other members of the testing team might be drawn from the Interactive-dialogue writers and/or the Lexicographer.

The Tools

> In the center of the chair was an enormous Head, without a body to support it or any arms or legs whatever. There was no hair upon this head, but it had eyes and a nose and mouth, and was much bigger than the head of the biggest giant.
> — L. Frank Baum

The minimum for tools, of course, is zero, but automation can help achieve a higher fidelity simulation, and therefore more reliable data.

An audio channel is critical. The participant should have a way of hearing and talking to the Wizard that approximates the implementation method: a phone or a headset.

2: The problems concern familiarity with the design. Usability requires some detachment from the product being inspected. In drafting tasks, for instance, someone who knows the design and its intentions will tend inadvertently to play to its strengths, where someone with more distance isn't even aware of those attractors and will set tasks that reflect user goals more than system expectations. Similarly, with data analysis, someone close to the design often has trouble recognizing flaws that the test exposes, faulting the user or the technology. These are very real dangers, but a good Prime will control and filter the task list, and manage the analysis accordingly.

A speech recognizer, especially one that closely resembles the implementation recognizer, can be invaluable. The Wizard and his assistants should still get the audio input directly, so that they can rapidly choose the system response, but a screen log of the recognition can help decide when and how to deploy repair strategies, help monitor for out-of-vocabulary words, give a general sense of problem areas for later analysis, and consistently reinforce for the Wizard the limitations of the technology, to help govern the performance.

Other input filters are also useful. For instance, depending on the capabilities of the underlying system, the Wizard may have to follow rules like "Ignore everything after the first ten words/phrases." A counter spliced into the recognizer could play a tone or flash a light after ten sequential recognitions, or even dampen the input audio.

A speech synthesizer is also potentially very useful. Synthesis has a number of liabilities, but it provides an extremely convincing filter between the Wizard and the user, to help enforce the illusion of talking to a machine, and provides some flexibility for "age," "gender," and other vocal characteristics. If the design incorporates synthesis, the Oz test should definitely incorporate it as well. If a synthesizer is used, the output utterances must be largely prepared and stored; on-demand, in-test typing will not sufficiently approach real-time interaction.

Alternatively, **a vocoder** can be used, something which just tweaks spoken input in the direction of flatter, more mechanical speech.

The test will also require a number of computers — how many depends on the set up, but between one and three, representing the following functions:

A computer dedicated to the database, for search and display. An assistant should perform this activity and route the information to the Wizard, preferably in a form that requires little adaptation before responding to the participant. The assistant, using this computer, should also track the information slots and their fillers (for instance, departure time, departure location, arrival time, and arrival location, for flight information), and route it to the Wizard as appropriate.

A computer dedicated to the interaction. This machine is the Wizard's, and should have a large display. A portion of the screen real estate should be reserved for the information that goes back to the participant (that is, it must be networked to the database computer, receiving messages from the assistant), and a portion for any recognition data that is generated (if the test includes a recognizer). But the bulk of the screen is for a representation of the interaction model: timing information, lists of legitimate utterances (global and local), assignments of agent responsibility, lists of recognition/understanding rules, and packaged dialogue routines. (Note, the computer should be able to display this information, but how and whether any of it is displayed should be configurable by the Wizard; keeping it all active at the same time would be very noisy.)

If elements of the soundscape are being tested, they also need to be represented on the Wizard's computer, and likewise deployable from there.

A computer dedicated to the capture and logging of data, for later analysis. This one is for the principal observer. It captures the audio, captures any attendant recognition data (including n-best lists with respect to established thresholds), supports the generation of transcripts, and allows for on-the-fly commentary by the observer.

Audio-monitoring equipment is essential. The audio data should feed into the project corpus, to aid the software developers and to feed the omnivorous maw of the interface lexicon. The sessions should be transcribed, for the evaluators and designers to study and refer back to.

Video monitoring equipment is relatively expendable in voice interface testing, for both Oz and usability research, but if it is available and does not represent an undue draw on resources, it should always be used. Video data can uncover aspects of the interaction that might otherwise be unavailable (frustration, distraction, physical problems with the audio channel equipment, and so on), and can augment or contextualize results gathered by other logging methods.

The Process

> *She also made many magical experiments, hoping to discover something that would aid her.*
> — L. Frank Baum

Like a usability test, an Oz test is fundamentally an experiment. The controls are looser, and the statistical analyses less rigorous, than in a prototypical scientific experiment. But it's an experiment nonetheless. It requires a dependent variable (the interaction model) to be tested against a range of independent variables (or conditions): the participants (or, more properly, the participants' speech behaviors, the task(s), and the environment. It is also, as this dependent-to-independent-variable ratio should make clear, what is technically known in the experimental sciences as "a fishing expedition" — in the sense that specific hypotheses cannot be investigated with any rigor and the exercise is mostly about generating as rich a body of data for analysis as is manageable.

The Interaction Model

The interaction model is the set of design considerations, utterance constraints, and flow possibilities that define the (projected) cooperative speech activities of the system and the user. It is represented in the test by the Wizard, along with the assistants, and the support tools (by what Bernsen, Dybkjær, and Dybkjær, 1998: 131ff, call "the Wizard interface"). Effectively, this is a design specification, as defined in Chapter 14.

The Participants

The participants should approximate end users as closely as possible. Ideally, they should be drawn from the projected end-user population. An Oz test can get by with 5–7 participants, provided the user population is fairly homogeneous, and the discourse field is well understood (as, for instance, with a warehouse inventory or distribution tool). If the projected user population is heterogeneous, as many voice-interface services are (as, for instance, with a traffic service), the participant group should sample this population more widely. The defining characteristics (for instance, someone who operates a vehicle regularly in a manner that implicates traffic information) are necessary filters; after that, age and gender are probably the most important characteristics to include, since these are the traits that condition discourse and technology adoption most clearly. The marketing folks may wish to include other demographic variables (like profession and income); in general, this push should be resisted, or accommodated politically, unless those characteristics genuinely reflect significant end-user characteristics. In any case, the more user characteristics the test accommodates, the larger the participant group will be (and the more resources will be expended). Including age and gender pushes the group size to the neighborhood of a dozen.[3]

In general, however, unless you have very firm reasons to believe that the system will be used in different ways by different categories of users (in particular, by "novices" and "experts," two notorious fictions in interaction design, but sometimes workable fictions all the same), you should not attempt to test multiple groups. For one thing, you likely won't have a sufficient sample for any generalizations to stand. For another, in a new product you would have to create the experts yourself (and a new product is the only one you would be doing an Oz test on; if it existed already, you would test the product directly, not a simulation of it); and you can always test the same group later, if you want to look for an expert effect.

You should always try to include a few of what Rubin (1994: 129) calls *LCUs*, for "least competent users" — participants whose experience level is relatively low. They provide a good worst-case for the system; if *they* can get through, that bodes well for everyone else. And they often supply useful insight into the naïve strategies that at least some of the end users will employ, suggesting, for instance, directions for progressive assistance.

Recruiting participants is almost always problematic, because everybody is busy, or at least has priorities other than testing software. You can get them through fiat, of course, horse-trading with their managers for something or other, but people who are ordered to

---

3: You never achieve the homogeneity of variance necessary for robust statistical reliability, of course, with sample sizes at this level. But, as Nielsen and others have repeatedly pointed out in related work (that is, in usability research), the statistical requirements of product testing are far less severe than for psychological or sociological research that investigates general principles of human behavior. See, for instance, Nielsen (1994).

participate are not always the most cooperative, and hostile participants do not always engage in the most representative behaviors. The only rule of thumb here is to try and ensure they get something back, even if it is only an afternoon off, some coffee, and a donut. Psychologists and sociologists have been living off the ability to grant academic credit for decades; there is no equivalent in corporate circumstances, but key chains and screensavers are sometimes acceptable rewards. And a thank-you note afterwards is always a good idea (copying their boss, if their participation is in any way connected to corporate concerns).

Always have a few back-up participants ready, people you can call at a moment's notice to fill in for someone who doesn't show up; don't let the setup go to waste.

The Tasks

The tasks should represent projected system use. For instance, testing a travel-information and booking system involves setting tasks for users like finding routes between Salt Lake City, Utah, and Torino, Italy; cross-checking times, availability, costs; booking hotel rooms and ground transportation; and so on. Testing a weather service involves setting tasks like finding relevant weather for various locations, perhaps correlated with activities like sailing, skiing, and traveling (for instance, checking airport conditions). Testing a movie-information service involves setting tasks like finding out where *Throne of Blood* or *Ran* is playing, as well as prices, times, theater addresses, perhaps directions, and other relevant information.

In general, you want the encounter to be fairly natural. But you shouldn't be unduly concerned with "naturalness" and "artificiality." The tasks are necessarily artificial. The whole situation is artificial. And, while Oz tests are a species of usability, they are more exploratory than usability tests. So, if there are aspects of the design you want to put special pressure on, you shouldn't worry that they are unnatural. For instance, you might be particularly interested in a registration procedure, which users may only have to engage in once. Work it into the test in a couple of places, anyway (with other intervening activities).

Designing the tasks means giving the participants representative problems, which they have to solve by using the system. Lightly embedding these problems within a scenario can give the participant a sense of structure and a clearer focus on the goals of the interaction, but the scenario only needs a hint of a plot — not character development or scene setting. Two examples from Bernsen, Dybkjær, and Dybkjær (1998) provide strong negative and positive examples of task depiction, as shown in Figures 15.1 and 15.2.

Figure 15.1 is a catalogue of what to avoid. There are too many extraneous details here, and not enough relevant ones (departure times, return times, any price concerns; all of which might be inferred, but it shouldn't be the participant's job to make off-task inferences). It's certainly a good idea to establish a problem space of this sort, where appropriate, but forcing the participant to pull dates and times out of the air distracts from the interactive task.

> You study at the Academy of Music in Aalborg. You are going
> to visit your parents in Copenhagen for Christmas. They have
> promised to pay for your ticket.

FIGURE 15.1   A poorly-designed usability task, from Bernsen, Dybkjær, and Dybkjær (1998: 148)

> Customer number 110.
>
> Travellers: Jens Høst (ID-number 27) and Anton Sigurdsen (ID-number 28).
>
> Jen and Anton want to fly return, Copenhagen to Aalborg (and back)
>   on Tuesday, 27th October
>   leaving at 7:00 or 7:30
>   coming back by 17:25.

FIGURE 15.2   A well-designed usability task, adapted from Bernsen, Dybkjær, and Dybkjær
(1998: 148)

Figure 15.2, on the other hand, from a later Oz test in the development of the same system, is much a better example (I have tinkered with Bernsen and the Dybkjærs's formatting and wording somewhat). All of the relevant information is here, in the way most travelers would conceive it, and the excess (motivations, personal circumstances) is gone. Note that the specifications needn't be this precise, departures "in the morning" or "in early evening" are fine, return dates "between Christmas and New Years" are okay, and introducing ranges (as well as variables like "prefers an aisle seat") promotes a more negotiative interaction that, in turn, generates richer data.

The importance of layout for task depiction is often underestimated. Tasks are regularly given in paragraph format which increases the participant's cognitive load notably, especially as the number of information slots he has to employ increases, forcing him to first read linearly and then often to search back and forth through the text as the dialogue unfolds.[4]

4: Here's a typical example (since I de-typified the Bernsen et al. example):

You want to know the price of a first class, round-trip ticket from Cherbourg to Paris. You wish to travel to Paris on Friday, September 22nd, leaving Cherbourg after 6:00 p.m. and to return to Cherbourg on Sunday, September 24th, arriving before 10:00 p.m.

(Life et al., 1996)

The tricky part about tasks is finding good problems, and then phrasing the tasks in such a way that you identify those problems adequately, but don't give directions for solving them. Presenting the users with the tasks is inherently problematic. The difficulty is related to task presentation in any usability work: you want to give the participants enough information to perform the task, but you don't want to tell them how to perform the task. In a document-navigation task, for instance, you want the participants to find certain pockets of information in the document, but you don't want to give them the key words the document uses, or you've just told them to go to the index or the table of contents and look for those words. What if the words are not the ones that would naturally occur to a user? You've biased the test heavily in favor of the document's self-image. Or, in a graphic-interface test, you want the participants to carry out some actions, but you need to tell them in a way that doesn't (explicitly or implicitly) give them the menu labels or the button terminology.

And, as above, the natural-language phenomenon of convergence exacerbates this problem almost beyond repair. *Any* language you use to describe the tasks will bias the participant toward its use, and therefore bias the test toward already-chosen system terminology. And the use of graphics has limited applicability. In the end, you have to trust that the discourse-field research has been thorough and the vocabulary and structures it has produced are representative, but it is still best to stick to the most general terminology and to follow a minimalist strategy for task directions.

Notice that you should not plant the terminology you prefer in the task description, but only the terminology that reflects common usage (uncovered in your discourse analysis). For instance, your recognizer might be happier with "departure time," but if the (projected) user group prefers "leave," that's how the task should be phrased. Any sociolinguistic variation you uncover in the discourse analysis should also be represented. For instance, if people usually say "come back" but sometimes say "return," then your task descriptions should reflect the alternation.

With a speech-system task list, too, you have to be especially careful that any resemblance to a script is minimized. Describe a task in a general way that focuses on a goal; words like "then" and "next" and "after" are good diagnostics that your task list is really

---

I don't mean to single out Life et al. — this example is very typical — but the typographical and information-mapping complications with this task description are substantial, starting with the italics (not having seen the original task lists, I don't know if it is just a weird quoting convention in the literature or if the testers really have so little understanding of readability as to italicize the text for users, but it is very common to find task examples in italics). *Cherbourg* is repeated three times, *Paris* twice, the dates and times are distended and embedded in a linear text — all of which complicates, rather than simplifies, the participant's activity (affecting interaction, attitude, and potential success). There is also no explicit task, just a fictional ascription of desires (*want*, *wish*), which adds an inferential layer to the participant's activities. Over the course of multiple tasks, these factors can add up.

a scenario, or a set of instructions for interaction, which participants have a tendency to follow quite faithfully. You are, in that case, testing more their ability to carry out a play script than to engage the system. Bernsen and the Dybkjærs (1998: 160), for instance, found that many of their participants were simply reading their task descriptions aloud to the Wizard.

It is often a good idea to make the early tasks relatively simple, even trivial, to put the participants at ease. The only argument against this arrangement is that it might not reflect the projected use of the product.

There will likely be a learning effect, so the responses to later questions are not always easy to compare to earlier responses (for instance, participants might rate the system poorly after a given task, more favorably after another task, but the rating may have less to do with the tasks themselves than with where those tasks are performed with respect to each other). The best approach is to use a different randomization of the tasks for each session, or to alternate groups of tasks (reversing the first ten and the last ten for half the participants, for instance).

Having a well-considered, representative, and structured set of tasks is critical to the success of an Oz test, but this period is still an exploratory stage in the development, and it is equally important to have at least one free-form task. It will never be truly free-form, of course, because you've effectively coached the participants by having them carry out a series of structured tasks. But you've got a representative user, so he shouldn't have too much trouble responding to tasks like the two in Figure 15.3.

There are, of course, all sorts of ways the open-ended approach exemplified by the Figure 15.3 tasks could go wrong (not the least of which are attentional detours into memory or time-management planning), but there's little to lose if you include one or two such tasks at the end of the session (you've already got lots of data), and the strategy will increase the realism of the test for the participant, while putting pressure on the interaction model. Even a hostile or mischievous participant, who puts excessive pressure on the model, might help turn up problems or spark new design ideas.

The task list should be drawn up by the Usability Prime. The Interaction Architect, the Technology Expert, the Subject-matter Expert, and any or all members of the design team

> Think of the last flight you took. Now, please use the AcmeVoice travel agent to book the trip.

> Think of somewhere you'd like to travel, with at least one friend or family member, and use the AcmeVoice travel agent to make the arrangements.

FIGURE 15.3   Open-ended usability tasks for testing a travel system

should provide suggestions for the problems. The experts should review the task list once it has been assembled, but should not (without very good reason) change the tasks. The Architect and the Quality Assurance Prime should have sign-off.

A score sheet should be spun off from the task list, including every task, but stripping out the unnecessary parts (like the scalar questions), and leaving sufficient space after each task to record comments. The observer can print it up and work by hand, or just use it directly on a computer.

The Testing Environment

The testing environment should approximate (or replicate) the end-use context to the extent that the context might impinge on behavior. For instance, the test of a warehousing tool should take place in the warehouse. The test of a hardware installation or repair tool should be in the presence of the hardware, in an environment in which it would be installed, and should be conducted as the participant engages in the relevant activity. For something like a traffic-information service, which is completely abstract and useless unless it is used to make driving decisions while behind the wheel of a vehicle, the environment is more problematic; the best solution (in terms of ethics, and possibly legal implications, which preclude testing someone who is actually behind the wheel of a car during a typical rush hour in a typical large city), is perhaps a driving simulator (Bernsen and Dybkjær, 2001, for example). The same considerations hold for vehicle-centered directions, though not for pedestrian-centered directions.

For most general-purpose information services, any room with a comfortable chair is appropriate, though you should always be careful to have the appropriate equipment available (for instance, some users may prefer a speaker phone, some a cordless, some a cell).

The Test

> *It pleased him to test the cleverness of his workmanship.*
> — L. Frank Baum

Once all the ingredients are in place — the participants, the Wizard and the Wizard-support network, the task prescriptions, and the attendant equipment — the actual test is usually the least of your worries.

Voice-interface tests are much easier to conduct remotely than tests for most other forms of interactive technology, but it is still best to bring people into a facility to run the tests in person. This procedure won't affect the interaction much, if at all, but setting up the test, being available for any necessary intervention, getting visual impressions of the participants' attitudes, and, especially, conducting a post-test interview, are all compromised significantly by remote testing. The advantages to remote testing are mostly for the participant, who has to invest less time, and to the testing budget.

You need to ensure the participants know you're testing the system, not them. Many people, if not most, in today's computer-saturated world are familiar with usability testing at some level, and more than a handful have been involved in such testing. But it is still a good idea to ease their potential anxiety by thanking them for helping you with the system, by generally treating them as software evaluators, not as test subjects, by answering any questions they might have, and by reassuring them about the confidentiality of their specific involvement and the anonymity which will cloak any reporting of their comments or actions.

For thoroughness and consistency, and to make the results more generalizable, these remarks should be written down and expressed to the participants in a uniform way. (You needn't read them off verbatim, as if you were giving them their Miranda rights, but neither should you ad lib extensively, or omit anything.)

You will need to get a consent form and nondisclosure agreement signed by the participants. That is, unless you or your legal advisors are really finicky, one form that covers both domains: permission to report their behaviors (under cover of anonymity), and assurance that they will keep the project confidential. If the participants are not onsite, fax or attach or courier the form(s).

The coordinator, who welcomes the participants and sets the tasks up for them, should also be available if the interaction breaks down before the tasks have been completed, to soothe frustrations and even provide hints if necessary (better to have compromised data than no data), and to get things rolling again. If the coordinator and the Wizard's assistant are the same person, this availability can cause notable delays.

It is imperative that the participants be firmly under the illusion that they will be interacting with a computer by voice. Bernsen, Dybkjaer, and Dybkjaer (1998:136) raise concerns about "lying" to the participants about the wizard/system, but their solution does not gain any ethical ground, and without this deception there is little point in conducting an Oz test. The important consideration is that the participants be informed before they leave the site.[5]

5: Bernsen, Dybkjær, and Dybkjær (1998: 136), citing Nielsen's discussion of usability testing ethics (1993:181–185), say "[participants] should not be told a direct lie for ethical reasons, ... Instead, they should be given vague information which may be interpreted as if the system is a real one." As they say on the immediately preceding page, "it is important that [participants] believe that the simulated system is a real system" (135). That's an understatement: it is critical they believe the system is real. The test is pointless otherwise. And purposely misleading someone is in exactly the same moral ball park as directly lying to them (if a tester's ethics preclude one, it is beyond me why they would include the other), while vagueness can only lead to an overall sense of uncertainty that would not be beneficial to the test. Moreover, Bernsen, Dybkjaer, and Dybkjaer also misread Nielsen, who quite clearly uses Wizard-of-Oz tests as an example of a method that involves "a deception that should be disclosed" *after* the post-test questionnaire (1993: 184). Nielsen does not advocate eliminating the deception, nor even trying to downgrade it to a strategic equivocation, just revealing it before the participants leave. Compare this to, for instance, the practice Sony Ericsson initiated in the summer of 2002, hiring actors to pose as tourists who asked

Make sure *everything* you need is on hand, in place, and functioning. Nothing frustrates people who have volunteered for your study more than needless delays because of technical problems or because there aren't enough pencils to go around. Keep a checklist, and run through it a good half-hour before the participants arrive.

Run the participants one at a time. For much usability work, it is advisable to have two people working together. Pairing participants generates richer data, mostly by encouraging the participants to verbalize their strategies as they plan out the task, suggest alternate moves to each other, express frustration, and so on. But that is for direct-manipulation tasks, which are ordinarily performed silently, and the verbal nature of cooperative behavior gives the tester much more to work with than furrowed brows and looks of sudden enlightenment. Speech-systems tests are already data rich, because the interaction itself reveals much about the cognitive processes behind the participants' behaviors.

The task list should chunk the tasks into groups, punctuated by a few quick, within-test ranking questions, like the following:

Using the system to carry out these tasks was

For these tasks, the system was

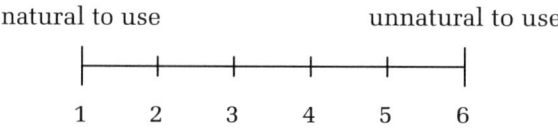

I enjoyed the interaction

---

people to take their pictures, extolling the virtues of the phone (T68i) and its camera add-on. Jon Maron, director of marketing communications for the company said the actors "didn't make any secret of the fact they work for the company, if asked" (Reuters News Wire, 5 August 2002). The difference represented by that "if asked" is ethically momentous; it puts the onus on the victim of the deception to uncover it. Nielson's recommendations are entirely correct and entirely ethical.

This strategy not only allows you to collect useful, mid-task responses, it ensures the participants are steadily thinking about the quality of the interaction, and provides a small buffer between tasks. This buffer also gives the participant a chance to shift gears for a moment, so that she is somewhat fresher for the next task, and fatigue doesn't set in so quickly. If the task-set is somewhat unnatural (if, for instance, you ask them to find out information on five different movies in different areas of town, when they would normally only check one or two movies in the same area), the buffer reduces the peculiarity somewhat by detaching the tasks.

Aside from completing these in-test evaluations, don't require anything further of the participants; just have them focus on the tasks. In particular, any kind of verbal protocol analysis (e.g., Ericsson and Simon, 1993) is clearly inappropriate. The principal interaction of the user is by voice. It would be weird in the extreme for a user to say "I'd like to find out about digital cameras" to the system, then tell the tester "I just told the system that I'd like to find out about digital cameras."[6]

Observers/data gatherers should observe and gather data; what are important are impressions, rather than numbers. The equipment is capturing durations, turns, slippages, and overlaps, but impressions — the natural byproduct of human observation — are much harder for the equipment. Impressions, of course, will be the principal output of the data analysis phase, but recording them on the fly can provide a rough and early road map to the transcripts and help feed the interview process. Gather them on a score sheet, task-by-task, watching for

- Vocabulary

   The transcripts will provide the hard data about this, but observers should note their impressions of the range of vocabulary (large variation or small, short words or long), and record the use of any noteworthy words or phrases.

- Utterance structures

   A general sense of completeness, fluency, and structure (Lots of questions? Lots of statements? Lots of multiple-argument utterances? Or a tendency toward single-argument utterances?).

6: There may actually be reasons to implement a think-aloud methodology of some sort, especially in research circumstances (as opposed to product-development circumstances). For instance, researchers might want to plumb the user's cognition by asking her to articulate her reasoning before and after utterances directed at the system. This sort of work would have to be implemented in a model, I would think, with a walkie-talkie style send/receive button, so the user (and the experimenters) can easily segregate her interactive talk and her talk-about-the-interactive-talk.

- Interactive patterns

  How the participants build their interactions, and respond to the Wizard's moves. For instance, their in-task data sequences might follow a certain sequence (departure < destination < date < time), or they might be largely ad hoc, or some admixture.

- Participant's reactions

  Points of success and failure are particularly important junctures, but overall attitudes are also significant. Do they get tired of regular confirmation subdialogues, for instance?

- A sense of the flow

  How the participants manage turns (or allow the Wizard to manage them). The numbers may tell one story about who has the initiative in a dialogue, for instance, and subjective perceptions may tell another about who is in control.

- Also watch generally for how these observations all correlate with furrowed or raised brows.

The Oz session should be followed by an interview, keyed to a questionnaire, and given by the observer. Self-administered questionnaires are a very spotty instrument in these circumstances, and free-form interviews are difficult to correlate with one another for usefully general findings (either among the participants of one iteration, or between the results of different iterations). Appendix 15.1 offers a slate of general-purpose, voice-interface scalar questions.

If the participant is remotely involved, then use the telephone; otherwise, it should be face-to-face, over some refreshments in a congenial atmosphere that encourages frank responses. The interview is a critical augmentation of the test. As Bernsen and the Dybkjærs note, "users tend to suffer in silence during the dialogue and complain afterwards" (1998: 216). Conversely, some exchanges that seem, externally, to be problematic or frustrating turn out not to have perturbed the participant much at all.

The second-last question of the interview should be:

Anything else?

Is there anything I neglected to ask that you think we should know about with respect to the system?

This question needs to be followed by a frank confession that the system was simulated, followed by the last (slated) question of the interview:

Anything else?

Now that you know that the system was simulated, is there anything you want to add to your comments?

This question not only might produce some useful suggestions, it also gives the participant a chance to react to the trick and express any concerns he may have about the test.

The audio from the test sessions has to be transcribed for discussion and analysis, something which is best done on the day of the session. That makes for a long day if you run two sessions per day for the iteration, and it can be delayed until after all the sessions are complete. If you've used recognition software (or if you now feed the audio into recognition software) you should keep the first-pass recognition transcript for interest and research, but for reporting and analyzing Oz tests, you need to edit them into a more coherent shape (that is, corresponding to what the wizard and the participant heard). All transcripts should be identified by a participant ID, and headed with a participant profile, and every utterance should be given a unique identifier (see Appendix 15.2).

The Analysis

> *"Then," said the Professor, "I will get out my famous magnifying-glass and throw the insect upon a screen in a highly-magnified condition, that you may all study carefully its peculiar construction and become acquainted with its habits and manner of life."*
> — L. Frank Baum

It is essential to get user evaluations, between-task and post-test. The most important quality a voice interface (indeed, any interface) can have is the ability to satisfy its users, and it is a truism in speech-system research that "objective" measures of the interface (overall duration, time-per-task, number of turns, number of repairs, you name it) don't correlate especially well with users' assessments of their own satisfaction. (That's not to say there are no design strategies which lead to, or enhance, user satisfaction — if there weren't, the whole enterprise would be a crap shoot — nor to say that the user's self-reports are infallibly reliable, just that satisfaction doesn't quantify easily on the basis of simple expectations, and that, for this particularly critical aspect of the interaction, the user is the *only* source.)

Data analysis falls roughly into two categories, which take the labels *quantitative* and *qualitative*: numbers and impressions. Most software engineers have more respect for the former; most usability specialists have more faith in the latter; both are necessary, and both intermingle (quantitative analysis relies on impressions drawn from the numbers, qualitative analysis relies on the amount of related impressions).

Quantitatively, the application of these results may often not be immediately apparent, but you should log a wide range of categories (sometimes these can be captured automatically, though much of it will have to be tabulated by hand, post-hoc). The measures become more important as the development cycle advances, so that Oz iterations can be tested against each other or against other usability inspection methods, and can be used to establish metrics for usability tests. Measurements should include:

- Turn length (in words: maximum, minimum, average, and mode)

- Number of turns per task (maximum, minimum, average, and mode)

- Task duration

- Number — and complete inventory — of out-of-vocabulary words (a low out-of-vocabulary percentage is a very good sign)

- Task success ratio (number of successfully completed tasks over number of tasks attempted)

- Incidents of meta-communication (repair, clarification, system-inquiries; type and number, per-task and overall)

- Interaction slippages (overall number, percentage)

- Task slippages (overall number, percentage)

- Timeouts (overall number, percentage)

- Assistance (overall number, percentage of system turns)

- Dysfluencies (number of hesitations, false starts, nonwords; type and number overall)

- Number of word types and tokens (totals, type/token ratio)

- Dialogue acts (number of each type, overall and per task)

- Initiative (percentage of system-initiative utterances per task; the percentage of user-initiative utterances per task will be the inverse)

- Cohesion phenomena (number and type per phenomenon: anaphora, ellipsis, connectives)

- Satisfaction responses, from the scalar questions (average and mode ratings per question)

- Cooperativity problems (number/type, average/type; see Table 15.1)

- Number and type of Wizard improvisations

These numbers give you snapshots of the data, which you then flip through looking for patterns. Are there lots of repairs? Lots of word types overall? Lots of tokens of a few types?

Problem		
Area	**Label**	**System (in)action**
Quantity of Information	Under-info	provides insufficient information.
	Over-info	provides superfluous information.
Quality of Information	Lie	provides false information.
	Guess	provides unwarranted information.
Relevance of Information	Irrelevant	provides irrelevant information.
Manner of Expression	Obscure	provides an obscure utterance.
	Ambiguous	provides an ambiguous utterance.
	Prolix	provides a too-lengthy utterance.
	Disorder	provides a poorly ordered utterance.
Grounding	Grounding-user/back	doesn't have relevant user background information.
	Grounding-user/cont	ignores user's grounding contributions.
	Grounding-self	ignores own grounding contributions.
	Grounding-domain	is missing domain knowledge.
	Grounding-feedback	fails to provide feedback.
Repair	Repair-initiation	fails to initiate repair or clarification (after slippage).
	Repair-response	fails to respond to a repair or clarification.
	Repair-request	fails to respond to a request for a repair or clarification.
Other	Other	falls down in some way that doesn't fit the other categories well (for instance, a user report of unnaturalness that might trace back to cohesion issues; or a failure to pick up or deploy turn-cues).

TABLE 15.1 Cooperativity problems, adapted from Bernsen, Dybkjær and Dybkjær (1998: 210–211)

Too many out-of-vocabulary words? Are the participants generally satisfied about ease of use? Bored and unhappy about the lack of naturalness?

To a limited extent, you also need to look for patterns between and among the snapshots; in particular, anything that correlates with satisfaction judgments. But also, are there more system inquiries for one task than another? More meta-communicative behaviors? More system initiative utterances (or the inverse)?

The real pattern and correlation hunting, though, comes with multiple Oz tests (and, later on, with the usability testing cycle), as the design grows and changes: compare turns/task and duration/task from tasks in early tests to tasks in later tests; compare meta-communicative behaviors and rates of dysfluency between tests; compare number and type of system inquiries between tasks; above all, compare everything to the scalar satisfaction ratings. Developmental patterns are powerful design instruments, for checking to see if things are getting better or worse (a possibility, especially when changes to one part of the interaction has ramifications on results for other parts of the interaction).

Again, you have to be fairly cautious interpreting any patterns you turn up, and interpreting the quantitative data generally. There simply isn't sufficient rigor in this work to warrant inferential statistical analysis (in the sense of reasonable confidence levels in the reliability of generalizations from them). But there is meaning in the numbers all the same, and combined with your design sensibilities, the interview results, the qualitative analysis results, and the prominence of the patterns, they can help you make decisions about directions to take.

Some additional factors to consider:

- Meta-communication does not necessarily mean there are problems; and, more particularly, it does not mean nonsatisfaction by the users. Meta-communication is a normal function of everyday talk. But monitoring it is still fruitful, because it is a deviation from the task, and can indicate (especially over multiple users at the same interactive juncture) user uncertainties.

- Similarly, dysfluencies are normal elements of everyday talk, and in themselves they don't tell you much at all. But if they pile up at the same or similar points in multiple user interactions, they can indicate user uncertainties.

- A single utterance, or turn, may represent multiple problems; don't think you have exhausted the diagnosis with one problem report.

- The ideal number of Wizard improvisations is zero, and the Wizard should stick to the design parameters closely. But compromised data is always better than no data (as long as the compromises are understood), and rather than suffer a total breakdown of the session, the Wizard may have to deviate from the interaction model in some unforeseen way. Not only does this keep the session alive and salvage some data, it could also point the way rather directly to design improvements. The Wizard is part of the design team.

- There will occasionally be recalcitrant participants. Bernsen and the Dybkjærs (1998: 224–5), for instance, report the use of a negated, agentless, passive sentence by a user who later confessed she was deliberately trying structures she thought the system "would be unable to handle." There may be stretches of tests involving such users that produce usable data, but artificially induced breakdowns tell you more about the participant than the system, and such data should just be turfed. It is, and will be for a very long time, ridiculously easy to blow-up a speech system. Users who do so intentionally are not your concern; it is the unintentional blow-ups that usability work is after.[7]

The qualitative data comes from what you perceive about the participants' engagement with the system — from observations during the test, and from what (and how) the participants report their views of the system to you.

During the test, take good notes. Catalogue everything that seems relevant. And rework those notes as soon after the session as possible, as if they will be read by someone else (they might be, but the primary point is just to preserve meaning for you; the cryptic notes you write in the heat of the test will look like Etruscan to you in a few days).

Watch for things that don't show up so well numerically. Are the users following particular strategies (resisting initiative opportunities, adopting word limits, avoiding pronouns)? Are there indications of satisfaction/frustration (expressions, tone of voice, pacing)? One important characteristic of speech system interactions that is very tough to catch quantitatively (except in scalar responses, which lack specificity) is "naturalness." Does the interaction seem smooth and unforced, or choppy and forced? Pay particular attention to sequences of utterances, especially ones that go on for a while or seem awkward.

During the interviews, prompt users regularly for elaborations, and, again, take comprehensive notes, reworking them later. Be redundant, but be somewhat subtle about it. Asking highly redundant questions (for instance, scalar questions about liking the system and enjoying the system) sometimes annoys respondents, but it also frequently elicits more information, by getting them to think about the issue in slightly different terms, or simply by revisiting an issue they might still have something to say about.

However, and by whomever, the data was collected, the Usability Prime takes responsibility for it, and for its deployment. The point of Oz testing is to probe design decisions early on; reporting the results therefore focuses on those decisions, and on the underlying conceptions about users and tasks that informed those decisions. Is the user profile accurate? Is the task model accurate? Pay careful attention to wizard improvisations; they indicate clear breakdowns in the interaction model, and may point in the direction of solutions.

7: It's a slightly different matter, however, if this sort of breakdown occurs in demos, betas, and the like, if the recalcitrant participant is a vendor who then uses the breakdowns to disparage the system or otherwise argue against its adoption. In such cases, you need to make the point that real users don't deliberately sabotage the system; that they, conversely, try to make it *work*, so they can perform the tasks they need to perform.

The flaws in the interaction model require close attention. Bernsen et al. (1998: 210–211) recommend a Gricean coding taxonomy for system contributions to dialogue breakdowns that is very thorough, a version of which I offer in Table 15.1.[8]

This group of problems can be expanded or contracted as the specific system requires. For instance, Bernsen et al. (1998: 210) include (in the information quantity category) "The system is not fully explicit in communicating to users the commitments they have made," which is relevant to systems where users make commitments, especially those with high grounding criteria.

In concert with this taxonomy (or another, but you need to be able to identify more of what's going on than "problem," "another problem," and "yet another problem"), you need a reporting structure that:

- Identifies the symptom

- Offers the diagnosis

- Indicates the prognosis (i.e., assigns a severity rating)

  and

- Recommends a treatment

With a usability test, the last step is often omitted (though this decision depends on the philosophy of the usability expert), in order to leave all treatment concerns explicitly to the design team; with an Oz test, the emphasis is on design, rather than on market-ready usability, and the more suggestions the better. And with an Oz test, the severity rating[9] is often omitted (again, depending on the philosophy of the testers), since the design is usually so fluid that ratings are not especially stable.

Take the following stretch of dialogue: (The numbers prefixed to each utterance are part of the Bernsen-Dybkjær-Dybkjær[2] transcription scheme, which I've omitted for the other stretches of dialogue from their study that I've used in this book. I include it here, because of its value in pinpointing exactly where in the transcript the problem occurs. S and U stand for System and User, respectively; the following number is the participant number; after the dash comes the turn number. So, for instance, S3-29 is the twenty-ninth turn by the system (i.e., the Wizard) in its dialogue with participant 3.)

8: I've made a few changes, but they are not drastic; the Bernsen-Dybkjær-Dybkjær taxonomy is a good one. Mostly, I have just edited out their specific problems as a bit too local for all speech systems, and I have made the Gricean/ Conversational groupings more explicit with my terminology. The specific labels in the tables are mnemonics I developed, but they become quickly unworkable as a coding scheme, and they need to be put into some order of shorthand (e.g., I-und, I-ov, Lie, Gu, Irr, Obs, Amb, Pro, Dis, G-u/b, G-u/c, G-d, D-f, R-in, R-res, R-req).

9: I use Nielsen's (1994: 49) ratings: 0 (no problem at all), 1 (cosmetic), 2 (minor usability problem), 3 (major usability problem), 4 (catastrophic usability problem).

S3-29   System: Please state the customer number.

U3-29   Caller: Number four.

S3-30   System: Customer number four.

How many people will travel?

U3-30   Caller: Two adults and two children.

S3-31   System: Two people.

Please state the ID number of the first person.

U3-31   Caller: Change.

S3-32   System: Two people.

Is this correct?

U3-32   Caller: No.

S3-33   System: How many people will travel?

U3-33   Caller: Four.

S3-34   System: Four people.

Bernsen et al. (1998: 179)

The problem report on a slippage of this sort, will look something like Figure 15.4.

Severity	3 – Major usability problem	
Symptom	S3-30	System: How many people will travel?
	U3-30	Caller: Two adults and two children.
	S3-31	System: Two people
Diagnosis	Grounding-domain	
Treatment	People often divide up the travelers according to ticketing conventions. Listen for more than one number in the utterance, correlated with different ticket categories, and do the math.	

FIGURE 15.4   A problem report on an interaction slippage

Usability Testing

User testing with real users is the most fundamental usability method and is in some sense irreplaceable, since it provides direct information about how people use computers and what their exact problems are with the concrete interface being tested.
— Jakob Nielsen

There is so much overlap between an Oz test and a usability test — in terms of participants, tasks, team, set-up, and analysis — that this section almost comes down to "ditto, minus the Wizard."

This section *is* decidedly briefer because of all the ground covered in the Oz section, and anything I don't take up explicitly in this discussion (for instance, the use of between-task scalar questions, or the treatment of subjects, or the handling of data), you can assume is simply imported wholesale from the Oz section. (If this arrangement is inconvenient for you, my apologies. I'm a big fan of redundancy, but the reams of verbatim repetition it would mean for me to cover the same topics again in this section gives even me pause; it would amount to a near-fraudulent exploitation of the money you spent on this book, and wouldn't be very fair to the trees, either.)

But there are three very significant, closely related differences that affect, in particular, the planning and the results of usability testing, in contrast to Oz testing.

The first difference is the screechingly obvious minus-the-Wizard part, with its complement: plus-the-system. The interface isn't carved in stone by the usability-testing stage, or there would be no point in the tests, but it is much further along. An important consideration on this front is that now the recognition errors are real, and measuring them has meaning (unlike with Oz tests). Jennifer Lai and Nicole Yankelovich (2002) stress this factor eloquently:

> *usability studies are particularly important for uncovering problems due to recognition errors, which are difficult to simulate effectively in a Wizard-of-Oz study, but are a leading cause of usability problems. The effectiveness of an application's error recovery functionality must be tested in the environments in which real users will use the application.*

The second difference is the timing. Usability testing occurs much later in the development cycle than Oz testing. That means the stakes are correspondingly higher.

The third difference concerns the objectives. A usability test is not, as an Oz test is, for probing elemental design choices. It is for measuring the nearly-finished product against users' behaviors, to finalize it for market release.

The primary implication of these three factors is the necessity for firm, consensual usability objectives.

From the design work, Oz testing, other inspection methodologies you may have used, and competitor evaluations, you should derive a set of metrics and scores that establish the criteria by which the system's usability will be judged. As Wixon and Wilson point out, having hard numerical objectives confers authority:

> One of the reasons attributes like "time to market" and "reliability" (eliminating bugs) play such a major role in the development process is that they have "metrics" that are clearly defined. Usability "bugs" can be just as severe as reliability bugs and should be treated with the same quantitative respect.
> (Wixon and Wilson, 1997)

Amorphous goals like "user understands most system utterances" or "few inappropriate responses" can be satisfied virtually by fiat; precise goals like "user responds appropriately to system utterances ninety percent of the time, or better" can only be satisfied by measurement.

You should make all of the measurements outlined above for Oz testing. Additionally, you should measure, and set usability thresholds for, the following metrics:[10]

- Dialogue quality metrics

    - Mean recognition scores

    - Task success ratio

    - Timeouts (percentage)

    - Recognition scores

        - Rejections

        - Deletions

        - Insertions

        - Substitutions

10: These metrics are strongly influenced by the evaluation model developed by Marilyn Walker and her colleagues at the AT&T research labs (see, e.g., Walker, Kamm, and Litman, 2000), the weirdly named PARADISE (from *PARAdigm for DIalogue System Evaluation*). I have, however, eliminated their "dialogue efficiency metrics," which concern task duration and number of turns. Those measurements should be gathered, but their relation to usability is too tenuous to use them as evaluation metrics. A few people seem to think that speed is an important characteristic of speech systems. Kate Dobroth, for instance, says that "more than anything, people want spoken interactions to be fast" (Weinschenk and Barker, 2000: 217). But the strong consensus among human-factors specialists working with voice systems is that speed is not a usability issue at all. Indeed, another paper headed by Walker, using data gathered under the PARADISE model, argues that timing data and a satisfactory user experience don't correlate well at all (Walker, Boland, and Kamm, 2000). The satisfaction scores in my proposed framework are also different from the PARADISE model.

- Interaction slippages (overall number, percentage of turns)

- Task slippages (overall number, percentage of turns)

- Out-of-vocabulary words (percentage)

- System inquiries (percentage)

- Assistance (percentage)

- Repairs (percentage)

- Clarifications (percentage)

- Early terminations (percentage)

- User satisfaction ratings

  - In-test

    - Average rating overall

    - Average rating per question

  - Interview

    - Average rating overall

    - Average rating per question

- Number of cooperativity problems tolerable (by type)

- Number of usability problems tolerable (by severity)

The specific thresholds set for any given iteration of usability testing depend on too many factors to set out any numbers here.

While the structure of the tests may be highly similar, then, the specific motivations are quite different (the general motivation — matching the interface effectively to the users and their tasks — is the same).

Other Usability Inspection Methods

Involving users in the design process throughout the lifecycle of a speech application is crucial. A natural, effective interface can only be achieved by understanding how, where, and why target users will interact with the application.
  — Jennifer Lai and Nicole Yankelovich

Usability has evolved into a set of considerations strongly influenced by the idea of empirical testing with real users. And with this set of considerations has come a variety of non-user-testing methodologies which nevertheless try to get at what users would find most conducive to successful use of the product (or, in some cases, least obstructive to successful use). Often these are considered as cheaper alternatives to testing; they developed hand-in-hand with the discount usability movement. But there is no substitute for user testing, in either its Oz or its formal-usability guises; and any new voice interface that goes to market without iterations of both is foolhardy.

There is, however, augmentation for user testing, in both its Oz and formal-usability guises. They can be cost-reduction measures, if they are successful enough to reduce the need for further user tests, but they are best seen as part of an alternating cycle of user testing and other inspection methods, rather than as stripped-down procedures which substitute for user tests. It might all come down to the same series in the end (say two user tests interspersed with two other inspections). But looking at them as ways to avoid user testing — rather than as ways which support user testing and which contribute to the overall triangulation on the users needs, goals, and behaviors — is, simply, the wrong attitude.

Of the familiar usability inspection methods, the two most promising for voice interaction design are:

- Heuristic evaluation

- Pluralistic talkthrough

These two inspection methods do not exhaust the possibilities for usability engineering in voice-interface development. Virtually all the canonical inspection methods, though primarily developed for graphic-interfaces, are adaptable for speech system inspection (see the seminal Mack and Nielsen article for an overview, 1994: 5–6, and the seminal Nielsen and Mack book, 1994, for practical explications of those methods). But heuristic evaluations and pluralistic talkthroughs both represent very efficient ways to complement Oz and usability testing in the voice-interface development cycle. Heuristic evaluations work especially well with the Oz iterations — before it starts, between tests, or at the end of the Oz cycle, before hard prototyping begins. Pluralistic talkthroughs work best with the usability test iterations, either on a prototype before the testing begins, or between tests.

Heuristic Evaluation

Jakob Nielsen treats heuristic evaluation in a range of works; the definitive one is his chapter in Nielsen and Mack (1994: 25–61). In a heuristic evaluation, a usability expert examines the design according to a preset body of criteria (that is, the heuristics; from Greek *heuriskein,* 'to discover').

This inspection method is anchored in a set of usability heuristics Nielsen advocates — principles that focus an inspection on those design issues which are especially responsible to the user. The heuristics have an obvious graphic bias, and have been widely adopted (or, in any case, lip service to them is widespread) among web designers, but the bulk of the principles are highly applicable to voice interaction usability as well. Here is a relevant subset (adapted slightly from Nielsen 1994: 30, and correlated where appropriate with the Gricean maxims and with voice interface best-practice lore):

- Clarity of system state

  The system should always keep users informed about what is going on, through appropriate feedback within reasonable time.

- Use of real-world language

  The system should speak the users' language, with words, phrases, and concepts familiar to the user, rather than system-oriented terms. Follow real-world conventions, making information appear in a natural and logical order ($\approx$ Grice's maxim of manner: Be orderly).

- User control and freedom

  Users can end up routed into some exchange they do not want to participate in, and should always have a natural and obvious escape route. Support undo and redo.

- Consistency and standards

  Users should not have to wonder about the meaning of the system's utterances or elements of the soundscape ($\approx$ Grice's maxims of manner: Avoid ambiguity, Avoid obscurity). Follow platform conventions.

- Slippage prevention[11]

  Even better than good repair messages is a careful design that prevents a communicative slippage from occurring in the first place.

- Flexibility and efficiency of use

  Accelerators may often speed up the interaction for the expert user such that the system can cater to both inexperienced and experienced users. Support turn-overlap. Support customization. Support mixed initiative.

- Aesthetic and minimalist design

11: This one is effectively Nielsen's Error prevention, a label I've changed because of (1) a philosophical disposition against the idea of user errors, and (2) the importance in speech systems of reserving the word *error* for recognition failures.

System utterances should not contain information that is irrelevant or rarely needed. ($\approx$ Grice's maxims of quantity, maxim of relation). Every extra unit of information in a dialogue competes with the relevant units of information and diminishes their relative clarity. Support progressive assistance.

- Help users recognize, diagnose, and recover from slippages

Repairs should be expressed in plain language, and precisely indicate the problem.

It's not surprising that Neilsen's heuristics, based on usability experience and refined through factor analysis, should correlate both with the specifics of Grice's Cooperative principle and with voice interface good practice, since all interfaces are dialogic at heart and all good interfaces maximize cooperation. But when they are drawn together in this fashion, they provide a reassuring conspiracy of principles for usable voice interface design that illustrates how valuable a heuristic review can be.

Pluralistic Talkthrough

This inspection method is, of course, adapted from the pluralistic *walk*through (Mack and Nielsen, 1994: 5; Bias, 1994), though my accommodation of it to voice interface development also involves crossing it somewhat with a feature inspection (Mack and Nielsen, 1994: 6). The notion of feature is almost wholly irrelevant for voice interaction (at least when seen in terms of the scope of graphic interface features), but the primary focus of a feature inspection is on the number and type of steps necessary to perform specific tasks: change *steps* to *turns*, and this focus is at least as important for voice interfaces as for graphic, perhaps more.

A pluralistic talkthrough involves bringing the design team and representatives of all the stakeholders together to talk through the interaction possibilities together, with particular sensitivity to the number and type of turns implicated. It is a sort of usability-test-by-committee method.

In its general form, the talkthrough is a staple of voice interaction design. Designers and writers constantly work through stretches of dialogue. A pluralistic talkthrough is (1) a more formalized routine; (2) staged expressly as a diagnostic procedure; which (3) involves representatives of the design team, the technology team, the users, and any other interested stakeholders (marketing people, for instance, and upstream or downstream vendors). It has some parallels to participatory design, focus-group research, and brainstorming.

The benefits and the liabilities of this method are both related to its usability-by-committee format. We all know the derogatory definition of a camel (or, if it's a sneer that some generations of readers haven't heard, here it is: "a horse designed by a committee"). But if the committee was designing a horse for the desert, they got it exquisitely right. Committees, and collectivities of humans generally, can go very wrong. Antagonisms and closed-mindedness can lead to ego- or ideology-driven compromises that doom their decisions. They can also, however, go very right. Collectivities wrote the Magna Carta and the American Declaration of Independence and the skits of Monty Python's Flying Circus. Generosity and open-mindedness (along with well-placed tenacity) can lead to synergistic compromises that are far greater than the sum of their parts. People can do remarkable things together, with the right attitude and a good moderator.

What that means is that assembling and preparing the stakeholders for the talkthrough is a major part of the job, and conducting the meeting is equally important. The designers and technologists need to come in with thick skins and receptive ears. The vendors and the marketers need to come in ready to contribute their expertise about the customers and the components, but not to pursue any short-cut-to-the-profit agendas. The users (whom, if you have the luxury, you should screen for constructive assertiveness) should be treated by everyone with courtesy and respect (and, as in Oz and usability testing, incentives are also important). And everyone needs to realize that they all win if the focus is kept resolutely on usability.

The coordinator/moderator of the meeting (the Quality Assurance and Usability Primes are the best candidates) needs to prepare all the representatives individually, before the talkthrough meeting, and collectively, at the start of the meeting. She also needs to keep the meeting moving forward, not letting it lose momentum by trying to solve every single issue to everyone's satisfaction; to broker arguments; to turn all commentary in constructive directions; and to know when it's time for more doughnuts. These meetings can take the better part of a day.

The talkthrough has in common with all usability inspection methods a fundamentally diagnostic character: looking always for points of potential breakdown, for slippages, for bad routings. The participants probe each system utterance for ambiguities, obscurities, and other sources of cognitive friction. But there is room for more active creativity in a talkthrough than in most inspections; the presence of multiple constituencies, working through the same designs, provides a valuable opportunity to search for alternative strategies or wordings.

There must be a hard prototype ready, with one technical requirement in place, for a pluralistic talkthrough to proceed.[12] The technical requirement is for an offline or paused mode of some kind. Whatever the ultimate interactive mode of the design (moded or unmoded), the prototype has to be able to withstand long delays between inputs (without shutting down, offering assistance, or the like). Also, whatever the final form of the design,

---

12: The presence of a hard prototype is the primary way in which a pluralistic talkthrough differs from a pluralistic (and other forms of) walkthrough, which are often linked to storyboarding or table-topping, techniques that utilize paper prototypes. The chief implication of this difference is that talkthroughs don't require a single "right" answer for each interaction.

it is best (faster, cheaper, more easily altered) to use a text-to-speech engine. (If you really feel you need natural speech for part of the system, you might consider what is sometimes called a *bionic wizard* methodology: that is, a mixed wizard-and-synthetic-speech tandem.)

The pluralistic talkthrough proceeds like this. Once there is a hard prototype to work with, the design team chooses several representative tasks. All of the participants come together in a board room (or other suitable meeting place). The moderator sets out the objectives: finding and offering tentative solutions to usability problems, for a specific set of tasks, in a collaborative environment. The moderator establishes the ground rules: there are no wrong moves, just ones the system might not (yet) be able to deal with; there are no dumb questions; there should be no aggression. The moderator establishes the procedure (a cheat-sheet for each participant is also a good idea here):

1. For each task, the user goals are stipulated ahead of time.

2. For each system utterance, (while the system goes into an inactive mode) every participant writes down their response (which might be silence, an utterance, or even hanging up).

3. After all the responses are committed to paper, each participant reads their response, with or without explanation (as appropriate).

4. After all responses have been read, a general discussion ensues on the matters of:

   • How the system induced each response.

   • How the system might have induced a better response (especially from those participants whose responses represent slippages, breakdowns, or potential bad routings). Notice that the solutions generated here are not constrained to the previous system turn; they might go back all the way to a welcoming message, for instance.

   • How the system might best respond to some specified participant utterances (especially those that the designers or technologists foresee as leading to problems).

   • What the response to the system should be. (It is important that one response be chosen from among the participants' suggestions, so that the inspection can continue.)

5. The system is brought back into active mode, and the moderator speaks the chosen utterance to the system.

6. Go back to 2 (until the task is completed).

7. After the task is completed, the participants go into free-form discussion about the design implications of the talkthrough.

For purposes of subsequent analysis, as well as appeals to the common ground, the moderator is tasked with recording all system and "user" utterances (that is, all inputs and outputs), as well as collecting all problem reports and design suggestions.

Beta Tests, Field Studies

There's just nothing like trying it out with real people to find out all the ways you blew it.
— Nicole Yankelovich

Field studies have two general variants: local and remote; or, being there and not.

Local Field Studies

Much like task analysis, local field studies are a species of ethnography: participant observation in the user's stomping grounds.

The difference between field studies and task analyses is primarily one of purpose, although that difference implicates major differences in equipment and interaction, and sometimes small differences in methodology as well. With task analysis, whatever the technology users are working with, that's the starting point, the basis for a first-order change. With field studies, unless the whole development phase has gone desperately awry, the technology is stable, usable, and effective, but requires further tuning or monitoring. In task analysis, the target technology is only partially known, and the point of the study is to gain sufficient purchase on a set of tasks to develop a product that supports users in the performance of those tasks. Field studies evaluate an existing product in vivo, hoping that it works pretty darn well already but still striving to better accommodate it to the user and his task.

Despite these rather striking divergences, all of the observation/interview methodologies of task analysis treated in Chapter 10 translate very smoothly to local field studies (with perhaps less emphasis in the latter on constraining the interaction to predetermined scenarios).

Remote Field Studies

Remote field studies have more in common with market research than with anthropology. Analysts don't visit the user's village. They phone, or write, or just eavesdrop from afar; typical instruments in these studies are the telephone survey, the questionnaire, and monitoring software. Of these, the last is the most important for design purposes, and where the other instruments are used it is often just to explore more closely findings the monitoring software has exposed.

Surveys and questionnaires are primarily tools for investigating attitudes and inclinations, and can help identify the general sources of those attitudes (annoyance at an agent, or satisfaction with the greeting), as well as charting those inclinations (preference for web-interaction, because of visual display; an interest in additional voice services). The phone interview, in particular, can also work very well in connection with service monitoring (calling up a sample of people who abandoned transactions before completion, for instance, and finding out why).

The possibility exists, too, of using phone-based systems to gather data on their own; this technique, of course, is one of the banes of the Web, where virtually all requests for information are naked grabs for marketing data, but if used judiciously and sincerely it can be an effective way to carry out remote field studies: at random intervals, or predetermined key points (quarterly, yearly, after hitting numeric call thresholds), or in contextually defined ways (sampling early terminators, for instance) you can deploy a system-initiated survey, querying satisfaction and soliciting commentary.

In any case, the user/system interactions should be monitored statistically quite closely, generating at least the following data:

- Number of early terminations per hundred calls

- Number of zero-outs per hundred calls

- The average duration of the calls

- The peaks and valleys in usage

- Repairs per call

  - Nonrecognitions

    - Rejections

    - Timeouts

  - Interaction slippages

  - Task slippages

Specific systems will also have specific aspects that might be monitored. A banking application, for instance, might want to see how often the bill-payment function is used versus the transfer function versus the balance function, as well as in what order those functions are most frequently engaged.

Because of the linear and temporal dimensions of speech systems, voice interfaces can be monitored on a scale that is unheard of for graphic interfaces; call centers have banks of tools for data capture of this sort, though their motives are not always pure. The potential for speech systems to incorporate intensive monitoring as a design tool is incredibly rich. The system could, in principle, catch all out-of-vocabulary errors, or the last five turns of early-termination calls, or entire calls that involved more than two error cascades, and route them to files for analysis. The callers, of course, need to be informed that the data is being captured, and that it is for quality assurance reasons.

Beta Tests

Beta tests (field tests of the limited release of an imminently market-ready product) should be thorough. Members of the design team should be present at first deployment and should conduct follow up visits after users have lived with the product for a while. With a general-purpose, phone-based product, the "site" you visit might be a commuter train or a park bench or an airport.

Phone-based voice interfaces, especially for general-purpose information products, have an incredible beta-test advantage over most products: open a line, give out the phone number, and log every call meticulously. This opportunity provides a very cheap test bed for the product, gigabytes of training-input for recognizers, and hours of raw data (audio and transcribed) for designers. And even if it is released directly to the market, without a beta test, the nature of the system provides for the sort of on-going quality assurance monitoring that few other products can hope for.

Summary

> *Testing always works.*
> — Steve Krug

This chapter outlines a number of extremely useful methodologies for investigating, testing, and monitoring user satisfaction.

We looked in considerable detail at **Wizard-of-Oz testing**, a method that sets prototypical tasks for prototypical users to perform, using a simulated speech system in which a tester pretends to be the computer system. It is a lo-fi methodology, but still has some clear **resource demands** — calling for **human and computational assistance**, a specific **testing environment**, and a well-designed testing **process**. The test should provide multiple opportunities for user-mediated evaluation — primarily through observing their successes, failures, and repertoire of strategies, but also through **scalar questions** interspersed with tasks, and integrated into **post-test interviews**. The **analysis** of the data Oz-testing generates must be both **quantitative** and **qualitative**, though the latter is generally more helpful for design.

More briefly, we also took up **usability testing**. The principal ways in which usability testing diverges from Oz testing concern the use of a prototype (not a wizard), the later placement in the development cycle, and the much firmer set of evaluation **metrics**. We also reviewed two different **usability inspection methods**, one involving just a usability expert, the other a parliament of stakeholders. In the **heuristic evaluation**, a usability expert inspects the voice interface (in a design specification, a prototype, or market-ready system) according to a number of preset criteria, such as clarity of system state, flexibility, and efficiency. In the **pluralistic talkthrough**, representatives from the various constituencies involved in the development, deployment, and utilization of the product interact collectively with the system to perform a pre-established set of tasks. We also briefly sketched out **beta-testing** and **field-study** methods and considerations.

Appendix 15-1

Sample Scalar Questions

This appendix lists some sample scalar questions for a post-test interview (Oz or usability).

Note: All questions should be followed by pursuit of explanation or expansion ("Why?" "In what way?"), and negative assessments should always be followed by pursuit of suggestions for improvement.

I like this system

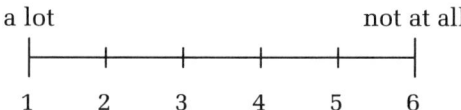

This system behaved the way I expected it to.

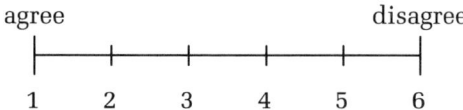

This system behaved the way I wanted it to.

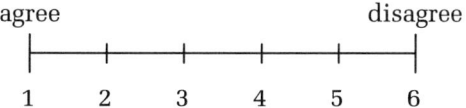

This system is

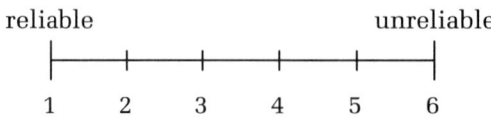

This system is

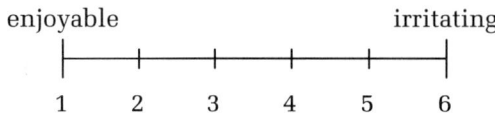

This system is

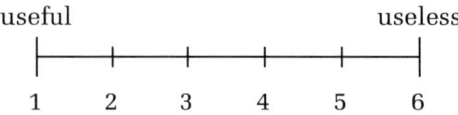

This system is

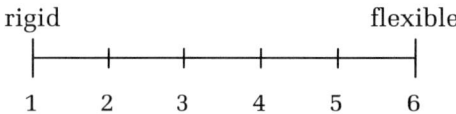

This system is

This system is

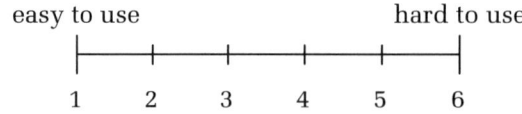

Using this system was

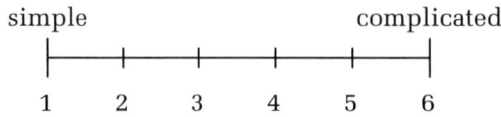

Using this system, I felt

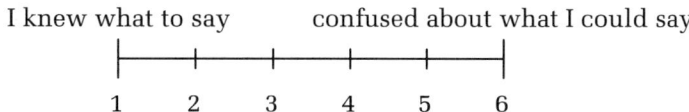

This system's responses were

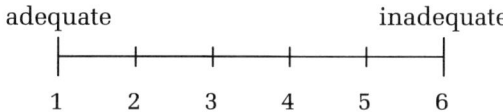

This system's responses were

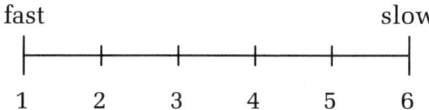

This system's responses were

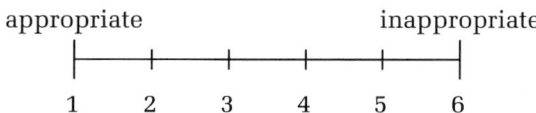

This system's responses were

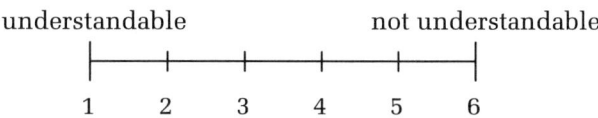

The voice(s) is (are)

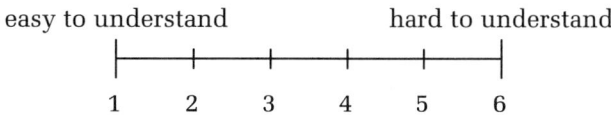

The voice(s) is (are)

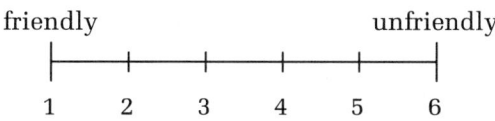

I would use this system

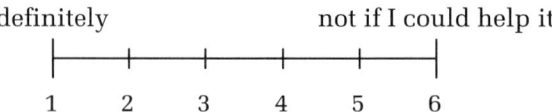

Appendix 15-2

Transcription Sample

The dialogue is from Walker (2000: 392), but all the particulars (transcript number, date, participant profile) are fictional. This sample is a minimalist transcript for a test session; many elaborations are possible, but you need to code the participant, session, and date; provide participant description; and identify all turns. It is also helpful to have the basic linguistic information available at a glance (additional user data, along with this information, should be tabulated in a separate file).

Participant #10, session 1   Native language: English
Female, 32   Birth country: USA
Computer professional, very experienced with e-mail, moderately experienced with speech systems.

1	ELVIS	Hi, ELVIS here . . . You have 5 new or unread messages in your inbox. Say Read, Summarize, or say Help for more options
2	Caller	Read.
3	ELVIS	Select by Sender, Subject or in order of arrival?
4	Caller	Sender.
5	ELVIS	Which Sender?
6	Caller	Kim.
7	ELVIS	In the messages from Kim, there's one message about "Inerviewing Antonio" and one message about "A meeting today". The first message is titled "Interviewing Antonio". It says, "I'd like to interview him. I could also go on to lunch. Kim."
8	Caller	Next.
9	ELVIS	This is a message about meeting today it said the meeting this morning is at 10:00 in 2d516. See you there, Kim.
10	Caller	I'm done here.

16

Conclusion — Pursuing Habitability

At least half a dozen laboratories are hard at work in search of an electronic system for recognizing speech and coding it, no matter who talks, whether the speaker has a cold, is excited, or is underwater. When it succeeds, we shall be able to talk to a typewriter which has no stenographer in front of it or yell a number at a telephone and have it do its own dialing.
    — David O. Woodbury, 1959

Two truisms of speech-system development:

Truism #1   Computational speech and language technologies are progressing at tremendous speed.

Truism #2   Predictions of an imminent breakthrough in these fields have entered their sixth decade.

Or, put another way, the year 2001 has come and gone without the HAL-9000. The problems of speech interaction are far more difficult than have been imagined, even by many linguists, who should have known better. In particular, building a real-time, goal-driven, speech-activated, speech-directed, speech-generating system — a voice interface — is not a matter of building a bigger, better, faster, pattern matcher.

From the earliest days of voice interfaces, the say-*yes*-for-yes days, speech-system folk have always "assumed that speech technology would 'improve' until applications became better behaved" (Balentine, 1999: 206). That assumption — or, call it what it is, wishful thinking — continues, and while the technological improvements over the last decade have been genuinely staggering, given the starting point and the complexity of the problems, this wishful thinking is founded on misguided, engineering-based reasoning.

Engineers have always tended to view speech communication in a Shannon-and-Weaver computational model, where there is a message with some value, a channel for it to cross, noise along the way, redundancy to combat the noise, and so on. In this model "A speaker is a mapping from messages onto wave forms and a hearer is a mapping from wave forms onto messages" (Fodor, 1975: 103). That's good, that's right, that's fruitful, if you're an engineer. Patterns do need to be matched, lots of them, very quickly. But designers need to (and users do) view communication in a more dynamic, interactive frame.

It was misguided from the beginning to put so much faith in the technology basket, but it is now unmistakable that improved service will depend much more heavily on human-factors concerns (anticipation, guidance, repair) than on chips and code. As Norman Fraser, a European pioneer in conversational systems, puts it, "the success of a spoken dialogue system is only slightly related to technical matters." Specifically citing ethos and natural-ness in error handling, he added that

> I have, for example, seen trial systems with a disgracefully low word accuracy score
> receiving a user satisfaction rating of around 95%. I have also seen technically excellent
> systems being removed from service due to negative user attitudes.
> (Bernsen et al., 1998: 21)

Technical excellence is far from incidental, but it is also far from sufficient. It is, at best, half the story.

Voice interfaces are expert systems, and expertise is "an ongoing collaborative and discursive construction of tasks, solutions, visions, breakdowns, and innovations," not the "stable individual mastery of well-defined tasks" many people presume (Engeström and Middleton, 1996: 4). Expertise is flexible, responsive, and shifting. The caller already has such expertise. Her success in a given speech-system interaction depends on how much of that expertise the system can deploy, and even more on how much it can allow her to deploy; it depends, that is, on the system's ability to collaborate through language. The system's success too, therefore, both in an immediate interaction, and long-term in the marketplace, depends on that ability.

We first worked through a broad cross-section of knowledge in this book that supports crafting such an expert system, from linguistics through pragmatics, conversation analysis, social psychology, and lexicography, under the constant influence of human–computer interaction research. Rooted in that knowledge, we then turned to strategies in all aspects of the development of voice interaction projects, from user and discourse analysis, through agent design, repair planning and initiative management, legacy and pre-existing source issues, writing-for-speaking and turn design, planning the call flow and specifying the design, testing, testing, and testing. But there is one last point in all this to take from Aristotle, the Grand Old Man of Everything — an indispensable point.

"Everyone who makes," The Philosopher says, speaking in the context of *techne*, "makes for an end" (*Nicomachean Ethics*, 6.2). Crafting is purposeful, both in the immediate sense of building exemplary artifacts ("virtuous," Aristotle would call them), and in the ultimate sense of pursuing the goals of the craft. The purpose of voice interaction design is to build exemplary interaction models, in the pursuit of habitability.

Discussions of habitability are often couched in terms of "keeping" the user within the limits of the system, and that's a crucial element in the crafting of individual exemplary voice interfaces. But more generally, the pursuit of habitability is the crafting of a system the user *can* inhabit. The difference is perspective, but it's an important perspective. On the one hand, we focus on constraining the user; on the other hand, it's accommodating the user. You have to do both, but the driving goal of voice interaction design, as it is with graphic interfaces, and with ergonomics generally, should be accommodation.

The technology is finally here, not for a HAL-9000, who has to operate the full array of systems required by a space ship going all the way to Saturn, but for a host of his more single-minded cousins, who need rather to supply information, manage sales, arrange for planes, trains, and automobiles. But technology cannot provide habitability on its own, not without informed, inspired, thoroughly integrated design work.

Glossary

802.11: IEEE standard for short-range wireless technology linking digital appliances (e.g., linking a camera and a printer, or a phone and a PDA). 802.11 is in partial competition with Bluetooth, but many appliances support both protocols.

AAC: Augmentative and Alternative Communication, a field of technology that makes extensive use of speech recognition and synthesis to assist people with disabilities.

Acceptance: See Contribution model.

Accommodation: See Convergence.

Adjacency pair: See Dialogic pair.

ADSI: Analog Display Services Interface, the standard protocol enabling data services, such as visual display, over analog telephone networks.

Affordance: An HCI design term adapted, chiefly by Donald Norman, from James Gibson's ecological model of psychology — in Norman's terms, affordances are "the perceived and actual properties of [a] thing, primarily those fundamental properties that determine just how a thing could possibly be used" (1988). For instance, one affordance of a telephone handset is that it is graspable, another it that its buttons are pushable, a third is that its earpiece produces audible signals, and so on.

Agent: 1. A computerized, human-like character that forms part or all of the output for a voice interface. They respond to and respond in language (textual or verbal).
2. A software element that carries out some processes (web searching, file management, dialogue evaluation, . . .), with some autonomy, but at the direction of a user or another agent.

Alarm, false: See Insertion.

Alias: An alternate acceptable word for the preferred target word. There is a widespread, tendency to use the word *alias* for *synonym* (indeed, as a synonym for *synonym*) in computational linguistics. But a synonym is not an alias, even though the system coding may be forced to treat it as such. *Alias* entails that there is one true term that the alias (or aliases) can stand in for, whereas synonymy is a relationship between two or more terms with very substantial referential overlap.

Amalgam, Dialogue act: See Dialogue act.

Anaphora: Words that get their semantic specification from the presence of other words. Mostly anaphora is a matter of pronominal reference (as in "Homer thinks he has the flu," where the meaning of *he* depends on the presence of another word, *Homer*), though it has other manifestations (as in "There were a thousand HCI professionals at the CHI conference. The researchers loved Las Vegas." In this case, "the researchers" gets its full specification from the presence of "a thousand HCI professionals").

Anthropomorphism: The trope whereby nonhuman entities are given human attributes, almost invariably including speech, for rhetorical effect. Within HCI research, there are sporadic debates about whether computers should be anthropomorphic. In this book, *anthropomorphism* (along with the issues surrounding it) is collapsed into the term *personification*. See also Personification.

Articulatory synthesis: A type of wave-form synthesis based on digital models of the vocal tract.

ASR: Automatic Speech Recognition (system).

Assertion: Often confounded with the related but distinct term, *assertive*.
1. Syntactically, a form that asserts something (as opposed to an interrogative or an imperative). For instance, "The cat is on the mat".
2. A dialogue act that asserts something directly (as opposed to assuming, insisting, guessing, and so on). A prototypical example of the dialogue act category, Assertive.

Assertive acts: Informative dialogue acts which provide information (for instance, statements, claims, and descriptions).

Assertive: Often confounded with the related but distinct term, *assertion*.
The category of dialogue acts concerned with conveying information (of which the assertion is a prototypical member).

Association ratio: A measure of nonchance probability that a word collocates with another, which controls for corpus size and linear order (see Church and Hanks, 1989).

ASU: Automatic Speech Understanding.

ATD: Automatic Telephone Dialogues. A phone-only voice interface.

Audio interface: See Voice interface.

Audio logo: A short sound sequence (usually music, but sometimes notes or tones) used by a company to brand its products, often in conjunction with a voice recording of the company name.

Auditory icon: See Earcon.

AUI: Auditory User Interface; see voice interface.

Backchannel: A dialogue act uttered not to take or keep the turn, but to signal a shared understanding with the agent who has the floor (e.g., murmuring "uh-huh" or "right").

Background: See Ground.

Bailout: When a user exits a system or subsystem (e.g., terminating a dialogue before the natural completion point). See also Zero-out.

Barge-in: A term indicating that the caller can speak (that the system will listen) even during system output. The preferred term is the one used in ordinary language when someone interrupts and takes the floor, *Turn overlap* (or just *Overlap*).

Bluetooth: Short-range wireless technology for linking digital appliances (e.g., linking a camera and a printer, or a phone and a PDA). Named for the 10th century Viking King, Herald Bluetooth, who united several Nordic Fiefdoms. Bluetooth is in partial competition with 802.11, but many appliances support both protocols.

C&C: See Command-and-control.

CA: See Conversation analysis.

Call flow: The architecture of a voice interface, the blueprint for script structure; for instance, the possible paths a dialogue can follow instantiating the goal of booking a ticket, including the tasks and sequence. See also Script.

Candidates: The guesses that recognition engines return for what the input pattern was.

Cascading errors: See Error spiral.

Cluster, Dialogue act: See Dialogue act.

Cocktail party effect: The phenomenon of focused acoustic attention, such that in a cacophony of sounds (like cocktail party chatter), people can carry on conversations (or eavesdrop on them) by selecting what they will listen to and filtering other input down to a background hum.

Coherence, referential: The conceptual linkages among discourse elements (sentences, utterances, turns) by virtue of evoking the same objects or ideas — for instance, the way the following sentence evokes the same person twice: "John fell down because Jane tripped him." It is distinct from, but patterns with, associative cohesion.

Coherence, relational: The conceptual linkages among discourse elements (sentences, utterances, turns) by virtue of standard patterns (such as causation, elaboration, or contrast) — for instance, the relation between the two clauses in "John fell down because Jane tripped him." It is distinct from, but patterns with, connective cohesion.

Cohesion, associative: The formal linkages among discourse elements (sentences, utterances, turns) by virtue of using linked expressions referring the same object or idea (most commonly noun phrases and pronouns) — for instance, the use of *John* and *him* in "John fell down because Jane tripped him." It is distinct from, but patterns with, referential coherence.

Cohesion, connective: The formal linkages among discourse elements (sentences, utterances, turns) by virtue of using words and/or syntactic structures that evoke standard conceptual patterns (such as causation, elaboration, or contrast) — for instance, the use of *because* in "John fell down because Jane tripped him." It is distinct from, but patterns with, relational coherence.

Colligate: The noun form of the term describes a word which is frequently contiguous to another word (e.g., *dale* is a colligate of *hill*, in "over hill and dale"). The verb form of the term describes the activity of words falling into contiguous relationships with other words (e.g., *dale* colligates with *hill*, in "over hill and dale"). See also Colligation.

Colligation: A lexical co-occurrence relation (distinct from collocation) with two aspects. One aspect, sometimes called "the idiom principle," is the tendency for certain groups of words to stick together, like "kick the bucket" and "hold the phone." The other is the tendency for certain words to correlate with certain syntactic locations, like the frequency with which the noun, *deposit*, shows up as a direct object ("I would like to make a deposit.").

Colligation, textual: The tendency for certain words or phrases to occur in specific textual locations ("Once upon a time" at the beginning of fairy tales, for instance).

Collision (error): Miscommunication that results from the user and the system talking simultaneously, usually triggered by a system-initiated turn after a time-out, just as the user begins input. Even systems with first-rate turn-overlap capacities can be subject to collisions.

Collocate: The noun form of the term signals a word which has the tendency to show up in proximity to another word; for instance, *make* is a strong collocate of *call*. The verb form signals that tendency; for instance, *make* strongly collocates with *call*. See also Collocation.

Collocation: The tendency of words to show up in proximity to each other (to be collocates); for instance, *cell* tends to show up proximally to *phone*.

Command-and-control system: A system (usually with speech input) that follows simple utterances; for instance, in the movie *Blade Runner*, Decker interacts with ESPER, an appliance that can probe a digital image with commands like "Move in," "Stop," "Enhance," and "Track 45 right." Also "C&C."

Commissive acts: Obligative dialogue acts which oblige the speaker (promises, bets, refusals, . . .).

Common ground: See Ground.

Competence error: See Task slippage.

Concatenation synthesis: See Concatenative synthesis.

Concatenative synthesis: A type of speech synthesis created by recording words, phrases, and syllables from human speakers and assembling them into utterances. See also Speech synthesis.

Conceptual habitability: See Habitability.

Conceptual category strategy: A strategy for building system vocabularies organized by cross-associations among categories at top levels of interactions, especially to prepare for resolving ambiguities. For instance, with a general consumer-products information site, someone selecting "music" as the main category for an exchange might be looking for information on audio players and other playback electronics, but the system should also be prepared for musical instruments, compact disks, and DVDs, in case the user's construal of "music" is different from the site's.

Conceptual pact: A possibly short-term name or description for the mutual referential coherence established by agents in a dialogue — a "flexible and temporary agreement . . . to conceptualize an object in a particular way" (Brennan, 1998).

Concordance: As a noun, the term describes the collection of uses for specific words in a given corpus (everywhere Shakespeare had someone say "mayhaps," for instance). As a verb, the term describes the act of assembling such a collection. In voice interaction research, it is tremendously valuable for investigating how often and (especially) in what contexts a word is used in a corpus ("fuel-injection" in automotive repair discourse, for instance). Also "KWIC" (keyword in context).

Confidence: The commitment with which a recognizer returns a word-candidate for an input pattern; for instance, the pattern might be "recognize speech", and the system returns "recognize speech" with a high rating, "wreck a nice beach" with a lower confidence rating. Also "Confidence score."

Confirmation: A groundskeeping dialogue act particularly important in voice interfaces because of recognition fragility (e.g., the second of the following two utterances: "North Dakota?" "Yes, North Dakota"). See also Feedback(2).

Connected word recognition: A mode which takes speech input only if the words are spoken with noticeable silence between them.

Continuous word recognition: A mode which takes speech input in natural streams.

Consonant: A class of speech sounds characterized either by interruptions to the sound stream, such as stops and fricatives, or by short sonorant sounds, which mark syllable boundaries — in contrast to the more sustained sonorance of vowels, which occupy the centers of syllables.

Constitutive acts: Dialogue acts, like apologies, that are generally recognized socially as more formal. Frequently, they have a good deal of social machinery around them, like declaring marriage or christening a ship.

Content words: Nouns, verbs, adjectives, adverbs (and, in some theories, prepositions): the words that carry the semantic freight, the content, of an utterance (in comparison to function words, like *the*, *a*, or *please*).

Contribution graph: See Contribution model.

Contribution model: A model of conversation, developed by Herbert Clark and Edward Schaefer, which views every utterance of a dialogue in terms of two phases: the presentation phase, the part that introduces concepts and intentions to the dialogue; and the acceptance phase, the part that indicates how successful the previous utterances' presentation phase was. Janet Cahn and Susan Brennan (1999) have formalized this model, providing a mechanism for graphing out the presentation and acceptance phases of utterances in a dialogue, called a *Contribution graph*.

Convergence: The tendency for speakers to talk like one another (adopt the same terms, speak in roughly the same durations, even begin to speak in one another's dialects). A very useful (but not infallible) principle for voice interface design. Also called *Accommodation*. When it is deliberately induced (in a psychological experiment or interaction design) it is known as *Priming* or *Entrainment*.

Conversation Analysis: A field of study (and a general philosophy of interaction), developing out of sociology, that focuses on the formal properties of conversation: who speaks when, how the change of speakers is managed, what types of utterances pattern with what other types of utterances, and so on. Also "CA."

Conversational ground: See Ground.

Conversational maxims: A body of practices observed by cooperative conversants, researched and codified by H. Paul Grice (hence, they are often called *Gricean maxims*), in the research field of pragmatics (hence, their study is often called *Conversational pragmatics*). They concern the Relevance, Quantity, Quality, and Manner of conversational contributions.

Conversational pragmatics: See Conversational maxims.

Corpora: Plural of *Corpus*.

Corpus: A large-to-vast structured set of texts and/or audio records of speech (now digitally stored and processed).

CRM: Customer Relationship Management. A field dedicated to maintaining customer relationships, critically involving such matters as the flow of information, efficiency of transaction, and protection of privacy. Voice interfaces are often touted as a way to help manage customer relationships, especially in these three areas.

CSp: Computer Speech.

CTI: Computer-Telephony Integration. The field of bringing computers and telephones together.

CUI: Conversational User Interface. See Voice interface.

Cut-through: See Turn overlap.

DARPA: (U.S.) Defense Advanced Research Projects Agency, which has sponsored several speech-system initiatives, including the Communicator program, which funds conversational voice interface research.

DBCI: Dialogue-Based Collection of Information (questionnaires and other form-filling activities).

Declarative acts: Constituitive dialogue acts in which the speaker, serving a specifically sanctioned social role, declares a new social reality: pronouncing marriage, sentencing a criminal, christening a ship or a baby.

Deletion error: When the recognizer fails to detect a legitimate portion of a user's utterance (possibly the entire utterance).

Density strategy: A strategy of always populating a vocabulary with the largest number of terms compatible with good retrieval performance. Also "lexical density."

Dialect: A language variety defined more by the properties of its speakers than by the purposes or contexts of its use (as for instance, registers and genres are); properties which are often used to subdivide the notion of dialect. For instance, the property of geography during formative language development can lead to specific regiolects (like Southern American English); the property of socioeconomic status can lead to specific sociolects (like Cockney); and the property of a linguistic impact from prior or parallel languages spoken by an ethnic group can lead to specific ethnolects (like Yiddish- or African-influenced varieties of English).

Dialogic pair: An dialogic pair-unit (i.e., not just any two given utterances, but two utterances that function together; for instance, a question and an answer, or an invitation and an acceptance). Known in the Conversation analysis literature as an Adjacency Pair, a term rejected here because (1) *adjacency* suggests space, not time, and utterances occur in time; (2) the pair needn't occur in strict adjacency (or contiguity); and (3) it is opaque.

Dialogue act: The force of an utterance, what it *does* (asks a question, makes a request, declines an offer, . . .). People act by speaking (for instance, the utterance "What are you doing Sunday?" might perform the action of pre-inviting; that is, laying the groundwork for an invitation), a phenomenon traditionally called a "speech act." However, these utterances usually require another utterance ("Nothing, why?"), and the best perspective on them for voice interface design is as part of a sequence, rather than in isolation; hence, the term "dialogue act".[1]

Dialogue acts often come in clusters, such as an answer to one question followed by an offer of more help, or the greeting + identification + offer with which most commercial phone calls are answered.

Dialogue acts always come as amalgams, usually of task management and dialogue management. "Would you like a ham sandwich?" for instance, amalgamates the offer of a sandwich with a turn-assignment (since you are now called upon to take a turn and accept or decline).

Dialogue flow: The way the turns alternate and the initiative shifts in a dialogue.

Dialogue-management acts: The category of dialogue acts concerned with coordinating and grounding the dialogue (as opposed to task-management acts, which are concerned with accomplishing goals).

Dialogue Manager (DM): The software module most responsible for the flow of the exchange. It usually analyzes each sentence, decides the next utterance(s), and identifies whatever action should be taken. Also DMS (Dialogue Management System).

Dialogue: A linguistic exchange between agents.

1: This usage has plenty of precedent in the fields of computational pragmatics and dialogue studies, but it was introduced (by Bunt, 1979) in a much more restricted way, as a subset of speech acts which regulate or ground dialogues. See Bunt and Black (2000) for some discussion.

Dialogue initiative: See Initiative.

Dialogue-tree system: A voice interface which is built around branching hierarchies of choices; also "Menu system."

Diction: Word choice. The ancient rhetorical concern with diction was with choosing the optimal word for a given audience, purpose, and context, which is the core of voice interaction design. See Vocabulary, the Vocabulary problem.

Dictionary: An instrument which lists words, their spelling, their pronunciation, and their meaning. In voice interaction design, it sometimes means the collection of words available for a system to use (produce and recognize), a usage rejected in this book. See also Vocabulary.

Directed dialogue: A fixed-initiative dialogue fixed on the system. See Initiative.

Directive acts: Obligative dialogue acts which (attempt to) induce the obligation of the hearer (requests, invitations, forbiddings, . . .).

Directive prompts: System utterances that tell the user exactly what the available (or sometimes preferred) vocabulary is at the point of prompt ("Please say 'yes' or 'no.'").

Disambiguation: Resolving the meaning of an utterance when there are two or more candidates (is it *Mary*, *merry*, or *marry*?). The disambiguation might be accomplished by the natural-language system, checking what the most appropriate candidate is with respect to a given syntactic frame, or what best fits the discourse context. Or an appeal might be made to the user, via a *disambiguation query* ("Do you want Cisco Systems or SysCo foods?"). More specifically, in speech recognition, it refers to the process through which a captured signal is matched to one word or phrase.

Discourse analysis: 1. Academically, a field that grows out of social psychology and studies discourse with a range of conceptual tools (chiefly those borrowed from conversation analysis and pragmatics). 2. In voice-interface development, an early design phase that involves the analysis of discourse used in the application domain, to build vocabulary, understand register, and investigate interaction scenarios. See also Natural dialogue studies.

Discourse: A body of utterances: (1) in abstract terms, the type of utterances available, the preferred words, structures and moves, in a given domain, as in "the discourse of shopping;" (2) in concrete terms, the actual utterances used within certain established bounds (a time period, the covers of a book, and so on), as in Descartes' *Discourse on Method*.

Discrete word recognition: See Isolated word recognition.

DMS: Dialogue Management System. See Dialogue Manager.

DTMF: Dual Tone Multiple Frequency. See Keypad interface.

Dysfluencies: Interruptions to the (ideal) speech flow: repetitions, false starts, ums, ahs, prolonged hesitations. . . . All of these events are very common in regular speech, but we tend to hear right past them most of the time; recognizers are relatively poor at hearing past dysfluency.

Earcon: A punning analogy to the graphic interface *icon*, earcons are sounds designed to trigger associations with familiar objects, for the purpose of exploiting metaphorical implications. More generally, they are semantically/functionally specified output sounds, such as a beep to signal to the user it's now her turn to speak. Also "auditory icons."

Echo cancellation: A procedure that filters off from an incoming signal the echo of an outgoing signal, thereby (among other virtues) facilitating caller-on-system turn overlap. In the telephony world, turn overlap is sometimes called echo cancellation. See also Turn overlap.

ECU: Environmental Control Unit, a centralized speech-enabled processor for controlling domestic systems and appliances (lights, heating and cooling, stereo, etc.). Designed to enhance general home marketability, they have also proven valuable as an assistive technology for people with restricted mobility.

Ellipsis: Omitting words or phrases because they're predictable. Usually, this term concerns linguistic ellipsis (in which the source of the prediction is previous words or phrases: "Would you like ham with your eggs? Or bacon?" leaving out the underlined material "Would you like ham with your eggs? Or would you like bacon with your eggs?" But equally important is the notion of pragmatic ellipsis (in which the source of the prediction is the utterance context). "Two for *Throne of Blood*, please," someone might say to a ticket agent at a theater, for instance, leaving out the underlined material "Sell me two tickets for *Throne of Blood*, please" because the context makes it predictable.

Endpoint: As a noun, it refers to the point at which the input ends. As a verb, the act of identifying the (presumed) end of input.

Endpointer: The silence-detection device with respect to user input: it decides when the caller has finished speaking.

Enrollment: See Training.

Entailment: A relation between sentences such that the truth of one sentence necessitates the truth of another. "Fred is a bachelor" entails "Fred is male" (but notice that "Fred is male" does not entail "Fred is a bachelor").

Entrain: See Priming.

Entrainment: See Priming.

Error cascade: See Error spiral.

Error, competence: See Task slippage.

Error spiral: The situation where one error triggers more errors (for instance, a system response causes users to speak more loudly, which in turn throws off the recognizer). Also known as an "Error cascade", and an "Error amplification." Common related phrases are "Cascading errors" and "Spiraling errors."

Error, recognition: See Recognition errors.

Escalating detail: Also known as "Incremental prompt", "Expanded prompt." See Expansion.

Ethnolect: See Dialect.

Exchange: A subdialogue related to a specific topic or subtask.

Expanded prompt: Also known as "Incremental prompt" and "Escalating detail." See Expansion.

Expansion: A design strategy that adds specificity and direction to subsequent system utterances based on context and caller response. See also Progressive assistance.

Expressive acts: Constituitive dialogue acts in which the speaker presents her feelings (apologizing, for instance, conveys regret; thanking conveys gratitude).

Explicit feedback: See Feedback.

Explicit prompt: See Prompt.

Failure, recognition: See Recognition error.

False acceptance: When the recognizer accepts an illegal word ("hearing" it as a legal word); a species of Substitution. Also "false recognition."

False alarm: See Insertion.

False mapping: See Substitution.

False recognition: See False acceptance.

False rejection: See Rejection.

Feedback: A systematically ambiguous term in voice-interface development.
1. An indication by one agent about his processing of the other's previous utterance(s) — hearing, understanding, agreeing or disagreeing with, and so on. It might be "positive" in the sense of indicating successful processing, or "negative," indicating difficulties in processing.
2. A specific subclass of system utterances that probe the system's understanding by presenting it to the user in degrees of explicitness. Feedback may be explicit ("Did you just say 'Good-bye'?"), implicit ("Calling Susan . . ."), inferential ("The weather in Boston is . . ."), or open ("OK"). Often called "confirmation" in speech system research and design.

Felicity conditions: The conditions which determine the appropriateness and success of a dialogue act (for instance, for an apology, that a potential offense has occurred; for a sentencing, that a conviction has occurred and the speaker is a judge).

Field: One of the three aspects of Register, *field* signals the activity the communicative event supports and suffuses — inquiring, buying, carrying out a bank transaction. See also Register, Mode, Tenor.

Filtering through the phone: An early-phase, data-gathering technique for analyzing the discourse of tasks that do not ordinarily have a significant verbal dimension, a species of Natural dialogue study. For instance, people normally use scheduling calendars visually. Filtering that task through the phone would involve giving one person visual access to such a calendar, connecting her by phone to someone who cannot see it, and having them converse about appointments, dates, and the like. The vocabulary, utterance structure, and interaction patterns are then harvested to feed the design a voice interface for a scheduling application.

Finite State Machine: A computational model with a finite number of states and defined transitions among them. For instance, one state might be WORD, another VERB, with a transition defined between them as "is inflected for tense." The model would then pass the item *jumped* from WORD to VERB because the transition criterion is met by *jumped* (i.e., the model would recognize *jumped* as a verb).

Fixed initiative: See Initiative.

Floor: When an agent has her turn in an exchange, when she is speaking, she is said to have the floor.

Flow: See Call flow, Dialogue flow.

Flow-regulating acts: Dialogue acts which manage turn taking.

Formant synthesis: A type of wave-form synthesis based on modeling vowel formants.

Formant: The frequency ranges of a vowel with the greatest amplitudes. Acoustically, formants are what discriminate vowels from each other, and their transition curves help discriminate neighboring consonants.

Fricative: A consonant distinguished (articulatorly) by a vocal constriction through which air is forced and (acoustically) by a dominant hissing element, like the first sound of *ship*.

Function words: Words that perform syntactic or discursive functions and link content words — the lexical connective tissue of language; effectively, every word type except a noun, verb, adjective, or adverb (for instance, articles, particles, demonstratives, qualifiers, quantifiers, intensifiers, pronouns, . . .).

Functional habitability: See Habitability.

Fusion: A recognition error where the input is segmented into an insufficient number of constituents (e.g., "wreck a nice beach" is heard as "recognize speech"). See also Split.

Gain: A speech-system attribute concerning volume of input. If the gain is set too low, the system can hear quiet speech, but also tends to accept background noise as legal input. If the gain is set high, background noise is effectively filtered, but some legal input might also be lost.

Goat: Unfortunate telephony term for a person whose utterances a speech recognizer has trouble recognizing.

Grammar: In linguistics, *grammar* means the model of language knowledge, including (at least) phonology, morphology, syntax, and semantics. In ordinary language, *grammar* means the body of rules associated with proper speaking and writing. Among many speech professionals, *grammar* often means the body of acoustic models for a speech recognition system, a usage rejected in this book. See Vocabulary.

Grapheme: The abstract form of a letter.

Graphic interface: An interface by which a user performs tasks on a computer, inputting by direct manipulation (keyboards and/or various pointing devices) of objects depicted on a screen (buttons, fields, menus). The default computer interaction method at the beginning of the 21st century. Also, "GUI."

Greeting: 1. A dialogue act marking initial social contact ("Hi, how are ya?"). 2. The initial message a caller gets upon engaging a voice interface: ("Hi, this is AT&T Amtrak Schedule System. This is TOOT. How may I help you?").

Gricean maxims: See Conversational maxims.

Gricean pragmatics: See Conversational maxims.

Ground: The collective assumptions supporting a discourse event. The ground for this glossary entry, for instance, includes the title of the book, the term-definition template that is part of the register of glossaries, the ordinary-language use of *ground*, and so on. The collective assumptions behind any discourse event are of course limitless, so discussions of the ground tend toward a very small subset of those assumptions. The Common ground is what the conversants share as a function of culture, profession, and so on, prior to the interaction. The Conversational ground is what they establish as they talk. The Background is the Common ground + the Conversational ground.

Grounding criteria: The thresholds set for the system to govern the grounding of words, events, or agreements in a human–computer dialogue. For instance, a credit card number would generally have grounding criteria set very high (in fact, usually 100%), so that the system would in all cases repeat the number back for the caller to verify. But the grounding criteria for the word "account" would be set quite low if the recognizer confidently had "transfer" "300 dollars" and "checking."

Grounding: The act of establishing and confirming and calibrating the discourse, especially as it concerns the introduction of new topics, actions, or functions.

Groundskeeping acts: Dialogue acts which build and maintain the conversational ground between speakers (confirmations, rejections, repairs, . . .).

GUI: Graphic(al) User Interface; see Graphic interface.

Habitability: The accommodation of a computer system to the language people use in a given domain. Watt (1968: 338) says a habitable computer language is "one in which its users can express themselves without straying over the language's boundaries into unallowed sentences." Ogden and Bernick (1996) see four components: lexical habitability (having all the necessary words), syntactic habitability (parsing all the necessary utterance constructions), functional habitability (allowing all the necessary interactive procedures), and conceptual habitability (having all the necessary knowledge in the domain).

Habitable: See Habitability.

HCI: Human Computer Interaction, the field that studies the physical (ergonomic), cognitive, and perceptual elements of computer use.

Headword: An abstract word heading a dictionary or thesaurus entry, collapsing much variation into one form. The headword *bank* (the verb), for instance, collapses the variation in words like *bank, banks, banking,* and *banked.* Lexicographers prefer the term *lemma* (plural *lemmata*) for headword.

Heteronyms: Words which are spelled the same, but pronounced differently — "She will *lead*, but you'll have to drive fast. She has a *lead* foot."

HLT: Human Language Technology. A generic term for the study of natural language from the perspective of building machines that can interact with humans via natural language, and machines that can help humans deal with natural language in various ways (such as automatic translation and information extraction from texts). The term is largely used in interdisciplinary contexts. See also NLE.

HMM: Hidden Markov Model. A Finite State Machine in which (1) the transitions have ordained probabilities (the Markov part), and (2) the states are not observable from the outside (the Hidden part), only the outcome. This architecture is very common for speech recognizers and parsers.

Holonym: See Metonym.

Homonym: A word that sounds like another, but is very distinct in meaning, and often spelled differently as well — *reel* and *real* are homonyms.

Homonymous: The quality of being a homonym to some other word. Also "Homophonous."

Homophonous: See Homonymous.

HTML: HyperText Mark-up Language, a subset of SGML that primarily tags for appearance, rather than function. It is the base language of most web sites. See also SGML and XML.

Hyperarticulate: To exaggerate pronunciations (making them longer, louder, or more varied in pitch), usually to resist noise.

Hyperarticulation: A hyperarticulated pronunciation.

Hypernym: See Metonym.

Hyponym: See Metonym.

ICT: Information and Communication Technologies (a largely British English term).

Idiom: A routine and familiar sequence of words that has a unique meaning distinct from its component parts or its expected syntax (*kick the bucket* for "to die").

Idiom principle: See Colligation.

Imperative: A command ("Wash the car!").

Implicit prompt: See Prompt.

Incremental feedback: See Feedback.

Incremental prompt: Also known as "Escalating detail," "Expanded prompt." See Prompt, Expansion.

Inferential feedback: See Feedback.

Inferential prompt: See Prompt.

Infirm results: Words that the recognizer hypothesizes from input, but has not fully committed to yet. Also, "Infirm words."

Informative acts: Dialogue acts which concern information (statements, claims, questions, . . .).

Initiative: There is a systematic ambiguity with this word in voice-interface design. 1. The first utterance of a dialogic pair; for instance, a question that puts "reactive pressure" on the following utterance to be an answer in response. See Dialogic pair. 2. The flow-control of a dialogue. Whichever agent has control of the conversational flow has initiative. Note that initiative is not just who is talking. In an interview, for instance, the interviewer usually maintains the initiative. While both parties speak regularly, and the interviewee often speaks for longer durations, the interviewer controls the flow by setting the agenda, asking the questions, requesting elaborations, and so on. Speech systems can be fixed-initiative (invariably fixed with the system), in which one agent maintains all the control, or mixed-initiative (also variable initiative), in which either agent can take control at any time. Some researchers distinguish between task initiative and dialog initiative, encapsulating each, so that a caller might have task initiative overall, but the system might have, at a given point, dialogue initiative. Among the synonyms for *system initiative* are *directed initiative*, *directed dialogue*, and *system-directed interaction*.

Initiative manager: A software module in some voice-interface designs that weights the initiative toward the system or the user depending on the state of repair, the experience of the user, the requirements of interaction script, and the need for specific dialogue management acts.

Insertion: A recognition error: when noise (or possibly an out-of-vocabulary word) is "recognized" as a legitimate vocabulary item. Also "False alarm", "False recognition", "Intrusion, Misfire."

Insertion sequence: A sequence of dialogic pairs occurring between the initiative and response of another dialogic pair. For instance, a boy asks a girl "Would you like to go to the prom with me?" (an initiative), to which she replies with another initiative, "What kind of car do you drive?" He responds, "An 1899 Duryea," and she then responds to his initiative, "Certainly. I would be honored."

Interaction slippage: A communicative breakdown between the user and the speech system that is the fault of the interaction design.

Interrogative (act): There is an unfortunate but systematic ambiguity with this word. Syntactically, an interrogative is a question ("What kind of car do you drive?"). Pragmatically, it is an informative dialogue act which seeks information (constituent questions, yes/no questions, inquiries). One might use a syntactic interrogative that does not primarily seek information (for instance, the offer, "Would you like the salt?"). One might also use an interrogative dialogue act that is not syntactically an interrogative (for instance, the utterance, "My son is a doctor" in a circumstance designed to elicit the interloctor's son's occupation).

Intonation: See Prosody.

Intrusion: See Insertion.

Isolated word recognition: A system (or mode) which takes speech-only input if the words are spoken one at a time; that is, in isolation.

IVR: Interactive Voice Response (system); effectively, a synonym of *Voice interface*, though its early adoption has led it to pick up connotations of primitiveness; often, too, it is used for voice-output/keypad-input systems.

Jargon: Terminology characteristic of specific practices, professions, or interests, often baffling to people outside those areas.

Keypad interface: An interface whereby user input is via the telephone keypad ("Press 1 for checking, 2 for savings, . . ."). Also DTMF (system), which is not a particularly good term because it is hopelessly opaque to nontelephony people, and Touch-tone interface, which is problematic because Touch Tone® is a registered trademark of AT&T.

Keyword: A salient word in a discourse — in particular, one that a voice recognition engine is tuned to. The term is mostly associated with voice-response systems, but all speech systems function in critical ways upon keyword recognition. See also Wordspotting.

KWIC: Key Word In Context, a rapidly aging term for corpus-driven Concordance research.

Legacy: The constellation of concerns and expectations that a new design or implementation faces when previous products or services were already in place — typically, the legacy for a voice interface is either a human-mediated interaction or a keypad system.

Lemma: See Headword.

Lemmata: The plural of *lemma*; see Headword.

Lexical accommodation: See Convergence.

Lexical density: See Density strategy.

Lexical entrainment: See Priming.

Lexical habitability: See Habitability.

Lexical priming: See Priming.

Lexical token: See Token, lexical.

Lexical type: See Token, lexical.

Lexicography: The study of the forms, meanings, and uses of words, almost always with respect to dictionary making. In this book, I take lexicography to stand for the study of words generally.

Lexicology: The study of the forms, meanings, and uses of words, for an abstract understanding of their cognitive representation and use; that is, to understand how the brain stores and deploys them. The lexicological issues discussed in this book are subsumed under the label *Lexicography*.

Lexicon: See Dictionary, Vocabulary.

Linguistic ellipsis: See Ellipsis.

LVCSR: Large Vocabulary Continuous Speech Recognition; also Large Vocabulary *Conversational* Speech Recognition.

Manner, maxims of: The conversational maxims that call for cooperative conversants to use utterances that are clear, to the point, and orderly. See also Conversational maxims.

Man–Machine interface: An early term for User Interface, strongly associated with command-line interfaces and the early years of graphic interfaces. Also "MMI."

Mapping, false: See Substitution error.

Maxims, conversational: See Conversational maxims.

Menu system: A voice interface that presents the user with lists of options to follow (usually, the lists give way to additional lists of options, so that a dialog tree results). See also Dialogue-tree system.

Meronym: Also "partonym." See Metonym.

Metonym: A word related to another through some conceptual relationship (other than near-identity, which is covered by *synonym*). In this book, the term is used as an umbrella label for four specific lexicographic technical terms: *meronym* (also *partonym*) which means a part-relationship; *holonym*, its opposite, which means a whole-relationship (in the pair of words, *wheel/car*, *wheel* is the meronym, *car* is the holonym); *hyponym*, which indicates a subset relationship; and *hypernym*, its opposite, which indicates a superset relationship (in the pair of words *girl/female*, *girl* is the hyponym, *female* is the hypernym).

MI: 1. For conversational systems, Mixed Initiative; see Initiative. 2. For collocation, in corpus linguistics, Mutual Information; see Mutual Information.

Misfire: See Insertion.

Misrecognition: See Substitution error.

Mixed initiative: See Initiative.

Mixed mode: Systems that utilize both keypad- and voice-interaction styles.

MMI: See Man–machine interface.

Modal: 1. In linguistics, an auxiliary verb which correlates with obligation or possibility (e.g., *must*, *can*). In philosophy, a type of logic that includes obligation and possibility. 2. In interaction design, an adjectival form of Mode.

Mode: 1. In interaction design, mode is (1) the input/output channel (vision, sound, touch, etc.), or (2) a system state. 2. In systemic-functional linguistics, it is one of the three aspects of Register. *Mode* as an aspect of Register is much the same as in interaction design: the communicative channel. In a voice-only interface there is one input mode, speech; however, when background noise is high, or in other times of desperation, such systems may have to fall back on manipulative input. It has two output modes, speech and nonspeech sound (which might be further divided into music and representational sounds). See also Field, Tenor, Register.

Moded interaction: The division of a voice interface into distinct, user-cued states of listening and speaking. For instance, the caller can speak (legitimately) only after hearing a tone, or the caller can provide legitimate input only by pressing and holding a button (like a walkie-talkie or CB radio).

Modeless interaction: The ability in a voice interface for either agent to speak at any time, without special tones, button presses, or the like. See Moded interaction.

Move: A conversational turn.

Mutual Information: A measure of nonchance probability that a word collocates with another. It is rendered somewhat differently by different analysts. Church and Hanks (1989: 77) define it thusly:

$$\log_2 \frac{P(x,y)}{P(x)*P(y)}$$

where P(x,y) is the frequency of word x occurring within a given span of word y, P(x) is the overall frequency of x in the corpus (ditto for Pf(y)).

Natural dialogue studies: An early discourse-analysis design phase in which (sometimes simulated) dialogues are studied for information about vocabulary, utterance structure, and interaction patterns. See also Filtering through the phone.

Natural language: What we humans speak and hear and read and write — in particular, as opposed to artificial languages (as used in programming, logical analysis, and the like) and artificially constrained language (as used in keyword systems).

N-best (list): A list of matching patterns for some acoustic input (i.e., the system's N best guesses [candidates] as to what word the user has spoken, where N = some preset number).

N-gram (model): A statistical recognition protocol which calculates the probabilities for candidates of an Nth word given the N − 1 preceding words. For example, where N = 3, the probability that the 3rd (N^{th}) word is *picked* is very high if the preceding two words are (N − 2) *Peter* and (N − 1) *Piper*.

NLE: Natural Language Engineering. A generic term for deploying a range of technologies that can enable the use of natural language for and by machinery (interfaces, translation, information retrieval, and the like); it is largely a term used in computational linguistics. See also HLT.

NLI: Natural Language Interface, the subset of Voice interfaces that accommodate continuous, speaker-independent spoken input, and which are designed in accordance with empirical results from the study of human–human conversation. See also Voice interface.

NLP: Natural Language Processing. See NLU.

NLU: Natural Language Understanding. The field of computational linguistics dedicated to the comprehension of natural (as opposed to artificial, or artificially constrained) language. Also "Natural Language Processing."

Noise: Anything that degrades the message — unrelated sounds, signal cut-outs, even fatigue or foreign-language interference.

Nomenclature: A term for the specialized lexical resources of a specific domain; used mostly in the field of terminography.

Obligative acts: Dialogue acts that concern obligations (promises, invitations, bets).

Onomasiology: From lexicography, the study of the meaning of words, involving such matters (for instance) as synonymy and antonymy. This book, primarily uses the ordinary-language term, *conceptual*, for onomasiological issues. See also the complementary lexicographical term, Semasiology.

OOG: Out of Grammar. See Out-of-Vocabulary slippage.

OOV: See Out-of-Vocabulary slippage.

Open prompt: See prompt.

Open feedback: See Feedback.

Orthography: Spelling conventions (British orthography, for instance, includes *cheque* and *neighbour*; American orthography includes *check* and *neighbor*).

Ortholect: A definable spelling system — the British ortholect is different from the American, for instance; the text messaging ortholect has marked differences from both.

Out-of-Vocabulary slippage: The recognition engine fails to recognize a word because it is not in the current vocabulary; the word might be out of the system vocabulary completely (out-of-vocabulary, out-of-system), or only out of the working vocabulary at the time of utterance (out-of-vocabulary, in-system). Also called *out-task vocabulary* (or *error*), *out-of-grammar error*, and *correct rejection*. *Out-of-vocabulary* dates to isolated-word recognition systems. *Out-of-grammar* is a highly similar term (but somewhat more opaque, and a bit misleading; I ignore the difference in this book), which is linked to natural-language understanding routines and may also extend to unrecognized sequences of words.

Out-task vocabulary: See Out-of-Vocabulary-slippage.

Overlap: Short for Turn overlap; a design feature that allows the caller to speak during system output.

Over-Verification: See Verification.

OVW: Out-of-Vocabulary Word. See Out-of-Vocabulary slippage.

PACKET network: A data-oriented, connectionless, wideband network whose principal function is routing and switching of data packets. Such networks can handle voice, but not efficiently (due to routing delays).

Pact, conceptual: See Conceptual pact.

Paralinguistic: Aspects of communication that are not part of the primary linguistic signal, but which influence its character (and sometimes its meaning): for instance, how loudly something is said, or how quickly, or at what pitch.

Parametric coding: Speech synthesis from scratch; also called *rule-based synthesis* and *wave-form synthesis*. See also Speech synthesis.

Parse: To figure out the syntax of an utterance.

Parser: A device or routine which assigns a syntactic structure to a string of words.

Participatory design: A variety of usability in which the tester role plays as a user and works through various tasks. Also "PD."

Partonym: See Metonym.

Path: The "route" a user takes "through" a voice interaction. The term exploits a linear metaphor, which suits the serial nature of speech, but it can also be misleading if designers start to think of *the* path through an interaction. Good, habitable design frequently requires providing for many paths through the interaction.

PD: See Participatory design.

PDA: Personal Digital Assistant. A handheld computer for managing personal and business information, such as contacts and calendars, often with telephone, Internet, and other networking features.

Perplexity: A word-based measure of the branching factor in parsing (i.e., the number of words which can be chosen at a given juncture in the vocabulary); or, more generally, too many competing word choices. For instance, given the word "ice," any number of words might come next ("cube," "age," "rink," "cream," "fishing," . . .), which gives it a high perplexity measure.

Personification: The trope whereby abstractions are given human attributes for rhetorical effect. In this book, *personification* is used in a broader way that includes the sense of (and issues implicated by) the term *anthropomorphism*.

Phatic communion: "A mere exchange of words," as Branislaw Malinowski defined it (1923: 315), whose primary function is to effect "a tie of social sentiment or other;" utterances whose main job is to link people socially, the verbal equivalent of a wave or handshake.

Phone: 1. A speech sound. 2. Short form of *telephone*.

Phone-filtering: See Filtering through the phone.

Phoneme: A set of speech sounds (phones) regarded by speakers of a language as the "same sound" — in the way, for instance, that Granny Smiths, Spartans, and so on are regarded as the "same fruit," namely, an apple.

Phonetics: The study of speech sounds.

Phonology: The study of speech sounds as they are used by specific languages.

Plosive: See Stop.

POS: Part of Speech (i.e., noun, verb, preposition, and so on). This acromymn is popular among lexicographers and corpus linguists. POS tags, for instance, are tags embedded in corpora after words to identify their part of speech.

POTS network: Plain Old Telephone Service. A voice-oriented, connection-driven, narrowband network whose principal function is to transmit 3 kHz voice signals. Such networks can handle data, but not efficiently (due to bandwidth).

Pragmatic ellipsis: See Ellipsis.

Pragmatics: The study of language (in philosophy and linguistics), especially meaning and function in language, with specific reference to context (as opposed to semantics, in particular, which studies meaning abstracted from context).

Pragmatics, conversational: See Conversational maxims.

Predesign studies: See Natural dialogue studies.

Presentation: See Contribution model.

Prime: See Priming.

Priming: Like a pump or an internal combustion engine, a mind can be made more ready for input, and will more readily generate output, by getting a little dose of the critical substance; in this case, relevant language. A person who hears the word *number* is more prepared to hear and process some numbers; a person who hears the word *hooligan* is more likely to use it within the ensuing discourse than otherwise. This phenomenon has important consequences for voice interaction design, as it suggests that one can plant, and then listen, for specific words (see Boyce et al., 1996). Within the context of speech systems, the term *lexical entrainment* is often used for the strategy of priming speakers to respond with the system's preferred terms. See also Convergence.

Progressive assistance: Providing more and more detailed direction to the caller, when communication failures occur; a type of Expansion. The first failure might trigger just "Sorry, I missed that," the next "Please speak more briefly," and so on. Also known as "Escalating detail." See also Expansion.

Prompt: A system utterance that guides user input. It may be explicit ("For weather, say *weather*"), implicit ("Which type of restaurant would you like?"), inferential ("I can answer questions about Strindberg, the Royal Institute of Technology, and Stockholm"), or open ("How may I help you?"). In some usages (though not in this book), *prompt* refers to all system utterances. See also System utterance.

Prosody: The rhythm of continuous speech (questions, for instance, have a different prosody from assertions, because the pitch rises expectantly at the end). Also, "intonation", "suprasegmental phonology."

Prototype: 1. An early instantiation of a design (usually mocked up in some way, sometimes just on paper, usually on machinery); an important element of the design process, since it allows for testing to begin early in the development cycle. 2. A sort of speech-sample donor, someone who talks to a system, which then merges and averages the prototype samples to arrive at recognition values.

Quality, maxims of: The conversational maxims that call for cooperative conversants to use utterances which are true and accurate. See also Conversational maxims.

Quantity, maxims of: The conversational maxims that call for cooperative conversants to use utterances with the contextually appropriate quantity of information — for instance, using professional jargon where the user knows it, but using fuller ordinary-language expressions when the user doesn't know the jargon. See also Conversational maxims.

Q-zone: Quiet-zone. An electronically enforced area of nonactivity for cell phones, enabled by Bluetooth technology.

Random access: Not truly random access, of course, just nonlinear information access (something other than the neighboring items in a list): absolutely crucial in navigating an aural information space.

Reactive pressure: See Dialogic pair.

Reach envelope: The physical extent of the interface; in voice design, reach envelope concerns the physical interaction devices (headset, hand-held, tied to a hardware configuration, microphone-and-speaker, and so on).

Recognition error: Misrecognition of user input. See also Deletion, Insertion, Substitution, Split, and Fusion (subcategories of recognition errors).

Recognition failure: See Recognition error.

Recognition, false: See False acceptance.

Reflexivity: One of Charles F. Hockett's "design features" of natural language — a characteristic that helps differentiate it from other natural forms of communication: the ability of humans to use language to refer to language. Designers need to capitalize on this feature, to plan out the interface language, and anticipate the users' language. And speech systems need to encode reflexivity, to help them clarify, query, contextualize, and negotiate.

Regiolect: See Dialect.

Register: Language use within a specific domain, characterized by specialized terms and structures, and analytically comprised of Field, Tenor, and Mode. See also Sublanguage.

Rejection: When input cannot be deciphered and the system can only reject it. It also makes sense to talk about a class of rejection slippages, since time-outs and spoke-too-soons pattern much like classic rejections in terms of system response.

Relevance, maxim of: The conversational maxim that calls for cooperative conversants to use utterances which are suited to the topic, context, and purpose of the exchange. See also Conversational maxims.

Repair: Correcting problems that result from errors or slippages; for instance, soliciting another token from the user. See also Recognition error, Interaction slippage, Task slippage.

Reprompt: Generic term for asking the caller to repeat herself after a rejection. See System utterance.

Reserve synonym: A synonym not stored in the primary interaction vocabulary, but held in reserve and swapped into the dialogue manager for repair.

Response: There is a systematic ambiguity with this word in voice-interface design; it can have the following meanings: 1. The second utterance of a dialogic pair; for instance, an answer, which responds to the "reactive pressure" of a preceding question (an initiative). See Dialogic pair. 2. The specific second utterance of a summons/response dialogic pair; in telephone exchanges, the summons is usually the ringing of the phone, and the response manifests as a greeting and/or identification.

Rule-based synthesis: Speech synthesis from scratch; also called *parametric coding* and *waveform synthesis*. See also Speech synthesis.

Salience: For a system word or phrase, the property of standing out from the crowd, usually through some form of physical alteration (increased volume or duration).

SALT: Speech Application Language Tags, an extension of HTML and other markup languages (cHTML, XHTML, WML, etc.) for adding a voice interface to web sites, for both voice-only (e.g., telephone) and multimodal browsing.

Scaffolding prompts: Tapering in a specific exchange — as Kotelly puts it, prompts "in a series become progressively shorter based on the knowledge callers have gained as they go along" (Kotelly, 2003: 167). See also Tapering.

Scenario: A hypothetical or generic situation. Scenarios are important throughout the design and development of voice interfaces — primarily in the invention stage, where designers attempt to forecast possible interaction patterns; and in Wizard-of-Oz tests and other usability methods, where testers evaluate the design in situation-based encounters with representative users. See also Wizard of Oz.

Schema: A "conventional activity" involved in the performance of a task, specifically in terms of voice-interface interactive structure (Rudnicky, 1999). For instance, the task of booking a flight involves schemas such as "establishing the departure time," "establishing the destination," and so on.

Schemata: The irregular plural of *schema* (this book uses the regularized plural, *schemas*).

Script: 1. The instantiated design of a voice interface, derived from drama (in terms of dialogue) and artificial intelligence (in terms of "episodic structure"). See also Call flow.

2. The system utterances, written down for voice talent to record.

SDS: Spoken Dialogue System. See Voice interface.

Semantics: The study of linguistic meaning in the narrow sense of word definitions, truth conditions, and logical syntax — though this word is often generalized to a broader sense of meaning (including, for instance, pragmatic aspects of meaning).

Semasiology: From lexicography, the study of the formal properties of words — spelling and pronunciation. It involves such notions as homonymy. This book primarily uses the ordinary-language term, *formal*, for semasiological issues. See also the complementary term in lexicography, Onomasiology.

SGML: Standard Generalized Markup Language, a formatting language for tagging document elements. Its greatest value is in managing textual and graphic data that is either subject to frequent revision or needs to be published in different formats. See also HTML and XML.

Sheep: Telephony term for a person whose utterances a speech recognizer finds very easy to recognize.

Simplest systematics model: In Conversation Analysis, the basic two-rule, turn-distribution model (Sacks, Schegloff, and Jefferson 1978):
1. a) If the concluding speaker selects a next speaker, then that speaker should take the next turn.
   b) If the concluding speaker doesn't make (or cue) a choice, then anybody can weigh in.
   c) If no one takes the floor, then the concluding speaker can start another turn.
2. Whatever option is taken, at the next relevant point, 1 a–c kick in again.

SLDS: Spoken Language Dialogue System. See Voice interface.

SLI: Spoken Language Interface. See Voice interface.

Slippage, interaction: See Interaction slippage.

Slippage, task: See Task slippage.

SLS: Spoken Language System (the preferred DARPA term). See Voice interface.

SLU: Spoken Language Understanding; Natural Language Understanding when the input is verbal. See also NLU.

Sociolect: See Dialect.

Sociolinguistics: The study of how social relations determine, afford, or constrain communication; and (somewhat less centrally) the way in which communicative patterns and tendencies construct, shape, and maintain social relations.

Sonorance: The musical, resonant aspect of speech sounds, characterized especially by vowels (like [a] and [o]). Most consonants are not sonorant, though there is a special class of sonorant consonants that have vowel-like properties. Compare, for instance, the hissing sound of [s] (a nonsonorant consonant) to the more musical sound of [l] (a sonorant consonant).

Sound and feel: An analogy to "look and feel" used in graphic interface design: the overall tone and structure of a voice interface, enforced by thematic consistency in soundscape and interaction style.

Soundness: The interface quality of disallowing all invalid actions. A graphic interface, for instance, can gray-out menu items. In a voice interface, soundness is impossible to achieve, and all soundness considerations must be handled by repair designs.

Soundscape: The overall use of sound in the interface, usually confined to nonspeech sounds (music, tones, samples, etc.), but more broadly including the vocal qualities of the agent(s) as well.

Speaker-dependent recognition: The recognition capacity of a system which takes speech input only from specific individual speakers, on whom the system has been "trained."

Speaker-independent recognition: The recognition capacity of a system which takes speech input from any speaker, without the need for speaker-dependent recognition training.

Speech act: An action accomplished by an utterance (for instance, the sentence "Get out of Dodge by sundown" performs the action of Threatening). See Dialogue act, the preferred term from the voice-interface perspective.

Speech synthesis: Generating speech output from a computer (particularly as distinct from playing back fully recorded human utterances). It might be wave-form synthesis (also "parametric coding" or "rule-based synthesis"), in which the speech is generated from scratch through mathematical models of speech sounds, or concatenative synthesis, in which small speech elements from human recordings are spliced together.

SpeechActs: An early conversational voice interface designed and implemented at Sun Microsystems for calendar and email functions; named for the pragmatic notion of "speech act."

Speech-enabled: See Voice.

Spiraling errors: See Error spiral.

Split: A recognition error where the input is segmented into too many constituents (e.g., "recognize speech" is heard as "wreck a nice beach.") See also Fusion.

Spoke-too-soon slippage: A communicative breakdown because the user says something when the system is not prepared to listen; that is, when the system does not support, or has temporarily disabled, turn overlap. See Slippage, Turn-overlap.

SR: Speech Recognition.

SRI: Speech Recognition Interface. See Voice interface.

SS: Speech Synthesis.

Stop: A type of consonant, in which the air is completely blocked from leaving the mouth, and then released with a small burst of energy; also called a *plosive*.

STS: Spoke Too Soon. See Spoke-too-soon.

STT: Speech To Text.

State machine: See Finite state machine.

Sublanguage: A specialized subset of a natural language, used within a particular domain, including a particular vocabulary and particular syntactic tendencies. The term is favored by computational linguists, disfavored by voice interaction designers. See Register.

Substitution error: A recognition error: when a legitimate vocabulary item is "recognized" as another legitimate vocabulary item. Also known as "False mapping", "Misrecognition." It also makes sense to talk of a substitution class of errors, since deletions and insertions pattern very much like classic substitutions in terms of system response.

Subtask: See Task.

SUI: Speech User Interface. See Voice interface.

Suitcase strategy: The design strategy of retaining rejected user utterances for later analysis, which is particularly successful with out-of-vocabulary, in-system utterances — that is, for situations where an utterance cannot be recognized in one context, but might be at another point in the interaction.

SUNDIAL: Speech UNderstanding and DIALogue.

Suprasegmental phonology: See Prosody.

SV: Speaker Verification.

SWTS: Spoke Way Too Soon.

Syllable: A concentration of sustained sonorant energy in a word or morpheme, a vowel-centered unit of speech; *vowel* has two syllables, for instance, *speech* has one.

Synonym: A word that is very similar in meaning to another word, often said to have the "same meaning" as the other word. For instance, *attorney* is a synonym for *lawyer* (and vice versa).

Synonym, reserve: See Reserve synonym.

Synonymy: The relationship holding among or between words with near-perfect referential overlap; that is, among Synonyms.

Syntactic habitability: See Habitability.

Syntax: The principles of correct word assembly: "chickens over red three fell" is ill-formed, "three red chickens fell over" is well-formed, according to English syntactic principles.

Synthesis, speech: See Speech synthesis.

System-directed: See Initiative.

System-directive acts: Obligative dialogue acts which (attempt to) induce specific behaviors from the system — providing help, pausing, resuming, and the like.

System utterance: Any speech output from a voice interface. Also called, in some usages, Prompt. See also Utterance.

System vocabulary: See Vocabulary.

Talk ahead: A design feature that allows (experienced) callers to trigger actions that have not yet been presented. For instance, the caller might start a credit-card dialogue by saying "Mastercard" (at an appropriate juncture) even before the system prompts for credit card information.

Talk-over: See Turn overlap.

Talk-through: See Turn overlap.

Tapering: A design strategy which reduces the length of subsequent system utterances based on context and dialogue history. See also Ellipsis, Scaffolding utterances.

Task: Those user activities that constitute goal-directed interactive behavior; for instance, booking a flight or purchasing a book. This book does not distinguish between *subtask* and *task*, except to the extent that a subtask must be a component of some larger, goal-directed behavior. But they are both tasks. With voice systems, tasks may be as brief as one turn, or as long as an entire dialogue.

Task-management acts: The category of dialogue acts concerned with accomplishing goals (as opposed to dialogue-management acts, which are concerned with coordinating and grounding the dialogue).

Task initiative: See Initiative.

Task slippage: Problems that arise not because of the recognition engine (errors) or the interface design (interaction slippages), but because of specific functionality issues related to the task. For instance, the user of a voiced banking system might request information for which she doesn't have authorization (someone else's loan balance). Also "competence errors" (Wolf and Zadrozny, 1998).

Telematics: The field of vehicle-based information systems, representing the convergence of four technologies: the automobile, computing, wireless communications, and the Global Positioning System.

Tenor: An aspect of Register, *tenor* concerns the agents in an interaction, but not strictly as individuals (Tom or Sally or SpeechActs), though that plays a part. Rather, tenor focuses on the role of participants in a communicative event — as information seekers and information providers, as shoppers and sellers, as travelers and travel agents. See also Register, Field, Mode.

Textual colligation: See Colligation, textual.

Timeout: A predetermined period of (caller) silence that triggers some system behavior.

Token, lexical: Lexical token and the closely related notion, lexical type, mark an important distinction in thinking about word usage and vocabulary space (mental or mechanical). A type is a word as we commonly conceptualize it: a sound sequence paired with a meaning. A token is a given instance of a type. So, in the string, "Run, Lola, Run," there are three tokens but only two types (*run* and *Lola*). When your word processor counts words, it counts every word-token (multiple occurrences of the same word type qualifies as multiple words). But counting words in terms of vocabulary space means counting word types (multiple occurrences of the same word type qualifies as one word).

Token, word: See Token, lexical.

Touch-tone interface: See Keypad interface.

Training: The act of speaking to a speech-recognition system so that it becomes familiar with the cadences and idiosyncrasies of that speaker. Also "enrollment."

Transparent: The interface property of being "invisible," unnoticed, so that the user's awareness is fully on the task and not on the medium at all.

TTS: Text To Speech. The process whereby written text is turned into (synthetic) speech. Also TtS.

TUI: Touch-tone User Interface; see Keypad interface.

Turn distribution: The way in which speaking agents share the Floor. See Floor, Simplest systematics.

Turn: The utterances from one speaker to the next when the previous speaker has ended and the next speaker begins (with conversational beginnings being a special case with no previous speaker, and endings being a special case with no next speaker).

Turn overlap: Also Overlap. When one agent in a dialogue begins speaking before the other has finished. For most voice interfaces, this feature is highly desirable, letting the caller control the flow of the interaction by interrupting the system. The most common term for this feature is *Barge-in*. Other terms include *Talk-over*, *Talk-through*, *Echo cancellation*, and *Cut-through*.

Type, lexical: See Token, lexical.

Type, word: See Token, lexical.

UCD: User-Centered Design.

Under Verification: See Verification.

USI: Universal speech interface, a standardized speech-interface style, with established primitives and interaction patterns, still in the very provisional stages. Cf. Rosenfeld et al., 2000.

Utterance: The natural unit of interactive speech, but a loosely defined and highly dependent notion. It might be a word, a phrase, a sentence, a small cluster of sentences, even a single morpheme or a grunt — whatever the conversants regard as a unit.

Utterance, System: See System utterance.

VAD: Voice-Activated Dialing. See Voice dial.

Variable initiative: A synonym of Mixed initiative. See Initiative.

Verification: The act of ensuring the system has the right values for some data-slot (numbers for a credit card). Underverification is the situation where verifications are necessary but not executed; oververification is where no verification is required, but it is executed (Smith and Gordon, 1997).

Vocable: A word form, independent of its meaning and function (*truck*, for instance, just as a shell which has some unspecified function and meaning, in a sentence like "*Truck* has five letters"). Often used of words like "oh" and "ah" and "er," which are (erroneously) thought not to have meaning or function.

Vocabulary: The word-hoard; all the words known to an individual or a system, or a linguistic domain. People have vocabularies. Speech recognition systems have vocabularies. English has a vocabulary. Air-traffic control discourse has a vocabulary. This book, or this page, or this entry, can even be said to have a vocabulary.

This book uses *vocabulary* in a more specific, slightly technical sense than it usually has in speech-system development, using it to designate the body of acoustic models the speech recognition engine employs. Vocabularies in this sense might be swappable, in that different collections of acoustic models can be available at different points in the interaction. That is, there may be a system vocabulary (the complete collection of available models) and any number of working vocabularies (the collection available at any given point in the interaction). The traditional word in speech-system development for this module is *grammar*, but that usage distorts both the technical linguistic and the ordinary-language sense of *grammar* so dramatically that it should be abandoned. (*Dictionary* is also sometimes used this way in speech-system development, a usage avoided in this book for similar reasons.)

Vocabulary Problem: The relationship between the number of acoustic patterns a speech system can store and how quickly it can match any one of those patterns to the input: an increase in vocabulary size means an increased demand for processing. (Cf. Furnas et al., 1983, 1987; Brennan, 1998.)

Vocabulary, system: See Vocabulary.

Vocabulary, working: See Vocabulary.

Vocoder: A blend of *voice* + *coder*, it is a device for modifying speech signals, particularly helpful in Wizard of Oz tests for making a human sound machine-like.

Voice: As a verb, the process of giving a voice interface to an artifact (pre-existent or new), or of giving a nontrivial voice interactive modality to such an artifact, as in "SunMicrosystems was one of the earliest companies to voice their email program," or "MIT has been voicing applications for over a decade." Also "speech-enable", "voice-enable."

Voice dial: To dial numbers by saying words ("Call Elroy Jetson").

Voice-enable: See Voice.

Voice interface: The subject of this book: a set of voice-based interaction processes for humans to perform tasks with a computer. A voice interface for some application, X, is effectively all the human-factor concerns associated with the way speech is used to interact with X. While there are sometimes subtle differences among some of these terms, for some researchers, the following words are effectively synonyms of voice interface: *Audio Interface, Conversational User Interface* (*CUI*), *Interactive Voice Response* (*IVR*) *System, Natural Language Interface* (*NLI*), *Spoken Dialogue System* (*SDS*), *Spoken Language Dialogue System* (*SLDS*), *Spoken Language Interface* (*SLI*), *Spoken Language System* (*SLS*), *Speech Recognition Interface* (*SRI*), *Speech User Interface* (*SUI*), *Voice Response Unit* (*VRU*), *Voice User Interface*

(*VUI*). In this book, the default for *Voice interaction* is a system modeled on human–human conversation, and the term *Voice-response system* is used for discrete-command speech applications.

Voice-response system: The term adopted in this book (adapted from VRU) for phone-based speech applications built on a discrete command paradigm, not a conversational paradigm. See also Voice interface.

VoiceXML: A voice-enabling markup language based on XML, providing for speech input and output. Designed largely to support voice browsing of the Web, it (and other XML-based meta-languages) also has considerable value for rapid prototyping and deploying a range of voice applications. Developed and promoted by the VoiceXML Forum, sponsored by AT&T, IBM, Lucent Technologies, and Motorola. Also "VXML."

Vowel: A class of speech sounds characterized by sustained sonorance, which holds syllables together — in contrast to the interruptive sounds or shorter sonorance of consonants.

VoxML: A voice-enabling markup language developed and promoted by Motorola, based on XML. See also VoiceXML.

VXML: See VoiceXML.

VRU: Voice-Response Unit. See Voice interface.

VUI: Voice User Interface. See Voice interface.

Wave-form synthesis: Speech synthesis from scratch; also called *rule-based synthesis* and *parametric coding*. See also Speech synthesis.

WAP: Wireless Application Protocol. A secure protocol for accessing information through portable digital appliances — mobile phones, PDAs, and the like; it was developed especially to provide for Internet content on such devices, accommodating HTML and XML, and optimized for a tailored language, WML (Wireless Markup Language).

WER: Word Error Rate; a measure of speech recognition success.

Wi-Fi: Wireless Fidelity, used generically of any 802.11 network. See 802.11.

WIMP: Windows, Icons, Mouse, and Pointer (also "Windows, Icons, Menus, Pointer" and Windows, Icons, Mouse, Pull-down menus"): effectively, a graphic interface. The term invariably refers to a human–computer interaction style that includes more than the named elements (buttons, fields, keyboards, . . .). Originally coined by command-line aficionados to denigrate graphic interfaces, it has been adopted by interface designers.

Wizard: See Wizard of Oz.

Wizard of Oz: A methodology for interface research and for the design and testing of specific interfaces. Its defining characteristic is that a human (the Wizard) pretends to be a machine. While applicable to any domain in principle (with appropriate technological aids), it has come to be associated strongly with voice interaction design because of its widespread use in that domain as a research and usability instrument.

WML: Wireless Markup Language.

Word type: See Token, lexical.

Word, content: See Content words.

Word, function: See Function words.

Wordspotting: A recognition/understanding strategy of looking for words and phrases in an acoustic signal (rather than trying to recognize the entire signal), and reconstituting the intended meaning of the entire signal on that basis. See also Keyword.

Working vocabulary: See Vocabulary.

WOZ, WOz: Wizard of Oz; a methodology for the design and evaluation of interfaces; see Wizard of Oz.

XML: Extensible Markup Language, a subset of SGML adapted especially for web documents, developed by the W3C. Its principal virtues are that it allows functional tags on information (e.g., labeling the title or author or publisher) rather than HTML's formatting tags (e.g., labeling italic or bold or center-aligned text), and customizability (so that a digital-camera site designer could define the tag "pixel size" and "media").

Zero-out: The act of getting out (bailing out) of a voice system in order to talk to a human (usually by keying a zero). See also Bailout.

Bibliography

Adams, D. 1979. *The Hitchhiker's Guide to the Galaxy*. New York: Random House.

Aldrich, T. B. 1903. "Pongapong papers."

Allen, J., Ferguson, G., and Stent, A. 2001. "An architecture for more realistic conversational systems." Paper read at Intelligent User Interfaces (IUI-01), 22:27–38 at Santa Fe, NM.

Allen, J. F., Byron, D. K., Dzikovska, M., Ferguson, G., Galescu, L., and Stent, A. 2001. "Towards conversational human-computer interaction." *AI Magazine*.

Allen, J. F., Miller, B. W., Ringger, E. K., and Sikorski, T. 1996. "A robust system for natural spoken dialogue." Paper read at Annual Meeting of the Association for Computational Linguistics, June 1996.

Allwood, J. 1994. "Obligations and options in dialogue." *Think* 3:9–18.

Andrews, R., Biggs, M., and Seidel, M., eds. 1996. *The Columbia World of Quotations*. New York: Columbia University Press.

Ardissono, L., Boella, G., and Damiano, R. 1998. "A plan-based model of misunderstandings in cooperative dialogue." *International Journal of Human-Computer Studies* 48:649–679.

Aristotle. 1984. *The Complete Works*, edited by J. Barnes. 2 volumes. Princeton: Princeton University Press.

Arons, B. 1991. "Hyperspeech: Navigating in speech-only hypermedia." Paper read at Hypertext, 91.

Atkinson, J., and Drew, P. 1979. *Order in Court: The Organisation of Verbal Interaction in Judicial Settings*. London: Macmillan.

Atkinson, J. M. 1990. "Low-budget usability testing." *STC Intercom*, 4, February, 1990, 1.

Attwater, D. J., Edgington, M., Durston, P., and Whittaker, S. 2000. "Practical issues in the application of speech technology to network and customer service applications." *Speech Communication* 31:279–291.

Bailey, R. W. 1996. *Human Performance Engineering: Designing High Quality Professional User Interfaces for Computer Products, Applications and Systems*. Upper Saddle River, NJ: Prentice Hall.

———. 2000. "Human interaction speeds." *UI Design Update Newsletter* (August 2000). http://www.humanfactors.com/downloads/aug00.asp.

————. 1999. "Speed reading." *UI Design Update Newsletter* (February 1999). http://www.humanfactors.com/downloads/feb99.asp.

Bakhtin, M. M. 1986. *Speech Genres and Other Late Essays*. Translated by V. McGee. Edited by C. Emerson and M. Holquist. Austin: University of Texas Press.

Balentine, B. 1999. "Re-engineering the speech menu." In *Human Factors and Voice Interactive Systems*, edited by D. Gardner-Bonneau. Cambridge, MA: Kluwer Academic Publishers.

Balentine, B., and Morgan, D. P. 1999. *How to Build a Speech Recognition Application: A Style Guide for Telephony Dialogues*. San Ramon, CA: Enterprise Integration Group, Inc.

Balkanski, C., and Hurault-Plantet. 2000. "Cooperative requests and replies in a collaborative dialogue model." *International Journal of Human-Computer Studies* 53:915–968.

Ball, G., and Breese, J. 2000. "Emotion and personality in a conversational agent." In *Embodied Conversational Agents*, edited by J. Cassell, J. Sullivan, S. Prevost, and E. Churchill. Cambridge: The MIT Press.

Ball, L. J., and Ormerod, T. C. 2000. "Applying ethnography in the analysis and support of expertise in engineering design." *Design Studies* 21 (4):403–421.

Bannon, L. J., and Bødker, S. 1991. "Beyond the interface: Encountering artifacts in use." In *Designing Interaction: Psychology at the Human-Computer Interface*, edited by J. Carroll. New York: Cambridge University Press.

Barnett, J., and Singh, M. 1996. "Designing a portable spoken dialogue system." In *Dialogue Processing in Spoken Language Systems, ECAI '96 Workshop. Lecture Notes in Computer Science* 1236 (1997):156–70

Barry, W. A. 1970. "Marriage research and conflict: An integrative review." *Psychological Bulletin* 73:41–54.

Baum, L. 1900. *The Wonderful Wizard of Oz*. Chicago: George M. Hill.

————. 1904. *The Marvelous Land of Oz*. Chicago: George M. Hill.

Bell, L., and Gustafson, J. 1999. "Utterance types in the August database." Paper read at Third Swedish Symposium on Multimodal Communication, October 15–16, at Lingköpings Universitet.

Bénjoint, H. 1983. "On field work in lexicography." In *Lexicography: Principles and Practice*, edited by R. R. K. Hartmann. New York: Academic Press.

Bernsen, N. O. 2000. "What is natural interactivity?" Paper read at Workshop from Spoken Dialogue to Full Natural Interactive Dialogue. Theory, Empirical Analysis and Evaluation, May 29, 2000, at Athens, Greece.

Bernsen, N. O., Dybkjær, H., and Dybkjær, L. 1997. "Elements of speech interaction." Paper read at Third Spoken Language Dialogue and Discourse Workshop, September 1997, at Vienna.

Bernsen, N. O., and Dybkjær, L. 2000a. "A methodology for evaluating spoken-language dialogue systems and their components." Paper read at 2nd International Conference on Language Resources and Evaluation (LREC '2000), 31.5.-2.6.2000, at Athens, Greece.

————. 2000b. "Is that a good spoken language dialogue system?" Paper read at Workshop Using Evaluation within HLT Programs: Results and Trends," 31.5.-2.6.2000, at Athens, Greece.

Bernsen, N. O., and Dybkjaer, L. 2001. "Combining multi-party speech and text exchanges over the Internet." Paper read at Eurospeech 2001.

Bernsen, N. O., Dybkjaer, L., and Dybkjaer, H. 1998. *Designing Interactive Speech Systems: From First Ideas to User Testing*. London: Springer-Verlag.

Bias, R. G., and Mayhew, D. J., eds. 1994. *Cost Justifying Usability*. West Tisbury, MA: Deborah J. Mayhew and Associates.

Bickmore, T., and Cassell, J. 2000. "How about the weather?: Social Dialogue with embodied Conversational agents." *Proceedings of the AAAI all Symposium on Socially Intelligent Agents*. North Falmouth, MA.

Bilange. 1991. *A Task-Independant Oral Dialogue Model*. Paris: Cap Gemini Innovation.

Bird, S., and Harrington, J. 2001. "Editorial: Speech annotation and corpus tools." *Speech Communication* 33:1–4.

Blakemore, D. 1992. *Understanding Utterances*. Oxford: Blackwell.

Bloomfield, L. 1933. *Language*. New York: Holt, Rinehart & Winston.

Boehm, B. 2000. *Spiral Development: Experience, Principles, and Refinements* (Special Report CMU/SEI-00-SR-08, ESC-SR-00–08), June 2000. Available from http://www.sei.cmu.edu/cbs/spiral2000/february2000/BoehmSR.html.

Bonvillain, N. 1997. *Women and Men: Cultural Constructs of Gender*. Upper Saddle River, NJ: Prentice Hall.

Borruso, S. 1995. "Voice of the revolution." *Wired* 3. http://www.wired.com/wired/archives/3.09/voices.html.

Boyce, S. 1999. "Spoken natural language dialogue systems: User interface issues for the future." In *Human Factors and Voice Interactive Systems*, edited by D. Gardner-Bonneau. Cambridge, MA: Kluwer Academic Publishers.

———. 2000. "Natural spoken dialogue systems for telephony applications." *Communications of the ACM* 43 (9):29–34.

Brennan, S. E. 1998a. "The grounding problem in conversations with and through computers." In *Social and Cognitive Psychological Approaches to Interpersonal Communication*, edited by S. R. Fussell and R. J. Kreuz. Mahwah, NJ: Lawrence Erlbaum Associates.

———. 1998b. "The vocabulary problem in spoken dialogue systems." http://www.psychology.sunysb.edu/sbrennan-/papers/luperfoy.pdf.

Brennan, S. E. 2000. "Processes that shape conversation and their implications for computational linguistics." Paper read at 38th Annual Meeting of the Association for Computational Linguistics, at Hong Kong.

Brennan, S. E., and Hulteen, E. A. 1994. "Interaction and feedback in a spoken language system: A theoretical framework." *Knowledge-based Systems* 8 (2–3):143–150.

Brennan, S. E., and Ohaeri, J. O. 1999. "Why do electronic conversations seem less polite? The costs and benefits of hedging." Paper read at International Joint Conference on Work Activities, Coordination, and Collaboration, at San Francisco, CA.

Bretan, I., Ereback, A. L., MacDermid, C., and Waern, A. 1995. "Simulation-based dialogue design for speech-controlled telephone services." CHI '95 Proceedings. Denver, Colorado.

Brown, G., and Yule, G. 1983. *Discourse Analysis*. Cambridge: Cambridge University Press.

Brown, N. R., and Vorsburh, A. M. 1989. "Evaluating the accuracy of a large vocabulary speech recognition system." Paper read at The Human Factors Society 33rd Annual Meeting. Denver, CO.

Brown, T., Simard, C., and Takhteyev, Y. *Mixing TTS and recorded speech* [website]. Stanford University 2000 [cited date. Available from http://www.stanford.edu/nass/comm369/MixingTTSandRecorded.ppt.]

Bunt, H. C. 1991. "Dynamic interpretation and dialogue theory." In *Papers from the Second Symposium on Logic and Language*, edited by L. Kalman and L. Polos. Budapest: Akademiai Kiado.

———. 1995. "Semantics and pragmatics in the ELTA dialogue project." Paper read at ELSNET Workshop on Dialogue and Discourse, April 1995, at Dublin.

Bunt, H. C., and Black, B. 2000. "The ABC of Computational Pragmatics." In *Abduction, Belief and Context in Dialogue*. Edited by H. Bunt and W. Black. Philadelphia, PA: John Benjamins, 1–47.

Burnard, L. 1995. "Using SGML for linguistic analysis: The case of the BNC." *Markup Languages* 1 (2):31–51.

———. *The Text Encoding Initiative's recommendations for the Encoding of Language Corpora: Theory and Practice*. 1997. http://users.ox.ac.uk/~lou/wip/Soria/.]

Butler, S. 1872. *Erewhon*. London: Trubner.

Button, G., and Sharrock, W. 1995. "On simulacrums of conversation: Towards a clarification of the relevance of conversation analysis for human-computer interaction." In *The Social and Interactional Dimensions of Human-Computer Interfaces*. Cambridge: Cambridge University Press.

Buxton, W. 1995. "Speech, language and audition." In *Readings in Human-Computer Interaction: Toward the Year 2000*, edited by R. M. Baecker, J. Grudin, W. Buxton, and S. Greenberg. San Francisco: Morgan Kaufmann Publishers.

Byrne, B. 2001. "Turning GUIs into VUIs: Dialog design principles for making web applications accessible by telephone." *Voice XML Review*, June 2001.

Byrne, J. H. 1994. *Mrs. Byrne's Dictionary of Unusual, Obscure, and Preposterous Words*. New York, NY: Birch Lane.

C. & G. Merriam Company. 1913. *Webster's Revised Unabridged Dictionary*: Springfield, MA: Mirriam-Webster.

Cahn, J. 1990. "The generation of affect in synthesized speech." *Journal of the American Voice Input/Output Society* 8 (1990):1–19.

Cahn, J. E., and Brennan, S. E. 1999. "A psychological model of grounding and repair in dialog." Paper read at AAAI Fall Symposium on Psychological Models of Communication in Collaborative Systems, at North Falmouth, MA.

Calisher, H. 1964. *Extreme Magic*. Boston: Little, Brown.

Carbonell, J., and Collins, A. 1970. *Papers A/B*. Cambridge, MA: Bolt, Beranek and Newman.

Carroll, J., ed. 1991. *Designing Interaction: Psychology at the Human-Computer Interface*. New York: Cambridge University Press.

Carroll, L. 1871. *Alice Through the Looking Glass*. New York: Bloomsbury Publishing.

———. 1962. *Alice's Adventures in Wonderland;* and, *Through the Looking Glass*: Harmondsworth, Puffin.

Carter, R. 1999. *The rise of the customer* [website]. The Brisbane Institute. http://www.brisinst.org.au/papers/carter_rob_customer/

Cassell, J., Sullivan, J., Prevost, S., and Churchill, E. 2000. *Embodied Conversational Agents*. Cambridge, MA: The MIT Press.

Cassell, J., Bickmore, T., Campbell, L., Vilhjálmsson, H., and Yan, H. 2000. "Human conversation as a system framework: Designing embodied conversational agents." In *Embodied Conversational Agents*, edited by J. Cassell, J. Sullivan, S. Prevost, and E. Churchill. Cambridge, MA: The MIT Press.

Cassell, J. 2000. "Nudge nudge wink wink: Elements of face-to-face conversation for embodied conversational agents." In *Embodied Converational Agents*, edited by J. Cassell, J. Sullivan, S. Prevost, and E. Churchill. Cambridge, MA: The MIT Press.

Cassell, J., Stocky, T., Bickmore, T., Gao, Y., Nakano, Y., Ryokai, K., Tversky, D., Vaucelle, C., and Vilhjálmsson, H. 2002. "MACK: Media lab autonomous conversational kiosk." Paper read at IMAGINA '02, January 12–15, 2002, at Monte Carlo.

Cavazza, M. O. 2003. "Conversational characters: 'Talking heads' as the user interface of the future." *BCS Review*. http://www.bes.org/review03/articles/multimedia/cavazza.htm.

Chafe, W. 1994. *Discourse, Consciousness, and Time: The Flow and Displacement of Conscious Experience in Speaking and Writing*. Chicago: University of Chicago Press.

Chu-Carroll, J. 1996. "A plan-based model for response generation in collaborative consultation dialogs." Ph.D. Thesis, University of Delaware, Washington, DC.

Chu-Carroll, J., and Brown, M. K. 1998. "An evidential model for tracking initiative in collaborative dialogue interactions." *User Modeling and User-adapted Interaction* 8 (3–4):215–253.

Church, K. W., and Hanks, P. 1990. "Word association norms mutual information and lexicography." *Computational Linguistics* 16 (1).

Churcher, G. E., Atwell, E. S., and Souter, C. 1997. *Dialogue Management Systems: A Survey and Overview*. Report 97.6. University of Leeds; School of Computer Studies.

Churchill, E. F., Cook, L., Hodgson, P., Prevost, S., and Sullivan, J. W. 2000. " 'May I help you?' Designing embodied conversational agent allies." In *Embodied Conversational Agents*, edited by J. Cassell, J. Sullivan, S. Prevost, and E. Churchill. Cambridge: The MIT Press.

Clark, H. 1992. *Arenas of Language Use*. Chicago: University of Chicago Press.

Clark, H., and Brennan, S. E. 1991. "Grounding in communication." In *Perspectives on Socially Shared Cognition*, edited by L. B. Resnick, J. Levine, and S. D. Teasley. Washington, DC: American Psychological Association.

Clark, H., and Haviland, S. E. 1977. "Comprehension and the Given-new Contract." In *Discourse Production and Comprehension*, edited by R. Freedle. Mahwah, NJ: Lawrence Erlbaum.

Clark, H., and Schaefer, E. 1987. "Concealing one's meaning from overhearers." *Journal of Memory and Language* 26:209–225.

———. 1989. "Contributing to discourse." *Cognitive Science* 13:259–294.

Clark, H., and Wilkes-Gibbs, D. 1986. "Referring as a collaborative process." *Cognition* 22:1–39.

Coggle, P. 1993. *Do You Speak Estuary*. London: Bloomsbury.

Cohen, M. H., Giangola, J. P., and Balogh, J. 2004. *Voice User Interface Design*. Boston: Addison-Wesley Professional.

Cohen, P. R., Morgan, J., and Pollack, M. E., eds. 1990. *Intentions in Communication*. Cambridge, MA: The MIT Press.

Colby, K. M. 1999. "Human-computer conversation in a cognitive therapy program." In *Machine Conversations*, edited by Y. Wilks. Dordrecht: Kluwer.

Cooper, A. 1995. *About Face: The Essentials of User Interface Design*. Foster City, CA: IDG Books Worldwide.

———. 1999. *The Inmates Are Running the Asylum: Why High-tech Products Drive Us Crazy and How to Restore the Sanity*. Indianapolis, IN: Sams.

Cooper, J. W., Viswanathan, M., Byron, D., and Chan, M. 2001. "Building searchable collections of enterprise speech data." *Proceedings of the First ACM/IEEE-CS Joint Conference on Digital Libraries*. Roanoke, VA.

Coulthard. 1985. *An Introduction to Discourse Analysis*. 2nd Edition. Boston, MA: Addison-Wesley.

Crowley, S., and Hawhee, D. 1999. *Ancient Rhetorics for Contemporary Students*. Toronto: Allyn and Bacon.

Dahlbäck, N., Swamy, S., Nass, C., Arvidsson, F., and Skågeby, J. 2001. "Spoken interaction with computers in a native or non-native language — Same or different?" Proceedings of INTERACT 2001. Tokyo, Japan.

Danieli, M., and Gerbino, E. 1996. "Metrics for evaluating dialogue strategies in a spoken language system." Torino, Italy: CSELT — Centro Studi e Laboratori Telecomunicazioni.

Day, M. C., and Boyce, S. 1993. "Human factors in human-computer system design." *Advances in Computers* 36 (1993):333–430.

De Sola Pool, I. 1977. *The Social Impact of the Telephone.* Cambridge, MA: The MIT Press.

Delomier, D., Meunier, A., and Morel, M. A. 1989. "Linguistic features of human machine oral dialog." Paper read at Eurospeech '89, at Paris.

Dennett, D. 1997. *Kinds of Minds: Towards an Understanding of Consciousness.* Boulder, CO: Perseus Publishing.

Derriks, B., and Willem, D. 1998. "Negative feedback in information dialogues: Identification, classification and problem-solving procedures." *International Journal of Human-Computer Studies* 48:577–604.

Dirven, R., and Verspoor, M. 1998. *Cognitive Exploration of Language and Linguistics.* Philadelphia, PA: John Benjamins.

Doney, P., and Cannon, J. 1997. "An examination of the nature of trust in buyer-seller relationships." *Journal of Marketing* 61:35–51.

Dray, S., and Mrazek, D. 1996. "A day in the life of a family: An international ethnographic study." In *Field Methods Casebook for Software Design*, edited by D. Wixon and J. Ramey. New York: John Wiley & Sons.

Drexler, K. E. 1990. *Engines of Creation.* Garden City, NJ: Anchor Press.

Drexler, K. E., Peterson, C., and Pergamit, G. 1991. *Unbounding the Future: The Nanotechnology Revolution.* New York, NY: Quill Books.

Dumas, J. S., and Redish, J. C. 1993. *A Practical Guide to Usability Testing.* Westport, CT: Ablex.

Dunbar, R. 1997. *Grooming, Gossip, and the Evolution of Language.* Cambridge, MA: Harvard University Press.

Dunne, J. 1993. *Back to the Rough Ground: Phronesis and Techne in Modern Philosophy and in Aristotle.* Notre Dame: University of Notre Dame Press.

Dvorak, J. C. 2002. "Voice recognition: Another dead end." *PC Magazine*, January 21, 2002.

Dybkjær, L., and Dybkjær, H. 1993. "Wizard-of-Oz experiments in the development of the dialogue model for P1." Copenhagen, Denmark: Spoken Language Dialogue Systems, STC Aalborg University, CCI Roskilde University, CST University of Copenhagen, Denmark.

Dybkjaer, L., Bernsen, N. O., and Dybkjaer, H. 1998. "A methodology for diagnostic evaluation of spoken human–machine dialogue." *International Journal of Human-Computer Studies* 48:605–625.

Eagley, A. H. 1983. "Gender and social influence: A social psychological analysis." *American Psychologist* 38:971–981.

Eckert, W. 1995. "Gesprochener mensch maschine dialog." Ph.D. Thesis, Universitat Erlangen, Nurnberg.

Eisenberg, A. 2000. "Sounding off: Talking computers will imitate male and female stereotypes with real and synthesized voices." *Houston Chronicle*, October 20, 2000.

Eisenberg, D. 2001. "Dial tone 2.0: The phone talks back." *Time* (19 March).

Elhadad, M. 1990a. "Constraint-based text generation." New York, NY: Dept. of Computer Science, Columbia University.

Engestrom, Y., and Middleton, D., eds. 1996. *Cognition at Work.* Cambridge: Cambridge University Press.

———, eds. 1998. *Cognition and Communication at Work.* Cambridge: Cambridge University Press.

Enkvist, N. E. 1978. "Stylistics and text linguistics." In *Current Trends in Text Linguistics*, edited by W. Dressler. Berlin: de Gruyter.

Ericsson, K. H., and Simon, H. A. 1993. *Protocol Analysis: Verbal Reports as Data*. Revised ed. Cambridge, MA: The MIT Press.

Farley, M. 2001. "User-centered design for VoiceXML application." *VoiceXML Review*, June 2001.

Firth, J. R. 1957. "A synopsis of linguistic theory, 1930–1955." In *Studies in Linguistic Analysis*. Oxford: Blackwell.

———. 1968. *Selected Papers of J. R. Firth 1952–59*, edited by F. R. Palmer. London: Longmans.

Fishman, P. 1978. "Interaction: The work women do." *Social Problems* 25:397–406.

Flammia, G. 1998. "Discourse segmentation of spoken dialogue: An empirical approach." Ph.D. Thesis. Laboratory for Computer Science. MIT.

Fodor, J. 1992. "The big idea: Can there be a science of mind?" *Times Literary Supplement*, 3rd July, 1992.

Foley, J. D., and Wallace, V. L. 1974. "The art of natural graphic man-machine conversation." *Proceedings of the IEEE* 62:462–471.

———. 1973. *Howards End*. London: Edward Arnold.

Frank, G. et al. 2002. "JUST-TALK: An application of responsive virtual human technology." Paper read at 24th Interservice/Industry Training, Simulation and Education Conference. Orlando, FL.

Fraser, N. M., and Gilbert, N. 1991. "Simulating speech systems." *Computer Speech and Language* 5:81–89.

Furnas, G., Landauer, T. K., Gomez, L. M., and Dumais, S. T. 1987. "The vocabulary problem in human-system communication: An analysis and a solution." Bell Cmmunications Research. *CACM* 30 (11):964–971.

Furnas, G. W., Landauer, T. K., Gomez, L. M., and Dumais, S. T. 1983. "Statistical semantics: Analysis of the potential performance of keyword information systems." *Bell Systems Technical Journal* 62:1753–1806.

Gardner, R. 2002. *When Listeners Talk: Response Tokens and Listener Stance*. Philadelphia: John Benjamins.

Gardner-Bonneau, D. 1999. "The future of voice interactive applications." In *Human Factors and Voice Interactive Systems*, edited by D. Gardner-Bonneau. Cambridge, MA: Kluwer Academic Publishers.

———. 1999. "Guidelines for speech-enabled IVR application design." In *Human Factors and Voice Interactive Systems*, edited by D. Gardner-Bonneau. Cambridge, MA: Kluwer Academic Publishers.

———, ed. 1999. *Human Factors and Voice Interactive Systems*. Boston: Kluwer Academic Publishers.

Gazdar, G. 1979. *Pragmatics: Implicature, Presupposition and Logical Form*. New York: Academic Press.

Geis, M. I. L. 1995. *Speech Acts and Conversational Interaction*. Cambridge: Cambridge University Press.

General Magic. *zach_sarchastic_email_demo.ram* [website]. General Magic 2002a [cited August 30, 2002. Available from http://www.generalmagic.com/demos/htmlperson.shtml; site no longer available online.]

———. *zach_sarchastic_male_self-intro.ram* [website]. General Magic 2002b [cited August 30, 2002. Available from http://www.generalmagic.com/demos/htmlperson.shtml.]

———. *jen_pure_friendly_cd_dial_demo.ram.* [website]. General Magic 2002c [cited August 30, 2002. Available from http://www.generalmagic.com/demos/htmlperson.shtml.]

Givón, T. 1990. *Syntax: A Functional-Typological Introduction.* Vol. II. Amsterdam: John Benjamins.

Glass, J., Hazen, T. J., and Hetherington, L. 1999. "Real-time telephone-based speech recognition in the JUPITER domain." Phoenix, A2: *Proceedings of the International Conference on Acoustics, Speech, and Signal Processing.*

Glass, J. R., and Hazen, T. J. 1998. "Telephone-based conversational speech recognition In the Jupiter domain." Paper read at 98, at Sydney, Australia.

Gong, L., and Nass, C. 2000. "Does adding a synthetic face always enhance speech interfaces?". Department of Communication, Stanford University, Stanford, CA.

Gong, L., Nass, C., Simard, C., and Takhteyev, Y. 2001. "When non-human is better than semi-human: Consistency in speech interfaces." In *Usability Evaluation and Interface Design: Cognitive Engineering, Intelligent Agents, and Virtual Reality*, edited by M. J. Smith, G. Salvendy, D. Harris, and R. Koubek. Mahwah, NJ: Lawrence Erlbaum.

Gorin, A. L., Riccardi, G., and Wright, J. H. 1997. "How may I help you?" *Speech Communication* 23 (1997):113–127.

Gould, J., and Lewis, C. 1985. "Designing for usability — Key principles and what designers think." *Communications of the ACM* 28 (1985):300–311.

Green, T. 2004. *Chasing HAL: In pursuit of a human-computer interface* [website]. *The MITRE* Digest, August 23, 2001 2001 [cited 2004]. Available from http://www.mitre.org/news/digest/archives/2001/chasing_hal.html.

Grice, H. P. 1957. "Meaning." *Philosophical Review* 66:377–388.

———. 1968. "Utterer's meaning, sentence meaning, and word meaning." *Foundations of Language* 4:225–242.

———. 1989. *Studies in the Way of Words.* Cambridge, MA: Harvard University Press.

Grosz, B. J., and Sidner, C. L. 1986. "Attention, intention, and the structure of discourse." *Computational Linguistics* 12 (3):175–204.

———. 1990. "Plans for discourse." In *Intentions in Communication*, edited by P. R. Cohen, J. Morgan, and M. E. Pollack. Cambridge, MA: The MIT Press.

Guyomard, M., and Siroux, J. 1987. "Experimentation in the specification of oral dialog." *NATO-ASI Recent Advances in Speech Understanding and Dialogue Systems* July 1987.

———. 2002a. "The future of cellphones is here. Sort of." *New York Times*, February 14, 2002.

———. 2002b. "A tyranny of digital controls invades the comfort of home." *New York Times*, April, 28, 2002.

Hahn, K., Olveda, S., Baesman, R., and Liu, S. *Fact and Opinion in the World of TTS and Recorded Speech.* [website]. Stanford University 2000 [cited 2004.] Available from http://www.stanford.edu/~nass/comm369/Fact_Opinion_Study.ppt.

Halliday, M. A. K. 1978. *Language as Social Semiotic: The Social Interpretation of Language and Meaning.* London: Edward Arnold.

———. 1989. *Spoken and Written Language.* Edited by F. Christie. Oxford: Oxford University Press.

Halliday, M. A. K., and Hasan, R. 1976. *Cohesion in English.* London: Longmans.

Harris, Z. 1968. *Mathematical Structures of Language.* New York: Wiley.

———. 1982. "Discourse and sublanguage." In *Sublanguage*, edited by R. Kittredge and J. Lehrberger. Berlin: Walterde Gruyter, 231–236.

———. 1985. "Distributional structure." In *The Philosophy of Linguistics*, edited by J. J. Katz. New York: Oxford University Press.

———. 1988. *Language and Information*. New York: Columbia University Press.

Harwood, J. 1986. *The Rhetorics of Thomas Hobbes and Bernard Lamy (Landmarks in Rhetoric and Public Address)*. Carbondale. IL: Southern Illinois University Press.

Hauser, M. D. 1997. *The Evolution of Communication*. Cambridge, MA: The MIT Press.

Hayes, P. J. 1987. "Session III: Language and displays for human-computer communication change in human-computer interfaces on the space station: Why it needs to happen and how to plan for it." In *Human Factors in Automated and Robotic Space Systems: Proceedings of a Symposium*, edited by T. B. Sheridan, D. S. Kruser, and S. Deutsch: Committee On Human Factors, National Research Council.

Hayes, P. J., Ball, E., and Reddy, R. 1981. "Breaking the man-machine communication barrier." *Computer* 14 (3):19–30.

Hayes, P. J., and Reddy, R. 1983. "Steps toward graceful interaction in spoken and written man-machine communication." *International Journal of Man-Machine Studies* 19 (3):231–284.

Helander, M., ed. 1996. *Handbook of Human-Computer Interaction*. North Holland: Elsevier Science Publishers.

Hellweg, E. 1999. "Are you talking to me?" *Business 2.0*, February 1999.

Hempelmann, C. F. 2000. "Incongruity and resolution of humorous narratives—Linguistic humor theory and the medieval bawdry of Rabelais, Boccaccio, and Chaucer." M.A. Thesis, English, Youngstown State University.

Heritage, J. 1984. "A change of social token." In *Structure of Social Action*, edited by J. M. Atikinson and J. Heritage. Cambridge: Cambridge University Press.

HeyAnita. 2004. *HeyAnita* [website] 2004 [cited April 2004]. Available from http://www.heyanita.com.

Hirschberg, J., Litman, D., and Swerts, M. 2001. "Identifying user corrections automatically in spoken dialogue systems." Florham Park, NJ: AT&T Labs-Research.

Hobbes, Thomas. See Harwood, J. 1986.

Hobbs, J. R. 1979. "Coherence and coreference." *Cognitive Science* 3:67–90.

Hockett, C. F. 1960. "The origin of speech." *Scientific American* 203:88–96.

Hoey, M. 1998. "The hidden lexical clues of textual organisation: A preliminary investigation into an unusual text from a corpus perspective." Paper read at 3rd International Conference on Teaching and Learning Corpora, at Keble College, Oxford.

———. 2003. "What's in a word?" *MED Magazine*, August 2003.

Hone, K., and Baber, C. 1999. "Modelling the effects of constraint upon speech-based human-computer interaction." *International Journal of Man-Machine Studies* 50:85–107.

Horton, A. 1994. *Writing the Character-centered Screenplay*. Berkeley, CA: University of California Press.

Huang, A., Lee, F., Nass, C., Paik, Y., and Swartz, L. 2001. "Can voice user interfaces say 'I'? An experiment with recorded speech and TTS." Stanford, CA: Department of Communication, Stanford University.

Hubal, R. C. et al. 2000. "The virtual standardized patient-simulated patient-practitioner dialogue for patient interview training." In *Envisioning Healing: Interactive Technology and the Patient-Practitioner Dialogue*, edited by J. D. Westwood, H. M. Hoffman, G. T. Mogel, R. A. Robb, and D. Stredney. Amsterdam: IOS Press.

Hulstijn. 2000. "Dialogue models for inquiry and transaction." Ph.D. Thesis, Universiteit Twente, Enschede, The Netherlands.

Hurston, Z. N. 1942. *Dust Tracks on a Road*. Philadelphia, PA: J. B. Lippincott Company.

Hutchby, I., and Woofit, R. 2001. *Conversation and Technology*. New York, NY: Basil Blackwell.

————. 1998. *Conversation Analysis*. New York, NY: Basil Blackwell.

Hutchins, E. J., Hollan, J. D., and Norman, D. A. 1986. "Direct manipulation interfaces." In *User-centered System Design: New Perspectives on Human-Computer Interaction*, edited by D. A. Norman and S. W. Draper: L. Erlbaum Associates.

IBM. 2001. *IBM Websphere Voice Server Software Developers Kit (SDK) Programmes Guide*. Version 1.5. Second edition.

Isbister, K., and Nass, C. 2000. "Consistency of personality in interactive characters: Verbal cues, nonverbal cues, and user characteristics." *International Journal of Human-Computer Studies* 53 (2):251–267.

Isocrates. 1980. *Isocrates with an English Translation*. Translated by G. Norlin. Vol. Three. Cambridge, MA: Harvard University Press.

Jacobs, T. E., Sukkar, R. A., and Burke, E. R. 1992. "Performance trials of the Spain and United Kingdom Intelligent Network automatic speech recognition systems." Paper read at IEEE Workshop on Interactive Voice Technology for Telecommunications Applications, at Piscataway, NJ.

Jaffrey, M. 1989 *Madhur Jaffrey's Cookbook*. New York, NY: Harper & Row.

James, H. 1975. *The Portrait of a Lady*. New York, NY: WW. Norton.

Jannedy, S., and Mobius, B. 1997. Name pronunciation in German text-to-speech synthesis. *Fifth Conference on Applied Natural Language Processing*. Washington, DC.

Jefferson, G. 1974. "Error correction as an interactional resource." *Language in Society* 2:181–199.

Johnson, S. 1997. *Interface Culture: How New Technology Transforms the Way We Create and Communicate*. San Francisco: Harper.

Johnston, D. C. 1996. "Newsrooms: Dialing the city desk—Get human!" *Columbia Journalism Review* (September/October).

Jones, D. M., Hapeshi, K., and Frankish, C. 1990. "Design guidelines for speech recognition interfaces." *Applied Ergonomics* 20 (1):40–52.

Jones, M. L. R. 1999. "A new agenda for HCI research: Rethinking user needs." Paper read at ACM Conference on Human Factors in Computing Systems (CHI 99), at Pittsburgh, PA.

Kalish, D. 1967. "Semantics." In *The Encyclopedia of Philosophy*, edited by P. Edwards. New York, NY: Macmillan Publishing.

Kamm, C. 1994. "User interfaces for voice applications." In *Voice Communication Between Humans and Machines*, edited by D. Roe and J. Wilpon. Washington, DC: National Academy Press.

Kamm, C., Walker, M., and Lawrence, R. 1997. "The role of speech processing in human–computer intelligent communication." *Speech Communication* 23 (1997):263–278.

Kampelman, M. M. 1985. *Covert Action Information Bulletin,* May 6, 1985.

Karat, C. M., Halverson, C., Horn, D., and Karat, J. 1999. "Patterns of entry and correction in large vocabulary continuous speech recognition systems." Paper read at ACM Conference on Human Factors in Computing Systems (CHI 99), at Pittsburgh, PA.

Karat, J., Lai, J., Danis, C., and Wolf., C. n.d. *Speech User Interface Evolution*: IBM T. J. Watson Research Center.

Karis, D., and Dobroth, K. M. 1995. "Psychological and human factors issues in the design of speech recognition systems." In *Applied Speech Technology*, edited by A. Syrdal, R. Bennett, and S. Greenspan. Toronto: CRC Press.

Kestenbaum, D. *Analysis: Computerized voice machines with a more humanlike sound.* (2002) [website]. NPR 2002 [cited August 2004]. Available from http://www.npr.org/programs/atc/features/2002/apr/computervoices/.

Kieras, D. E. 1997. "Task analysis and the design of functionality." In *The Computer Science and Engineering Handbook*, edited by A. B. Tucker. Boca Raton, FL: CRC Press.

Knott, A., and Sanders, G. A. 1997. "The classification of coherence relations and their linguistic markers: An exploration of two languages." *Journal of Pragmatics* 30:135–175.

Kolzer, A. 1999. "Universal dialogue specification for conversational systems." Ulm, Germany: DaimlerChrysler AG—Research and Technology.

Kotelly, B. 2003. *The Art and Business of Speech Recognition: Creating the Noble Voice.* Boston, MA: Addison-Wesley.

Kowtko, J. C., and Price, P. J. 1989. "Data collection and analysis in the air travel planning domain." In *DARPA Speech and Natural Language Workshop*. Los Altos, CA: Morgan Kaufmann Publishers.

Kreutel, J. 2000. "Reconstructing conversational games in an obligation-driven dialogue model." Paper read at Text, Speech and Dialogue: Third International Workshop, TSD 2000, September 2000, at Brno, Czech Republic.

Kreutel, J., and Matheson, M. 1996. "Obligations, intentions, and the notion of conversational games." Paper read at Gotalog 2000, at Goteborg University.

Krug, S. 2000. *Don't Make Me Think! A Common Sense Approach to Web Usability.* Indianapolis, IN: New Riders.

Labov, W. 1972. *Language in the Inner City: Studies in the Black English Vernacular-conduct and Communication.* Philadelphia: University of Pennsylvania Press.

Lacohee, H., and Anderson, B. 2001. "Interacting with the telephone." *International Journal of Human-Computer Studies* 54 (2001):665–699.

Lai, J., and Yankelovich, N. 2002. "Conversational speech interfaces." In *Human-Computer Interaction Handbook*, edited by J. Jacko and A. Sears. Mahwah, NJ: Lawrence Erlbaum, 698–713.

Lakoff, G., and Johnson, M. 1980. *Metaphors We Live By.* Chicago: University of Chicago Press.

Lamel, L., Bennacef, S., Rosset, S., Devillers, L., Foukia, S., Gangolf, J. J., and Gauvain, J. L. 1997. "The LIMSI RailTel System: Field trial of a telephone service for rail travel information." *Speech Communication* 23 (1997):67–82.

Lamel, L., Rosset, S., Gauvain, J. L., Bennacef, S., Garnier-Rizet, M., and Prouts, B. 2000. "The LIMSI Arise System." *Speech Communication* 31 (2000):339–353.

Langkilde, I., Walker, M. A., Wright, J., Gorin, A., and Litman, D. 1999. "Automatic prediction of problematic human-computer dialogues in how may I help you?" Paper read at ASRU 99.

Laurel, B. 1986. "Interface as mimesis." In *User Centred System Design: New Perspectives in Human-Computer Interaction*, edited by D. A. Norman and S. W. Draper. Mahwah, NJ: Lawrence Erlbaum Associates.

———. 1997. "Interface agents: metaphors with character." In *Software Agents*, edited by J. M. Bradshaw. Cambridge, MA: The MIT Press.

Lebowitz, F. 1981. *Social Studies.* New York, NY: Random House.

Lee, E. J., Nass, C., and Brave, S. 2000. "Can computer-generated speech have gender? An experimental test of gender stereotypes." Paper read at CHI 2000, Apr. 1–6 2000, at The Hague, The Netherlands.

Leeuwen, T. van. 1999. *Speech, Music, Sound*. New York, NY: Macmillan Press.

Lender, W. 1991. "What's in a lexical entry?" In *Linguistica Computazionale*, edited by L. Cignoni and C. Peters. Pisa: Instituto di linguistica computazionale.

Levow, Cina Anno. 1999. "Understanding recognition failures in spoken corrections in human-computer dialog": *Proceedings of ESCA Workshop on Dialogue and Prosody*. Eindhoven, The Netherlands.

Lewis, J. R. 1999. "Effect of error correction strategy on speech dictation throughput." Paper read at *Human Factors and Ergonomics Society*, 1999.

Liberman, Mark. *Liguistics 101* [website (course notes)]. University of Pennsylvania 2000 [cited April 2004.] Available from http://www.ling.upenn.edu/courses/Fall_2000/ling001/syntax.html.

Life, A., Salter, I., Temem, J. N., Bernard, F., Rosset, S., Bennacef, S., and Lamel, L. 1996. "Data collection for the MASK kiosk: WOz vs prototype system. London, U.K.; Paris, France: Ergonomic Unit, UCL.

Litman, D. J., Marilyn, A. W., and Kearns, M. S. 1999. "Automatic detection of poor speech recognition at the dialogue level." Paper read at ACL-99.

Lochbaum, K. 1994. "Using collaborative plans to model the intentional structure of discourse." Ph.D. Thesis, Center for Research in Computer Technology, Applied Sciences, Harvard University, Cambridge, MA.

Lockhart, G. 1988. *The Weather Companion: An Album of Meteorological History, Science, and Folklore*. New York, NY: John Wiley & Sons.

Louw, B. 1993. "Irony in the text or insincerity in the writer? The diagnostic potential of semantic prosodies." In *Text and Technology: In Honour of John Sinclair*, edited by M. Baker, G. Francis, and E. Tognini-Bonelli. Philadelphia/Amsterdam: John Benjamins.

Mack, R. L., and Nielsen, J. 1994. "Executive summary." In *Usability Inspection Methods*, edited by J. Nielsen and R. L. Mack. New York, NY: John Wiley & Sons.

MacKay, D. M. 1962. "The use of behavioural language to refer to mechanical processes." *The British Journal for the Philosophy of Science* 50:89–103.

Maier, E., and Sitter, S. 1992. "An extension of rhetorical structure theory for the treatment of retrieval dialogues." Paper read at CogSci '92, The 14th Annual Conference of the Cognitive Science Society, July, 1992, at Bloomington, N.

Malinowski, B. 1923. "The problem of meaning in primitive languages." In *The Meaning of Meaning: A Study of the Influence of Language upon Thought and the Science of Symbolism*, edited by C. K. Ogden and I. A. Richards. London: Kegan Paul, Trench, Trubner & Co.

Mané, A., Boyce, S., Karis, D., and Yankelovich, N. 1996. "Designing the user interface for speech recognition applications." A CHI 96 workshop. *SIGCHI Bulletin*, October 1996.

Mann, and Thompson. 1987. "Antithesis: A study in clause combining and discourse structure." In *Language Topics: Essays in Honour of Michael Halliday*, edited by R. Steele and T. Threadgold. Amsterdam: John Benjamins.

Marchand, R. 1998. *Creating the Corporate Soul: The Rise of Public Relations and Corporate Imagery in American Big Business*. Berkeley, CA: University of California Press.

Marcu, D. 2000. "The rhetorical parsing of unrestricted texts: A surface-based approach." *Computational Linguistics* 26 (3):395–448.

Marti, S. J. W. 1999. "Active messenger: Email filtering and mobile delivery. M.Sc. Thesis, Media Arts and Sciences, School of Architecture and Planning, MIT, Cambridge, MA.

Martin, P., Crabbe, F., Adams, S., Baatz, A., and Yankelovich. N. 1996. "SpeechActs: A spoken-language framework." *IEEE Computer* 29 (7):33–40.

Marx, M. 1995. "Toward effective conversational messaging." Master's Thesis, Program in Media, Arts and Sciences, MIT, Cambridge.

Marx, M., and Schmandt. C. 1994. "Putting people first: Specifying names in speech interfaces." Paper read at ACM Symposium on User Interface Software and Technology, November 2–4, 1994, at Marina del Rey, CA.

———. 1996. "MailCall: Message presentation and navigation in a nonvisual environment." Cambridge, MA: MIT Media Lab.

Massaro, D. W., Cohen, M. M., Beskow, J., and Cole, R. A. 2000. "Developing and evaluating conversational agents." In *Embodied Conversational Agents*, edited by J. Cassell, J. Sullivan, S. Prevost, and E. Churchill, Cambridge, MA: The MIT Press.

Mayhew, D. J. 1999. *The Usability Engineering Lifecycle.* San Francisco, CA: Morgan Kaufmann Publishers.

McArthur, T. 1998. *Living Words: Language, Lexicography, and the Knowledge Revolution.* Exeter, UK: University of Exeter Press.

McCawley, J. D. 1986. "What linguists might contribute to dictionary-making if they could get their act together." In *Real World Linguistics*, edited by P. C. Bjarkman and V. Raskin. Norwood, NJ: Ablex.

McIlvenny, P., and Raudaskoski, P. 1996. "The mutual relevance of conversation analysis and linguistics: A discussion in reference to interactive discourse." In *Proceedings of the Thirteenth Scandinavian Conference of Linguistics*, edited by L. Heltoft. Roskilde, Denmark.

McLaughlin, M., and Cody, M. 1982. "Awkward silences: Behavioural antecedents and consequences of the conversational lapse." *Human Communication Research* 8:299–316.

Microsoft. 1998. *Guidelines for designing character interaction* [website]. Microsoft Corporation 1998 [cited Aug 2004]. Available from http://msdn.microsoft.com/library/default.asp?url=/library/en-us/msagent/guidln_6qpa.asp.

Miller, L. A., and Thomas, J. C. 1977. "Behavioral issues in the use of interactive systems." *International Journal of Human-Computer Studies* 51 (1999):169–196.

Minker, W. 1999. "Design considerations for knowledge source representations of a stochastically-based natural language understanding component." *Speech Communication* 28 (1999):141–154.

Minsky, M. 1986. *The Society of Mind.* New York: Simon and Schuster.

Moeschler, J. 2001. "Speech act theory and the analysis of conversations." In *Essays in Speech Act Theory*, edited by D. Vanderveken and S. Kubo. Philadelphia: John Benjamins Publishing Company.

Möller, Jens-Uwe. 1995. "Towards learning dialogue structures from speech data and domain knowledge: Challenges to conceptual clustering using multiple and complex knowledge." [cited July 2004]. Available from www.cs.cmu.edu/~dgroup/papers/tlds.pdf.

Moon, Y. 1998. "Impression management in computer-based interviews: The effects of input modality, and distance." *Public Opinion Quarterly* 62:610–622.

Moore, J., and Walker, M. 1995. "Proceedings of the 1995 AAAI workshop on empirical methods in discourse interpretation and generation." Technical report. Paper read at 1995 AAAI workshop.

Morris, C. W. 1938. "Foundations of the theory of sign." In *International Encyclopedia of Unified Science*, edited by O. Neurath, R. Carnap, and C. Morris. Chicago: University of Chicago Press.

Muller, M., and Daniel, J. 1990. "Toward a definition of voice documents." Paper read at COIS.

Mulligan, K., ed. 1987. *Speech Act and Sachverhalt: Reinach and the Foundations of Realist Phenomenology*. Dordrecht, Holland and Boston: Nijhoff.

Murray, R., and Arnott, J. L. 1993. "Toward the simulation of emotion in synthetic speech: A review of the literature of human vocal emotion." *Journal of the Acoustic Society of America* 93 (2):1097–1108.

Nass, C., Moon, Y., and Green, C. 1997. "Are machines gender-neutral? Gender-stereotypic responses to computers with voices." *Journal of Applied Social Psychology* 27:864–876.

Nass, C., Isbister, K., and Lee, E. J. 2000. "Truth is beauty: Researching embodied conversational agents." In *Embodied Conversational Agents*, edited by J. Cassell, J. Sullivan, S. Prevost, and E. Churchill, Cambridge: The MIT Press.

Nass, C. 2004. "Etiquette equality: Exhibitions and expectations of computer politeness." *Communications of the ACM*, 47:35–37.

Nass, C., Simard, C., and Takhteyev, Y. 2004. *Should recorded and synthesized speech be mixed?* 2004 [cited July 2004]. Available from http://www.stanford.edu/~nass/comm369/pdf/MixingTTSandRecordedSpeech.pdf.

Neilsen, J. 1989. "Usability engineering at a discount." In *Designing and Using Human-Computer Interfaces and Knowledge Based Systems*, edited by G. Salvendy and M. J. Smith, Amsterdam: Elsevier Science Publishers.

———, ed. 1990. *Designing User Interfaces for International Use*. Boston: Elsevier Science Publishers.

———. 1993. *Usability Engineering*. Boston: Academic Press.

———. 1994. "Guerrilla HCI: Using discount usability engineering to penetrate the intimidation barrier." In *Cost-Justifying Usability*, edited by R. G. Bias and D. J. Mayhew. San Francisco, CA: Morgan Kaufmann.

Neilsen, J., and Mack, R. L., eds. 1994. *Usability Inspection Methods*. New York: John Wiley & Sons.

Nicholl, J. *Subject: The King's English Newsgroups: rec.arts.sf-loversView: Complete Thread (39 articles)* [Online Newsgroup]. The King's English Newsgroups, 1990-05-15 11:53:58 PST 1990 [cited].

Norman, D. A. 2000. "Making technology invisible [An interview with E. Bergman]." In *Information Appliances and Beyond: Interaction Design for Consumer Products*, by E. Bergman. San Francisco: Morgan Kaufmann Publishers.

Novick, D. 1999. "Using the cognitive walkthrough for operating procedures." *Interactions* 6:31–37.

Noyes. 2001. "Talking and writing—how natural in human-machine interaction?" *International Journal of Human-Computer Studies* 55 (2001):503–519.

Nuance communications. *Nuance communications* [website]. Nuance 2002 [cited July 2002]. Available from www.nuance.com.

O'Shaughnessy, D. 2000. *Speech Communications: Human and Machine*. Second ed. Piscataway, NJ: IEEE Press.

O'Toole, Kathleen. 2000. Virtual voice technology raises real questions about ethics. *Stanford Report* (July 26, 2000). Available: http://news-service.stanford.edu/news2000/virtual voice726-a.html. [cited sept. 2004.]

Oettinger, Anthony G. 1965. "An Essay in Information Retrieval or The Birth of a Myth." *Information and Control* 8:64–79.

Ogden, C. K., and Richards, I. A., eds. 1923. *The Meaning of Meaning: A Study of the Influence of Language upon Thought and the Science of Symbolism*. London: Kegan Paul, Trench, Trubner & Co.

Ogden, W. C., and Bernick, P. 1997. "Using natural language interfaces." In *Handbook of Human-Computer Interaction*, edited by M. Helander. Second edition. North Holland: Elsevier Science Publishers.

Omoigui, N., He, L., Gupta, A., Grudin, J., and Sanocki, E. 1999. "Time-compression: Systems concerns, usage, and benefits." Paper read at CHI 99.

Ooi, V. 1998. *Computer Corpus Lexicography*. Edinburgh: Edinburgh University Press.

Opitz, K. 1983. "On dictionaries for special registers." In *Lexicography: Principles and Practice*, edited by R. R. K. Hartmann. New York, NY: Academic Press.

Orr, W. D., ed. 1968. *Conversational Computers*. New York, NY: John Wiley & Sons.

Oviatt, S., Cohen, P., and Wang. M. 1994. "Toward interface design for human language technology: Modality and structure as determinants of linguistic complexity." *Speech Communication* 15:3–4.

Oviatt, S. L., MacEachern, M., and Levow, G. 1998. "Predicting hyperarticulate speech during human-computer error resolution." *Speech Communication* 24 (2):1–23.

Oviatt, S., and Cohen, S. 2000. "Multimodal interfaces that process what comes naturally." *Communications of the ACM* 43 (3):45–53.

Peckam, J. 1993. "A new generation of spoken dialogue systems: Results and lessons from the SUNDIAL project." Paper read at Eurospeech '93, at Berlin.

Penn, W. 1909. *Fruits of Solitude*. New York, NY: P. F. Collier.

Peterson, G. E., and Barney, H. L. 1952. "Control methods used in a study of the identification of vowels." *Journal of Acoustical Society of America* 24:175–184.

Pinker, S. 1994. *The Language Instinct*. New York, NY: W. Morrow and Co.

Plato. 1961. *Collected Dialogues*, edited by E. Hamilton and H. Cairns. Princeton, NJ: Princeton University Press.

Polifroni, J., and Seneff, S. 2000. "Galaxy-II as an architecture for spoken dialog evaluation." Paper read at 2nd International Conference on Language Resouces and Evaluation (LREC).

Polifroni, J., Seneff, S., Glass, J., and Hazen, T. J. 1998. "Evaluation methodology for a telephone-based conversational system." Paper read at LREC '98.

Pomerantz, A. 1986. "Extreme case formulations: A way of legitimizing claims." *Human Studies* 9:219–229.

Pope, A. 1963. *Collected Poems*. London: Dent.

Potamianos, A., Ammicht, E., and Kuo, H. J. 2000. "Dialogue management in the Bell Labs communicator system." Murray Hill, NJ: Bell Labs, Lucent Technologies.

Powell, W. G. *Commentary: It's the content, dummy: Why most web sites fail* (June 2001) [website]. Hanson News 1.3 2001 [cited Feb 2002]. Available from http://www.hansoninc.com/newsletter/news-i3.asp.

Prince, E. 1988. "Discourse analysis: A part of the study of linguistic competence." In *Linguistics: the Cambridge Survey*, edited by F. J. Newmeyer. Cambridge, MA: Cambridge University Press.

Raggett, D. *A general purpose architecture for intelligent tutoring systems* [website]. The CLASS project Collaboration in Language and Speech. Science and Technology 2002 [cited Feb 2003]. Available from http://www.class-tech.org/events/NMI_workshop2/papers/.

Raman, T. V. 1997. *Auditory User Interfaces: Towards the Speaking Computer*. Boston, MA: Kluwer Academic Publishers.

Raskin, J. 2000. *The Humane Interface*. Boston, MA: Addison-Wesley.

Raskin, J. 1993. "Down with GUIs!" *Wired*, December 1993, 122.

———. 1994. "Intuitive equals familiar." *Communications of the ACM* 37 (9):17.

Reeves, B., and Nass, C. 1996. *The Media Equation: How People Treat Computers, Television and New Media Like Real People and Places*. New York, NY: Cambridge University Press/CSLI.

Rich, A. 1973. *Diving into the Wreck: Poems 1971–1972*. New York, NY: W. W. Norton & Company.

Ringle, M., and Bruce, B. C. 1981. "Conversation failure." In *Knowledge Representation and Natural Language Processing*, edited by W. G. Lehnert and M. H. Ringle. Mahwah, NJ: Lawrence Erlbaum.

Robles, E., Bienstock, H., Treinen, M., Heenan, C., and Nass, C. 2001. "Using design to increase disclosure: A study of the effects of modality and gender of prompt on willingness to disclose." Stanford, CA. Unpublished paper.

Rodman, R. D. 1999. *Computer Speech Technology*. Boston, MA: Artech House.

Roget, P. M. 1925. *Roget's Thesaurus*.

Rohrer, T. 2003. *Metaphors we compute by: bringing magic into interface design* [website]. University of Oregon 1995 [cited February 2003]. Available from http://philosophy.uoregon.edu/metaphor/gui4web.htm.

Rosenfeld, R., Olsen, D., and Rudnicky, A. 2000. "A universal human-machine speech interface." Pittsburgh, PA: School of Computer Science, Carnegie-Mellon University.

Rosset, S., Bennacef, S., and Lamel, L. 1999. "Design strategies for spoken language dialog systems." Paper read at European Conference on Speech Technology, EuroSpeech, September 1999, at Budapest.

Rubin, J. 1994. *Handbook of Usability Testing: How to Plan, Design, and Conduct Effective Tests*. New York: John Wiley & Sons.

Rudnicky, A. I. 1995. "The design of spoken language interfaces." In *Applied Speech Technology*, edited by A. Syrdal, R. Bennett, and S. Greenspan.

Rudnicky, A. I., Bennett, C., Black, A. W., Chotomongcol, A., Lenzo, K., Oh, A., and Singh, R. 2000. "Task and domain specific modellng in the Carnegie-Mellon Communicator System." Pittsburgh, PA: School of Computer Science, Carnegie-Mellon University.

Rudnicky, A. I., Thayer, E., Constantinides, P., Tchou, C., Shern, R., Lenzo, K., Xu W., and Oh, A. 1999. "Creating natural dialogs in the Carnegie-Mellon Communicator System." Paper read at Eurospeech 1999.

Rudnicky, A. I. 1995. "The design of spoken language interfaces." In *Applied Speech Technology*, edited by A. Syrdal, R. Bennett, and S. Greenspan. Boca Raton, FL: CRC Press.

Sacks, H. 1992. *Lectures on Conversation*. Edited by G. Jefferson. 2 vols. Oxford: Basil Blackwell.

Sacks, H., Schegloff, E. A., and Jefferson, G. 1974. "A simplest semantics for the organization of turn-taking in conversation." *Language* 50 (4):696–735.

Sadek, D., and de Mori, R. 1998. "Dialogue systems." In *Spoken Dialogues with Computers*, edited by R. de Mori. New York: Academic Press.

Salvucci, D. D. 2001. "Predicting the effects of in-car interface use on driver performance: An integrated model approach." *International Journal of Human-Computer Studies* 55 (2001):85–107.

Sanders, G. A., and Scholtz, J. 2000. "Measurement and evaluation of embodied conversational agents." In *Embodied Conversational Agents*, edited by J. Cassell, J. Sullivan, S. Prevost, and E. Churchill. Cambridge, MA: The MIT Press.

Schank, R. C., and Abelson, R. P. 1977. *Scripts, Plans, Goals and Understanding: An Inquiry into Human Knowledge Structures.* Hillsdale, NJ: L. Erlbaum.

Schegloff, E. A. 1968. "Sequencing in conversational openings." *American Anthropologist* 70:1075–1095.

———. 1972. "Notes on a conversational practice: Formulating place." In *Studies in Social Interaction*, edited by D. N. Sudnow. New York, NY: MacMillan, The Free Press.

———. 1979. "The relevance of repair to syntax-for-conversation." In *Discourse and Syntax*, edited by T. Givon. New York, NY: Academic Press.

———. 1986. "The routine as achievement." *Human Studies* 9:111–151.

———. 1988. "On an actual virtual servo-mechanism for guessing bad news." *Social Problems* 32:442–457.

———. 1992. "Repair after next turn: The last structurally provided defense of intersubjectivity in conversation." *American Journal of Sociology* 98:1295–1345.

Schegloff, E. A., Jefferson, G., and Sacks, H. 1977. "The preference for self-correction in the organization of repair in conversation." *Language* 53 (2):361–382.

Schegloff, E. A., and Sacks, H. 1973. "Opening up closings." *Semiotica* 8:289–327.

Schillo, M. 1996. "Working while driving: Corpus-based language modelling of a natural English voice user interface to the in-car personal assistant." Master Thesis, Department of Computer Studies, Leeds University.

Schmandt, C. 1984. "Speech synthesis gives voiced access to an electronic mail system." *Speech Technology*, 66–68.

———. 1994. *Voice Communication with Computers: Conversational Systems.* New York, NY: Van Nostrand Reinhold.

Schroeter, J. 2001. "The fundamentals of text-to-speech." *VoiceXML Review* 1 (3).

Schwartz, E. 2004. *WAP-enabled phones to get voice interface.* CNN.com, December 27, 1999 1999 [cited August 2004]. Available from http://www.cnn.com/1999/TECH/computing/12/27/wap.voice.idg/.

Scientific American. 1871. 24:32.

Screven, C. G. 2000. "Information design in informal settings: Museums and other public spaces." In *Information Design*, edited by R. Jacobson. Cambridge, MA: The MIT Press.

Searle, J. R. 1979. *Expressions and Meaning.* Cambridge, MA: Cambridge University Press.

Searle, J. R., and Vanderveken, D. 1985. *Foundations of Illocutionary Logic.* Cambridge: Cambridge University Press.

Seuss, Dr. 1957. *The Cat in the Hat.* Boston, MA: Houghton Mifflin.

Seyfarth, R. M., and Cheney, D. L. 1992. "Meaning and mind in monkeys." *Scientific American* 267:122–129.

Shaw, M., and Garlan, D. 1996. *Software Architecture—Perspectives on an Emerging Discipline.* Upper Saddle River, NJ: Prentice Hall.

Shephard, A. 1989. "Analysis and training in information technology tasks." In *Task Analysis for Human-Computer Interaction*, edited by D. Diaper. Chichester: Ellis Horwood.

Shneiderman, B. 1998. *Designing the User Interface: Strategies for Effective Human-Computer Interaction.* Third ed. Reading, MA: Addison-Wesley.

Sikorski, T., and Allen, J. F. 1997. "A task-based evaluation of the TRAINS-95 dialogue system." In *Dialogue Processing in Spoken Language Systems*, edited by E. Maier et. al. Berlin: Springer.

Sinclair, J. 1987. *Introduction to Collins Cobuild English Language Dictionary.* London: Collins.

———. 1991. *Corpus Concordance Collocation.* Oxford: Oxford University Press.

Sinclair, R. C., Mark, M. M., Moore, S. E., Lavis, C. A., and Soldat, A. S. 2000. "Psychology: An electoral butterfly effect." *Nature* 208:665–666.

Smith, A. 1904. *An Inquiry into the Nature and Causes of the Wealth of Nations.* Fifth ed. London: Methuen and Co., Ltd.

Smith, M. J., Salvendy, G., Harris, D., and Koubek, R., eds. 2001. *Usability Evaluation and Interface Design: Cognitive Engineering, Intelligent Agents, and Virtual Reality.* Mahwah, NJ: Lawrence Erlbaum.

Smith, R. W. 1997. "An evaluation of strategies for selective utterance verification for spoken natural language dialog." Paper read at 5th Applied Natural Language Processing Conference, at Washington, USA.

Smith, R. W., and Hipp, D. R. 1993. *Spoken Natural Language Dialog Systems: A Practical Approach.* New York, NY: Oxford University Press.

Sperber, D., and Wilson, D. 1986/1995. *Relevance. Communication and Cognition.* Second edition ed. Oxford: Basil Blackwell.

SRI/Amex. *Amex travel agent data* [website]. SRI International 1989 [cited. Sept. 2004] Available from http://www.ai.sri.com/~communic/amex/amex.html.

Stalnaker, R. C. 1972. "Pragmatics." In *Semantics of Natural Language*, edited by D. Davidson and G. Harman. Dordrecht: D. Reidel.

Stark, T. 2002. *Something in the air: An interface only a mother could love* [website]. IBM Developer Works, November 2001 [cited Sept. 2002].

Stellmann, P., and Brennan, S. E. 1993. "Flexible perspective-setting in conversation." In *Abstracts*, 34th Annual Meeting of the Psychonomic Society. Washington, DC.

Stent, A. 2000. "Rhetorical structure in dialog." Paper read at 2nd International Natural Language Generation Conference (INLG '2000), June 2000.

Stent, A. J., and Allen, J. F. 2000. "Annotating argumentation acts in spoken dialog." Rochester, NY: University of Rochester, Computer Science Department.

Steuten, A. A. G. 1997. "Structure and cohesion in business conversations." *WEB-SLS The European Student Journal of Language and Speech* 96 (3).

Stevens, W. 1982. *The Collected Poems of Wallace Stevens.* New York, NY: Vintage Books.

Stifelman, L. J. 1993. "User repairs of speech recognition errors: An intonational analysis." Cambridge, MA: MIT Media Laboratory.

Stubbs, M. 1995. "Collocations and semantic profiles: On the cause of the trouble with quantitative studies." *Functions of Language* 2 (1):23–55.

Stucker, H. 1999. "Personality typing." *Wired*, July 1999.

Sun-Microsystems. 1998. *Java Speech API Programmer's Guide.* Palo Alto, CA: Sun-Microsystems.

Tannen, D. 1990. *You Just Don't Understand: Women and Men in Conversation.* New York, NY: Ballantine Books.

Thompson, H. S. 1988. *Generation of Swine.* New York, NY: Vintage Books.

Tognazinni, B. 1990. "User testing on the cheap." *Apple Direct*, 21–27.

———. 1996. *Tog on Software Design.* Reading, MA: Addison-Wesley.

———. 2001. *Ask Tog: Apple squandering the advantage* [website]. Nielsen Norman Group 2001 [cited Sept. 2004]. Available from http://www.asktog.com/columns/035SquanAdv.html.

———. *The butterfly ballot: Anatomy of a disaster.* Ask Tog, January, 2001. Available from http://www.asktog.com/columns/042ButterflyBallot.html.

Tourangeau, R., Couper, M., and Steiger, D. 2001. "Social presence in web surveys." Paper read at 2001 FCSM Conference.

Traum, D., and Hinkelman, E. 1992. "Conversation acts in task-oriented spoken dialogue." *Computational Intelligence* 8 (3):575–599.

Traum, D. 1999. "Speech acts for dialogue agents." In *Foundations of Relational Agency*, edited by M. Wooldride and A. Rao, Dordrecht: Kluwer.

Trench, R. C. 1859. *On the Study of Words.* New York, NY: Redfield.

Turner, R. M. 1998. "Context-mediated behavior for intelligent agents." *International Journal of Human-Computer Studies* 48:307–330.

Turunen, M., and Hakulinen, J. 2000. "Mailman—a multilingual speech-only e-mail client based on an adaptive speech application framework." Paper read at Workshop on Multilingual Speech Communication.

———. 2001. "Agent-based adaptive interaction and dialogue management architecture for speech applications." Paper read at Fourth International Conference TSD 2001.

Tweney, D. 2002. *HAL 9000 is ready to take your order* [website]. Business 2.0 2001 [cited Sept. 2004].

Vanhoucke, W. V., Neeley, L., Mortati, M., Sloan, M. J., and Nass, C. 2001. "Effects of prompt style when navigating through structured data." Paper read at INTERACT 2001, Eighth IFIP TC.13 Conference on Human-Computer Interaction, at Tokyo, Japan.

Walker, A. G. 1999. *Handbook on Questioning Children: A Linguistic Perspective. 2nd Edition.* Washington, DC: American Bar Association.

Walker, M. A., Kamm, C., and Litman, D. J. 2000. "Towards developing general models of usability with PARADISE." *Natural Language Engineering* 1 (1):1–16.

———. Fromer, J., Fabbrizio, G., Mestel, C., and Hindle, D. 1998. "What can I say?: Evaluating a spoken language interface to email." Paper read at ACM CHI '98.

———. Boland, J., and Kamm, C. 2000. "The utility of elapsed time as a usability metric for spoken dialogue systems." ASRU '99.

———. Fromer, J. C., and Narayanan, S. 1997. "Learning optimal dialogue strategies: A case study of a spoken dialogue agent for email." Paper read at 36th Annual Meeting of the Association of Computational Linguistics, COLING/ACL 98.

Wallace, V., and Irani, K. B. 1971. "On network linguistics and the conversational design of queueing networks." *Journal of the ACM* 18:616–629.

Wardaugh, R. 1985. *How Conversation Works.* New York, NY: Basil Blackwell.

Waterworth, J. A. 1984. "Interaction with machines by voice: Human factors issues." *British Telecom Technology Journal* 2 (4).

———. and Talbot, M. 1987. *Speech and Language-based Interaction with Machines: Towards the Conversational Computer.* Chichester: Ellis Horwood.

Watt, S. 1998. "Seeing this as people: Anthropomorphism and common-sense psychology." Ph.D. Dissertation, Cognitive Science, The Open University.

Watt, W. C. 1968. Habitability. *American Documentation* 19 (3):338–351.

Weiner, E. 1990. "The federation of English." In *The State of the Language*, edited by C. Ricks and L. Michaels. Berkeley: University of California Press.

Weinreich, U. 1962. "Lexicographic definition in descriptive semantics." In *Problems in Lexicography*, edited by F. W. Householder and S. Saporta. Bloomington, IN: Indiana University Press.

Weinschenk, S., and Barker, D. T. 2000. *Designing Effective Speech Interfaces*. New York, NY: Wiley & Sons.

Whitcut, J. 2000. *Better Wordpower*. New York, NY: Oxford University Press.

Whitney, W. D. 1867. *Language and the Study of Language: Twelve Lectures on the Principles of Linguistic Science*. New York, NY: Charles Scribner & Co.

Widdowson, H. G. 1979. *Explorations in Applied Linguistics*. New York, NY: Oxford University Press.

Williams, J. R. 1998. "Guidelines for the use of multimedia in instruction." Paper read at Human Factors and Ergonomics Society 42nd Annual Meeting.

Willinsky, J. 1994. *Empire of Words. The Reign of the OED*. Princeton, NJ: Princeton University Press.

Wilpon. 1994. "Applications of voice-processing technology in telecommunications." In *Voice Communication between Humans and Machines*, edited by D. Roe and J. Wilpon. Washington DC: National Academy Press.

Wilson, K. G. 1993. *The Columbia Guide to Standard American English*. New York: Columbia University Press.

Wittgenstein, L. 1968. *Philosophical Investigations*. Translated by G. E. M. Anscombe. 3rd ed. Oxford: Blackwell.

Wixon, D., and Wilson, C. 1997. "The usability engineering framework for product design and evaluation." In *Handbook of Human-Computer Interaction*, edited by M. Helander. Amsterdam: North-Holland.

Wolf, C., and Zadrozny, W. 1998. "Evolution of the conversation machine: A case study of bringing advanced technology to the marketplace." Paper read at Conference on Human Factors in Computing Systems (CHI 98).

Woodbury, D. O. 1959. "The translating machine." *The Atlantic Monthly* 204 (2):60–64.

Woodson, W. E. 1981. *Human Factors Design Handbook*. Columbus, OH: McGraw-Hill.

Woofit, R., Fraser, N. M., Gilbert, N., and McGlashan, S. 1997. *Humans, Computers and Wizards: Conversation Analysis and Human Computer Interaction*. London: Routledge.

Yankelovich, N. 1994. "Talking vs. Taking: Speech access to remote computers." Paper read at ACM CHI '94, April 24–28, 1994, at Boston, MA.

———. 1996. "How do users know what to say?" *ACM Interactions*, November/December 1996.

———. *Using natural dialogs as the basis for speech interface design* 1997 [cited August 2004.] Available from http://research.sun.com/speech/publications/mit1998/ MITPressChapter.v3.html.

———. 1998. "Using natural dialogs as the basis for speech interface design." Unpublished http://research.sun.com/speech/publications/mit-1998/ MITPressChapter.v3.html. [cited Sept. 2004]

———. Levow, G. A., and Marx, M. 1995. "Designing SpeechActs: Issues in speech user interfaces." Paper read at Conference on Human Factors in Computing Systems (CHI 95).

Yeats, W. B. 1989. *The Collected Poems*. Edited by R. Finneran. Revised 2nd ed. New York, NY: Macmillan.

Yee, P., and Harris, R. 1989. "The pragmatics of usability." Paper read at Technicom '89.

Yngve. 1970. "On getting a word in edgewise." In *Papers from the Sixth Regional Meeting*, edited by M. A. Campbell, J. Lindholm, A. Davison, W. Fisher, L. Furbee, J. Lovins, E. Maxwell, J. Reighard, and S. Straight. Chicago, IL: Chicago Linguistics Society.

Yule, G. 1996. *The Study of Language*. Cambridge: Cambridge University Press.

Zadrozny, W., Wolf, C., Kambhatla, N., and Ye, Y. 1998. "Conversation machines for transaction processing." Hawthorne, NY: IBM T. J. Watson Research Center.

Ziefle, M. 1998. "Effects of display resolution on visual performance." *Human Factors* 40 (4):555–568.

Zue, V. R. 1997. "Conversational interfaces: Advances and challenges." Paper read at Eurospeech, at Rhodes, Greece.

Index

Page numbers followed by n indicate footnotes.